TWO-PHASE FLOW AND HEAT TRANSFER IN THE POWER AND PROCESS INDUSTRIES

KU-566-694

(left to right) J. G. Collier, A. E. Bergles, F. Mayinger, G. F. Hewitt, and J. M. Delhaye.

TWO-PHASE FLOW AND HEAT TRANSFER IN THE POWER AND PROCESS INDUSTRIES

A. E. Bergles
Iowa State University
Ames, Iowa, U.S.A.

J. G. Collier
Atomic Energy Technical Unit
U. K. Atomic Energy Authority
Harwell, England

J. M. Delhaye
Centre d'Etudes Nucléaires
Grenoble, France

G. F. Hewitt
Engineering Sciences Division and HTFS
U. K. Atomic Energy Authority
Harwell Laboratory, England

F. Mayinger
Institut für Verfahrenstechnik
Universität Hannover
Hannover, West Germany

● HEMISPHERE PUBLISHING CORPORATION

Washington   New York   London

McGRAW-HILL BOOK COMPANY

New York   St. Louis   San Francisco   Auckland   Bogotá
Hamburg   Johannesburg   London   Madrid   Mexico
Montreal   New Delhi   Panama   Paris   São Paulo
Singapore   Sydney   Tokyo   Toronto

055944745

TWO-PHASE FLOW AND HEAT TRANSFER IN THE POWER AND PROCESS INDUSTRIES

Copyright © 1981 by Hemisphere Publishing Corporation. All rights reserved. Printed in the United States of America. No part of this publication may be reproduced, stored in a retrieval system, or transmitted, in any form or by any means, electronic, mechanical, photocopying, recording, or otherwise, without the prior written permission of the publisher.

2 3 4 5 6 7 8 9 0   B R B R   8 9 8 7 6 5 4 3 2 1

Library of Congress Cataloging in Publication Data

Main entry under title:

Two-phase flow and heat transfer in the power and
    process industries.

    Includes bibliographies and index.
    1. Two-phase flow.   2. Heat—Transmission.
3. Power-plants—Equipment and supplies.   4. Chemical
plants—Equipment and supplies.   I. Bergles, A. E.,
date
TA357.T94   621.402'2   80-22025
ISBN 0-07-004902-5

D
621.4022
TWO

Contents

Preface

This book grew out of discussions held during the NATO Advanced Study Institute on Two-Phase Flows and Heat Transfer held in Istanbul, Turkey in 1976. At that time a short course was proposed that would emphasize not only the fundamentals of two-phase flow and heat transfer but also industrial applications. Under cosponsorship of Hemisphere Publishing Corporation and Brookhaven National Laboratory, "Two-Phase Flow and Heat Transfer in the Power and Process Industries" was presented by AEB, JGC, JMD, and GFH in October 1978 and October 1980 at Brookhaven in Upton, New York. In March 1980, FM joined forces to present a revised version of the course in Hannover, under cosponsorship of Hemisphere and Universität Hannover.

An extensive series of lecture notes was developed for these courses. After consultation with Hemisphere Publishing Corporation, it was decided to publish the latest version of these notes as a book. While recognizing that this book is part of the continued "exponential" growth of literature on this subject, it is felt that the material will be of interest and value to students, researchers, and practitioners. The collection of material appears to be unique in that a complete overview is presented, ranging from fundamental methods of analysis to practical plant design and operational safety.

Since all of us are extensively involved in teaching and writing on two-phase flow and heat transfer, it is inevitable that we have drawn on material that has been presented elsewhere. This includes lectures at several NATO Advanced Study Institutes, Von Karman Institute for Fluid Dynamics Lecture Series, International Heat Transfer Conferences, and Multi-Phase Flow and Heat Transfer Symposium–Workshops—all of which have been published by Hemisphere. Additionally, some of the material by JGC is based on *Forced Convection Boiling and Condensation*, published by McGraw-Hill (UK). The material by GHF is partially derived from *Annular Two-Phase Flow* (Pergamon Press), *Two-Phase Flow and Heat Transfer* (Oxford University Press), and *Measurement of Two-Phase Flow Parameters* (Academic Press).

AEB wishes to acknowledge the support given for this project by the Alexander von Humboldt Foundation and the Institut für Verfahrenstechnik der Universität Hannover from 1979 to 1980.

The material in this book was typed by Mrs. Barbara Granito, and we would like to thank her for her careful and patient work. Our thanks are also due to Mr. William Begell, president of Hemisphere Publishing Corporation, and his staff for their support in this endeavor. We must also thank the students in our courses whose intelligent comments and questions have led to the better development of this material.

Finally, we would like to thank our wives (Penny, Ellen, France, Shirley, and Franziska) who bear so nobly their lonely lives as "two-phase flow widows."

A. E. Bergles
J. G. Collier
J. M. Delhaye
G. F. Hewitt
F. Mayinger

TWO-PHASE FLOW AND HEAT TRANSFER IN THE POWER AND PROCESS INDUSTRIES

Chapter 1

Two-Phase Flow Patterns

J. M. DELHAYE

1.1 Introduction

A two-phase mixture flowing in a pipe can exhibit several
interfacial geometries (bubbles, slugs, film,...). But this
geometry is not always clearly defined, which prevents the flow
patterns from being precisely and objectively described.

In single-phase flow, laminar and turbulent flows are
differently modelled. Laminar flows can be described by instantaneous quantities, the solutions of the Navier-Stokes equations,
whereas turbulent flows are described by time- or statistical-
averaged quantities which are the solutions of a system involving
the Reynolds equations and some closure equations.

Likewise, in two-phase flows, the flow patterns must be known
in order to model the physical phenomenon as closely as possible.
It is obviously impossible to describe bubbly flows and annular
flows with a good accuracy by means of the same model. It is far
better to adopt two different models, each one fitting the individual
flow description. Nevertheless, this approach is still difficult
because of the transition zone between two flow patterns and the
lack of physical knowledge to describe these buffer zones.

In addition to the random character of each flow configuration,
two-phase flows are never fully-developed. In fact the gas phase
expands due to the pressure drop along the pipe. This may lead to
a modification of the flow structure such as an evolution from
bubbly flow to slug flow. The flow pattern depends also upon the
singularities occurring along the pipes such as bends, junctions,
air-injection devices, etc...

The parameters which govern the occurrence of a given flow
configuration are numerous and it seems hopeless to try to represent
all the transitions on a two-dimensional flow chart. Among these
various parameters, one can select, (1) the volumetric flow-rates
of each phase, (2) the pressure, (3) the heat flux at the wall,
(4) the densities and viscosities of each phase, (5) the surface
tension, (6) the pipe geometry, (7) the pipe characteristic
dimension, (8) the angle of the pipe with respect to the horizontal
plane, (9) the flow direction (upward, downward, cocurrent,
countercurrent), (10) the inlet length, (11) the phase-injection
devices. The last two points are of the utmost importance and one

always has to keep in mind that a flow map is proposed for given
conditions and that it must be used for the same conditions.

The purpose of this chapter is to describe the different
configurations of gas-liquid pipe flows, to examine the transition
phenomena between the flow patterns and to propose a set of flow
maps allowing the determination of the flow pattern, given the
system parameters. This chapter starts with the definitions of
the basic quantities describing two-phase pipe flows and with a
survey of the experimental techniques which can be used to
recognize the flow structure.

1.2 Describing Parameters of Two-Phase Pipe Flows

Given the fluctuating character of two-phase flows, averaging
operators have to be introduced. These operators which act on
space or time domains are studied in detail in Chapter 2 of this
book. Only definitions and fundamental properties of these
operators are given hereunder.

1.2.1 *Phase Density Function*

The presence or absence of phase $k(k=1,2)$ at a given point $\mathbf{x}$
and a given time t is characterized by the unit value (or zero
value) of a phase density function $X_k(\mathbf{x},t)$ defined as follows:

$$X_k(\mathbf{x},t) \triangleq \begin{cases} 1 \text{ if point } \mathbf{x} \text{ pertains to phase k} \\ 0 \text{ if point } \mathbf{x} \text{ does not pertain to phase k} \end{cases}$$

$$(1.1)$$

The phase density function is a binary function analogous to the
intermittency function used in single-phase flow.

1.2.2 *Instantaneous Space-Averaging Operators*

Instantaneous field variables may be averaged over a line, an
area or a volume, i.e. over a n-dimension domain (n=1,2,3 for a
segment, an area or a volume). For instance, in a pipe flow, the
field variables can be averaged over a diameter, a chord, a plane
cross section or a finite control volume. At a given time, this
n-dimension domain $\mathcal{D}_n$ can be divided into two sub-domains $\mathcal{D}_{kn}$
pertaining to each phase (k=1,2),

$$\mathcal{D}_1 = \mathcal{D}_{11} + \mathcal{D}_{21}$$

Consequently two different *instantaneous space-averaging operators*
are introduced,

$$\{\ \}_n \triangleq \frac{1}{\mathcal{D}_n} \int_{\mathcal{D}_n} \mathrm{d}\mathcal{D}_n \qquad (1.2)$$

and

$$< >_n \overset{\Delta}{=} \frac{1}{\mathcal{D}_{kn}} \int_{\mathcal{D}_{kn}} d\mathcal{D}_n \qquad (1.3)$$

The *instantaneous space-fraction* R_{kn} is defined as the average over $\mathcal{D}_n$ of the phase density function $X_k(\mathbf{x},t)$,

$$R_{kn} \overset{\Delta}{=} \{X_k\}_n = \frac{\mathcal{D}_{kn}}{\mathcal{D}_n} \qquad (1.4)$$

This definition leads directly to the usual instantaneous space-fraction,
(i) over a segment

$$R_{k1} = L_k / \sum_{k=1,2} L_k \qquad (1.5)$$

where L_k is the cumulated length of the segments occupied by phase k
(ii) over a surface,

$$R_{k2} = A_k / \sum_{k=1,2} A_k \qquad (1.6)$$

where A_k is the cumulated area occupied by phase k
(iii) over a volume,

$$R_{k3} = V_k / \sum_{k=1,2} V_k \qquad (1.7)$$

where V_k is the volume occupied by phase k.

The *instantaneous volumetric flowrate* Q_k through a pipe cross section of area A is defined by,

$$Q_k \overset{\Delta}{=} \int_{A_k} w_k \, dA = A \, R_{k2} <w_k>_2 \qquad (1.8)$$

where w_k is the axial component of the velocity of phase k.

The *instantaneous mass flowrate* M_k is given by,

$$M_k \overset{\Delta}{=} \int_{A_k} \rho_k w_k dA = A \, R_{k2} <\rho_k w_k>_2 \qquad (1.9)$$

where ρ_k is the density of phase k.

1.2.3 Local Time-Averaging Operator

Local field variables can be averaged over a time interval $[t-T/2; t+T/2]$. As for single-phase turbulent flow, this time interval of magnitude T must be chosen large enough compared with the time scale of the turbulence fluctuations, and small enough compared with the time scale of the overall flow fluctuations. This is not always possible and a thorough discussion of this delicate question can be found in papers by Delhaye and Achard (1977, 1978).

If we consider a given point **x** in a two-phase flow, phase k passes this point intermittently, and a field variable f_k (**x**,t) associated with phase k is a piecewise continuous function. Denoting by T_k(**x**,t) the cumulated residence time of phase k within the interval T, we can define two different *local time-averaging operators*,

$$\overline{} \triangleq \frac{1}{T} \int_{[T]} dT \tag{1.10}$$

and

$$\overline{}^x \triangleq \frac{1}{T_k} \int_{[T_k]} dT \tag{1.11}$$

The *local time-fraction* α_k is defined as the average over T of the phase density function X_k,

$$\alpha_k(\mathbf{x},t) = \overline{X_k(\mathbf{x},t)} = \frac{T_k(\mathbf{x},t)}{T} \tag{1.12}$$

1.2.4 Commutativity of Averaging Operators

Considering all the definitions given previously, one can easily derive the following identity,

$$\overline{R_{kn} <f_k>_n} \equiv \{\overline{\alpha_k f_k}^x\}_n \tag{1.13}$$

A particular case for equation (1.13) is obtained by taking $f_k \equiv 1$, which leads to

$$\overline{R_{kn}} \equiv \{\alpha_k\}_n \tag{1.14}$$

Note that identities (1.13) and (1.14) are valid for segments (n=1), areas (n=2) or volumes (n=3).

As a consequence, the time-averaged volumetric and mass flow-rates can be expressed in the following ways,

$$\overline{Q_k} = A \; \overline{R_{k2} <w_k>_2} \equiv A \; \{\overline{\alpha_k w_k}^x \}_2 \qquad (1.14$$

$$\overline{M_k} = A \; \overline{R_{ke} <\rho_k w_k>_2} \equiv A \; \{\overline{\alpha_k \rho_k w_k}^x \}_2 \qquad (1.16)$$

1.2.5 *Qualities*

The *mass-velocity* $\overline{G}$ is defined by,

$$\overline{G} \overset{\Delta}{=} \frac{\overline{M}}{A} \qquad (1.17)$$

where $\overline{M}$ is the time-averaged total mass flowrate.

The (*true*) *quality* x is defined as the ratio of the gas mass-flowrate to the total mass-flowrate,

$$x \overset{\Delta}{=} \frac{\overline{M_G}}{\overline{M_G} + \overline{M_L}} = \frac{\overline{M_G}}{\overline{M}} \qquad (1.18)$$

It is currently impossible to measure or to calculate with a high precision the quality of a liquid-vapor mixture flowing in a heated channel and withstanding a phase change. Nevertheless, a fictitious quality, the so-called *equilibrium* or *thermodynamic quality*, can be calculated by assuming that both phases are flowing under saturation conditions, i.e. that their temperatures are equal to the saturation temperature corresponding to their common pressure.

1.2.6 *Volumetric Quantities*

The *volumetric quality* β is defined as the ratio of the gas volumetric flowrate to the total volumetric flowrate,

$$\beta \overset{\Delta}{=} \frac{\overline{Q_G}}{\overline{Q_G} + \overline{Q_L}} \qquad (1.19)$$

The *local volumetric flux* j_k is a local time-averaged quantity defined by,

$$j_k \overset{\Delta}{=} \overline{X_k w_k} = \alpha_k \; \overline{w_k}^x \qquad (1.20)$$

Its area-average J_k over the total cross section area A is a space/time-averaged quantity called the *superficial velocity*. This quantity is defined by,

$$j_k \overset{\Delta}{=} \{\overline{X_k w_k}\}_2 = \{j_k\}_2 \qquad (1.21)$$

This quantity is directly related to the volumetric flowrate. In
fact we have, taking identity (1.13) into account,

$$J_k = \{\overline{\alpha_k w_k}\}_2 = \overline{R_{k2}} \ \overline{<w_k>_2} = \overline{Q_k}/A \qquad (1.22)$$

If the density of phase k is constant, the superficial velocity J_k
can be expressed in terms of the quality x_k and the mass velocity
$\overline{G}$ as follows,

$$J_k = \overline{M_k}/\rho_k A = x_k \overline{G}/\rho_k \qquad (1.23)$$

The *mixture superficial velocity* J is defined as the sum of the
superficial velocities of each phase,

$$J \overset{\Delta}{=} J_1 + J_2 \qquad (1.24)$$

One can show that J is the velocity of a cross section plane throug
which the total volumetric flowrate is equal to zero.

1.3   Experimental Determination of Flow Patterns

Flow patterns can be recognized on an integral basis by
observing the flow itself or a picture or by looking at movies.
These techniques are necessary but not sufficient since they rarely
provide quantitative information. Very often they are supplemented
by an examination of the statistical properties of a local quantity
such as the phase density function, the instantaneous line fraction
or the instantaneous wall pressure.

1.3.1 *Integral Techniques*

If the pipe walls are transparent, the flow pattern can be
determined by taking pictures or normal or high-speed movies. When
the walls are opaque one can use X-ray photography or cinematograph
These optical techniques are described in detail by Delhaye (1979).

When motion pictures are taken, one can focus on a given zone
of the test section or one can track a specific item of the flow
such as a bubble, a gas plug or a roll wave. This is performed by
using a mirror which can rotate at an adjustable speed (Fig. 1.1).
The internal structure of the flow appears clearly and it has
even been possible to measure the speed of waves in an annular flow
(Hewitt and Lovegrove, 1970).

Flow patterns in forced convection boiling can be observed
directly by using specific geometries (Fig. 1.2) or transparent
heated walls (Fig. 1.3). If the walls are opaque X-rays or neutron
techniques can be employed (Bennett et al., 1965).

When it is based on a visual examination, the determination
of the transition between two flow configurations is a qualitative
concept which depends on the observer. This is why other methods

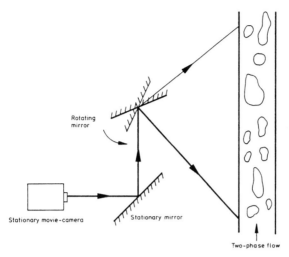

Figure 1.1: Bubble Tracker used by Roumy (1969)

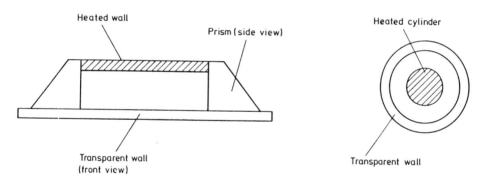

Figure 1.2: Visualization of Forced Convection Boiling in Channels of Particular Geometries.

are used, which are based on the statistical properties of a local fluctuating quantity.

1.3.2 *Local Techniques*

More specifically, flow patterns can be classified according to the properties of the probability density functions or power spectrum densities of some variables.

At a given point, the *phase density function* X_k can be

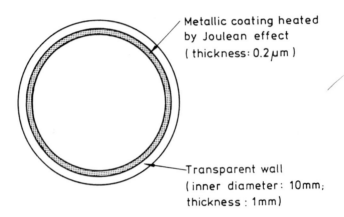

Metallic coating heated
by Joulean effect
(thickness: 0.2 μm)

Transparent wall
(inner diameter: 10mm;
thickness : 1mm)

Figure 1.3: Visualization of Forced Convection Boiling in Heated Wall Channels

determined by means of a small sized probe located at that point.
This probe must be sensitive to the phase which surrounds its
extremity, hence to the changes of a physical property between one
phase and the other. When the liquid phase is electrically
conducting one can use the change in the electrical conductivity
(Lackme, 1967; Raisson, 1968; Serizawa et al., 1975). When the
liquid phase is not electrically conducting, e.g. a Freon, one
can use the changes in the heat transfer coefficient or in the
optical index.

The *line gas-fraction* R_{G1} can be measured by an X-ray attenuation technique. Jones and Zuber (1975) give probability density
functions of R_{G1} as a function of the flow pattern for an air-water
mixture flowing upward in a vertical channel with a rectangular
cross section. Some of their results appear in Figure 1.4.

The fluctuations of the *pressure* at the wall also enable the
flow patterns to be classified. Hubbard and Dukler (1966) looked
at the power density spectra of the pressure signals for horizontal
air-water flow at atmospheric pressure. They classified the flow
structures according to the distribution of the energy versus the
frequency. Figure 1.5 shows some typical results.

1.4   Upward Cocurrent Flow in a Vertical Pipe

If the pipe axis is positively oriented in the upward direction
an upward cocurrent flow will be such that

$$\overline{Q_G} > 0 \qquad \overline{Q_L} > 0$$

1.4.1 *Description*

The main flow patterns encountered in a vertical pipe are
shown in Figure 1.6. *Bubbly flow* is certainly the most widely
known configuration, although at high velocity its milky appearance
prevents it from being easily recognized. Bubbles are spherical
only if their diameters do not exceed one millimeter, whereas,
beyond one millimeter, their shape is variable. Roumy (1969)
distinguishes two bubbly flow patterns. In the independent bubble

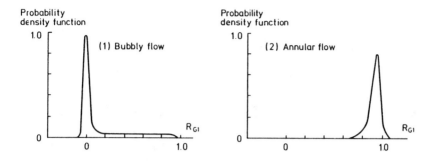

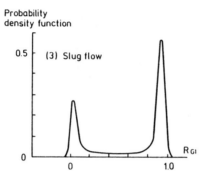

Figure 1.4: Probability Density Functions for R_{Gl} (Jones and Zuber, 1975)

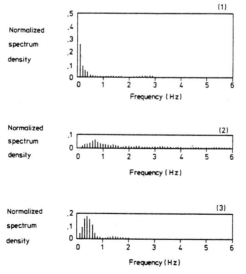

Figure 1.5: Power Density Spectra of Pressure Signals (Hubbard and Dukler, 1966)
(1) Separated Flow, (2) Dispersed Flow, (3) Intermittent Flow.

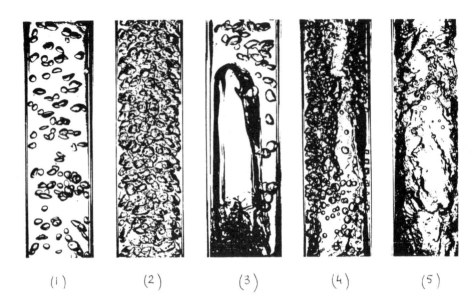

Figure 1.6: Upward Cocurrent Flow in a Vertical Pipe Air-water Flow Patterns
            (Roumy, 1969) (1) Independent bubbles, (2) Packed bubbles, (3) Slug
            flow, (4) Churn flow, (5) Annular flow. Pipe diameter : 32 mm

configuration, bubbles are spaced and non-interacting with each
other. On the contrary, in the packed configuration bubbles are
crowded together and strongly interacting with each other.

     Slug flow is composed of a series of gas plugs. The head of a
gas plug is generally blunt whereas its end is flat with a bubbly
wake. A simple visual observation reveals that the liquid film whic
surrounds a gas plug, moves downward with respect to the pipe wall.

     Given a constant liquid flowrate, an increase of the gas flow-
rate leads to a lengthening and a breaking of the gas plugs. The
flow pattern evolves toward an annular flow in a chaotic way.
This transition configuration is called a *churn flow.*

     Dispersed annular flow is characterized by a central gas core
loaded with liquid droplets and flowing at a higher velocity than
the liquid film which clings to the wall. Droplets are torn off
from the crest of the waves which propagate on the surface of the
liquid film. They diffuse in the gas core and can eventually
impinge onto the film surface. Hewitt and Roberts (1969) evidenced
the existence of a *wispy annular flow* where the liquid droplets
gather into clouds within the central gas core (Fig. 1.7).

     Finally, if the wall temperature is high enough to vaporize
the film, the droplets will constitute a *mist flow.*

     Figure 1.8 shows the configurations taken by a liquid-vapor
flow in a heated pipe as a function of the wall heat flux. The
liquid enters the pipe at a constant flowrate and at a temperature

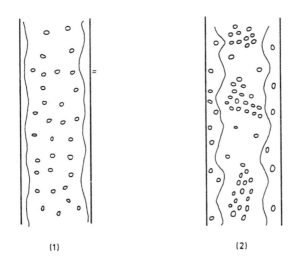

(1)   (2)

(1) Dispersed, (2) Wispy annular

Fig. 1.7: Annular flows

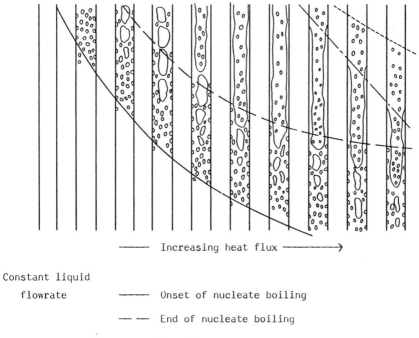

——— Increasing heat flux ———→

Constant liquid
flowrate   ——— Onset of nucleate boiling

                        — — End of nucleate boiling

                        — . — Dryout

                        ----- Limit of the superheated vapor region

Fig. 1.8: Convective boiling in a heated channel (Hewitt and Hall Taylor, 1970)

lower than the saturation temperature. When the heat flux increases
the vapor appears closer and closer to the pipe inlet. The *local
boiling length* is the extent of pipe where the bubbles are
generated at the wall and condense in the liquid core where the
liquid temperature is still lower than the saturation temperature.
The vapor is produced by two mechanisms, (1) wall nucleation, (2)
direct vaporization on the interfaces located in the flow itself.
The latter increases in importance as one proceeds along the pipe.
There is progressively less and less liquid between the wall and
the interfaces. Consequently the thermal resistance decreases.
So does the wall temperature, which brings the wall nucleation to
a stop. In annular flow, the liquid film flowrate decreases
through evaporation and entrainment of droplets, although some
droplets are redeposited. In heat flux controlled systems, when
the film is completely dried-out, the wall temperature rises
very quickly and can exceed the melting temperature of the wall.
This phenomenon has received several names such as dryout, burnout,
boiling crisis or critical heat flux. Note that the film dryout
in an annular flow does not provide the only explanation for a
boiling crisis, i.e. a sudden increase of the wall temperature.

1.4.2 *Flow Maps*

A flow map is a two dimensional representation of the flow
pattern existence domains. The coordinate systems are different
according to the authors and so far there is no agreement on the
best coordinate system. Selecting a series of flow maps is not easy
and no recommendation can be made since no method has proved
entirely adequate so far. Nevertheless the following charts can
give some expectations which, however, may be contradictory from
one method to the other.

Hewitt and Roberts' map (1969) is the most widely used chart
for air-water and steam-water flows. It was originally established
for an air-water mixture flowing in a pipe 31.2 mm in diameter at
a pressure varying from 1.4 to 5.4 bar. The coordinate system is
the following (Fig. 1.9),

(i) Abscissa:

$$\rho_L J_L^2 = \overline{G}^2 (1-x)^2/\rho_L \tag{1.25}$$

(ii) Ordinate:

$$\rho_G J_G^2 = \overline{G}^2 x^2/\rho_G \tag{1.26}$$

These coordinates have to be evaluated at the pressure of the
observed zone. The steam-water results of Bennett et al. (1965)
are also well represented in the Hewitt and Roberts' diagram.
They deal with steam-water mixtures flowing in a pipe 12.7 mm in
diameter at a pressure varying from 34.5 to 69 bar. In the
coordinate calculation the mass quality x is approximated by the
equilibrium quality.

Taitel and Dukler (1977) compared Hewitt and Roberts' techniqu
to several other methods, among them those developed by Govier and
Aziz (1972) and Oshinowo and Charles (1974). The comparison

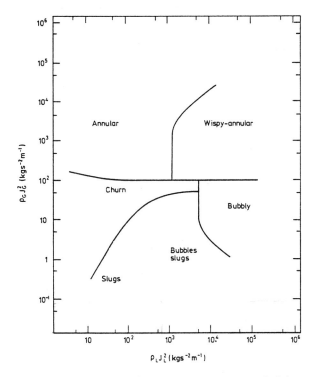

Fig. 1.9: Hewitt and Roberts' map (air-water flow up to 5.9 bar; steam-water flow up to 69 bar)

revealed several discrepancies not only on the quantitative aspects but also on the general trends of the transition curves. This can be easily explained due to the subjective definitions of the flow patterns and to the single representation for several transitions. Consequently, Taitel and Dukler (1977) went back to a physical analysis of the transitions and came up with the following conclusions,

(i) Transition between bubbly flow and slug flow or between bubbly flow and churn flow (Fig. 1.10)

The ordinate is the ratio J_L/J_G of the liquid superficial velocity to the gas superficial velocity. The abscissa is the quantity

$$\frac{J_G \rho_L^{\frac{1}{2}}}{[g(\rho_L - \rho_G)\sigma]^{\frac{1}{4}}}$$

where σ is the surface tension. The equation of the transition curve follows from the assumption of a change in the flow pattern occurring at a constant value of the void fraction. It reads,

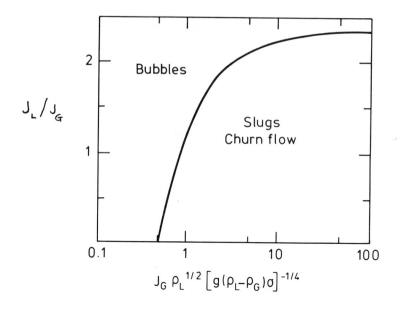

Fig. 1.10: Transition between Bubbly Flow and Slug Flow or between Bubbly Flow
and Churn Flow (Taitel and Dukler, 1977)

$$\frac{J_L}{J_G} = 2.34 - 1.07 \frac{\left[g(\rho_L - \rho_G)\sigma\right]^{\frac{1}{4}}}{J_G \rho_L^{\frac{1}{2}}} \qquad (1.27)$$

(ii)   Transition between slug flow and churn flow (Fig. 1.11)

The volumetric quality β is the ordinate whereas the abscissa
is the quantity $J/\sqrt{gD}$ where J is the mixture superficial
velocity (Eq. 1.24), g the acceleration due to gravity and D
the pipe diameter. The transition line has a complex equation
which involves the liquid Reynolds number $J_L D/\nu_L$ where ν_L is
the liquid kinematic viscosity. This equation follows from
an analysis of the behavior of the liquid slugs between two
Taylor bubbles.

(iii) Transition between slug flow and annular flow or between churn
flow and annular flow (Fig. 1.12)

According to the authors, the dispersed-annular flow breaks
down if the droplets are no longer sustained in the core by
the gas flow. As a result the transition line is represented
in a diagram where the ordinate number is the Kutateladze
number,

$$\frac{J_G \rho_G^{\frac{1}{2}}}{\left[g(\rho_L - \rho_G)\sigma\right]^{\frac{1}{4}}}$$

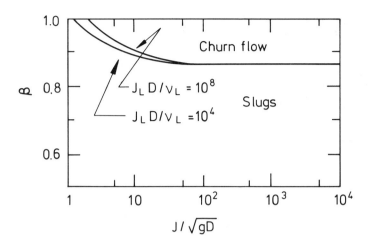

Figure 1.11: Transition between Slug Flow and Churn Flow (Taitel and Dukler, 1977)

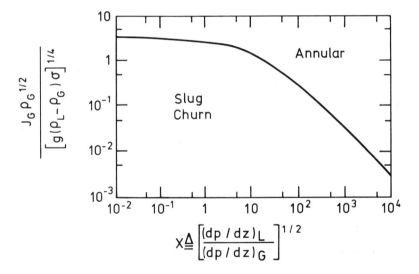

Figure 1.12: Transition between Slug Flow and Annular Flow or between Churn
Flow and Annular Flow (Taitel and Dukler, 1977)

whereas the abscissa is the Lockhart-Martinelli parameter X defined
by the following relation,

$$X \triangleq \left[\frac{(dp/dz)_L}{(dp/dz)_G} \right]^{\frac{1}{2}} \tag{1.28}$$

where $(dp/dz)_L$ and $(dp/dz)_G$ are the frictional pressure drops of

the liquid and the gas that would have been observed if these
fluids were flowing alone in the same pipe. The equation of the
transition line was obtained from theoretical considerations and
reads,

$$\frac{J_G \rho_G^{\frac{1}{2}}}{[g(\rho_L - \rho_G)\sigma]^{\frac{1}{4}}} = 3.09 \frac{(1+20X+X^2)^{\frac{1}{2}} - X}{(1+20X+X^2)^{\frac{1}{2}}} \qquad (1.29)$$

which simplifies to

$$\frac{J_G \rho_G^{\frac{1}{2}}}{[g(\rho_L - \rho_G)\sigma]^{\frac{1}{4}}} - 3.09 \text{ for } X \ll 1 \qquad (1.30)$$

and to

$$\frac{J_G \rho_G^{\frac{1}{2}}}{[g(\rho_L - \rho_G)\sigma]^{\frac{1}{4}}} = \frac{30.9}{X} \text{ for } X \gg 1 \qquad (1.31)$$

In order to use Taitel and Dukler's method one needs first to use
Figure 1.10, then Figure 1.12 if the flow is not a bubbly flow and
Figure 1.11 if the flow is neither bubbly nor annular.

The above technique for flow pattern discrimination has been
recently improved and simplified by Taitel and Dukler (1979).
According to a private communication from the authors, the agreement
with the experimental data is still better.

1.5 Downward Cocurrent Flows in a Vertical Pipe

If the pipe axis is positively oriented in the upward
direction a downward cocurrent flow will be such that

$$\overline{Q_G} < 0 \qquad \overline{Q_L} < 0$$

1.5.1 Description

So far the most comprehensive studies of downward cocurrent
flow patterns are due to Oshinowo and Charles (1974). These
authors distinguish six different flow configurations (Fig. 1.13).

The downward *bubbly flow* structure is quite different from
the upward bubbly flow configuration. In the latter case, bubbles
are spread over the entire pipe cross section whereas in the
downward flow bubbles gather near the pipe axis. This coring
effect is similar to the phenomenon observed in a flow of liquid
loaded with solid particles whose density is smaller than the
liquid density.

When the gas flowrate is increased, the liquid flowrate being
held constant, the bubbles agglomerate into large gas pockets.

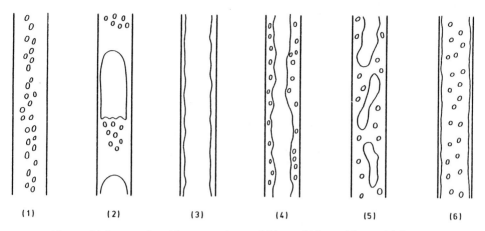

(1) Bubbles, (2) Slugs, (3) Falling film, (4) Bubbly
falling film, (5) Churn, (6) Dispersed annular

Figure 1.13: Downward Cocurrent Flow in a Vertical Pipe. Air-water Flow
Patterns (Oshinowo and Charles, 1974)

The top of these gas plugs is dome-shaped whereas the lower
extremity is flat with a bubbly zone underneath. This *slug flow* is
generally more stable than in the upward case.

The annular configuration can take several aspects. For
small liquid and gas flowrates, a liquid film flows down the wall
(*falling film flow*). If the liquid flowrate is higher bubbles are
entrained within the film (*bubbly falling film*). When liquid and
gas flowrates are increased a *churn flow* appears and can evolve
into a *dispersed annular flow* for very high gas flowrate.

1.5.2 *Flow Maps*

Oshinowo and Charles (1974) proposed a chart (Fig. 1.14) which
was obtained from their own experimental data. They studied two-
component mixtures of air and different liquids flowing in a pipe
25.4 mm in diameter at a pressure of around 1.7 bar. They chose
as abscissa and ordinate the quantities $Fr/\sqrt{\Lambda}$ and $\sqrt{\beta}(1-\beta)$ which are
calculated at the test section pressure and temperature. The Froude
number Fr is defined by the following relation,

$$Fr \stackrel{\Delta}{=} (J_G+J_L)^2/gD \tag{1.32}$$

where g is the acceleration due to the gravity and D the pipe
diameter. Λ is a coefficient which takes account of the liquid
physical properties. It is defined as follows,

$$\Lambda \stackrel{\Delta}{=} (\mu_L/\mu_W)\left[(\rho_L/\rho_W)\ (\sigma/\sigma_W)^3\right]^{-\frac{1}{4}} \tag{1.33}$$

μ, ρ, and σ designate respectively the liquid viscosity, density

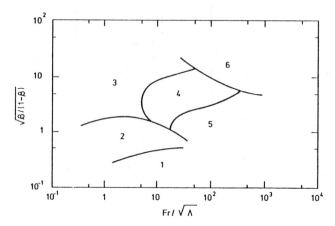

(1) Bubbles, (2) Slugs, (3) Falling film,
(4) Bubbly falling film, (5) Churn, (6) Dispersed
annular

Figure 1.14: Oshinowo and Charles Flow Map (1974)

and surface tension whereas W is a subscript referring to water at
20°C and 1 bar.

1.6   Cocurrent Flows in a Horizontal Pipe

The number of possible flow patterns in a horizontal pipe is
higher than in a vertical pipe. This is due to the effect of
gravity which tends to separate the phases and to create a
horizontal stratification.

1.6.1   *Description*

Alves (1954) proposed the classification shown in Figure 1.15.
In the *bubbly flow* configuration bubbles are moving in the upper
part of the pipe. When the gas flowrate is increased bubbles
coalesce and a *plug flow* takes place. For low liquid and gas
flowrates a *stratified flow* appears with a smooth interface. At
higher gas rates waves propagate along the interface (*wavy flow*)
and can reach the top wall of the pipe giving rise to a *slug flow*.
Finally at high gas flowrates and low liquid flowrates an *annular
flow* can exist with a thicker film in the lower part of the pipe.
These flow pattern names differ depending on the author. As an
example, Taitel and Dukler (1976) class plug flows and slug flows
under the same denomination (*intermittent flow*).

Figure 1.16 shows the evolution of a *vaporizing flow* in a
horizontal pipe. The liquid enters the heated pipe with a low
flowrate and at a temperature slightly lower than the saturation
temperature. An important remark concerns the upper part of the
tube which can dry out periodically and then suffer a sudden
increase in wall temperature. If the wall temperature is high
enough, the wall dries out completely and the liquid droplets form
a mist flow.

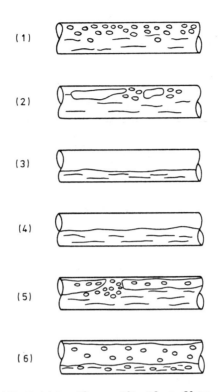

(1) Bubbly flow, (2) Plug flow,
(3) Stratified flow, (4) Wavy flow,
(5) Slug flow, (6) Annular flow

Figure 1.15: Flow Patterns in a Horizontal Pipe

Flow regimes in horizontal *condensing flows* were studied by
Soliman and Azer (1971) and also by Traviss and Rohsenow (1974)
who showed that the flow regimes could be predicted by the maps
used for adiabatic two-phase flow. However this conclusion was
questioned by Palen et al. (1977) who recommended additional work.

1.6.2 *Flow Maps*

Baker's diagram is still commonly used especially in petroleum
industries and for condenser design. The coordinates used by
Baker were simplified by Bell et al. (1970) and the resulting map
is shown in Figure 1.17. Quantities $\overline{G_G}$ and $\overline{G_L}$ are the superficial
gas and liquid mass velocities defined by the following relations,

$$\overline{G_G} \triangleq \overline{M_G}/A \qquad \overline{G_L} \triangleq \overline{M_L}/A \qquad\qquad (1.34)$$

where $\overline{M_G}$ and $\overline{M_L}$ are the gas and liquid mass flowrates and A the
pipe cross section area. Coefficients λ and ψ depend upon the
physical properties of the fluids and are defined as follows,

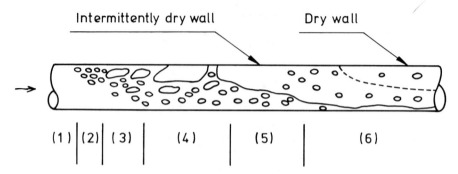

Intermittently dry wall   Dry wall

(1) | (2) | (3) | (4) | (5) | (6)

(1) Liquid single phase flow, (2) Bubbly flow, (3) Plug flow,
(4) Slug flow, (5) Wavy flow, (6) Annular flow

Figure 1.16: Flow Pattern Evaluation in a Horizontal Evaporator Tube

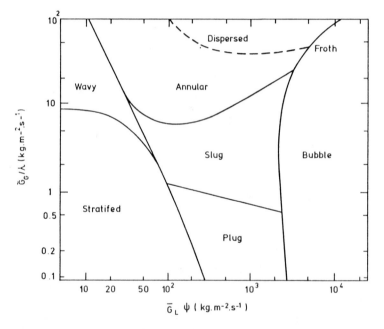

Figure 1.17: Modified Baker's Diagram (Bell et al., 1970)

$$\lambda \triangleq \left[\left(\frac{\rho_G}{\rho_A} \right) \left(\frac{\rho_L}{\rho_W} \right) \right]^{\frac{1}{2}} \tag{1.35}$$

$$\psi \triangleq \left(\frac{\sigma_{WA}}{\sigma} \right) \left[\left(\frac{\mu_L}{\mu_W} \right) \left(\frac{\rho_W}{\rho_L} \right)^2 \right]^{\frac{1}{3}} \tag{1.36}$$

Subscripts A and W refer respectively to the physical properties of air and water at 1 bar and 20°C. Consequently, for air-water flows at 1 bar and 20°C we have, $\lambda \equiv 1$ and $\psi \equiv 1$. For steam-water flows λ and ψ are given in Figure 1.18 as functions of the saturation pressure (Collier, 1972).

Mandhane et al. (1974) came up with a map based upon 5935 data points, 1178 of which concern air-water flows. Its coordinates are the superficial velocities J_L and J_G calculated at the test section pressure and temperature. This map (Fig. 1.19) is valid for the parameter ranges given in Table 1.1.

TABLE 1.1

Parameter Ranges for the Map Proposed by Mandhane et al. (1974)

Pipe inner diameter	12.7	– 165.1	mm
Liquid density	705	– 1009	$kg.m^{-3}$
Gas density	0.80	– 50.5	$kg.m^{-3}$
Liquid viscosity	$3 \ 10^{-4}$	– $9 \ 10^{-2}$	$kg.m^{-1}s^{-1}$
Gas viscosity	10^{-5}	– $2.2 \ 10^{-5}$	$kg.m^{-1}s^{-1}$
Surface tension	24	– 103	$mN.m^{-1}$
Liquid superficial velocity	0.09	– 731	$cm.s^{-1}$
Gas superficial velocity	0.04	– 171	$m.s^{-1}$

The general trends of Mandhane's map were found by *Taitel and Dukler* (1976) by means of a theoretical analysis of the flow pattern transitions. The theoretical chart proposed by Taitel and Dukler (Fig. 1.20) uses different coordinate systems according to the transitions which are considered.

(i) F versus X between stratified or wavy flow and intermittent flow with,

$$F \stackrel{\Delta}{=} \left(\frac{\rho_G}{\rho_L - \rho_G} \right)^{\frac{1}{2}} \frac{J_G}{(Dg \ \cos\alpha)^{\frac{1}{2}}} \qquad (1.37)$$

$$X \stackrel{\Delta}{=} \left[\frac{(dp/dz)_L}{(dp/dz)_G} \right]^{\frac{1}{2}} \qquad (1.38)$$

where D is the pipe diameter, α the pipe inclination angle with respect to the horizontal plane (positive for downward flow). The quantities $(dp/dz)_L$ and $(dp/dz)_G$ are the frictional pressure drops of the liquid and gas that would have been observed if these fluids were flowing alone in the pipe.

(ii) X constant between bubbly or intermittent flows and annular flows

(iii) K versus X between stratified flow and wavy flow, with K

J M Delhaye

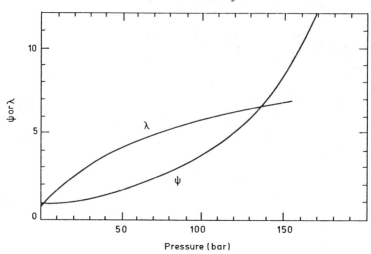

Figure 1.18: Values of λ and ψ for Steam-Water Flows (Collier, 1972)

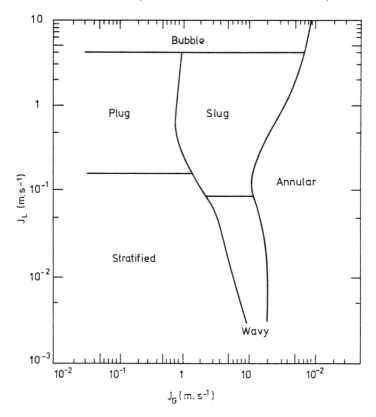

Figure 1.19: Flow Map Proposed by Mandhane et al. (1974)

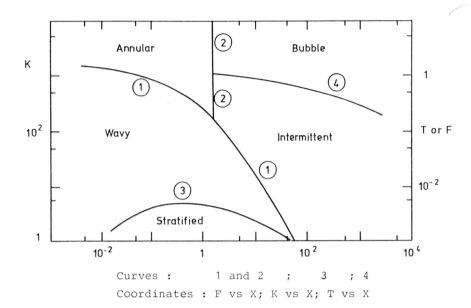

Curves :   1 and 2   ;   3   ; 4

Coordinates : F vs X; K vs X; T vs X

Figure 1.20: Theoretical Chart Proposed by Taitel and Dukler (1976)

defined by,

$$K \overset{\Delta}{=} \left[\frac{\rho_G J_G^2 J_L}{(\rho_L - \rho_G) g \nu_L \cos\alpha} \right]^{\frac{1}{2}} \qquad (1.39)$$

where ν_L is the liquid kinematic viscosity.

(iv) T versus X between bubbly flow and intermittent flow, with T defined by,

$$T \overset{\Delta}{=} \left[\frac{|dp/dz|_L}{(\rho_L - \rho_G) g \cos\alpha} \right]^{\frac{1}{2}} \qquad (1.40)$$

     The theoretical transitions obtained by Taitel and Dukler are compared in Figure 1.21 with Mandhane's map for an air-water mixture flowing in a 25 mm pipe at 1 bar and 25°C. Finally, we can note that Taitel and Dukler's method takes the pipe diameter into account.

1.7 Flow Patterns in a Rod Bundle

     Fuel elements of boiling water reactors and pressurized water reactors are generally assembled into rod bundles. Heat is extracted from the rods by water flowing parallel to their axes. Rod bundle geometries are described in detail by Lahey and Moody (1977) for boiling water reactors and by Tong and Weisman (1970) for pressurized water reactors. To our knowledge, only one paper

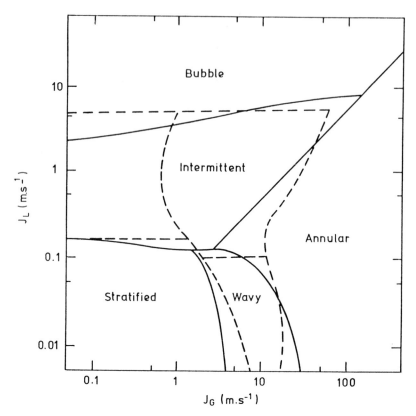

Theoretical transitions according to Taitel and
Dukler (1976)

------ Experimental diagram of Mandhane et al. (1974)

Figure 1.21: Horizontal Air-Water Flow at 1 bar and 25°C. Pipe inner diameter :
          25 mm

deals with flow patterns in rod bundles (Bergles, 1969). This is
not surprising, given the difficulty in introducing measurement
devices into the limited space between the rods.

     Bergles (1969) used resistive probes (see paragraph 1.2.2) to
determine the flow patterns at different points in a given cross
section of a four-rod bundle (Figure 1.22). Each rod was composed
of a 46 cm long non-heated zone and a 61 cm long heated zone.
Spacers located at 76 mm upstream and 25 mm downstream of the
heated portion enabled the rods to be positioned with respect to
the housing. Three resistive probes were placed in the interior
subchannel and in the corner subchannel as indicated in Figure
1.22 at 12.5 mm upstream of the end of the heated length. The
flow patterns were determined for a steam-water mixture at 69 bar.
Two cases were considered, (1) without heating (Fig. 1.23, 1.24 and

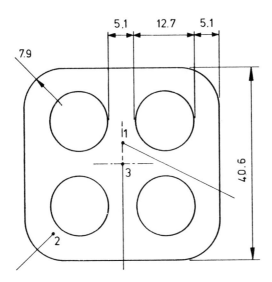

Figure 1.22: Dimensions in mm of the Four-Rod Bundle tested by Bergles (1969)

1.25), (2) with heating of the rods. Figure 1.23 shows a comparison between the flow pattern limits in the interior subchannel
(probes 1 and 3). Results are given in a $\bar{G}$ vs x_{eq} diagram where $\bar{G}$
is the mass velocity (Eq. 1.17) and x_{eq} the thermodynamic quality
calculated for the rod bundle as a whole. Actually, this representation is doubtful since $\bar{G}$ and x_{eq} are integral quantities
which do not refer to the particular subchannel under consideration. Nevertheless, this representation was adopted due to the
impossibility of obtaining subchannel quantities directly. On the
axis of the rod bundle (probe 3) changes in flow patterns occur at
lower qualities than within the gap between two rods (probe 1).
This indicates that the liquid tends to accumulate in the restricted
area. The interior subchannel flow regimes were compared with the
flow patterns obtained in a circular tube of the same hydraulic
diameter, at the same pressure. If the interior subchannel is
assumed to be limited by the rod surface and by the planes connecting their axes, the hydraulic diameter is found to be 12.6 mm.
Figure 1.24 shows a comparison between the flow pattern transitions
in the interior subchannel and those obtained in a tube, 10.2 mm in
diameter for the same length to hydraulic diameter ratio. The
agreement is good but should be considered as a coincidence since
the rod bundle results were obtained from integral values of the
mass velocity and quality instead of subchannel values. Finally,
flow patterns in the interior subchannel (point 3) and in the
corner subchannel (point 1) are compared in Figure 1.25. All the
transitions in the corner subchannel appear at higher qualities,
which indicates a liquid accumulation in this area. Hence, Figure
1.23 and 1.25 show clearly that different flow patterns can coexist
in a given rod bundle cross section. Limited data obtained with
heated rods tend to support this conclusion.

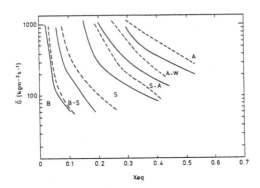

Figure 1.23: Comparison of Regimes at Two Points in Interior Subchannel
(------ Probe 1; ——— Probe 3) (B: bubbles; B-S: bubble-slug;
S: slug; S-A: slug-annular; A-W: annular waves; A: annular)

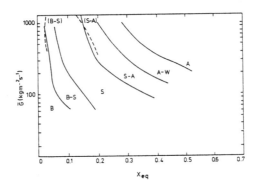

Figure 1.24: Comparison of Regimes for Interior Subchannel and Circular Tube
(——— Probe 3; ------ tube) (B: bubbles; B-S: bubble-slug; S:
slug; S-A: slug-annular; A-W: annular-waves; A: annular)

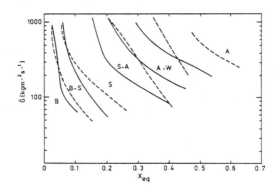

Figure 1.25: Comparison of Regimes for Interior and Corner Subchannel
(——— Probe 3; ------ Probe 2) (B: bubbles; B-S: bubble-slug;
S-A: slug-annular; A-W: annular-waves; A: annular)

1.8 **Flooding and** Flow Reversal

Flooding and flow reversal are basic mechanisms encountered in several phenomena occurring in nuclear reactor thermohydraulics, e.g. (i) rewetting from the top or countercurrent flow limiting conditions (Chan and Grolmes, 1975; Tien, 1977), (ii) motion of molten cladding material (Grolmes et al., 1974), (iii) downcomer flow pattern during an emergency core cooling (Block and Schrock, 1977).

Although more and more experiments are carried out on flooding and flow reversal, no adequate theories have been provided so far. However some empirical correlations have been developed but they can be used only in their range of validity. This seems a trivial statement but it appears that these correlations are occasionally applied to operating conditions far beyond their range of validity. Generally the ensuring results are not good, which is not surprising given the lack of physical bases for such empirical correlations.

1.8.1 *Description*

Figure 1.26 shows a vertical tube with a liquid injection device made up of a porous wall. At the beginning of the experiment there is no gas flow and a liquid film flows down the tube at a constant flowrate. If gas is flowing upward at a low rate, a countercurrent flow takes place in the tube (Fig. 1.26). When the gas flowrate is increased the film thickness remains constant and equal to the Nusselt thickness (Hewitt and Wallis, 1963). This thickness e is calculated by assuming that the flow is laminar and reads,

$$e = \left[\frac{3 \nu_L Q_L}{\pi D g} \right]^{\frac{1}{3}} \tag{1.41}$$

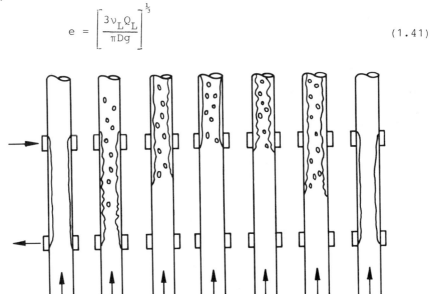

Figure 1.26: Flooding and Flow Reversal Experiment

where ν_L is the liquid kinematic viscosity, Q_L the liquid volumetric
flowrate, D the pipe diameter and g the acceleration due to gravity.
This relation is valid for liquid Reynolds numbers Re_L up to 4000
with,

$$Re_L \triangleq \frac{4Q_L}{\pi D \nu_L} \qquad\qquad (1.42)$$

For a higher gas flowrate the liquid becomes unstable and waves of
large amplitudes appear. Droplets are torn off from the crests of
the waves and are entrained with the gas flow above the liquid
injection level. These droplets diffuse in the gas core; some of
them impinge onto the wall and give rise to a liquid film (Fig.
1.26.2). Meanwhile the pressure drop in the tube above the liquid
injection level increases sharply as shown in Figure 1.27 (Dukler
and Smith, 1976). This constitutes the best experimental evidence
of the *flooding* phenomenon. When the gas flowrate is increased
further the tube section below the liquid injection level dries out
progressively (Fig. 1.26.3). No net liquid downflow appears for a
certain gas flowrate. However this transition value is not well
defined. At higher gas flowrates churn flow or annular flow appears
in the upper tube (Fig. 1.26.4).

    Suppose we decrease the gas flowrate. For a given value the
liquid film becomes unstable and large amplitude waves appear on
the interface. The pressure drop increases and the liquid film
tends to fall below the liquid injection device. This is called
the *flow reversal* phenomenon (Fig. 1.26.5). When the gas flowrate
decreases, the liquid film flows downward (Fig. 1.26.6) and the
upper tube dries out for a not-well defined value of the gas flowrate.

    The whole experiment shown in Figure 1.26 can be summarized
in the diagram of Figure 1.28.

1.8.2 *Experimental Correlations*

1.8.2.1 Flooding phenomenon. Wallis (1969) defines the following
non-dimensional superficial velocities,

$$J_G^* \triangleq \frac{J_G \rho_G^{\frac{1}{2}}}{\left[gD(\rho_L - \rho_G)\right]^{\frac{1}{2}}} \qquad\qquad (1.43)$$

$$J_L^* \triangleq \frac{J_L \rho_L^{\frac{1}{2}}}{\left[gD(\rho_L - \rho_G)\right]^{\frac{1}{2}}} \qquad\qquad (1.44)$$

where J_G and J_L are the superficial velocities given by Eq. (1.21).
Note that J_G^* and J_L^* are analogous to Froude numbers and represent
the ratio of the inertial forces to the gravity forces. The liquid
viscosity μ_L is taken into account by Wallis through a Grashof
number defined as

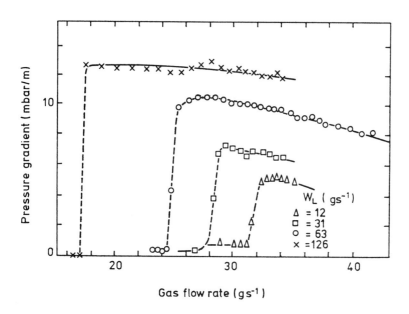

Figure 1.27: Pressure Drop versus Gas Flowrate (Dukler and Smith, 1976)

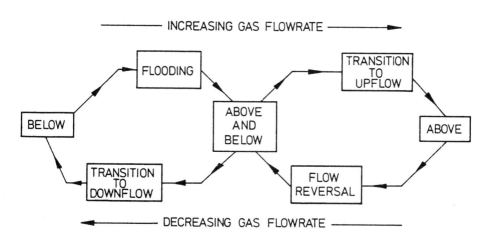

Figure 1.28: Flooding and Flow Reversal. Location of the Liquid Phase Above and/or Below the Liquid Injection Zone

$$N_L \overset{\Delta}{=} \left[\frac{\rho_L g D^3 (\rho_L - \rho_G)}{\mu_L^2} \right]^{\frac{1}{2}} \tag{1.45}$$

which represents the ratio of the gravity forces to the viscous forces. According to Wallis the flooding points can be correlated by the following formula,

$$J_G^{*\frac{1}{2}} + m\, J_L^{*\frac{1}{2}} = C \tag{1.46}$$

where m and C are given in Figure 1.29 and 1.30.

(i) When gravity forces are far more important than viscous forces, N_L is high and we have,

$$\begin{cases} m = 1 \\ 0.88 < C < 1 \quad \text{for round-edged tubes (Fig. 1.31.1)} \\ C = 0.725 \text{ for sharp-edged tubes (Fig. 1.31.2)} \end{cases}$$

Parameter C depends on the inlet and outlet conditions (Shires and Pickering, 1965), a fact which has been recognized by all the authors.

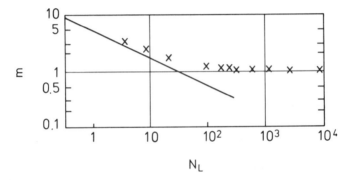

Figure 1.29: Value of m as a Function of N_L (Wallis, 1969)

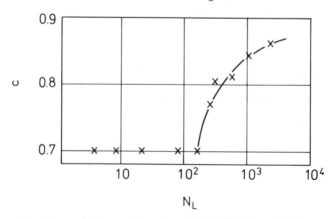

Figure 1.30: Value of C as a Function of N_L (Wallis, 1969)

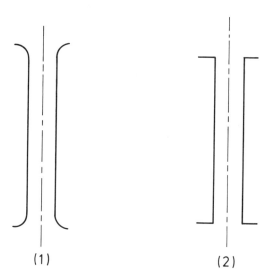

(1)   (2)

(1) Round-edged tubes   (2) Sharp-edged tube

Figure 1.31: Tube geometries

(ii) When gravity forces can be neglected with respect to viscous
forces, N_L is small and we have,

$$\begin{cases} m = 5.6 \ N_L^{-\frac{1}{2}} \\ C = 0.725 \text{ for round-edged tubes} \end{cases}$$

Despite its fairly good agreement with a relatively extended set
of data, the Wallis flooding correlation has a major drawback. It
does not take the length effect into account albeit this effect
can be very strong as shown in Figure 1.32.

     Suzuki and Ueda (1977) tried to overcome this snag but they
only came up with four different correlations which apply to four
different lengths!

1.8.2.2 Flow reversal phenomenon. Wallis (1969) proposed the
following correlation,

$$J_G^* = 0.5 \qquad\qquad\qquad\qquad (1.47)$$

The major disadvantage of this correlation lies in the diameter
effect which it involves. Actually the experimental results of
Pushkina and Sorokin (1969) did not show any diameter influence
as shown in Figure 1.33.

     Pushkina and Sorokin (1969) correlated their results fairly
well by the following expression,

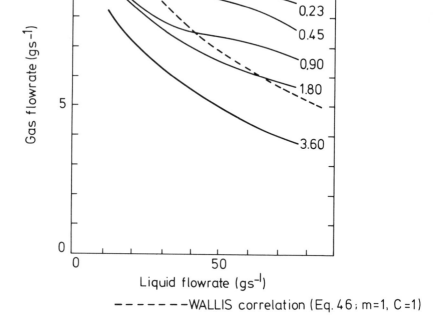

Figure 1.32: Length effect on flooding curve

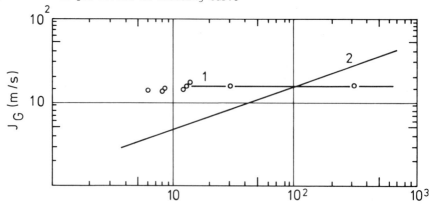

Tube diameter (mm)

(1) Pushkina and Sorokin results, (2) Wallis correlation (Eq. 1.47)

Figure 1.33: Diameter effect in flow reversal

$$Ku \overset{\Delta}{=} \frac{J_G \rho_G^{\frac{1}{2}}}{\left[g\sigma(\rho_L - \rho_G)\right]^{\frac{1}{4}}} = 3.2 \tag{1.48}$$

Ku is the Kutateladze number which represents the ratio of the gas inertial forces acting on capillary waves whose characteristic length is $\left[\sigma/(\rho_L - \rho_G)g\right]^{\frac{1}{2}}$. Pushkina and Sorokin (1969) ran their experiments with air and water flowing in tubes, 6 to 309 mm in diameter. In addition they checked their correlation against results obtained for other fluids by other authors. Note that for air and water, Eq. (1.48) yields a limiting air velocity of about 15.8 m/s at ambient pressure and temperature. In recent experiments with air and water flowing in large diameter tubes, up to 254 mm, Richter and Lovel (1977) confirmed the validity of the criterion proposed by Pushkina and Sorokin (Eq. 1.48).

Facing the diameter influence issue, Wallis tried to reconcile Pushkina and Sorokin criterion (Eq. 1.48) with his own correlation (Eq. 1.47). As a result, Wallis and Kuo (1976) proposed to correlate the Kutateladze number Ku as a function of a non-dimensional diameter D* defined by,

$$D^* \overset{\Delta}{=} D \left[\frac{(\rho_L - \rho_G)g}{\sigma}\right]^{\frac{1}{2}} \tag{1.49}$$

Note that D* is analogous to a Bond number which represents the ratio of gravity forces to surface tension forces. In the graph Ku vs D* the curves $J_G^* = $ const. are parabolas of equation

$$Ku = J_G^* \sqrt{D^*} \tag{1.50}$$

Figure 1.34 shows such a graph. However, Wallis and Kuo (1976) and Richter and Lovell (1977) pointed out that more experiments are needed to clear up the effects of length, diameter, surface tension, contact angle and viscosity.

1.8.3 Theoretical Approaches

The flooding phenomenon has been handled analytically by several authors. The onset of flooding has been explained by four different mechanisms.

(i)   occurrence of a standing wave on the liquid film (Shearer and Davidson, 1965)

(ii)   liquid film instabilities (Jameson and Cetinbudaklar, 1969; Cetinbudaklar and Jameson, 1969; Kusuda and Imura, 1974; Imura et al., 1977)

(iii) no net flow in the liquid film (Grolmes et al., 1974)

(iv)   droplet entrainment from the liquid film (Dukler and Smith, 1976)

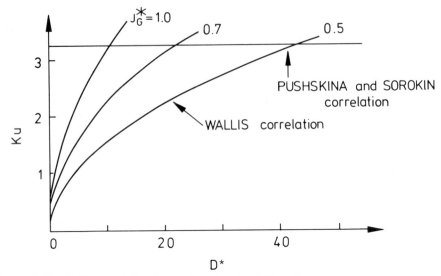

Figure 1.34: Wallis and Kuo Graph for Onset of Flow Reversal

The *flow reversal* phenomenon has been shown to be strongly connected to the hydrodynamics of hanging films which was studied by Wallis and Makkenchery (1974) and by Wallis and Kuo (1976).

Although a great effort has been made to understand these phenomena, no definite theoretical work has appeared so far and the issue is still open (Asahi, 1978; Bharathan, 1978; Duffey et al., 1978; Dukler and Smith, 1977; Dukler et al., 1978; Ueda and Suzuki, 1978).

Nomenclature

A	area
C	Wallis parameter (Eq. 1.41)
$\mathcal{D}$	domain
D	pipe diameter
D*	non-dimensional diameter (Eq. 1.44)
e	film thickness
F	Taitel and Dukler parameter (Eq. 1.32)
Fr	Froude number (Eq. 1.27)
f	arbitrary function
G	mass velocity (Eq. 1.17)
g	acceleration due to gravity
J	superficial velocity (Eq. 1.21)
J*	non-dimensional superficial velocity (Eqs. 1.38, 1.39)
j	local volumetric flux (Eq. 1.20)
K	Taitel and Dukler parameter (Eq. 1.34)

Ku Kutateladze number (Eq. 1.43)

L   length

M   mass flowrate (Eq. 1.9)

m   Wallis parameter (Eq. 1.41)

N   Grashof number (Eq. 1.40)

p   pressure

Q   volumetric flowrate (Eq. 1.8)

R   space-fraction (Eqs. 1.4, 1.5, 1.6, 1.7)

Re Reynolds number (Eq. 1.37)

T   time interval; Taitel and Dukler parameter (Eq. 1.35)

t   time

V   volume

w   axial velocity component

X   phase density function (Eq. 1.1); Lockhart and Martinelli parameter (Eqs. 1.23, 1.33)

x   quality

x   position vector

z   axial coordinate

α   local time-fraction (Eq. 1.12); pipe inclination angle

β   volumetric quality (Eq. 1.19)

Λ   Oshinowo and Charles parameter (Eq. 1.28)

λ   Baker parameter (Eq. 1.30)

μ   viscosity

ν   kinematic viscosity

ρ   density

σ   surface tension

ψ   Baker parameter (Eq. 1.31)

Subscripts

eq equilibrium

G   gas

k   phase index (k = 1, 2 or G, L)

L   liquid

n   dimension index

W   water at 20°C and 1 bar

Averaging operators

$\{\ \}_n$ space - averaging operator over $\mathcal{D}_n$ (Eq. 1.2)

$< >$ space - averaging operator over $\mathcal{D}_{kn}$ (Eq. 1.3)

———   time - averaging operator over $\lfloor T\rfloor$ (Eq. 1.10)

———x   time - averaging operator over $\lfloor T_k\rfloor$ (Eq. 1.11)

References

Alves, G E, (1954) "Cocurrent liquid-gas flow in a pipeline contactor". *Chem. Eng. Progr.*, **50** (9), 449-456.

Arnold, C R and Hewitt, G F, (1967) "Further developments in the photography of two-phase gas-liquid flow". *J. Photo Sci.*, **15**, 97-114.

Asahi, Y, (1978) "Flooding and flow reversal of two-phase annular flow". *Nuclear Engng. and Design*, **47**, 251-256.

Bell, K J, Taborek, J and Fenoglio, F, (1970) "Interpretation of horizontal in-tube condensation heat transfer correlations with a two-phase flow regime map". *Chem. Engng. Prog. Symposium Series*, **66** (102), 150-163.

Bennett, A W, Hewitt, G F, Kearsey, H A, Keeys, R K F and Lacey, P M C, (1965) "Flow visualization studies of boiling at high pressure". *Inst. Mech. Eng.*, *Proc. 1965-1966*, **180**, Part 3C, 260-270.

Bergles, A E, (1969) "Two-phase flow structure observations for high pressure water in a rod bundle". *Two-Phase Flow and Heat Transfer in Rod Bundles*, Schrock, V E, Ed., ASME, 47-55.

Bharathan, D, (1978) "Air-water countercurrent annular flow in vertical tubes". EPRI NP-786, Project 443-2, Interim Report.

Block, J A and Schrock, V E, (1977) "Emergency cooling water delivery to the core inlet of PWR's during LOCA". *Thermal and Hydraulic Aspects of Nuclear Reactor Safety*, **1**, *Light Water Reactors*, Jones, O C and Bankoff, S G, Eds., ASME, 109-132.

Cetinbudaklar, A G and Jameson, G J, (1969) "The mechanism of flooding in vertical countercurrent flow". *Chem. Eng. Sci.*, **24**, 1669-1680.

Chan, S H and Grolmes, M A, (1975) "Hydrodynamically-controlled rewetting". *Nuclear Engng. and Design*, **34**, 307-316.

Collier, J G, (1972) "Convective boiling and condensation", *McGraw-Hill*.

Cooper, K D, Hewitt, G F and Pinchin, B, (1964) "Photography of two-phase gas/liquid flow". *J. Photo. Sci.*, **12**, 269-278.

Delhaye, J M, (1979) "Optical methods in two-phase flows". *Proceedings of the Dynamic Flow Conference 1978, Dynamic Measurements in Unsteady Flow*, 321-343.

Delhaye, J M and Achard, J L, (1977) "On the use of averaging operators in two-phase flow modeling". *Thermal and Hydraulic Aspects of Nuclear Reactor Safety*, **1** : *Light Water Reactors*, Jones, O C and Bankoff, S G, Eds., ASME, 289-332.

Delhaye, J M and Achard, J L, (1978) "On the averaging operators introduced in two-phase flow modeling". *Transient Two-Phase Flow*, Proceedings of the CSNI Specialists Meeting, August 3-4, 1976, Toronto, Banerjee, S and Weaver, K R, Eds., AECL, **1**, 5-84.

Duffey, R B, Ackerman, M C, Piggott, B D G and Fairbairn, S A, (1978) "The effects of countercurrent single-and two-phase flows on the quenching rate of hot surfaces". *Int. J. Multiphase Flow*, **4**, 117-140.

Dukler, A E and Smith, L, (1976) "Two-phase interactions in countercurrent flow studies of the flooding mechanism", *US NRC Contract* AT (49-24) 0194, Summary report No 1.

Dukler, A E and Smith, L, (1977) "Two-phase interactions in countercurrent flow. Studies of the flooding mechanism". *US NRC Contract* AT (49-24) 0194, Summary report No 2.

Dukler, A E, Chopra, A, Moalim Maron, D and Semiat, R, (1978) "Two-phase interactions in countercurrent flow". *US NRC Contract* AT (49-24) 0194, Annual Report.

Govier, G W and Aziz, K, (1972) "The flow of complex mixtures in pipes". Van Nostrand-Reinhold.

Grolmes, M A, Lambert, G A and Fauske, H K, (1974) "Flooding in vertical tubes". *Symposium on Multiphase Flow Systems, Glasgow*, Paper A4.

Hewitt, G F and Hall-Taylor, N S, (1970) "Annular two-phase flow". Pergamon Press.

Hewitt, G F and Lovegrove, P C, (1970) "A scanning device for velocity measurement and its application to wave studies in annular two-phase flow". *J. of Physics E : Scientific Instruments*, **3**, 6-8.

Hewitt, G F and Roberts, D N, (1969) "Studies of two-phase flow patterns by simultaneous X-ray and flash photography". *AERE-M2159*.

Hewitt, G F and Wallis, G B, (1963) "Flooding and associated phenomena in falling film flow in a tube". *AERE-R4022*.

Hubbard, M G and Dukler, A E, (1966) "The characterization of flow regimes for horizontal two-phase flow : 1. Statistical analysis of wall pressure fluctuations". *Proceedings of the 1966 Heat Transfer and Fluid Mechanics Institute*, Saad, M A and Miller, J A, Eds., Stanford University Press, 100-121.

Imura, H, Kusuda, H and Funatsu, S, (1977) "Flooding velocity in a counter-current annular two-phase flow". *Chem. Engng. Sci.*, **32**, 79-87.

Jameson, G J and Cetinbudaklar, A, (1969) "Wave inception by air flow over a liquid film". *Cocurrent Gas-Liquid Flow*, Rhodes, E and Scott, D S, Eds., Plenum Press.

Jones, O C and Zuber, N, (1975) "The interrelation between void fraction fluctuations and flow patterns in two-phase flow". *Int. J. Multiphase Flow*, **2**, 273-306.

Kusuda, H and Imura, H, (1974) "Stability of a liquid film in a countercurrent annular two-phase flow (mainly on a critical heat flux in a two-phase thermosyphon)". *Bull. J.S.M.E.*, **17** (114), 1613-1618.

Lackme, C, (1967) "Structure et cinematique des ecoulements diphasiques a bulles". *CEA-R 3203*.

Lahey, R T and Moody, F J, (1977) "The thermal-hydraulics of a boiling water nuclear reactor". *American Nucl. Soc.*

Mandhane, J M, Gregory, G A and Aziz, K, (1974) "A flow pattern map for gas-liquid flow in horizontal pipes". *Int. J. Multiphase Flow*, **1**, 537-553.

Oshinowo, T and Charles, M E, (1974) "Vertical two-phase flow. Part 1. Flow pattern correlations". *The Canad. J. Chem. Engng.* **52**, 25-35.

Palen, J W, Breber, G and Taborek, J, (1977) "Prediction of flow regimes in horizontal tubeside condensation". *National Heat Transfer Conference, Salt Lake City.*

Pushkina, O L and Sorokin, Y L, (1969) "Breakdown of liquid film motion in vertical tubes". *Heat Transfer - Soviet Research.* **1** (5), 56-64.

Raisson, C, (1968) "Hydrodynamique de la double-phase a haute pression : evolution des configurations d'ecoulement jusqu'a l'echauffement critique". *These de docteur-ingenieur, Faculte des sciences, Universite de Grenoble.*

Richter, H J and Lovell, T W, (1977) "The effect of scale on two-phase counter-current flow flooding in vertical tubes". Final report on NRC *Contract No* AT (49-24) 0329, Thayer School of Engineering, Dartmouth College, Hanover, New Hampshire.

Roumy, R, (1969) "Structure des ecoulements dihasiques eau-air. Etude de la fraction de vide moyenne et des configurations d'ecoulement". CEA-R-3892.

Serizawa, A, Kataoka, I and Michiyoshi, I, (1975) "Turbulence structure of air-water bubbly flow - I. measuring techniques". *Int. J. Multiphase Flow*, **2**, 221-233.

Shearer, C J and Davidson, J F, (1965) "The investigation of a standing wave due to gas blowing upwards over a liquid film, its relation to flooding in wetted-wall columns". *J. Fluid Mech.*, **22**, Part 2, 321-335.

Shires, G L and Pickering, A R, (1965) "The flooding phenomenon in counter-current two-phase flow". *Proceedings on Symposium in Two-Phase Flow*, **2**, Exeter, B 501-B538.

Soliman, H M and Azer, N Z, (1971) "Flow patterns during condensation inside a horizontal tube". *ASHRAE Trans.* **77**, Part 1, No 2188, 210-224.

Suzuki, S and Ueda, T, (1977) "Behaviour of liquid films and flooding in counter-current two-phase flow. Part 1. Flow in circular tubes". *Int. J. Multiphase Flow*, **3**, 517-532.

Taitel, Y and Dukler, A E, (1976) "A model for predicting flow regime transitions in horizontal and near horizontal gas-liquid flow". *AICHEJ*, **22** (1), 47-55.

Taitel, Y and Dukler, A E, (1977) "Flow regime transitions for vertical upward gas-liquid flow : a preliminary approach through physical modelling". *AICHE 70th Annual Meeting, New York*, Session on Fundamental Research in Fluid Mechanics.

Taitel, Y and Dukler, A E, (1979) "Modelling flow pattern transitions for steady upward gas-liquid flow in vertical tubes". Submitted for publication.

Tien, C L, (1977) "A simple analytical model for counter-current flow limiting phenomena with vapor condensation". *Letters in Heat and Mass Transfer*, **4**, 231-238.

Tong, L S and Weisman, J, (1970) "Thermal analysis of pressurized water reactors". *American Nucl. Soc.*

Traviss, D P and Rohsenow, W M, (1973) "Flow regimes in horizontal two-phase flow with condensation". *ASHRAE Trans.*, No 2279, RP-63, 31-39.

Ueda, T and Suzuki, S, (1978) "Behaviour of liquid films and flooding in counter-current two-phase flow. Part 2. Flow in annuli and rod bundles". *Int. J. Multiphase Flow*, **4**, 157-170.

Wallis, G B, (1969) "One-dimensional two-phase flow". *McGraw-Hill.* {28,30,31}

Wallis, G B and Kuo, J T, (1976) "The behaviour of gas-liquid interfaces in vertical tubes". *Int. J. Multiphase Flow*, **2**, 521-536.

Wallis, G B and Makkenchery, S, (1974) "The hanging film phenomenon in vertical annular two-phase flow". *J. Fluids Engng.* 297-298.

Chapter 2

Basic Equations for Two-Phase Flow Modeling

J. M. DELHAYE

2.1 Introduction

This Chapter is essentially divided into two parts: Sections 2.2 to 2.5 deal with the different types of equations which can be used to describe a two-phase system whereas Sections 2.6 to 2.8 are devoted to a systematic classification of the most widely used models.

Section 2.2 provides the bases for *continuum mechanics of two-phase systems*. The local instantaneous equations are established and their direct utilization is exemplified for the dynamics of vapor bubbles and for the hydrodynamics of liquid films. Rational derivations of the instantaneous space-averaged and local time-averaged equations are given in Section 2.3 and 2.4 whereas the commutativity between time/space and space/time averaged equations is examined in Section 2.5.

Two-phase flow modeling is introduced via four wellknown models based on imposed velocity profiles. Section 2.6 deals with the presentation of the homogeneous, Bankoff, Wallis and Zuber-Findlay models. The nature of topological laws and the interaction laws required to close the set of equations of the two-fluid models is explained in Section 2.7. Finally, Section 2.8 gives a critical appraisal of some models based on specified evolutions.

2.2 Local Instantaneous Equations

Introduction

In single-phase flow, local balance laws at point x are expressed in terms of partial differential equations if point x does not belong to a surface of discontinuity. If point x belongs to a surface of discontinuity, the local balance laws are then formulated in terms of jump conditions which relate the values of the flow parameters on both sides of the surface of discontinuity (Truesdell and Toupin, 1960).

In two-phase flow, interfaces can be considered as surfaces of discontinuity. Consequently, the balance laws for each phase are expressed in terms of partial differential equations whereas on the interface they are formulated in terms of jump conditions.

Local instantaneous equations constitute the rational basis for almost all two-phase flow modeling procedures. They are used directly, e.g. in the study of bubble dynamics or film flows, or in an averaged form, e.g. in the study of pipe flows.

Jump conditions constitute a characteristic feature of two-phase flow analysis and provide relations between the phase inter-action terms which appear in the averaged equations (Delhaye 1974, Ishii 1975). Two kinds of jump conditions can be derived, the primary jump conditions and the secondary ones. The *primary jump conditions* are directly derived from the integral laws written for the following quantities: mass, linear momentum, angular momentum, total energy and entropy. The *secondary jump conditions* are established by combining the primary ones. They are written for the following quantities: mechanical energy, internal energy, enthalpy and entropy.

The interfacial entropy source, obtained from the entropy jump conditions, leads to the interfacial boundary conditions. If the interface transfers are assumed reversible it can thus be demonstrated that there is neither slip, nor temperature jump across the interface and that a certain relation exists between the free enthalpies of each phase.

The derivation of the local instantaneous equations starts with the integral balance laws written for a fixed control volume containing both phases. These integral laws are then transformed by means of the Leibniz rule and the Gauss theorems to obtain a sum of two volume integrals and a surface integral. The volume integrals lead to the local instantaneous partial differential equations valid in each phase whereas the surface integrals furnishes the local instantaneous jump conditions valid on the interface only.

Speed of Displacement of a Geometrical Surface (Truesdell and Toupin, 1960).

Consider a set of geometrical surfaces defined by the equation,

$$\boldsymbol{r} = \boldsymbol{r}(u,v,t) \tag{2.1}$$

where u and v are the coordinates of a point on this surface. The velocity of the surface point (u,v) is defined by,

$$\boldsymbol{v}_i = \left(\frac{\partial \boldsymbol{r}}{\partial t}\right)_{u,v \text{ const}} \tag{2.2}$$

The surface equation may also be expressed by,

$$f(x,y,z,t) = 0 \tag{2.3}$$

The normal unit vector $\boldsymbol{n}$ is related to the surface equation by,

$$n = \frac{\nabla f}{|\nabla f|} \qquad (2.4)$$

The speed of displacement of the surface is defined as the scalar product of v_i by n. We then deduce,

$$v_i \cdot n = \frac{-\partial f/\partial t}{|\nabla f|} \qquad (2.5)$$

This equation shows that the projection of v_i on the normal vector depends only on the surface equation expressed by (2.3). This equation is unique whereas expression (2.2) depends on the choice of parameters u and v.

Leibniz Rule (Truesdell and Toupin, 1960)

Let us consider a geometric volume $V(t)$ which is moving in space (Figure 2.1). This volume is bounded by a closed surface $A(t)$. At a given point belonging to this surface $A(t)$, let n be the unit normal vector, outwardly directed. The speed of displacement of the surface at that point is denoted $v_a \cdot n$. The Leibniz rule enables the time rate-of-change of a volume integral to be transformed into the sum of a volume integral and a surface integral,

$$\frac{d}{dt} \int_{V(t)} f(x,y,z,t)\, dV = \int_{V(t)} \frac{\partial f}{\partial t} dV + \oint_{A(t)} f v_A \cdot n dA \qquad (2.6)$$

Note that volume $V(t)$ is a geometric volume, not necessarily a material volume. If $V(t)$ is a material volume the Leibniz rule reduces to the Reynolds transport theorem.

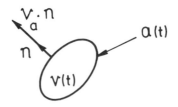

Figure 2.1

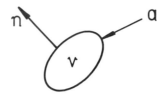

Figure 2.2

Gauss Theorems

Let us consider a geometric volume (Figure 2.2). This volume, bounded by surface A, is material or not, moving or not. At a given point of surface A, the unit normal vector $\mathbf{n}$ is outwardly directed. Let $\mathbf{B}$ and $\mathbf{M}$ be some vector and tensor fields. The Gauss theorems enable a surface integral to be transformed into a volume integral according to the following relations,

$$\oint_A \mathbf{n}\cdot\mathbf{B}\,dA = \int_V \nabla\cdot\mathbf{B}\,dV \tag{2.7}$$

$$\oint_A \mathbf{n}\cdot\mathbf{M}\,dA = \int_V \nabla\cdot\mathbf{M}\,dV \tag{2.8}$$

Integral Balances

Let us consider a fixed control volume V cut by an interface $A_i(t)$. This interface divides the control volume V into subvolumes $V_1(t)$ and $V_2(t)$, respectively bounded by surfaces $A_1(t)$ and $A_i(t)$, $A_2(t)$ and $A_i(t)$ (Figure 2.3).

Balance of mass. The time rate-of-change of mass contained in the control volume V is equal to the net influx of mass into V through the boundary surface A,

$$\frac{d}{dt}\int_{V_1(t)} \rho_1\,dV + \frac{d}{dt}\int_{V_2(t)} \rho_2\,dV = -\int_{A_1(t)} \rho_1\mathbf{v}_1\cdot\mathbf{n}_1\,dA - \int_{A_2(t)} \rho_2\mathbf{v}_2\cdot\mathbf{n}_2\,dA \tag{2.9}$$

where ρ_k and $\mathbf{v}_k$ are the density and the velocity vector of phase $k\,(k=1,2)$.

Balance of linear momentum. For the sake of simplicity surface tension is ignored and we postulate the following balance law for the control volume V. The time rate-of-change of linear momentum in the control volume V is equal to the sum of (1) the net influx of momentum into V through the boundary surface A, (2) the resultant of the external forces acting on volume V and on its boundary A. These forces are composed of volume forces, i.e. gravity forces, and surface forces, i.e. the stress tensor.

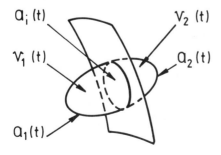

Figure 2.3

$$\frac{d}{dt} \int_{V_1(t)} \rho_1 \mathbf{v}_1 dV + \frac{d}{dt} \int_{V_2(t)}^{''} \rho_2 \mathbf{v}_2 dV =$$

$$- \int_{A_1(t)} \rho_1 \mathbf{v}_1 (\mathbf{v}_1 \cdot \mathbf{n}_1) dA - \int_{A_2(t)} \rho_2 \mathbf{v}_2 (\mathbf{v}_2 \cdot \mathbf{n}_2) dA$$

$$+ \int_{V_1(t)} \rho_1 \mathbf{F} dV + \int_{V_2(t)} \rho_2 \mathbf{F} dV$$

$$+ \int_{A_1(t)} \mathbf{n}_1 \cdot \mathbf{T}_1 dA + \int_{A_2(t)} \mathbf{n}_2 \cdot \mathbf{T}_2 dA \qquad (2.10)$$

where $\mathbf{F}$ is the external force per unit of mass and $\mathbf{T}$ the stress tensor.

Balance of angular momentum. The time rate-of-change of angular momentum in the control volume V is equal to the sum of (1) the net influx of angular momentum into V through the boundary surface A, (2) the resultant of the external torques acting on volume V and its boundary A,

$$\frac{d}{dt} \int_{V_1(t)} \mathbf{r} \times \rho_1 \mathbf{v}_1 \, dV + \frac{d}{dt} \int_{V_2(t)} \mathbf{r} \times \rho_2 \mathbf{v}_2 dV =$$

$$- \int_{A_1(t)} \mathbf{r} \times \rho_1 \mathbf{v}_1 (\mathbf{v}_1 \cdot \mathbf{n}_1) dA - \int_{A_2(t)} \mathbf{r} \times \rho_2 \mathbf{v}_2 (\mathbf{v}_2 \cdot \mathbf{n}_2) dA$$

$$+ \int_{V_1(t)} \mathbf{r} \times \rho_1 \mathbf{F} dV + \int_{V_2(t)} \mathbf{r} \times \rho_2 \mathbf{F} dV$$

$$+ \int_{A_1(t)} \mathbf{r} \times (\mathbf{n}_1 \cdot \mathbf{T}_1) dA + \int_{A_2(t)} \mathbf{r} \times (\mathbf{n}_2 \cdot \mathbf{T}_2) dA \qquad (2.11)$$

where $\mathbf{r}$ is the position vector.

Balance of total energy. The time rate-of-change of total energy (kinetic energy and internal energy) in the control volume V is equal to the sum of (1) the net influx of total energy into V through the boundary surface A, (2) the power of the external forces acting on volume V and on its boundary A, (3) the heat flux entering volume V through A,

$$\frac{d}{dt} \int_{V_1(t)} \rho_1 (\tfrac{1}{2} v_1^2 + u_1) dV + \frac{d}{dt} \int_{V_2(t)} \rho_2 (\tfrac{1}{2} v_2^2 + u_2) dV =$$

$$- \int_{A_1(t)} \rho_1 (\tfrac{1}{2} v_1^2 + u_1) \mathbf{v}_1 \cdot \mathbf{n}_1 dA - \int_{A_2(t)} \rho_2 (\tfrac{1}{2} v_2^2 + u_2) \mathbf{v}_2 \cdot \mathbf{n}_2 dA$$

$$+ \int_{V_1(t)} \rho_1 \mathbf{F} \cdot \mathbf{v}_1 dV + \int_{V_2(t)} \rho_2 \mathbf{F} \cdot \mathbf{v}_2 dV$$

$$+ \int_{A_1(t)} (n_1 \cdot T_1) \cdot v_1 dA + \int_{A_2(t)} (n_2 \cdot T_2) \cdot v_2 dA$$

$$- \int_{A_1(t)} q_1 \cdot n_1 dA - \int_{A_2(t)} q_2 \cdot n_2 dA \qquad (2.12)$$

where u is the internal energy per unit of mass and q the heat flux.

Balance of entropy. The time rate-of-change of entropy in the control volume V is equal to the sum of (1) the net influx of entropy into V through the boundary surface A due to the mass flow, (2) the net influx of entropy into V due to conduction through A, (3) the entropy source within volume V,

$$\frac{d}{dt} \int_{V_1(t)} \rho_1 s_1 dV + \frac{d}{dt} \int_{V_2(t)} \rho_2 s_2 dV$$

$$+ \int_{A_1(t)} \rho_1 s_1 v_1 \cdot n_1 dA + \int_{A_2(t)} \rho_2 s_2 v_2 \cdot n_2 dA$$

$$+ \int_{A_1(t)} \frac{1}{T_1} q_1 \cdot n_1 dA + \int_{A_2(t)} \frac{1}{T_2} q_2 \cdot n_2 dA$$

$$= \int_{V_1(t)} \Delta_1 dV + \int_{V_2(t)} \Delta_2 dV + \int_{A_i(t)} \Delta_i dA \geqslant 0 \qquad (2.13)$$

where the equality occurs when the evolution is reversible. In Eq. (2.13), s_k is the entropy per unit mass, T_k the absolute temperature, Δ_k the local entropy source per unit of volume and per unit of time within each phase and Δ_i the local entropy source per unit area of interface and per unit of time. As Eq. (2.13) is verified whatever $V_1(t)$ and $V_2(t)$ are, we then deduce that,

$$\Delta_k \geqslant 0 \qquad (k = 1,2) \qquad \Delta_i \geqslant 0 \qquad (2.14)$$

Generalized integral balance. Integral balances (2.9) to (2.13) can be rewritten under the following condensed form,

$$\sum_{k=1,2} \frac{d}{dt} \int_{V_k(t)} \rho_k \psi_k dV = - \sum_{k=1,2} \int_{A_k(t)} \rho_k \psi_k (v_k \cdot n_k) dA$$

$$+ \sum_{k=1,2} \int_{V_k(t)} \rho_k \phi_k dV - \sum_{k=1,2} \int_{A_k(t)} n_k \cdot J_k dA$$

$$+ \int_{A_i(t)} \phi_i dA \qquad\qquad (2.15)$$

For each balance law, the values of ψ_k, T_k, ϕ_k and ϕ_i are given in Table 1 where R is the antisymmetric tensor corresponding to vector $\mathit{\lambda}$ (Aris, 1962).

Table 2.1

Balance	ψ_k	J_k	ϕ_k	ϕ_i
Mass	1	0	0	0
Linear momentum	$\mathbf{v}_k$	$- T_k$	F	0
Angular momentum	$\mathit{\lambda} \times \mathbf{v}_k$	$- T_k \cdot R$	$\mathit{\lambda} \times F$	0
Total energy	$u_k + \frac{1}{2} v_k^2$	$q_k - T_k \cdot \mathbf{v}_k$	$F \cdot \mathbf{v}_k$	0
Entropy	s_k	$\frac{1}{T_k} q_k$	$\frac{1}{\rho_k} \Delta_k$	Δ_i

Transformation of the Integral Balance

The integral balance law (2.15) is transformed by means of the Leibniz rule (2.6) and the Gauss theorems (2.7, 2.8) into the following form,

$$\sum_{k=1,2} \int_{V_k(t)} \left[\frac{\partial}{\partial t} \rho_k \psi_k + \nabla \cdot (\rho_k \psi_k \mathbf{v}_k) + \nabla \cdot J_k - \rho_k \phi_k \right] dV$$

$$- \int_{A_i(t)} \left[\sum_{k=1,2} \dot{m}_k \psi_k + n_k \cdot J_k + \phi_i \right] dA = 0 \qquad (2.16)$$

with,

$$\dot{m}_k \triangleq \rho_k (\mathbf{v}_k - \mathbf{v}_i) \cdot n_k \qquad\qquad (2.17)$$

Eq. (2.16) has to be satisfied whatever $V_k(t)$ and $A_i(t)$ are. We thus deduce, (i) the local instantaneous phase equations

$$\frac{\partial}{\partial t} \rho_k \psi_k + \nabla \cdot (\rho_k \psi_k \mathbf{v}_k) + \nabla \cdot J_k - \rho_k \phi_k = 0 \qquad (2.18)$$

(ii) the local instantaneous jump condition

$$\sum_{k=1,2} (\dot{m}_k \psi_k + \mathbf{n}_k \cdot \mathbf{J}_k + \phi_i) = 0 \tag{2.19}$$

For each local instantaneous balance law, the values of ψ_k, $\mathbf{J}_k$, ϕ_k and ϕ_i are given in Table 2.1.

Local Instantaneous Equations in Each Phase

Primary equations.

- Mass:

$$\frac{\partial \rho_k}{\partial t} + \nabla \cdot (\rho_k \mathbf{v}_k) = 0 \tag{2.20}$$

- Linear momentum:

$$\frac{\partial \rho_k \mathbf{v}_k}{\partial t} + \nabla \cdot (\rho_k \mathbf{v}_k \mathbf{v}_k) - \rho_k \mathbf{F} - \nabla \cdot \mathbf{T}_k = 0 \tag{2.21}$$

- Angular momentum: This balance law does not supply any new local instantaneous equation. It simplifies to the relation,

$$\mathbf{T}_k = \mathbf{T}_k^{\ t} \tag{2.22}$$

where $\mathbf{T}_k^{\ t}$ is the transposed form of tensor $\mathbf{T}_k$. The stress tensor is thus symmetric.

- Total energy:

$$\frac{\partial}{\partial t} \left[\rho_k (\tfrac{1}{2} v_k^2 + u_k) \right] + \nabla \cdot \left[\rho_k (\tfrac{1}{2} v_k^2 + u_k) \mathbf{v}_k \right]$$

$$- \rho_k \mathbf{F} \cdot \mathbf{v}_k - \nabla \cdot (\mathbf{T}_k \cdot \mathbf{v}_k) + \nabla \cdot \mathbf{q}_k = 0 \tag{2.23}$$

- Entropy inequality:

$$\frac{\partial}{\partial t} (\rho_k s_k) + \nabla \cdot (\rho_k s_k \mathbf{v}_k) + \nabla \cdot \frac{\mathbf{q}_k}{T_k} = \Delta_k \geqslant 0 \tag{2.24}$$

Secondary equations.

- Mechanical energy equation: Multiplying the momentum equation (2.21) by $\mathbf{v}_k$ we get,

$$\frac{\partial}{\partial t}\left(\frac{1}{2}\rho_k v_k^2\right) + \nabla\cdot\left(\frac{1}{2}\rho_k v_k^2 \mathbf{v}_k\right)$$

$$- \rho_k \mathbf{F}\cdot\mathbf{v}_k - \nabla\cdot(\mathbf{T}_k\cdot\mathbf{v}_k) + \mathbf{T}_k : \nabla\mathbf{v}_k = 0 \qquad (2.25)$$

● Internal energy: Subtracting the mechanical energy equation (2.25) from the total energy (2.23) we obtain

$$\frac{\partial}{\partial t}(\rho_k u_k) + \nabla\cdot(\rho_k u_k \mathbf{v}_k) + \nabla\cdot\mathbf{q}_k - \mathbf{T}_k : \nabla\mathbf{v}_k = 0 \qquad (2.26)$$

● Enthalpy: We introduce now the enthalpy per unit of mass i_k and the viscous stress tensor τ_k. We have,

$$u_k = i_k - \frac{p_k}{\rho_k} \qquad (2.27)$$

$$\mathbf{T}_k = - p_k \mathbf{U} + \tau_k \qquad (2.28)$$

where p_k is the pressure within phase k and $\mathbf{U}$ the unit tensor.

Eq. (2.26) then reads,

$$\frac{\partial}{\partial t}(\rho_k i_k - p_k) + \nabla\cdot(\rho_k i_k \mathbf{v}_k) - \mathbf{v}_k\cdot\nabla p_k$$

$$+ \nabla\cdot\mathbf{q}_k - \tau_k : \nabla\mathbf{v}_k = 0 \qquad (2.29)$$

● Entropy: First we have to recall the fundamental thermodynamic equation (Callen, 1960),

$$u_k = u_k (s_k, \rho_k) \qquad (2.30)$$

and the definitions of pressure and temperature,

$$p_k \triangleq \rho_k^2 \left(\frac{\partial u_k}{\partial \rho_k}\right)_{s_k} \qquad (2.31)$$

$$T_k \triangleq \left(\frac{\partial u_k}{\partial s_k}\right)_{\rho_k} \qquad (2.32)$$

Eq. (2.30) and definitions (2.31, 2.32) lead directly to the Gibbs equations,

$$\frac{du_k}{dt} = T_k \frac{ds_k}{dt} - p_k \frac{d}{dt} (\frac{1}{\rho_k})$$ (2.33)

This equation is used to transform the internal energy equation (2.26) into an entropy equation. If we take into account the mass balance (2.20) and the definition of the Lagrangian derivative,

$$\frac{d}{dt} \triangleq \frac{\partial}{\partial t} + \mathbf{v}_k \cdot \nabla$$ (2.34)

we can transform the internal energy equation (2.26) into,

$$\rho_k \frac{du_k}{dt} + \nabla \cdot \mathbf{q}_k + p_k \nabla \cdot \mathbf{v}_k - \tau_k : \nabla \mathbf{v}_k = 0$$ (2.35)

On the other hand, the Gibbs equation can be rewritten by means of (2.20) and (2.34) as,

$$\frac{du_k}{dt} = T_k \frac{ds_k}{dt} - \frac{p_k}{\rho_k} \nabla \cdot \mathbf{v}_k$$ (2.36)

Combining (2.35) and (2.36) we obtain,

$$\rho_k \frac{ds_k}{dt} + \nabla \cdot \frac{\mathbf{q}_k}{T_k} - \mathbf{q}_k \cdot \nabla \frac{1}{T_k} - \frac{1}{T_k} \tau_k : \nabla \mathbf{v}_k = 0$$ (2.37)

Entropy source. By means of the mass balance (2.20) and of the Lagrangian derivative definition (2.34), the entropy inequality (2.24) becomes,

$$\rho_k \frac{ds_k}{dt} + \nabla \cdot \frac{\mathbf{q}_k}{T_k} = \Delta_k \geqslant 0$$ (2.38)

By comparing (2.37) and (2.38) we find immediately the expression for the entropy source,

$$\Delta_k = \mathbf{q}_k \cdot \nabla \frac{1}{T_k} + \frac{1}{T_k} \tau_k : \nabla \mathbf{v}_k \geqslant 0$$ (2.39)

This condition leads to restrictions on the constitutive laws written for the heat flux $\mathbf{q}$ and the viscous stress tensor τ. One can show that Eq. (2.39) implies that the heat conductivity and the viscosity are positive.

J M Delhaye

Local Instantaneous Equations at the Interface

Primary jump conditions.

- Mass

$$\rho_1 (v_1 - v_i) \cdot n_1 + \rho_2 (v_2 - v_i) \cdot n_2 = 0 \qquad (2.40)$$

or

$$\dot{m}_1 + \dot{m}_2 = 0 \qquad (2.41)$$

with

$$\dot{m}_k \overset{\Delta}{=} \rho_k (v_k - v_i) \cdot n_k \qquad (2.42)$$

If there is no mass transfer we have,

$$\dot{m}_1 \equiv \dot{m}_2 \equiv 0 \qquad (2.43)$$

and consequently,

$$(v_1 - v_i) \cdot n_1 = 0 \qquad (2.44)$$

$$(v_2 - v_i) \cdot n_2 = 0 \qquad (2.45)$$

Eliminating the interface displacement velocity we obtain,

$$(v_1 - v_2) \cdot n_1 = 0 \qquad (2.46)$$

Resolving v_1 and v_2 into their normal and tangential components,

$$v_1 = (v_1 \cdot v_1) n_1 + v_1^t \qquad (2.47)$$

$$v_2 = (v_2 \cdot n_2) n_2 + v_2^t \qquad (2.48)$$

If we admit no slip of one phase on the other at the interface (see Section 2.8), we have,

$$v_1^t \equiv v_2^t \qquad (2.49)$$

which, taking into account Eqs (2.46) to (2.49) gives

$$v_1 \equiv v_2 \qquad (2.50)$$

Linear Momentum

In the absence of surface tension we have,

$$\dot{m}_1 \mathbf{v}_1 + \dot{m}_2 \mathbf{v}_2 - \mathbf{n}_1 \cdot T_1 - \mathbf{n}_2 \cdot T_2 = 0 \tag{2.51}$$

or taking into account the mass balance (2.43),

$$\dot{m}_1 (\mathbf{v}_1 - \mathbf{v}_2) - \mathbf{n}_1 \cdot T_1 - \mathbf{n}_2 \cdot T_2 = 0 \tag{2.52}$$

Inviscid phases: The stress tensor simplifies to the pressure term only. Therefore Eq. (2.52) becomes,

$$\dot{m}_1 (\mathbf{v}_1 - \mathbf{v}_2) + (p_1 - p_2) \mathbf{n}_1 = 0 \tag{2.53}$$

Separating the normal and tangential components of $\mathbf{v}_k$, we have

$$\dot{m}_1 (\mathbf{v}_1{}^n - \mathbf{v}_2{}^n) + \dot{m}_1 (\mathbf{v}_1{}^t - \mathbf{v}_2{}^t) + (p_1 - p_2) \mathbf{n}_1 = 0 \tag{2.54}$$

with

$$\mathbf{v}_k = \mathbf{v}_k{}^n + \mathbf{v}_k{}^t \tag{2.55}$$

By resolving Eq. (2.54) into the normal and tangential components, we deduce,

$$\dot{m}_1 (\mathbf{v}_1{}^n - \mathbf{v}_2{}^n) + (p_1 - p_2) \mathbf{n}_1 = 0 \tag{2.56}$$

$$\mathbf{v}_1{}^t - \mathbf{v}_2{}^t$$

We can therefore conclude that if there is mass transfer between two inviscid fluids there is no slip at the interface.

Viscous phase: Consider a one dimensional flow in which the interface is perpendicular to the flow direction (Figure 2.4).

The momentum jump condition (2.52) reads,

$$m_1 (\mathbf{v}_1 - \mathbf{v}_2) + (p_1 - p_2) \mathbf{n}_1 + (\tau_2 - \tau_1) \cdot \mathbf{n}_1 = 0 \tag{2.58}$$

Multiplying scalarly by $\mathbf{n}_1$, we obtain,

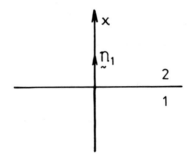

Figure 2.4.

$$\dot{m}_1 (\mathbf{v}_1 - \mathbf{v}_2) \cdot \mathbf{n}_1 + (p_1 - p_2) + \left[(\tau_2 - \tau_1) \cdot \mathbf{n}_1\right] \cdot \mathbf{n}_1 = 0 \qquad (2.59)$$

Since the flow is assumed to be one-dimensional, the mass balance in phase k, which is assumed to be incompressible is simplified to

$$\frac{du_k}{dx} \equiv 0 \qquad (2.60)$$

where u_k is the x-component of $\mathbf{v}_k$

Consequently,

$$\tau_k \equiv 0 \qquad (2.61)$$

The momentum jump condition (2.59) is therefore simplified to,

$$\dot{m}_1 (\mathbf{v}_1 - \mathbf{v}_2) \cdot \mathbf{n}_1 + p_1 - p_2 = 0 \qquad (2.62)$$

But since we have from Eq. (2.42),

$$\left(\frac{1}{\rho_1} - \frac{1}{\rho_2}\right) \dot{m}_1 = (\mathbf{v}_1 - \mathbf{v}_2) \cdot \mathbf{n}_1 \qquad (2.63)$$

Eq. (2.62) leads to

$$p_1 - p_2 = \frac{\rho_1 - \rho_2}{\rho_1 \rho_2} \dot{m}_1^2 \qquad (2.64)$$

The pressure is therefore higher in the denser fluid, whatever the transfer sense ($\dot{m}_1 \gtrless 0$). This conclusion is sometimes forgotten in the study of hydrodynamics of liquid films with interfacial change (Spindler et al. 1978).

Two-dimensional case with surface tension: If surface tension is taken into account the momentum jump condition is written

(Delhaye, 1974)

$$\dot{m}_1 v_1 + \dot{m}_2 v_2 - n_1 \cdot T_1 - n_2 \cdot T_2 + \frac{d\sigma}{d\ell} t - \frac{\sigma}{R} n_1 = 0 \qquad (2.65)$$

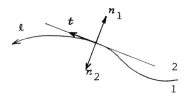

Figure 2.5.

where R is the radius of curvature, t the unit tangential vector and ℓ the curvilinear abscissa along the interface (Figure 2.5).

The fifth term in Eq. (2.65) expresses the mechanical effects due to surface tension gradients and is responsible for the Marangoni effect (Kenning, 1968; Levich and Krylov, 1969). If there is no mass transfer and if the phases are inviscid, Eq. (2.65) becomes

$$(p_1 - p_2) n_1 + \frac{d\sigma}{d\ell} t - \frac{\sigma}{R} n_1 = 0 \qquad (2.66)$$

This relation implies,

$$\frac{d\sigma}{d\ell} = 0 \qquad (2.67)$$

which is not the general case. Consequently, in the absence of phase change, the hypothesis of inviscid phases is not consistent with the presence of surface tension gradients. If we assume that the phases are viscous, then necessarily the surface tension gradients will induce fluid motions. If surface tension is constant, the momentum jump condition (2.66) simplifies to,

$$p_1 - p_2 = \frac{\sigma}{R} \qquad (2.68)$$

which is the well-known Young-Laplace law.

Angular Momentum

No new information is afforded by the angular momentum jump condition.

Total Energy

$$\dot{m}_1 (u_1 + \frac{1}{2} v_1^2) + \dot{m}_2 (u_2 + \frac{1}{2} v_2^2) + q_1 \cdot n_1 + q_2 \cdot n_2$$

$$- (n_1 \cdot T_1) \cdot v_1 - (n_2 \cdot T_2) \cdot v_2 = 0 \qquad (2.69)$$

Entropy Inequality

$$\Delta_i = - \dot{m}_1 s_1 - \dot{m}_2 s_2 - \frac{1}{T_1} q_1 \cdot n_1 - \frac{1}{T_2} q_2 \cdot n_2 \geqslant 0 \qquad (2.70)$$

Secondary jump conditions.

- Mechanical energy

  A mechanical energy jump condition can be derived by multiplying the momentum jump condition (2.51) by the velocity v_p of the interfacial fluid particles, whose normal component is the speed of displacement of the interface. We thus have,

$$v_p = (v_i \cdot n_k) n_k + v^t \qquad (2.71)$$

  where v^t is the tangential component of v_p.
  Eq. (2.51) becomes,

$$\dot{m}_1 v_1 \cdot v_p + \dot{m}_2 v_2 \cdot v_p - (n_1 \cdot T_1) \cdot v_p - (n_2 \cdot T_2) \cdot v_p = 0 \quad (2.72)$$

- Internal energy

  Subtracting the mechanical energy jump condition (2.72) from the total energy jump condition (2.69) we obtain,

$$\dot{m}_1 \left[u_1 + \frac{1}{2} (v_1 - v_p)^2 \right] + \dot{m}_2 \left[u_2 + \frac{1}{2} (v_2 - v_p)^2 \right]$$
$$+ q_1 \cdot n_1 + q_2 \cdot n_2$$
$$- (n_1 \cdot T_1) \cdot (v_1 - v_p) - (n_2 \cdot T_2) \cdot (v_2 - v_p) = 0 \qquad (2.73)$$

Unlike what occurs in the phase equation derivation, the kinetic energy of each phase is not eliminated. In fact, the method that we are using here enables the interface kinetic energy to be eliminated when the interfacial thermodynamic properties are taken into account (Delhaye, 1974).

- Enthalpy

  From the definition (2.42) of $\dot{m}_k$ we can derive the following identity,

$$v_k - v_p = \frac{\dot{m}_k}{\rho_k} n_k + v_k^t - v^t \qquad (2.74)$$

where v_k^t is the tangential component of v_k. Taking into account the definitions of the enthalpy (2.27) and of the stress tensor (2.28), Eq. (2.73), combined with Eq. (2.74), can be transformed into,

$$\dot{m}_1 \left[i_1 + \frac{1}{2}(v_1 - v_p)^2 - \frac{1}{\rho_1} (\tau_1 \cdot n_1) \cdot n_1 \right]$$

$$+ \dot{m}_2 \left[i_2 + \frac{1}{2}(v_2 - v_p)^2 - \frac{1}{\rho_2} (\tau_2 \cdot n_2) \cdot n_2 \right]$$

$$+ q_1 \cdot n_1 + q_2 \cdot n_2$$

$$- (\tau_1 \cdot n_1) \cdot (v_1^t - v^t) - (\tau_2 \cdot n_2) \cdot (v_2^t - v^t) = 0 \qquad (2.75)$$

• Entropy

In order to transform the enthalpy jump condition (2.75) into an entropy jump condition, we have to introduce the free enthalpies g_k of each phase,

$$g_k \overset{\Delta}{=} i_k - T_k s_k \qquad (2.76)$$

Eq. (2.75) then becomes,

$$\dot{m}_1 \left[T_1 s_1 + g_1 + \frac{1}{2}(v_1 - v_p)^2 - \frac{1}{\rho_1} (\tau_1 \cdot n_1) \cdot n_1 \right]$$

$$+ \dot{m}_2 \left[T_2 s_2 + g_2 + \frac{1}{2}(v_2 - v_p)^2 - \frac{1}{\rho_2} (\tau_2 \cdot n_2) \cdot n_2 \right]$$

$$+ q_1 \cdot n_1 + q_2 \cdot n_2$$

$$- (\tau_1 \cdot n_1) \cdot (v_1^t - v^t) - (\tau_2 \cdot n_2) \cdot (v_2^t - v^t) = 0 \qquad (2.77)$$

Interfacial entropy source. In order to combine the entropy inequality (2.70) with the entropy equation (2.77) we have to introduce an arbitrary temperature T_i which appears to be the interface temperature when the interface thermodynamic properties are taken into account (Delhaye, 1974; Ishii, 1975). This leads to the following expression for the entropy source

$$\Delta_i = \sum_{k=1,2} \left\{ \frac{\dot{m}_k}{T_i} \left[g_k + \frac{1}{2}(v_k - v_p)^2 - \frac{1}{\rho_k}(\tau_k \cdot n_k) \cdot n_k \right] \right.$$

$$\left. + (q_k \cdot n_k + \dot{m}_k s_k T_k) \left(\frac{1}{T_i} - \frac{1}{T_k} \right) \right.$$

$$- \frac{1}{T_i} (\tau_k \cdot n_k) \cdot (v_k^t - v^t) \bigg\} \geqslant 0 \qquad (2.78)$$

If we assume that the interface transfers are reversible, the entropy source Δ_i must be equal to zero whatever the mass flux, the viscous stress tensors and the heat fluxes are. We then deduce the following interfacial boundary conditions,

(i)   Thermal boundary condition

$$T_1 \equiv T_2 \equiv T_i \qquad (2.79)$$

(ii)   Mechanical boundary condition

$$v_1^t \equiv v_2^t \equiv v^t \qquad (2.80)$$

(iii) Phase change boundary condition

$$g_1 - g_2 = \left[\frac{1}{2}(v_2 - v_p)^2 - \frac{1}{2}(v_1 - v_p)^2 \right]$$
$$- \left[\frac{1}{\rho_2}(\tau_2 \cdot n_2) \cdot n_2 - \frac{1}{\rho_1}(\tau_1 \cdot n_1) \cdot n_1 \right] \qquad (2.81)$$

which can be written by means of Eqs (2.74) and (2.80)

$$g_1 - g_2 = \frac{1}{2}\dot{m}_k \left(\frac{1}{\rho_2^2} - \frac{1}{\rho_1^2} \right) - \left[\frac{1}{\rho_2}(\tau_2 \cdot n_2) \cdot n_2 - \frac{1}{\rho_1}(\tau_1 \cdot n_1) \cdot n_1 \right] \qquad (2.82)$$

Example 1: Dynamics of Vapor Bubbles

     In this Section we will search for the set of local instantaneous equations governing the dynamics of a spherical vapor bubble. For the sake of simplicity we make the following hypotheses,

(i)   on the bubble motion:

        (H1)   no gravity effect
        (H2)   spherical symmetry

(ii)   on the liquid behavior:

        (H3)   single-component fluid
        (H4)   Newtonian fluid
        (H5)   Constant viscosity μ_L
        (H6)   fluid obeying Fourier's law
        (H7)   Constant thermal conductivity k_L

(iii)   on the vapor behavior:

>   (H8)   single-component fluid
>
>   (H9)   Newtonian fluid
>
>   (H10) constant viscosity μ_V
>
>   (H11) fluid obeying Fourier's law
>
>   (H12) constant thermal conductivity k_V

(iv)   on the interface behavior:

>   (H13) no interfacial mass
>
>   (H14) no interfacial viscosity
>
>   (H15) constant surface tension σ

The local instantaneous equations are written in spherical coordinates (Bird, et al., 1960) taking into account the hypothesis of spherical symmetry (H2). They express the mass, momentum and energy balance for the liquid phase, the vapor phase and the interface. Finally, we will derive a special form of the liquid momentum equation called the Rayleigh equation. The hypotheses made will be given for each equation.

Mass balance.

- Liquid phase

  (H3) single-component fluid

$$\frac{\partial \rho_L}{\partial t} + \frac{1}{r^2} \frac{\partial}{\partial r} (\rho_L r^2 w_L) = 0 \qquad (2.83)$$

- Vapor phase

  (H8) single-component fluid

$$\frac{\partial \rho_V}{\partial t} + \frac{1}{r^2} \frac{\partial}{\partial r} (\rho_V r^2 w_V) = 0 \qquad (2.84)$$

- Interface

  (H13) no interfacial mass

  Jump condition (2.40) yields,

$$\rho_{vi}(w_{vi} - \dot{R}) = \rho_{Li}(w_{Li} - \dot{R}) \qquad (2.85)$$

  where the dot denotes a time derivative.

Note that if $\rho_{vi} \ll \rho_{Li}$ we obtain,

$$w_{Li} \simeq \dot{R} \qquad (2.86)$$

Momentum balance.

- Liquid phase

    (H1) no gravity effect

    (H3) single-component fluid

    (H4) Newtonian fluid

    (H5) constant viscosity μ_L

$$\rho_L \left(\frac{\partial w_L}{\partial t} + w_L \frac{\partial w_L}{\partial r} \right) = - \frac{\partial p_L}{\partial r} + \frac{4}{3} \mu_L \left(\frac{\partial^2 w_L}{\partial r^2} + \frac{2}{r} \frac{\partial w_L}{\partial r} - \frac{2w_L}{r^2} \right)$$

(2.87)

Note that for an incompressible liquid (ρ_L = constant) the viscous term cancels out due to the liquid mass balance (2.83).

- Vapor phase

    (H1)   no gravity effect

    (H8)   single-component fluid

    (H9)   Newtonian fluid

    (H10) constant viscosity μ_V

$$\rho_V \left(\frac{\partial w_V}{\partial t} + w_V \frac{\partial w_V}{\partial r} \right) = - \frac{\partial p_V}{\partial r} + \frac{4}{3} \mu_V \left(\frac{\partial^2 w_V}{\partial r^2} + \frac{2}{r} \frac{\partial w_V}{\partial r} - \frac{2w_V}{r^2} \right)$$

(2.88)

- Interface

    (H4)   Newtonian liquid phase

    (H9)   Newtonian vapor phase

    (H13) no interfacial mass

    (H14) no interfacial viscosity

When surface tension is taken into account, jump condition (2.51) can be generalized (Delhaye, 1974) and leads to the following equation,

$$P_{vi} - P_{Li} = \frac{2\sigma}{R} - \rho_{vi} (w_{vi} - \dot{R}) w_{vi} - \rho_{Li} (\dot{R} - w_{Li}) w_{Li}$$

$$+ \frac{4}{3} \mu_V \left(\frac{\partial w_V}{\partial r} \bigg|_i - \frac{w_{vi}}{R} \right) - \frac{4}{3} \mu_L \left(\frac{\partial w_L}{\partial r} \bigg|_i - \frac{w_{Li}}{R} \right) \qquad (2.89)$$

Internal energy balance.

● Liquid phase

    (H3) single-component fluid

    (H4) Newtonian fluid

    (H5) constant viscosity μ_L

    (H6) fluid obeying Fourier's law

    (H7) constant thermal conductivity k_L

$$\rho_L c_L^v \left(\frac{\partial T_L}{\partial t} + w_L \frac{\partial T_L}{\partial r} \right) = \frac{k_L}{r^2} \frac{\partial}{\partial r} \left(r^2 \frac{\partial T_L}{\partial r} \right)$$
$$- T_L \left(\frac{\partial p_L}{\partial T_L} \right)_{\rho_L} \frac{1}{r^2} \frac{\partial}{\partial r} (w_L r^2) + \frac{4}{3} \mu_L \left(\frac{\partial w_L}{\partial r} - \frac{w_L}{r} \right)^2 \qquad (2.90)$$

In this equation c_L^v denotes the specific heat at constant volume and the last term of the R.H.S. represents the viscous dissipation. Note that for an incompressible liquid (ρ_L = constant) the second term of the R.H.S. drops out.

● Vapor phase

    (H8) single-component fluid

    (H9) Newtonian fluid

    (H10) constant viscosity

    (H11) fluid obeying Fourier's law

    (H12) constant thermal conductivity k_v

$$\rho_v c_v^v \left(\frac{\partial T_v}{\partial t} + w_v \frac{\partial T_v}{\partial r} \right) = \frac{kv}{r^2} \frac{\partial}{\partial r} \left(r^2 \frac{\partial T_v}{\partial r} \right)$$
$$- T_v \left(\frac{\partial p_v}{\partial T_v} \right)_{\rho_v} \frac{1}{r^2} \frac{\partial}{\partial r} (w_v r^2) + \frac{4}{3} \mu_v \left(\frac{\partial w_v}{\partial r} - \frac{w_v}{r} \right)^2 \qquad (2.91)$$

In this equation c_v^v denotes the specific heat at constant volume and the last term of the R.H.S. represents the viscous dissipation.

● Interface

    (H4) Newtonian liquid phase

    (H6) liquid phase obeying Fourier's law

    (H9) Newtonian vapor phase

    (H11) vapor phase obeying Fourier's law

    (H13) no interfacial mass

    (H14) no interfacial viscosity

(H15) constant surface tension

When surface tension is taken into account, enthalpy jump
condition (2.75) can be generalized (Delhaye, 1974).
Nevertheless when surface tension is constant, this
generalized equation simplifies to equation (2.75), which
yields,

$$
k_L \left.\frac{\partial T_L}{\partial r}\right|_i - k_v \left.\frac{\partial T_v}{\partial r}\right|_i
$$

$$
= \rho_{vi}(\dot{R} - w_{vi}) \left[i_{vi} + \frac{1}{2}(w_{vi} - \dot{R})^2 - \frac{4}{3}\frac{\mu_v}{\rho_{vi}}\left(\left.\frac{\partial w_v}{\partial r}\right|_i - \frac{w_{vi}}{r}\right)\right]
$$

$$
- \rho_{Li}(\dot{R} - w_{Li}) \left[i_{Li} + \frac{1}{2}(w_{Li} - \dot{R})^2 - \frac{4}{3}\frac{\mu_L}{\rho_{Li}}\left(\left.\frac{\partial w_L}{\partial r}\right|_i - \frac{w_{Li}}{r}\right)\right]
$$

$$(2.92)$$

Rayleigh equation. In the studies of vapor bubble dynamics a
special form of the liquid momentum equation, the so-called
Rayleigh equation, is often used.

In addition to the hypotheses required to ensure the validity
of the liquid continuity equation (2.83), the mass jump condition
(2.85) and the liquid momentum equation (2.87), we assume that,

(H16) the liquid is incompressible (ρ_L = constant)

(H17) the vapor density ρ_v is negligible with respect to the
      liquid density ρ_L ($\rho_v \ll \rho_L$)

Taking into account hypothesis (H16), the liquid continuity
equation (2.83) can be directly integrated, which leads to

$$
w_L = \frac{A(t)}{r^2}
$$

$$(2.93)$$

As a result the liquid momentum equation (2.87) takes the following
form, whatever the liquid viscosity,

$$
\rho_L \left(\frac{\partial w_L}{\partial t} + w_L \frac{\partial w_L}{\partial r}\right) = - \frac{\partial p_L}{\partial r}
$$

$$(2.94)$$

Taking into account Eq. (2.93) and integrating from r = R to r
infinite, we find,

$$
\dot{w}_{Li}R + 2\dot{R}w_{Li} - \frac{1}{2}w_{Li}^2 = \frac{p_{Li} - p_{L\infty}}{\rho_L}
$$

$$(2.95)$$

where the dot denotes a time derivative.

Finally, hypothesis (H17) enables the mass jump condition to be written as equation (2.86). Consequently, Eq. (2.95) is written,

$$\ddot{R}R + \frac{3}{2}\dot{R}^2 = \frac{p_{Li} - p_{L\infty}}{\rho_L} \qquad (2.96)$$

which is called the *Rayleigh equation*.

Example 2: Hydrodynamics of Liquid Films

In this section we will derive the mass and momentum jump conditions at the free surface of an isothermal viscous liquid film flowing down an inclined plane. As we will deal only with values at the interface we will omit the subscript i. For the sake of simplicity we make the following hypotheses,

(i) on the film motion:

   (H1) two dimensional flow

(ii) on the liquid behavior:

   (H2) single component fluid
   (H3) Newtonian fluid

(iii) on the gas behavior:

   (H4) the gas phase intervenes only by its pressure p_G which is assumed constant

(iv) on the interface behavior:

   (H5) no interfacial mass
   (H6) no interfacial viscosity
   (H7) constant surface tension σ
   (H8) no phase change at the interface

The local instantaneous jump conditions are written in rectangular coordinates (Bird, et al. 1960). The x-axis is along the plane whereas the y-axis is perpendicular to it. The origin of the coordinates is at the wall.

Interfacial mass balance.

   (H5) no interfacial mass
   (H8) no phase change at the interface

Condition (2.43) yields,

$$\dot{m}_L \overset{\Delta}{=} \rho_L (\mathbf{v}_L - \mathbf{v}_i)\cdot\mathbf{n}_L \equiv 0 \qquad (2.97)$$

Let $y = \eta(x,t)$ be the thickness of the liquid film. Equations (2.4) and (2.5) lead to,

$$n_L = \begin{bmatrix} \dfrac{-\,\partial\eta/\partial x}{[1 + (\partial\eta/\partial x)^2]^{\frac{1}{2}}} \\[4mm] \dfrac{1}{[1 + (\partial\eta/\partial x)^2]^{\frac{1}{2}}} \end{bmatrix} \qquad (2.98)$$

$$v_L \cdot n_L = \dfrac{\partial\eta/\partial t}{[1 + (\partial\eta/\partial x)^2]^{\frac{1}{2}}} \qquad (2.99)$$

Consequently, Eq. (2.97) reads,

$$v_L = \dfrac{\partial\eta}{\partial t} + u_L \dfrac{\partial\eta}{\partial x} \qquad (2.100)$$

where u_L and v_L are the x and y components of the velocity vector v_L. Equation (2.100) is often called the *kinematic condition*.

Interfacial momentum balance.

  (H3) Newtonian liquid phase

  (H4) the gas phase intervenes only by its pressure p_G which is assumed constant

  (H5) no interfacial mass

  (H6) no interfacial viscosity

  (H7) constant surface tension σ

  (H8) no phase change at the interface

In these conditions the two-dimensional momentum jump condition reads,

$$- n_G \cdot T_G - n_L \cdot T_L = - \dfrac{\partial^2\eta/\partial x^2}{[1+(\partial\eta/\partial x)^2]^{3/2}} \, \sigma \, n_L \qquad (2.101)$$

where

$$T_G = \begin{bmatrix} -p_G & 0 \\[4mm] 0 & -p_G \end{bmatrix} \; ; \; T_L = \begin{bmatrix} -p_L+2\mu_L\dfrac{\partial u_L}{\partial x} & \mu_L\left(\dfrac{\partial u_L}{\partial y} + \dfrac{\partial v_L}{\partial x}\right) \\[4mm] \mu_L\left(\dfrac{\partial u_L}{\partial y} + \dfrac{\partial v_L}{\partial x}\right) & -p_L + 2\mu_L\dfrac{\partial v_L}{\partial y} \end{bmatrix}$$

• Normal projection of the momentum jump condition

   Taking the scalar product of Eq. (2.101) by n_L, we obtain,

$$P_L - P_G = - \sigma \frac{\partial^2 \eta / \partial x^2}{\left[1 + (\partial \eta / \partial x)^2 \right]^{3/2}}$$

$$+ 2\mu_L \frac{\dfrac{\partial v_L}{\partial y} - \left(\dfrac{\partial u_L}{\partial y} + \dfrac{\partial v_L}{\partial x} \right) \dfrac{\partial \eta}{\partial x} + \dfrac{\partial u_L}{\partial x} \left(\dfrac{\partial \eta}{\partial x} \right)^2}{1 + (\partial \eta / \partial x)^2} \qquad (2.102)$$

● Tangential projection of the momentum jump conditions

Taking the scalar product of Eq. (2.101) by a unit vector tangent to the interface, we obtain,

$$2 \frac{\partial \eta}{\partial x} \left(\frac{\partial u_L}{\partial x} - \frac{\partial v_L}{\partial y} \right) - \left(\frac{\partial u_L}{\partial y} + \frac{\partial v_L}{\partial x} \right) \left[1 - \left(\frac{\partial \eta}{\partial x} \right)^2 \right] = 0 \qquad (2.103)$$

2.3   Instantaneous Space-Averaged Equations

Mathematical Tools

Space-averaged equations for two-phase flow were derived by Delhaye (1968) and Vernier and Delhaye (1968). These calculations were developed following an article by Birkhoff (1964) concerning the errors to be avoided when setting up such equations for single-phase flow. The most thorough study of these equations and their application, within the frame of a two-fluid model, to a problem of liquid film hydrodynamics, is to be found in Kocamustafaogullari's thesis (1971).

Limiting forms of the Leibniz and Gauss theorems for volumes.
Given a fixed tube, axis Oz (unit vector $\boldsymbol{n}_z$) in which a volume V_k^* is cut by two cross section planes located a distance Z apart over areas A_{k1} and A_{k2} (Figure 2.6). Let V_k be the volume limited by A_{k1}, A_{k2}, and the portions A_i and A_{kW} of interface and wall enclosed between the two cross section planes. The unit vector normal to the interface and directed away from phase k is denoted by $\boldsymbol{n}_k$. The cross section planes limiting the volume V_k are not necessarily fixed and their speeds of displacement are denoted by $- \boldsymbol{v}_{Ak1} \cdot \boldsymbol{n}_z$ and $\boldsymbol{v}_{Ak2} \cdot \boldsymbol{n}_z$

(i) *Leibniz rule:*

The Leibniz rule applied to volume V_k leads to,

$$\frac{\partial}{\partial t} \int_{V_k(z,t)} f(x,y,z,t) \, dV = \int_{V_k(z,t)} \frac{\partial f}{\partial t} dV + \int_{A_i(z,t)} f \boldsymbol{v}_i \cdot \boldsymbol{n}_k dA$$

$$- \int_{A_{k1}(z,t)} f \boldsymbol{v}_{Ak1} \cdot \boldsymbol{n}_z dA + \int_{A_{k2}(z,t)} f \boldsymbol{v}_{Ak2} \cdot \boldsymbol{n}_z dA \qquad (2.104)$$

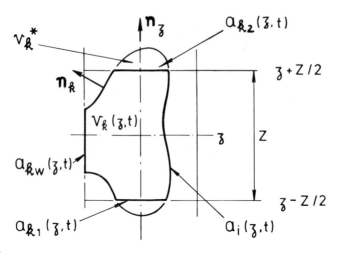

Figure 2.6.

where $\boldsymbol{v}_i \cdot \boldsymbol{n}_k$ is the speed of displacement of the interface A_i.

(ii) *Gauss theorems*:

The Gauss theorem applied to volume V_k leads to,

$$\int_{V_k(z,t)} \nabla \cdot \boldsymbol{B} \, dV = \int_{A_i(z,t)} \boldsymbol{n}_k \cdot \boldsymbol{B} \, dA + \int_{A_{kw}(z,t)} \boldsymbol{n}_k \cdot \boldsymbol{B} \, dA$$

$$- \int_{A_{k1}(z,t)} \boldsymbol{n}_z \cdot \boldsymbol{B} \, dA + \int_{A_{k2}(z,t)} \boldsymbol{n}_z \cdot \boldsymbol{B} \, dA \qquad (2.105)$$

which yields,

$$\int_{V_k(z,t)} \nabla \cdot \boldsymbol{B} \, dV = \int_{A_i(z,t)} \boldsymbol{n}_k \cdot \boldsymbol{B} \, dA + \int_{A_{kw}(z,t)} \boldsymbol{n}_k \cdot \boldsymbol{B} \, dA$$

$$+ \frac{\partial}{\partial z} \int_{V_k(z,t)} B_z \, dV \qquad (2.106)$$

where B_z is the z-component of $\boldsymbol{B}$.

For a tensor field we have,

$$\int_{V_k(z,t)} \nabla \cdot \boldsymbol{M} \, dV = \int_{A_i(z,t)} \boldsymbol{n}_k \cdot \boldsymbol{M} \, dA + \int_{A_{kw}(z,t)} \boldsymbol{n}_k \cdot \boldsymbol{M} \, dA$$

$$+ \frac{\partial}{\partial z} \int_{V_k(z,t)} \boldsymbol{n}_z \cdot M dV \tag{2.107}$$

Limiting forms of the Leibniz and Gauss theorems for areas. Given a tube, axis Oz (unit vector $\boldsymbol{n}_z$) in which a volume V_k is limited by a boundary A_i and cut by a fixed cross section plane over area A_k (Figure 2.7). The unit vector normal to the interface and directed away from phase k is denoted by $\boldsymbol{n}_k$. The intersection of interface A_i with the cross section plane is denoted by C. The unit vector normal to C, located in the cross section plane and directed away from phase k is denoted by $\boldsymbol{n}_{kC}$.

(i) Leibniz Rule:

$$\frac{\partial}{\partial t} \int_{A_k(z,t)} f(x,y,z,t) dA = \int_{A_k(z,t)} \frac{\partial f}{\partial t} dA + \int_{C(z,t)} f \boldsymbol{v}_i \cdot \boldsymbol{n}_{k} \frac{dC}{\boldsymbol{n}_k \cdot \boldsymbol{n}_{kC}}$$

$$\tag{2.108}$$

(ii) Gauss theorems

For vector fields we have,

$$\int_{A_k(z,t)} \nabla \cdot \boldsymbol{B} dA = \frac{\partial}{\partial z} \int_{A_k(z,t)} B_z dA + \int_{C(z,t)} \boldsymbol{n}_k \cdot \boldsymbol{B} \frac{dC}{\boldsymbol{n}_k \cdot \boldsymbol{n}_{kC}}$$

$$\tag{2.109}$$

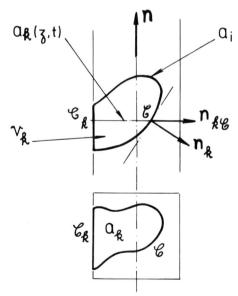

Figure 2.7.

If we take $\boldsymbol{B} = \boldsymbol{n}_z$ we obtain

$$\frac{\partial A_k(z,t)}{\partial z} = - \int_{C(z,t)} \boldsymbol{n}_k \cdot \boldsymbol{n}_z \frac{dC}{\boldsymbol{n}_k \cdot \boldsymbol{n}_{kC}} \tag{2.110}$$

For tensor fields,

$$\int_{A_k(z,t)} \nabla \cdot M dA = \frac{\partial}{\partial z} \int_{A_k(z,t)} \boldsymbol{n}_z \cdot M dA + \int_{C(z,t)} \boldsymbol{n}_k \cdot M \frac{dC}{\boldsymbol{n}_k \cdot \boldsymbol{n}_{kC}}$$

$$\tag{2.111}$$

Instantaneous Area-Averaged Equations

We will average the local instantaneous balance equation for phase k over the cross section area occupied by phase k (Figure 2.8).

The local, instantaneous balance equation is integrated over the area $A_k(z,t)$ limited by the boundaries $C(z,t)$ with the other phase and $C_k(z,t)$ with the pipe wall,

$$\int_{A_k(z,t)} \frac{\partial}{\partial t} \rho_k \psi_k dA + \int_{A_k(z,t)} \nabla \cdot (\rho_k \psi_k \boldsymbol{v}_k) dA$$

$$+ \int_{A_k(z,t)} \nabla \cdot J_k dA - \int_{A_k(z,t)} \rho_k \phi_k dA = 0 \tag{2.112}$$

Applying the limiting forms of the Leibniz rule (2.108) and of the Gauss theorems (2.109, 111) we obtain, since the pipe wall is supposed to be fixed and impermeable,

$$\frac{\partial}{\partial t} A_k < \rho_k \psi_k >_2 + \frac{\partial}{\partial z} A_k < \boldsymbol{n}_z \cdot (\rho_k \psi_k \boldsymbol{v}_k) >_2$$

$$+ \frac{\partial}{\partial z} A_k < \boldsymbol{n}_z \cdot J_k >_2 - A_k < \rho_k \phi_k >_2$$

$$= - \int_{C(z,t)} (\dot{m}_k \psi_k + \boldsymbol{n}_k \cdot J_k) \frac{dC}{\boldsymbol{n}_k \cdot \boldsymbol{n}_{kC}}$$

$$- \int_{C_k(z,t)} \boldsymbol{n}_k \cdot J_k \frac{dC}{\boldsymbol{n}_k \cdot \boldsymbol{n}_{kC}} \tag{2.113}$$

where

$$< f_k >_2 \overset{\Delta}{=} \frac{1}{A_k} \int_{A_k(z,t)} f_k(x,y,z,t) dA \tag{2.114}$$

$$\dot{m}_k \overset{\Delta}{=} \rho_k (\boldsymbol{v}_k - \boldsymbol{v}_i) \cdot \boldsymbol{n}_k \tag{2.115}$$

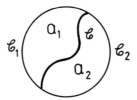

Figure 2.8.

By means of Table 2.1 given in Section 2.2, Eq. (2.113) gives the instantaneous area-averaged equations for the balance of mass, momentum, total energy and entropy.

Balance of Mass

$$\frac{\partial}{\partial t} A_k <\rho_k>_2 + \frac{\partial}{\partial z} A_k <\rho_k w_k>_2 = - \int_{C(z,t)} \dot{m}_k \frac{dC}{n_k \cdot n_{kC}} \qquad (2.116)$$

Balance of Momentum

Splitting the stress tensor T_k into the pressure term and the viscous stress tensor, we have,

$$\frac{\partial}{\partial t} A_k <\rho_k \mathbf{v}_k>_2 + \frac{\partial}{\partial z} A_k <\rho_k w_k \mathbf{v}_k>_2 - A_k <\rho_k F>_2$$

$$+ \frac{\partial}{\partial z} A_k <p_k \mathbf{n}_z>_2 - \frac{\partial}{\partial z} A_k <\mathbf{n}_z \cdot \mathbf{\tau}_k>_2$$

$$= - \int_{C(z,t)} (\dot{m}_k \mathbf{v}_k - \mathbf{n}_k \cdot T_k) \frac{dC}{n_k \cdot \mathbf{n}_{kC}} + \int_{C_k(z,t)} \mathbf{n}_k \cdot T_k \frac{dC}{n_k \cdot \mathbf{n}_{kC}}$$

$$\qquad (2.117)$$

If we project along the tube axis, we have,

$$\frac{\partial}{\partial t} A_k <\rho_k w_k>_2 + \frac{\partial}{\partial z} A_k <\rho_k w_k^2>_2 - A_k <\rho_k F_z>_2$$

$$+ \frac{\partial}{\partial z} A_k <p_k>_2 - \frac{\partial}{\partial z} A_k <(\mathbf{n}_z \cdot \mathbf{\tau}_k) \cdot \mathbf{n}_z>_2$$

$$= - \int_{C(z,t)} \mathbf{n}_z \cdot (\dot{m}_k \mathbf{v}_k - \mathbf{n}_k \cdot T_k) \frac{dC}{n_k \cdot \mathbf{n}_{kC}} + \int_{C_k(z,t)} \mathbf{n}_z \cdot (\mathbf{n}_k \cdot T_k) \frac{dC}{n_k \cdot \mathbf{n}_{kC}}$$

$$\qquad (2.118)$$

where w_k is the z-component of the velocity vector $\mathbf{v}_k$. If we assume that the pressure p_k is constant along C and C_k and equal to the averaged pressure $<p_k>_2$ over A_k, then by using Eq. (2.110),

J M Delhaye

Eq. (2.118) can be written

$$\frac{\partial}{\partial t} A_k < \rho_k w_k >_2 + \frac{\partial}{\partial z} A_k < \rho_k w_k^2 >_2 - A_k < \rho_k F_z >_2$$

$$+ A_k \frac{\partial p_k}{\partial z} - \frac{\partial}{\partial z} A_k < (\boldsymbol{n}_z \cdot \tau_k) \cdot \boldsymbol{n}_z >_2$$

$$= - \int_{C(z,t)} \boldsymbol{n}_z \cdot (\dot{m}_k \boldsymbol{v}_k - \boldsymbol{n}_k \cdot \tau_k) \frac{dC}{\boldsymbol{n}_k \cdot \boldsymbol{n}_{kC}}$$

$$+ \int_{C_k(z,t)} \boldsymbol{n}_z \cdot (\boldsymbol{n}_k \cdot \tau_k) \frac{dC}{\boldsymbol{n}_k \cdot \boldsymbol{n}_{kC}} \qquad (2.119)$$

Balance of total energy. If we introduce the enthalpy i_k per unit of mass, we obtain,

$$\frac{\partial}{\partial t} A_k < \rho_k (\tfrac{1}{2} v_k^2 + u_k) >_2 + \frac{\partial}{\partial z} A_k < \rho_k (\tfrac{1}{2} v_k^2 + i_k) w_k >_2$$

$$- A_k < \rho_k \boldsymbol{F} \cdot \boldsymbol{v}_k >_2 - \frac{\partial}{\partial z} A_k < (\tau_k \cdot \boldsymbol{v}_k) \cdot \boldsymbol{n}_z >_2$$

$$+ \frac{\partial}{\partial z} A_k < \boldsymbol{q}_k \cdot \boldsymbol{n}_z >_2$$

$$= - \int_{C(z,t)} \left[\dot{m}_k (\tfrac{1}{2} v_k^2 + u_k) - (\boldsymbol{T}_k \cdot \boldsymbol{v}_k) \cdot \boldsymbol{n}_k + \boldsymbol{q}_k \cdot \boldsymbol{n}_k \right] \frac{dC}{\boldsymbol{n}_k \cdot \boldsymbol{n}_{kC}}$$

$$- \int_{C_k(z,t)} \boldsymbol{q}_k \cdot \boldsymbol{n}_k \frac{dC}{\boldsymbol{n}_k \cdot \boldsymbol{n}_{kC}} \qquad (2.120)$$

Entropy inequality.

$$\frac{\partial}{\partial t} A_k < \rho_k s_k >_2 + \frac{\partial}{\partial z} A_k < \rho_k s_k w_k >_2$$

$$+ \frac{\partial}{\partial z} A_k < \frac{\boldsymbol{q}_k}{T_k} \cdot \boldsymbol{n}_z >_2 + \int_{C(z,t)} (\dot{m}_k s_k + \frac{\boldsymbol{q}_k}{T_k} \cdot \boldsymbol{n}_k) \frac{dC}{\boldsymbol{n}_k \cdot \boldsymbol{n}_{kC}}$$

$$+ \int_{C_k(z,t)} \frac{\boldsymbol{q}_k}{T_k} \cdot \boldsymbol{n}_k \frac{dC}{\boldsymbol{n}_k \cdot \boldsymbol{n}_k} = A_k < \Delta_k >_2 \geqslant 0 \qquad (2.121)$$

Instantaneous Volume-Averaged Equations

Transient two-phase flow modeling sometimes requires the solution of a set of partial differential equations written in terms of instantaneous area-averages. This set of partial differential equations is usually solved by using one of the following techniques,

(i)   a finite-difference method involving a discretization of the partial differential equations over finite sized meshes and leading to a *distributed* parameter model,

(ii) a profile method involving an integration over a finite length and leading to a *lumped* parameter model.

In essence, both resolution techniques deal with volume-averaged quantities.   It is the purpose of this Section to give the instantaneous volume-averaged equations with which the equations of both models must finally reconcile.

Assuming no mass flow through the wall, the integral balance over $V_k(z,t)$ reads,

$$\frac{\partial}{\partial t} \int_{V_k(z,t)} \rho_k \psi_k \, dV = - \int_{A_i(z,t)} \rho_k \psi_k (\mathbf{v}_k - \mathbf{v}_i) \cdot \mathbf{n}_k \, dA$$

$$+ \int_{A_{k1}(z,t)} \rho_k \psi_k (\mathbf{v}_k - \mathbf{v}_{Ak1}) \cdot \mathbf{n}_z \, dA - \int_{A_{k2}(z,t)} \rho_k \psi_k (\mathbf{v}_k - \mathbf{v}_{Ak2}) \cdot \mathbf{n}_z \, dA$$

$$- \int_{A_i(z,t)} \mathbf{n}_k \cdot \mathbf{J}_k \, dA - \int_{A_{kw}(z,t)} \mathbf{n}_k \cdot \mathbf{J}_k \, dA$$

$$+ \int_{A_{k1}(z,t)} \mathbf{n}_z \cdot \mathbf{J}_k \, dA - \int_{A_{k2}(z,t)} \mathbf{n}_z \cdot \mathbf{J}_k \, dA + \int_{V_k(z,t)} \rho_k \phi_k \, dV$$

$$(2.122)$$

Taking account of the following definitions,

$$<f_k>_3 \overset{\Delta}{=} \frac{1}{V_k} \int_{V_k(z,t)} f_k \, dV \qquad\qquad (2.123)$$

$$\dot{m}_k \overset{\Delta}{=} \rho_k (\mathbf{v}_k - \mathbf{v}_i) \cdot \mathbf{n}_k \qquad\qquad (2.124)$$

the integral balance (2.122) can be written as,

$$\underbrace{\frac{\partial}{\partial t} V_k <\rho_k \psi_k>_3}_{3} \quad - \quad \underbrace{V_k <\rho_k \phi_k>_3}_{2}$$

$$= \int_{A_{k1}(z,t)} \mathbf{n}_z \cdot \left[\rho_k \psi_k (\mathbf{v}_k - \mathbf{v}_{Ak1}) + \mathbf{J}_k \right] dA$$

$$- \int_{A_{k2}(z,t)} \mathbf{n}_z \cdot \left[\rho_k \psi_k (\mathbf{v}_k - \mathbf{v}_{Ak2}) + \mathbf{J}_k \right] dA \Bigg\}_3$$

$$- \int_{A_i(z,t)} (\dot{m}_k \psi_k + n_k \cdot J_k)\, dA \ - \int_{A_{kw}(z,t)} n_k \cdot J_k\, dA \qquad\qquad (2.125)$$

$$\underbrace{\phantom{- \int_{A_i(z,t)} (\dot{m}_k \psi_k + n_k \cdot J_k)\, dA}}_{4} \qquad \underbrace{\phantom{\int_{A_{kw}(z,t)} n_k \cdot J_k\, dA}}_{5}$$

The physical significance of Eq. (2.125) is straightforward: Term (1) is the storage of quantity ψ_k inside volume V_k, term (2) is the source of ψ_k inside volume V_k, terms (3) are the fluxes of ψ_k across the cross section planes. Terms (4) and (5) are the fluxes of ψ_k across the interface and the wall. They are of a particular importance since they are directly connected to the flow pattern through the interfacial area $A_i(z,t)$ and the wall area $A_{kw}(z,t)$ in contact with phase k.

An equivalent form of Eq. (2.125) is,

$$\frac{\partial}{\partial t}\, V_k <\rho_k \psi_k>_3 \ + \ \frac{\partial}{\partial z}\, V_k <\rho_k \psi_k w_k> \ + \ n_z \cdot J_k>_3 \ - \ V_k <\rho_k \phi_k>_3$$

$$= \int_{A_{k2}(z,t)} \rho_k \psi_k v_{Ak2} \cdot n_z\, dA \ - \int_{A_{k1}(z,t)} \rho_k \psi_k v_{Ak1} \cdot n_z\, dA$$

$$- \int_{A_i(z,t)} (\dot{m}_k \psi_k + n_k \cdot J_k)\, dA \ - \int_{A_{kw}(z,t)} n_k \cdot J_k\, dA \qquad\qquad (2.126)$$

where the integrals over A_{k1} and A_{k2} vanish if the cross section planes are fixed.

By means of Table 2.1 given in Section 2.2, Eqs (2.125) or (2.126) give the instantaneous volume-averaged equations for the balance of mass, momentum, total energy and entropy.

2.4 Local Time-Averaged Equations

Mathematical Tools

The use of the time-averaged local variables in two-phase flows was proposed by Teletov (1958) and reconsidered by Vernier and Delhaye (1968). Ishii (1975) in Chapter III of his book, traces the history of the subject in detail and compares the different types of averaged used.

The method used here to establish time-averaged equations differs from that used by Ishii. The interfaces we consider are density discontinuity surfaces. They have no thickness, unlike the interfaces considered by Ishii. Nevertheless, we find the same results but by using simpler calculations.

For solving two-dimensional or three-dimensional, transient problems, the local instantaneous equations can be time-averaged over a time-interval $[t-T/2; t+T/2]$. As for single-phase turbulent flow, this time interval $[T]$ must be carefully chosen, large enough compared to the turbulence fluctuations, and small enough compared to the overall flow fluctuations. This is not always possible, and a thorough discussion of this complex problem is to be found in the works of Delhaye and Achard (1977, 1978).

After recalling a few theorems on the derivatives of piecewise continuous functions, we will define the single time-averaging operators.

If we consider a given point in a two-phase flow, phase k passes this point intermittently and a function f_k associated with phase k will have the appearance shown on Figure 2.9. It will be a piecewise continuous function.

If we consider the time interval $[t-T/2;\ t+T/2]$, let $[T_k]$ be the subset of residence time intervals of phase k belonging to the interval $[T]$ and T_k the cumulated residence time of phase k in the interval $[T]$.

Limiting form of the Leibniz theorem.

$$\int_{[T_k]} \frac{\partial f_k}{\partial t}\, dt = \frac{\partial}{\partial t} \int_{[T_k]} f_k dt - \sum_{\substack{disc \\ \in [T]}} \frac{1}{|\mathbf{v}_i \cdot \mathbf{n}_k|}\, f_k \mathbf{v}_i \cdot \mathbf{n}_k \qquad (2.127)$$

If $f_k \equiv 1$, Eq. (2.127) reads,

$$\frac{\partial \alpha_k}{\partial t} = \frac{1}{T} \sum_{\substack{disc \\ \in [T]}} \frac{\mathbf{v}_i \cdot \mathbf{n}_k}{|\mathbf{v}_i \cdot \mathbf{n}_k|} \qquad (2.128)$$

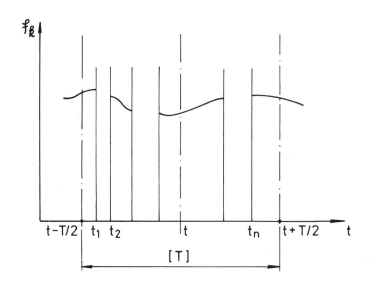

Figure 2.9.

with α_k the residence time fraction of phase k, defined by

$$\alpha_k \overset{\Delta}{=} \frac{T_k}{T} \tag{2.129}$$

Limiting forms of the Gauss theorems.

$$\int_{[T_k(\mathbf{x})]} \nabla \cdot \mathbf{B}_k(\mathbf{x},t)\,dt = \nabla \cdot \int_{[T_k(\mathbf{x})]} \mathbf{B}_k(\mathbf{x},t)\,dt$$

$$+ \sum_{\substack{\text{disc} \\ \in [T]}} \frac{1}{|\mathbf{v}_i \cdot \mathbf{n}_k|}\, \mathbf{n}_k \cdot \mathbf{B}_k(\mathbf{x},t) \tag{2.130}$$

The vector $\mathbf{B}_k$ can be replaced by a tensor $\mathbf{M}_k$. A relation identical to Eq. (2.130) is obtained. In particular, if

$$\mathbf{M}_k = \mathbf{U} \tag{2.131}$$

we have

$$\nabla \alpha_k = -\frac{1}{T} \sum_{\substack{\text{disc} \\ \in [T]}} \frac{\mathbf{n}_k}{|\mathbf{v}_i \cdot \mathbf{n}_k|} \tag{2.132}$$

Single Time-Averaged Equations

The local instantaneous balance law is integrated over the time interval $[T_k]$,

$$\int_{|T_k|} \frac{\partial}{\partial t}\, \rho_k \psi_k\,dt + \int_{|T_k|} \nabla \cdot (\rho_k \psi_k \mathbf{v}_k)\,dt$$

$$+ \int_{[T_k]} \nabla \cdot \mathbf{J}_k\,dt - \int_{|T_k|} \rho_k \phi_k\,dt = 0 \tag{2.133}$$

Taking into account the limiting forms of the Leibniz and Gauss theorems (2.127) and (2.130), Eq. (2.133) becomes

$$\frac{\partial}{\partial t}\,\overline{\alpha_k \rho_k \psi_k}^{x} + \nabla \cdot \overline{\alpha_k \rho_k \psi_k \mathbf{v}_k}^{x} + \nabla \cdot \overline{\alpha_k \mathbf{J}_k}^{x}$$

$$- \overline{\alpha_k \rho_k \phi_k}^{x} = -\sum_j \ell_j^{-1}\,(\dot{m}_k \psi_k + \mathbf{J}_k \cdot \mathbf{n}_k)_j \tag{2.134}$$

where

$$\overline{f_k}^x \triangleq \frac{1}{T_k} \int_{[T_k]} f_k dt \qquad (2.135)$$

$$\ell_j \triangleq T |\mathbf{v}_i \cdot \mathbf{n}_k|_j \qquad (2.136)$$

j denoting the j-th interface passing through **x** during the time interval [T].

Remarks on the definition of $\overline{f_k}^x$. A phase density function $X_k(\mathbf{x},t)$ is defined by,

$$X_k(\mathbf{x},t) \triangleq \begin{array}{ll} 1 & \text{if point } \mathbf{x} \text{ pertains to phase k} \\ & \\ 0 & \text{if point } \mathbf{x} \text{ does not pertain to phase k} \end{array} \qquad (2.137)$$

The time fraction α_k, defined by Eq. (2.129) is then equal to,

$$\alpha_k(\mathbf{x},t) \triangleq \frac{T_k}{T} = \frac{1}{T} \int_{[T]} X_k(\mathbf{x},t)dt \triangleq \overline{X_k}(\mathbf{x},t) \qquad (2.138)$$

On the other hand $\overline{f_k}^x$, defined by Eq. (2.135) is equal to,

$$\overline{f_k}^x = \frac{\frac{1}{T} \int_{[T]} X_k f_k dt}{\frac{1}{T} \int_{[T]} X_k dt} \triangleq \frac{\overline{X_k f_k}}{\overline{X_k}} \qquad (2.139)$$

Hence, we can see that the time-average of f_k over the time interval $[T_k]$ is the X_k-weighted average of f_k over [T]. This justifies the notation $\overline{f_k}^x$.

Primary balance equations. By means of Table 2.1 given in Section 2.2, Eq. (2.134) gives the local time-averaged equations for the balances of mass, momentum, total energy and entropy.

(i)   Mass

$$\frac{\partial}{\partial t} \alpha_k \overline{\rho_k}^x + \nabla \cdot \alpha_k \overline{\rho_k \mathbf{v}_k}^x = - \sum_j \ell_j \dot{m}_{kj} \qquad (2.140)$$

(ii)   Momentum

$$\frac{\partial}{\partial t} \alpha_k \overline{\rho_k \mathbf{v}_k}^x + \nabla \cdot \alpha_k \overline{\rho_k \mathbf{v}_k \mathbf{v}_k}^x - \nabla \cdot \alpha_k \overline{T_k}^x$$

$$- \alpha_k \overline{\rho_k F}^x = - \sum_j \ell_j^{-1} (\dot{m}_k \mathbf{v}_k - T_k \cdot \mathbf{n}_k)_j \qquad (2.141)$$

(iii) Total energy

$$\frac{\partial}{\partial t} \; \overline{\alpha_k \rho_k (u_k + \tfrac{1}{2} v_k^2)}^x \;+\; \nabla \cdot \overline{\alpha_k \rho_k (u_k + \tfrac{1}{2} v_k^2) \boldsymbol{v}_k}^x$$

$$+\; \nabla \cdot \overline{\alpha_k \boldsymbol{T}_k \cdot \boldsymbol{v}_k}^x \;+\; \nabla \cdot \overline{\alpha_k \boldsymbol{q}_k}^x \;-\; \overline{\alpha_k \rho_k \boldsymbol{F} \cdot \boldsymbol{v}_k}^x$$

$$=\; -\; \sum_j \ell_j^{-1} \left[\dot{m}_k (u_k + \tfrac{1}{2} v_k^2) - (\boldsymbol{T}_k \cdot \boldsymbol{v}_k) \cdot \boldsymbol{n}_k + \boldsymbol{q}_k \cdot \boldsymbol{n}_k \right]_j \qquad (2.142)$$

(iv) Entropy

$$\frac{\partial}{\partial t} \; \overline{\alpha_k \rho_k s_k}^x \;+\; \nabla \cdot \overline{\alpha_k \rho_k s_k \boldsymbol{v}_k}^x \;+\; \nabla \cdot \overline{\alpha_k \frac{1}{T_k} \boldsymbol{q}_k}^x$$

$$+\; \sum_j \ell_j^{-1} \; (\dot{m}_k s_k + \frac{1}{T_k} \boldsymbol{q}_k \cdot \boldsymbol{n}_k) \;=\; \overline{\alpha_k \Delta_k}^x \;\geqslant\; 0 \qquad (2.143)$$

Comments on the Single Time-Averaging Operators

In Eqs. (2.138) and (2.139) we used the single time-averaging operator defined by

$$\overline{g} \; (\boldsymbol{x}, t) \triangleq \frac{1}{T} \int_{[T]} g(\boldsymbol{x}, z) \, dz \qquad (2.144)$$

In two-phase flow, the g function is of the type,

$$g \; (\boldsymbol{x}, t) \;=\; X_k f_k \qquad (2.145)$$

and can be expanded into a Fourier series. The *message* $g(\dot{x}, t)$ can be split into two parts (1) the *signal* $g_s(\boldsymbol{x}, t)$ which is, in our case, the sum of the expansion terms whose angular frequencies are lower than an arbitrary cutoff frequency (2) the *noise* $g_n(\boldsymbol{x}, t)$ which is the sum of the remaining terms. Hence,

$$g(\boldsymbol{x}, t) \;=\; g_s(\boldsymbol{x}, t) \;+\; g_n(\boldsymbol{x}, t) \qquad (2.146)$$

The time averaging operator (2.144) is expected to low-pass filter the message $g(\boldsymbol{x}, t)$ on which it is acting,

$$\overline{g}(\boldsymbol{x}, t) \;\simeq\; g_s(\boldsymbol{x}, t) \qquad (2.147)$$

As the first time derivatives of the time-averaged variables appear in the single time-averaged balance equation (2.134), it would be worthwhile to have also,

$$\frac{\partial}{\partial t} \, \overline{g} \, (\mathbf{x}, t) \simeq \frac{\partial}{\partial t} \, g_s \, (\mathbf{x}, t) \qquad (2.148)$$

Delhaye and Achard (1977, 1978) showed that the single time-averaging operator (2.144) does not always fulfill conditions (2.147) and (2.148). Furthermore, the first time-derivative of $\overline{g}$ is discontinuous. For these reasons, Delhaye and Achard (1977, 1978) introduced a double time-averaging operator.

2.5 Composite-Averaged Equations

Commutativity of the Averaging Operators

The importance of composite, i.e. space/time- or time/space-averaged equations is considerable, since, at the present time, all the practical problems of two-phase flow in channels are dealt with using these equations.

These composite averaged equations can be obtained in two different ways, (1) by averaging the time-averaged local equations over a channel cross section area or over a slice, (2) by averaging over a time interval the instantaneous equations averaged over the cross section areas or slices occupied by each phase.

In order to verify the equivalence of the results obtained with one or the other of these methods, we will first establish or recall theorems concerning the commutativity of averaging operators. We will also introduce local and integral specific areas, since these quantities play an important part in mass, momentum and energy interface transfers. We will conclude by demonstrating the identity of the space/time- and time/space-averaging operators.

If we consider any scalar, vector or tensor function associated with phase k, the aim of our calculation is to find the group of variables on which a permutation of the time and space averaging operators can be carried out.

We have, by means of the averaging operator definitions,

$$\{\alpha_k \overline{f_k}^x\}_2 = \{\overline{X_k f_k}\}_2 \overset{\Delta}{=} \frac{1}{A} \int_A (\frac{1}{T} \int_{[T]} X_k f_k \, dt) \, dA \qquad (2.149)$$

Reversing the order of integration yields,

$$\{\alpha_k \overline{f_k}^x\}_2 = \frac{1}{T} \int_{[T]} (\frac{1}{A} \int_A X_k f_k \, dA) \, dt \qquad (2.150)$$

$$= \frac{1}{T} \int_T (\frac{1}{A} \int_{A_k} f_k \, dA) \, dt \qquad (2.151)$$

As a consequence we obtain the fundamental relation (Vernier and Delhaye, 1968),

$$\{\overline{\alpha_k f_k}^x\}_2 \equiv \overline{R_{k2} <f_k>_2}$$ (2.152)

where R_{k2} is the area fraction of phase k in the cross section defined by,

$$R_{k2} \overset{\Delta}{=} \frac{A_k}{A}$$ (2.153)

Eq. (2.152) plays an essential part in two-phase flow modeling. In effect, it is more or less convenient to formulate hypotheses on local time-averaged quantities (such as α_k or $\overline{f_k}^x$) than on instantaneous area-averaged quantities (such as R_{k2} or $<f_k>_2$), depending on the flow pattern (e.g. bubbly flow or slug flow). A particular case of Eq. (2.152) is obtained by taking,

$$f_k \equiv 1$$

which leads to,

$$\{\alpha_k\}_2 \equiv \overline{R_{k2}}$$ (2.154)

It is obvious that Eqs. (2.152) and (2.154) established for areas are also valid for segments and volumes. More generally, we have,

$$\{\overline{\alpha_k \overline{f_k}^x}\}_n \equiv \overline{R_{kn} <f_k>_n}$$ (2.155)

where,

    n = 1   for segments
    n = 2   for areas
    n = 3   for volumes

By writing Eq. (2.155) for segments (n = 1), probe measurements of local void fraction can be tallied with radiation attenuation measurements of segment void fractions.

Local and Integral Specific Areas

Fundamental identity. Given a pipe and a fixed control volume V limited by the pipe wall and the cross sections A' and A" located a distance Z apart (Figure 2.10). The area of the *moving* interfaces contained in volume V is denoted by $A_i(t)$.

For any arbitrary continuous vector field $B_k(x,t)$, the following identity is satisfied (Delhaye, 1976, Delhaye and Achard, 1978),

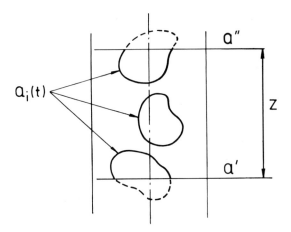

Figure 2.10.

$$\int_V \left[\sum_j \ell_j^{-1} (B_k \cdot n_k)_j \right] dV \equiv \overline{\int_{A_i(t)} B_k \cdot n_k \, dA} \qquad (2.156)$$

with,

$$\ell_j = T \left| v_i \cdot n_k \right|_j \qquad (2.157)$$

where T is the integration period and $v_i \cdot n_k$ the speed of displacement of $A_i(t)$. Subscript j denotes the j-th interface passing through x during the time interval $[T]$.

If the vector field B_k is chosen in such a way that on the interfaces,

$$B_k \equiv n_k \qquad (2.158)$$

identity (2.156) leads to a relation which connects the interfacial area to the speed of displacement of the interfaces,

$$\int_V (\sum_j \ell_j^{-1}) dV \equiv \overline{A_i(t)} \qquad (2.159)$$

Specific areas. The following definitions can then be set,

(i) the *local specific area* $\gamma(x)$ is a local quantity defined over the time interval $[T]$ by the following relation,

$$\gamma(x) \overset{\Delta}{=} \sum_j \ell_j^{-1} \qquad (2.160)$$

(ii) the *integral specific area* $\Gamma_3(t)$ is an instantaneous quantity defined on the volume V by the following relation,

$$\Gamma_3(t) \triangleq \frac{A_i(t)}{V} \tag{2.161}$$

Identity (2.159) can then be written,

$$\{\gamma(\boldsymbol{x})\} \equiv \overline{\Gamma_3(t)} \tag{2.162}$$

Identities (2.159) or (2.162) are fundamental as far as interfacial area measurements are concerned. They provide the link between the integral specific area which can be measured by chemical methods and the local specific area which could be determined by means of probe techniques.

Limiting form of the fundamental identity. If $Z \to 0$, identity (2.156) becomes,

$$\left\{ \sum_j \ell_j^{-1} (\boldsymbol{B}_k \cdot \boldsymbol{n}_k)_j \right\}_2 \equiv \frac{1}{A} \int_C \overline{\boldsymbol{B}_k \cdot \boldsymbol{n}_k \frac{dC}{\boldsymbol{n}_k \cdot \boldsymbol{n}_{kC}}} \tag{2.163}$$

which will appear as the link between the interaction terms occurring in the space/time or time/averaged equations. In Eq. (2.163), $\boldsymbol{n}_{kC}$ denotes the unit vector normal to C, located in the cross section plane and directed away from phase k. Curve C is the intersection of interface A_i with the cross section plane.

Space/Time- or Time/Space-Averaged Equations

If we time-average over $[T]$ the instantaneous area-averaged balance equation, we obtain,

$$\frac{\partial}{\partial t} \overline{A_k \langle \rho_k \psi_k \rangle_2} + \frac{\partial}{\partial z} \overline{A_k \langle \boldsymbol{n}_z \cdot (\rho_k \psi_k \boldsymbol{v}_k) \rangle_2}$$

$$+ \frac{\partial}{\partial z} \overline{A_k \langle \boldsymbol{n}_k \cdot \boldsymbol{J}_k \rangle_2} - \overline{A_k \langle \rho_k \phi_k \rangle_2}$$

$$= - \overline{\int_{C(z,t)} (\dot{m}_k \psi_k + \boldsymbol{n}_k \cdot \boldsymbol{J}_k) \frac{dC}{\boldsymbol{n}_k \cdot \boldsymbol{n}_{kC}}} - \overline{\int_{C_k(z,t)} \boldsymbol{n}_k \cdot \boldsymbol{J}_k \frac{dC}{\boldsymbol{n}_k \cdot \boldsymbol{n}_{kC}}}$$

$$\tag{2.164}$$

If we area-average the local time-averaged balance equation over the total cross section area, we obtain,

$$\frac{\partial}{\partial t} A \{ \alpha_k \overline{\rho_k \psi_k}^x \}_2 + \frac{\partial}{\partial z} A \{ \alpha_k \overline{\boldsymbol{n}_z \cdot (\rho_k \psi_k \boldsymbol{v}_k)}^x \}_2$$

$$+ \frac{\partial}{\partial z} A \{ \alpha_k \boldsymbol{n}_z \cdot \overline{\boldsymbol{J}_k}^x \}_2 - A \{ \overline{\alpha_k \rho_k \phi_k}^x \}_2$$

$$= - A \{ \sum_j \ell_j^{-1} \, (\dot{m}_k \psi_k + \boldsymbol{J}_k \cdot \boldsymbol{n}_k)_j \}_2 - \int_{C_1 + C_2} \alpha_k \boldsymbol{n}_k \cdot \overline{\boldsymbol{J}_k}^x \, \frac{dC}{\boldsymbol{n}_k \cdot \boldsymbol{n}_{kC}}$$

$$(2.165)$$

Thanks to Eqs. (2.152) and (2.163), Eqs. (2.164) and (2.165) are identical. Similar equations to (2.164) and (2.165) can be written for volume-averaged quantities.

2.6 Models Based on Imposed Velocity Profiles

Introduction

The word *model* denotes a set of equations describing a certain *picture* of an actual two-phase flow.

Consider for example an air-water bubbly flow in a vertical pipe. Such a two-phase flow is never truly axisymmetric and, in addition, the bubbles flow faster than the liquid due to their buoyancy. However, we can replace the actual flow by an idealized picture where the flow is supposed axisymmetric and where the bubbles are supposed to move with the liquid velocity. The mathematical *model* will use two or one space independent variables depending on the type of the averaged equations written: local time-averaged or instantaneous area-averaged.

The choice of a *picture* for a two-phase flow is essentially a choice of geometric properties (e.g. axisymmetry, cylindrical interfaces in annular flow), kinematic properties (e.g. no local relative velocity between the phases) or thermal properties (e.g. saturation conditions for one phase or both). These last groups of properties represent the kinematic or thermal *nonequilibria* between the phases.

The selection of a given kinematic and/or thermal pattern is equivalent to a choice of solutions for the mathematical model. As a result this implies some compatibility conditions on the set of equations. The first models used in two-phase flow were models where some parts of the solution were imposed. However the more and more complex applications of two-phase flows have given a strong impetus to the research concerning the unrestricted two-fluid models where the nonequilibria are no longer imposed in some way but are calculated as parts of the solution.

Figure 2.11 represents four models based on imposed velocity profiles. The *one-dimensional, one-velocity model* is the simplest model and corresponds to the homogeneous model. This model can be improved in two ways: by allowing a velocity distribution (*two-dimensional, one-velocity model or Bankoff model*) or by introducing a relative velocity between the phases (*one-dimensional, two-velocity model or Wallis model*). A combination of these two improvements leads to the *two-dimensional, two-velocity model or*

Zuber and Findlay model.

One-Dimensional, One-Velocity Model

If $\overline{w_G}^x$ and $\overline{w_L}^x$ are the local velocities of the gas and liquid phases averaged over their residence time T_G and T_L, we have,

$$\overline{w_G}^x \equiv \overline{w_L}^x \equiv \text{constant over the cross section of the pipe}$$

(2.166)

The time-average of the gas volumetric flowrate is written,

$$\overline{Q_G} \triangleq \overline{\int_{A_G} w_G dA} = A \overline{R_G <w_G>}$$

(2.167)

where A_G is the area occupied by the gas phase at a given time, A the total area of the pipe cross section, $<w_G>$ the instantaneous area-averaged velocity of the gas phase over A_G and R_G the instantaneous area fraction defined by,

$$R_G \triangleq \frac{A_G}{A}$$

(2.168)

One-dimensional one-velocity model
(Homogeneous model)

Two-dimensional, one-velocity model
(Bankoff model)

One-dimensional, two-velocity model
(Wallis model)

Two-dimensional, two-velocity model
(Zuber and Findlay model)

Figure 2.11: Models based on imposed velocity profiles.

According to Eq. (2.152) we have,

$$\overline{R_G <w_G>} \equiv \{\alpha_G w_G\}^{\overline{x}} \tag{2.169}$$

where $\{\ \}$ denotes an average over the pipe cross section and α_G the local time fraction defined by,

$$\alpha_G \overset{\Delta}{=} \frac{T_G}{T_L} \tag{2.170}$$

Eq. (2.167) now reads,

$$\overline{Q_G} = A \{\alpha_G w_G\}^{\overline{x}} \tag{2.171}$$

Taking into account Eq. (2.166) we obtain,

$$\overline{Q_G} = A \{\alpha_G\} \overline{w_G}^x \tag{2.172}$$

According to Eq. (2.154) we have,

$$\{\alpha_G\} = \overline{R_G} \tag{2.173}$$

As a result the time-average of the gas volumetric flowrate reads,

$$\overline{Q_G} = A\overline{R_G}\, \overline{w_G}^x \tag{2.174}$$

In a similar way we obtain for the time-average of the liquid volumetric flowrate,

$$\overline{Q_L} - A\ (1-\overline{R_G})\overline{w_L}^x \tag{2.175}$$

Taking into account the assumption expressed by Eq. (2.166) we obtain,

$$\boxed{\overline{R_G} = \beta} \tag{2.176}$$

where β is the *volumetric quality* defined by

$$\beta \overset{\Delta}{=} \frac{\overline{Q_G}}{\overline{Q_G}+\overline{Q_L}} \tag{2.177}$$

Two-Dimensional, One-Velocity Model. The Bankoff Model

    Bankoff (1960) assumed that the local liquid and gas
velocities are equal and that the radial distributions of this
velocity and of the time fraction are given by the following
equations,

$$\overline{w_G}^x \equiv \overline{w_L}^x = (\frac{y}{R})^{\frac{1}{m}} \overline{w_C}^x \tag{2.178}$$

$$\alpha_G = (\frac{y}{R})^{\frac{1}{n}} \tag{2.179}$$

where y denotes the distance from the wall, R the pipe radius,
$\overline{w_C}^x$ the velocity on the tube axis and where m and n are positive
constants.

    A calculation similar to the one done in Section 2.6 leads
to the following results,

$$\boxed{R_G = K\beta} \tag{2.180}$$

where K is a distribution parameter called the Armand parameter.
This parameter is given by,

$$K = \frac{2(m+n+mn)(m+n+2mn)}{(n+1)(2n+1)(m+1)(2m+1)} \tag{2.181}$$

and varies from 0.6 to 1 when m and n vary from 2 to 7 and from
0.1 to 7 respectively. Bankoff (1960) proposed the following
correlation,

$$K = 0.71 + 0.00145p \quad (p \text{ in bar}) \tag{2.182}$$

One-Dimensional, Two-Velocity Model. The Wallis Model

    Wallis (1963) assumed flat profiles for the velocities and the
time-fraction but he took into account a constant relative velocity
between the phases. One thus obtains a formula which can be
compared with Eq. (2.176) for the one-dimensional one velocity
model,

$$\boxed{\overline{R}_G = \frac{\beta}{1 + \frac{(1-\overline{R}_G)(\overline{w_G}^x - \overline{w_L}^x)}{J}}} \tag{2.183}$$

where J is the mixture volumetric flux defined by,

$$J \triangleq \frac{\overline{Q_G} + \overline{Q_L}}{A}$$ (2.184)

In gravity dominated two-phase flows, the relative velocity $(\overline{w_G}^x - \overline{w_L}^x)$ is a function of the area fraction $\overline{R_G}$ and of some physical properties of the fluids. For example in a bubbly flow we have,

$$\overline{w_G}^x - \overline{w_L}^x = w_\infty (1 - \overline{R_G})$$ (2.185)

where w_∞ is the rise velocity of a bubble in an infinite medium. This velocity depends on the physical properties of the fluids only.

In order to make his model more practical Wallis defined the following quantities,

$$j_{GL} \triangleq \alpha_G (\overline{w_G}^x - j)$$ (2.186)

where,

$$j \triangleq j_G + j_L \triangleq \alpha_G \overline{w_G}^x + \alpha_L \overline{w_L}^x$$ (2.187)

Accounting for the one-dimensional assumption, Eqs. (2.186) and (2.187) lead to,

$$J_{GL} \triangleq \{j_{GL}\} = \overline{R_G} (\overline{w_G}^x - J) = \overline{R_G} (1 - \overline{R_G}) (\overline{w_G}^x - \overline{w_L}^x)$$ (2.188)

$$J_{GL} = (1 - \overline{R_G}) J_G - \overline{R_G} J_L$$ (2.189)

where,

$$J_G \triangleq \overline{Q_G} / A$$ (2.190)

$$J_L \triangleq \overline{Q_L} / A$$ (2.191)

The correlation for the relative velocity given by Eq. (2.185) leads to a new expression for J_{GL},

$$J_{GL} = w_\infty \overline{R_G} (1 - \overline{R_G})^2$$ (2.192)

The *Wallis diagram* (Figure 2.12) represents the variations of J_{GL} as functions of $\overline{R_G}$. Eq. (2.192) is represented by a curve whereas Eq. (2.189) is represented by a straight line passing through points the coordinates of which are ($\overline{R_G} = 0$; $J_{GL} = J_G$) and ($\overline{R_G} = 1$; $J_{GL} = -J_L$).

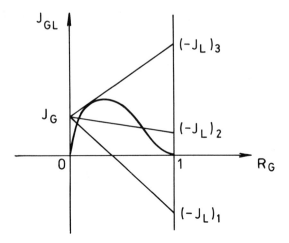

Figure 2.12: Wallis diagram for bubbly flow.

In *upward cocurrent flow* J_G and J_L are positive and the
straight line (Eq. 2.189) intersects the curve given by Eq. (2.192)
at a single point. The abscissa of this point corresponds to the
area fraction $\overline{R}_G$ observed in the pipe. In *countercurrent flow*
the gas flows upwards ($J_G>0$) and the liquid downwards ($J_L<0$). If,
in Figure 2.12, $|J_L|<|(J_L)_3|$, the straight line intersects the
curve at two points which correspond to the occurrence of a
double flow pattern within the same column (Figure 2.13). If the
liquid flowrate is further increased, the straight line becomes
tangent to the curve when $|J_L| = |(J_L)_3|$ and the flow becomes
cocurrent downward.

Two-Dimensional, Two-Velocity Model. The Zuber and Findlay Model

Zuber and Findlay (1965) defined a local drift velocity $\overline{w}_{Gj}^x$
by the following relation,

$$\overline{w}_{Gj}^x \triangleq \overline{w}_G^x - j = (1-\alpha_G)(\overline{w}_G^x - \overline{w}_L^x) \qquad (2.193)$$

This equation can also be written in the form,

$$\alpha_G \overline{w}_{Gj}^x = \alpha_G \overline{w}_G^x - \alpha_G j \qquad (2.194)$$

The Zuber and Findlay equation is obtained by taking the average
of Eq. (2.194) over the pipe cross section,

$$\overline{R}_G = \frac{J_G}{C_o J + \tilde{w}_{Gj}} \qquad (2.195)$$

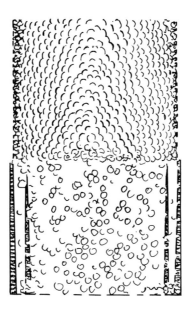

Figure 2.13: Loose packed bed and dense packed bed in the same column (Croix
        et al. 1973).

where the *distribution parameter* C_O and the *weighted drift
velocity* $\tilde{w}_{Gj}$ are defined by,

$$C_O \overset{\Delta}{=} \frac{\{\alpha_G j\}}{\{\alpha_G\}\{j\}} \qquad (2.196)$$

$$\tilde{w}_{Gj} \overset{\Delta}{=} \frac{\{\alpha_G \overline{w}_{Gj}^x\}}{\{\alpha_G\}} \qquad (2.197)$$

Recommended values for C_O and $\tilde{w}_{Gj}$ are given by Ishii (1977)
according to the flow pattern and the pressure range.

        Equation (2.195) can also be written in terms of the
volumetric quality β,

$$\boxed{\overline{R}_G = \frac{\beta}{C_O + \tilde{w}_{Gj}/J}} \qquad (2.198)$$

If velocity and void fraction profiles are flat then $C_O \equiv 1$ and
$\tilde{w}_{Gj} \equiv J_{GL}/\overline{R}_G$. The Zuber and Findlay model thus degenerates to
the Wallis model (Eq. (2.183). If there is no relative velocity
between the phases and if we let $C_O \equiv 1/K$, we recover Eq. (2.180)
of the Bankoff model. Finally if the profiles are flat and if
there is no relative velocity between the phases we end up with

Eq. (2.176) of the homogeneous model.

Eq. (2.198) also reads,

$$\frac{J_G}{\overline{R_G}} = C_o J + \tilde{w}_{Gj} \qquad\qquad (2.199)$$

If C_o and $\tilde{w}_{Gj}$ are constant for a given flow pattern the averaged gas velocity $J_G/\overline{R_G}$ is a linear function of the mixture superficial velocity J. Figure 2.14 gives an example of the Zuber-Findlay diagram applied to an air-water bubbly mixture flowing in a converging-diverging nozzle (Delhaye and Jacquemin, 1971).

2.7 Unrestricted Two-Fluid Models

Introduction

The models based on imposed velocity profiles are not adapted to the description of complex two-phase flows where nonequilibria play an important role. Such nonequilibria occur when a two-phase flow withstands strong variations in space or time. The mathematical description of such flows requires a two-fluid model where the kinematic or thermal nonequilibria are calculated as a part of the solution. The development of the two-fluid models has led to difficulties linked to the very nature of the averaging procedure and to the interaction laws needed to close the set of equations.

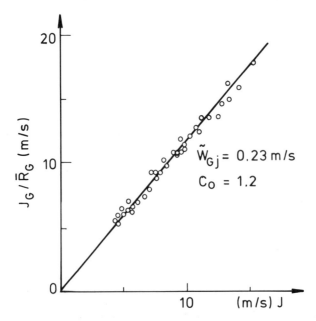

Figure 2.14: Zuber-Findlay diagram. Air-water bubbly mixture flowing in a converging-diverging nozzle (Delhaye and Jacquemin, 1971).

In the following we will explain the origin of the *topological laws* (Bouré, 1978) and we will emphasize the importance of the mathematical structure of the *interaction laws*.

Topological Laws

In this Section we will look for the origin of the topological laws as named by Bouré (1978). We will consider the instantaneous area-averaged equations governing the isothermal stratified flow of two inviscid, incompressible fluids in a horizontal channel. The calculation will concentrate on the interface behavior in the absence of phase change. For the sake of simplicity we will assume that the flow is two-dimensional and that the interface has no mass, no viscosity and no surface tension. Figure 2.15 shows the system geometry. The channel width is denoted by H whereas h and (H-h) denote the thickness of fluids 1 and 2.

In the following we will write the area-averaged equations directly, ignoring completely the two-dimensional details of the flow.

Mass balance.

    (H1)   incompressible fluids

    (H2)   no phase change at the interface

Eq. (2.116) simplifies to the following relations,

(i) for fluid 1:

$$\frac{\partial}{\partial t}\, h\rho_1 + \frac{\partial}{\partial x}\, h\rho_1 <u_1> = 0 \qquad\qquad (2.200)$$

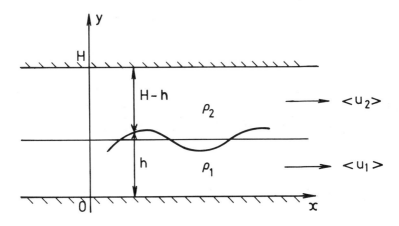

Figure 2.15.

(ii) for fluid 2:

$$\frac{\partial}{\partial t} (H-h)\rho_2 + \frac{\partial}{\partial x} (H-h)\rho_2 <u_2> = 0 \qquad (2.201)$$

Momentum balance along the x-axis.

    (H1)   incompressible fluids

    (H2)   no phase change at the interface

    (H3)   horizontal flow

    (H4)   inviscid fluids

Eq. (2.118) together with Eq. (2.110) yields,

(i) for fluid 1:

$$\frac{\partial}{\partial t} h\rho_1<u_1> + \frac{\partial}{\partial x} h\rho_1<u_1^2> + \frac{\partial}{\partial x} h<p_1> = p_{1i} \frac{\partial h}{\partial x} \qquad (2.202)$$

(ii) for fluid 2:

$$\frac{\partial}{\partial t} (H-h)\rho_2<u_2> + \frac{\partial}{\partial x} (H-h)\rho_2<u_2^2>$$

$$+ \frac{\partial}{\partial x} (H-h)<p_2> = - p_{2i} \frac{\partial h}{\partial x} \qquad (2.203)$$

Momentum jump condition.

    (H2)   no phase change at the interface

    (H4)   inviscid fluids

    (H5)   no surface tension

Jump condition (2.51) simplifies to the following relation,

$$p_{1i} = p_{2i} \stackrel{\Delta}{=} p_i \qquad (2.204)$$

Two-phase flow modeling. Considering Eqs. (2.200) to (2.204) we realize that, given H, ρ_1 and ρ_2 we have *four* partial differential equations (Eqs. 2.200 to 2.203), *five* dependent variables, i.e. h, $<u_1>$, $<u_2>$, $<p_1>$, $<p_2>$, and *three* supplementary variables, i.e. $<u_1^2>$, $<u_2^2>$, p_i.

(i) Space correlation coefficients:

    One can define a *space correlation coefficient* C_k by the following relation,

$$C_k \stackrel{\Delta}{=} \frac{<u_k^2>}{<u_k>^2} = C_k (x,t) \qquad (2.205)$$

Generally speaking, C_k should be expressed as a functional of the dependent variables. However, due to the absence of information concerning this correlation coefficient, we will *assume* that,

$$C_1 = C_2 = 1 \qquad (2.206)$$

(ii) Mechanical law:

We recall here that in the method based on the area-averaged equations, we completely ignore the two-dimensional equations which govern the flow. As a consequence, we have to *postulate* an equation which relates the pressure p_i at the interface to the dependent variables, knowing that we have also to reduce the number of these dependent variables by one. Physical intuition leads us to *assume* that we have a hydrostatic pressure distribution over the height of the channel,

$$<p_1> = p_i + \rho_1 g \frac{h}{2} \qquad (2.207)$$

$$<p_2> = p_i - \rho_2 g \frac{H-h}{2} \qquad (2.208)$$

As a result, we *postulate* the following *mechanical law*,

$$<p_2> = <p_1> - (\rho_1 - \rho_2) g \frac{h}{2} - \rho_2 g \frac{H}{2} \qquad (2.209)$$

(iii) Final system:

Thanks to *assumptions* (2.206) and (2.209) the four balance equations (2.200) to (2.203) can be written under the following matrix form,

$$A \frac{\partial X}{\partial t} + B \frac{\partial X}{\partial x} = 0 \qquad (2.210)$$

with matrices A and B and the solution vector X given by,

$$A \triangleq \begin{bmatrix} \rho_1 & 0 & 0 & 0 \\ -\rho_2 & 0 & 0 & 0 \\ \rho_1 <u_1> & h\rho_1 & 0 & 0 \\ -\rho_2 <u_2> & 0 & (H-h)\rho_2 & 0 \end{bmatrix} \qquad (2.211)$$

$$
B \triangleq \begin{bmatrix}
\rho_1 <u_1> & h\rho_1 & 0 & 0 \\
-\rho_2 <u_2> & 0 & (H-h)\rho_2 & 0 \\
\rho_1 <u_1>^2 + \rho_1 gh/2 & 2h\rho_1 <u_1> & 0 & h \\
-\rho_2 <u_2>^2 + \rho_2 g(H-h) & 0 & 2(H-h)\rho_2 <u_2> & H-h \\
\quad -\rho_1 g(H-h)/2 & & &
\end{bmatrix}
$$

$$(2.212)$$

$$
X \triangleq \begin{bmatrix} h & <u_1> & <u_2> & <p_1> \end{bmatrix}
\qquad (2.213)
$$

(iv) Interface stability:

The characteristic equation reads,

$$
\det \begin{bmatrix} B - cA \end{bmatrix} = 0 \tag{2.214}
$$

where c is the characteristic speed. After some calculation we obtain the following relation,

$$
\rho_1 (<u_1>-c)^2 (H-h) + \rho_2 (<u_2>-c)^2 h - gh(H-h)(\rho_1-\rho_2) = 0 \tag{2.215}
$$

which is fully consistent with the gravity long wave theory (Milne Thomson, 1955).

Stability is ensured if the characteristic speeds are real, i.e. if we have,

$$
(<u_1>-<u_2>)^2 < \frac{g(\rho_1-\rho_2)\left[\rho_1(H-h)+\rho_2 h\right]}{\rho_1 \rho_2} \tag{2.216}
$$

This condition means that the restoring effect due to the gravity should be high enough to overcome the destablizing effect due to the slip velocity.

If we assume a *constant* pressure over the channel height instead of a *hydrostatic* distribution, the gravity term no longer appears in the characteristic equation (2.215). As a result the solution is *always* unstable, which is physically unrealistic.

Other topological laws proposed. Topological laws depend on two-phase flow patterns. For dilute suspensions of bubbles, Van Wijngaarden (1976) obtained,

$$<p_G> = p_{Gi} \qquad\qquad (2.217)$$

$$<p_L> = p_{Li} + (\frac{1}{4} - R_G)(<w_G>-<w_L>)^2 \qquad\qquad (2.218)$$

This law is associated with the following momentum jump condition,

$$p_{Gi} - p_{Li} = \frac{2\sigma}{R} \qquad\qquad (2.219)$$

A similar law was given by Stuhmiller (1977).

Interaction Laws

Whatever the type of averaging operator used the two-fluid model requires the knowledge of *seven* interaction laws: three for the interfacial transfers of mass, momentum and energy between the phases and four for the momentum and energy transfer between each phase and the wall. These laws being known, one can calculate the void fraction, the velocities and the temperatures of both phases and, as a result, the kinematic and thermal nonequilibria. Therefore the interaction laws can be considered as the keystone of two-phase flow modeling.

Bouré (1978) has attempted to restrict the generality of the way the interaction laws could be formulated by studying their invariance properties in some changes of frame or origin. However the *mathematical nature* of these laws is not yet known rigorously. Algebraic or differential laws could be chosen but it is also possible to envisage some laws involving convolution integrals or interaction terms obeying a transport equation. The mathematical nature of the interaction laws has a drastic influence on the characteristic speeds calculated from the set of equations describing a given two-phase flow. In fact, if these laws do not contain any differential terms, the matrices of the system do not contain any terms linked to the interaction processes. However the occurrence of differential terms in interaction laws would not be surprising. Several authors have shown that such was the case in some particular situations: friction and heat transfer at the wall in transient laminar single-phase flow (Pham Dan Tam and Veteau, 1977; Achard, 1978), interfacial friction in transient dispersed flow (Achard, 1978), transient interfacial energy transfer (Banerjee, 1978).

Apart from the case of transient flows, friction and heat transfer between a two-phase mixture and a wall are rather well-known and are extensively studied in other chapters of this book. On the contrary, the interfacial mass, momentum and energy transfers are still being investigated. However some progress has been made recently concerning the interfacial friction in steady-state or transient dispersed flows (Ishii and Zuber, 1978; Achard, 1978).

2.8 Models Based on Specified Evolution

Although the two-fluid model is, by its very nature, the most consistent model, its use is difficult due to the seven interaction laws needed. In order to simplify the problem one has imagined *particular evolutions* of two-phase flows where the nonequilibria are either null or given by algebraic laws. We thus have,

(i) no local relative velocity (homogeneous model; Bankoff model, see Section 2.6)

(ii) correlations for the relative velocity given by means of,

- a slip ratio $\overline{<w_G>}/\overline{<w_L>}$ as a function of pressure and equilibrium quality

- a weighted drift velocity $\tilde{w}_{Gj}$ as a function of pressure and of the physical properties of the fluids (Zuber and Findlay model, see Section 2.6).

- a void fraction $\overline{R_G}$ as a function of pressure and equilibrium quality (Martinelli and Nelson, 1948)

(iii) thermal equilibria

- saturation conditions for the vapor phase

- saturation conditions for the liquid phase

It is of the utmost importance to understand that the choice of one or several nonequilibrium conditions is equivalent to the choice of a specific evolution of one or several dependent variables. This remark brings up two major consequences:

(i) a choice of a specific evolution for a given dependent variable yields some compatibility conditions on the interaction laws since the transfers impose the evolution of the flow.

(ii) The replacement of a partial differential equation by an algebraic equation changes the mathematical nature of the problem drastically. This change implies a modification in the propagation properties of the model with respect to the two-fluid model in which the nonequilibria are part of the solution.

Nomenclature

A	function, matrix (Eq. 2.211)
A	surface
B	vector, matrix (Eq. 2.212)
C_O	Zuber and Findlay parameter (Eq. 2.196)
c	characteristic speed
C	space correlation coefficient (Eq. 2.205)
C	line

c^V specific heat at constant volume

f   function

F   external force per unit of mass

g   free enthalpy per unit of mass, arbitrary function

h   height

H   channel height

i   enthalpy per unit of mass

j   local volumetric flux (Eqs. 2.186, 2.187)

J   superficial velocity (Eqs. 2.184, 2.190, 2.191)

J   flux term

k   thermal conductivity

K   Armand parameter (Eq. 2.181)

ℓ   line coordinate, specific length (Eq. 2.136)

$\dot{m}$ mass transfer per unit area of interface and per unit of time (Eq. 2.17)

M   tensor

m   exponent

n   exponent

n   normal unit vector

p   pressure (Eq. 2.31)

Q   volumetric flowrate (Eq. 2.167)

q   heat flux

r   position vector

r   radial coordinate

R   radius of curvature, pipe radius, area fraction (Eqs. 2.153, 2.168)

R   antisymmetric tensor

s   entropy per unit of mass

t   time

t   tangential unit vector

T   temperature (Eq. 2.32), time interval

T   stress tensor

u   surface coordinate; internal energy per unit of mass; x-component of the velocity vector

U   unit tensor

v   surface coordinate; y-component of the velocity vector

v   velocity vector

V   volume

w   radial component of the velocity vector, z-component of the velocity vector

$\tilde{w}_{Gj}$ Zuber and Findlay weighted drift velocity (Eq. 2.197)

x   coordinate

x   position vector

X   phase density function (Eq. 2.137), solution vector (Eq. 2.213)

y   coordinate

y   distance from the wall

z   coordinate

Z   distance

α   time fraction (Eq. 2.129) (2.170)

β   volumetric quality (Eq. 2.177)

γ   local specific area (Eq. 2.160)

Γ   integral specific area (Eq. 2.161)

Δ   entropy source per unit of volume or area and per unit of time

η   film thickness, time dummy variable

μ   viscosity

ρ   density

σ   surface tension

τ   viscous stress tensor

φ   source term

ψ   specific quantity

Subscripts

A   surface

c   pipe axis

G   gas

i   interface

j   interface number

k   phase index

L   liquid

n   dimension index, noise

p   interface velocity

s   signal

V   vapor

W   wall

Superscripts

n   normal

t   tangential, transposed

v   at constant volume

Operators

$<>_2$ area-averaging operator over A_k (Eq. 2.114)
$<>_3$ volume-averaging operator over V_k (Eq. 2.123)
———   single time-averaging operator (Eq. 2.139)
———^{x} x-weighted single time-averaging operator (Eq. 2.135)

References

Achard, J L, (1978) Contribution à l'Etude Théorique des Ecoulements Diphasiques Transitoires, Thèse de Docteur ès Sciences, Université Scientifique et Médicale et Institut National Polytechnique de Grenoble.

Aris, R, (1962) *Vectors, Tensors and the Basic Equations of Fluid Mechanics*, Prentice Hall.

Banerjee, S, (1978) "A surface renewal model for interfacial heat and mass transfer in transient two-phase flow." *Int. J. Multiphase Flow*, **4**, 571-573.

Bankoff, S G, (1960) "A variable density single-fluid model for two-phase flow with particular reference to steam-water flow." *J. Heat Transfer*, 265-272.

Bird, R B, Stewart, W E and Lightfoot, E N, (1960) *Transport Phenomena*, Wiley.

Birkhoff, G, (1964) "Averaged-conservation laws in pipes." *J. of Math. Anal. and Applic.*, **8**, 66-77.

Bouré, J A, (1978) Les Lois Constitutives des Modèles d'Ecoulements Diphasiques Monodimensionnels, à Deux Fluides: Formes Envisageables, Restrictions Résultant d'Axiomes Fondamentaux, CEA-R-4915.

Callen, H B, (1960) *Thermodynamics*, Wiley.

Croix, J M, Labroche, C, Lackmé, C and Merle, A, (1973) Une application des colonnes à lits denses: le traitement d'eaux phénolées industrielles par extraction liquide-liquide, *Bulletin d'Informations Scientifiques et Techniques du CEA*, No. 177, 11-24.

Delhaye, J M, (1968) Equations fondamentales des écoulements diphasiques, CEA-R-3429.

Delhaye, J M, (1974) "Jump conditions and entropy sources in two-phase systems. Local instant formulation." *Int. J. of Multiphase Flow*, 1, 395-409.

Delhaye, J M, (1976) Sur les surfaces volumiques locale et integrale en écoulement diphasique, *Comptes Rendus à l'Académie des Sciences*, Paris t. 283, Série A, 243-246.

Delhaye, J M and Achard, J L, (1978) "On the averaging operators introduced in two-phase flow modeling." *Transient Two-Phase Flow*, Proceedings of the CSNI Specialists Meeting, August 3-4, 1976, Banerjee, S and Weaver, K R, Eds, **1**, 5-84.

Delhaye, J M, and Achard, J L, (1977) "On the use of averaging operators in two-phase flow modeling." *Thermal and Hydraulic Aspects of Nuclear Reactor Safety, 1: Light Water Reactors*, Jones, O C and Bankoff, S G, Eds, ASME, 289-332.

Delhaye, J M and Jacquemin, J P, (1971) "Experimental study of a two-phase air-water flow in a converging-diverging nozzle." CENG/STT, Note TT No. 382.

Ishii, M, (1975) *Thermo-fluid dynamic theory of two-phase flow*, Eyrolles (Paris).

Ishii, M, (1977) "One dimensional drift-flux model and constitutive equations for relative motion between phases in various two-phase flow regimes." ANL-77-47.

Ishii, M and Zuber, N, (1978) "Relative motion and interfacial drag coefficient in dispersed two-phase flow of bubbles, drop or particles." Paper presented at AIChE 71st Annual Meeting at Miami Beach, Florida.

Kenning, D B, (1968) "Two-phase flow with non uniform surface tension." *Applied Mechanics Reviews*, **21**, No. 11.

Kocamustafaogullari, G, (1971) "Thermo-fluid dynamics of separated two-phase flow." Ph.D. Thesis, Georgia Institute of Technology, School of Mechanical Engineering.

Levich, V G and Krylov, V S, (1969) "Surface-tension-driven phenomena." *Annual Review of Fluid Mechanics*, **1**, 293-316.

Martinelli, A C and Nelson, D B, (1948) "Prediction of pressure drop during forced circulation boiling of water." *Trans. ASME*, 695-702.

Milne Thomson, L M, (1955) *Theoretical Hydrodynamics*, Mac Millan.

Phan Dan Tam and Veteau, J M, (1977) Lois constitutives en variables moyennées pour les écoulements de fluides: Etude de la loi de frottement instationnaire pour l'écoulement monophasique laminaire en conduite cylindrique, CEA-R-4816.

Spindler, B, Solésio, J N and Delhaye, J M, (1978) "On the equations describing the instabilities of liquid films with interfacial phase change." ICHMT 1978 International Seminar on Momentum, Heat and Mass Transfer in Two-Phase Energy and Chemical Systems, Dubrovnik, September 4-9, Hemisphere.

Stuhmiller, J H, (1977) "The influence of interfacial pressure forces on the character of two-phase flow model equations." *Int. J. Multiphase Flow*, **3**, 551-560.

Teletov, S G, (1958) "Two-phase flow hydrodynamics. 1-Hydrodynamics and energy equations" (in Russian), *Bulletin of Moscow University*, **2**.

Truesdell, C A and Toupin, R A, "The classical field theories." *Encyclopedia of Physics*, Flugge, S, Ed., **III/1**, Principles of Classical Mechanics and Field Theory, Springer-Verlag, 226-858.

Van Wijngaarden, L, (1976) "Some problems in the formulation of the equations for gas/liquid flows". *Theoretical and Applied Mechanics*, Koiter, W T, Ed., North Holland, 249-260.

Vernier, P H and Delhaye, J M, (1968) "General two-phase flow equations applied to the thermohydrodynamics of boiling water nuclear reactors." *Energie Primaire*, **4**, No. 1-2, 5-46.

Wallis, G B, (1963) "Some hydrodynamics aspects of two-phase flow and boiling." *International Developments in Heat Transfer*, ASME, 319-340.

Zuber, N, and Findlay, J A, (1965) "Average volumetric concentration in two-phase flow systems." *J. Heat Transfer*, 453-468.

Chapter 3

Frictional Pressure Drops

J. M. DELHAYE

3.1 Introduction

The calculation of two-phase pressure drop is an essential
issue in many industrial applications and quite a number of
studies and survey papers have been devoted to this area. It is
not the purpose of this Chapter to sum up existing reviews or add
another one to an already long list. Our goal is more oriented
toward a complete understanding of how the different models
proposed in the literature enter the general framework of two-
phase flow modelling as presented in Chapter 2. However, practical
considerations will not be left aside and design recommendations
will be given at the end of the Chapter.

As was shown in Chapter 2, the knowledge of the interaction
laws between the phases or between the two-phase mixture and the
wall basically relies on experimental evidence. For the determination of the frictional pressure drop, one must recall that this
quantity is not usually measured directly. The majority of
experimental data provides values of the total pressure drop only.
As a result, a void fraction measurement or evaluation is needed
to determine the acceleration and gravity pressure drops to be
subtracted from the total pressure drop to get the frictional
component. This has to be borne in mind whenever one uses a
pressure drop correlation. More precisely, if the void fraction
is not measured and if a correlation is used instead to establish
a correlation for the frictional pressure drop, the same void
correlation must be used when applying the frictional pressure
drop correlation. This obvious recommendation is often forgotten
and several extensive experimental works or comparative studies
cannot be used due to the absence of information concerning the
way the other pressure drop components are evaluated.

This Chapter is devoted to several wellknown models of
frictional pressure drop in *steady-state flow* within *single tubes*
or *channels*. For detailed studies of pressure drops in rod
bundles and tube banks, the reader is referred to the surveys by
Grand (1981) and Grant et al. (1974) respectively. The correlations
which will be presented in the following *do not depend on the flow
regime*. This is of course a major drawback since a correlation
for bubbly flow msut be different from a correlation used in
annular flow. However flow regime dependent correlations do exist
but their definite recommendation still requires further investi-

gation. The reader interested in these correlations for horizontal flow will find an interesting review in the paper by Mandhane et al. (1977).

Comparative studies will be mentioned in the last Section of the Chapter and will lead to a crude selection chart for frictional pressure drop correlation in steady-state flow.

3.2 Simplified Balance Equations

The two averaging operators, i.e. over a cross section area and over a time interval, give averages of products which can be expressed as a function of the product of averages by means of correlation coefficients. For instance

$$\{fg\} = A\{f\}\{g\}$$

$$\overline{fg}^x = B\overline{f}^x\ \overline{g}^x$$

where f and g are two arbitrary functions, A a space correlation coefficient and B a time correlation coefficient.

In the present state of knowledge, the following hypotheses are generally admitted,

(i)   The space correlation coefficients are equal to 1

(ii)   The time correlation coefficients are equal to 1

(iii) The equations of state valid for local quantities apply to averaged quantities.

(iv)   Longitudinal conduction terms in each phase as well as their derivatives, are negligible

(v)   The phase viscous stress derivatives and the power of these viscous stresses are negligible

(vi)   In *vertical* flow, the pressure is constant over a cross section.

Moreover, if the flow is assumed to be symmetrical with respect to a straight line, the vectorial momentum equation reduces to its projection on the symmetry axis. On the basis of these hypotheses, simplified phase and mixture equations are obtained as follows.

Balance of Mass

(i) Phase equation (k=G,L)

$$\frac{\partial}{\partial t}(R_k \rho_k) + \frac{\partial}{\partial z}(R_k \rho_k w_k) = -\frac{1}{A}\int_{C(z,t)} \overset{\circ}{m}_k \frac{dC}{\boldsymbol{n}_k \cdot \boldsymbol{n}_{kC}} \tag{3.1}$$

(ii) Mixture equation

$$\frac{\partial}{\partial t}\,(R_G\rho_G + R_L\rho_L) + \frac{\partial}{\partial z}\,(R_G\rho_G w_G + R_L\rho_L w_L) = 0 \tag{3.2}$$

Balance of Momentum

(i) Phase equation (k=G,L)

$$\frac{\partial}{\partial t}\,(R_k\rho_k w_k) + \frac{\partial}{\partial z}\,(R_k\rho_k w_k^2) - R_k\rho_k F_z + R_k\,\frac{\partial p}{\partial z}$$

$$= -\frac{1}{A}\int_{C(z,t)} n_z\cdot(\dot{m}_k v_k - n_k\cdot\tau_k)\,\frac{dC}{n_k\cdot n_{kC}}$$

$$+ \frac{1}{A}\int_{C_k(z,t)} n_z\cdot(n_k\cdot\tau_k)\,\frac{dC}{n_k\cdot n_{kC}} \tag{3.3}$$

(ii) Mixture equation

$$\frac{\partial}{\partial t}\,(R_G\rho_G w_G + R_L\rho_L w_L) + \frac{\partial}{\partial z}\,(R_G\rho_G w_G^2 + R_L\rho_L w_L^2)$$

$$- (R_G\rho_G + R_L\rho_L) F_z + \frac{\partial p}{\partial z}$$

$$= \frac{1}{A}\sum_{k=G,L}\int_{C_k(z,t)} n_z\cdot(n_k\cdot\tau_k)\,\frac{dC}{n_k\cdot n_{kC}} \tag{3.4}$$

Balance of Total Energy

(i) Phase equation (k=G,L)

$$\frac{\partial}{\partial t}\left[R_k\rho_k(\tfrac{1}{2}v_k^2 + i_k)\right] + \frac{\partial}{\partial z}\left[R_k\rho_k(\tfrac{1}{2}v_k^2 + i_k)w_k\right]$$

$$- R_k\,\frac{\partial p}{\partial t} - R_k\rho_k F\cdot v_k$$

$$= -\frac{1}{A}\int_{C(z,t)}\left[\dot{m}_k(\tfrac{1}{2}v_k^2 + i_k) - (\tau_k\cdot v_k)\cdot n_k + q_k\cdot n_k\right]\frac{dC}{n_k\cdot n_{kC}}$$

$$- \frac{1}{A}\int_{C_k(z,t)} q_k\cdot n_k\,\frac{dC}{n_k\cdot n_{kC}} \tag{3.5}$$

(ii) Mixture equation

$$\frac{\partial}{\partial t}\left[R_G\rho_G(\tfrac{1}{2}v_G^2 + u_G) + R_L\rho_L(\tfrac{1}{2}v_L^2 + u_L)\right]$$

$$+ \frac{\partial}{\partial z}\left[R_G\rho_G(\tfrac{1}{2}v_G^2 + i_G)w_G + R_L\rho_L(\tfrac{1}{2}v_L^2 + i_L)w_L\right]$$

$$- (R_G\rho_G\mathbf{F}\cdot\mathbf{v}_G + R_L\rho_L\mathbf{F}\cdot\mathbf{v}_L) = -\frac{1}{A}\sum_{k=G,L}\int_{C_k} C_k(z,t)\, q_k\cdot n_k\,\frac{dC}{n_k\cdot n_{kC}}$$

$$(3.6)$$

3.3 Homogeneous Model

In the homogeneous model we impose the following relations,

$$\overline{w_G}^x \equiv \overline{w_L}^x \triangleq w \tag{3.7}$$

$$\overline{T_G}^x \equiv \overline{T_L}^x \equiv T_{sat}(p) \tag{3.8}$$

where w is the velocity component along the axis of the pipe and
T the temperature, $T_{sat}(p)$ being the saturation temperature
corresponding to the pressure p, constant over a cross section.
If we further assume flat velocity and time fraction profiles we
have (Eq. 2.176),

$$R_G = \beta \tag{3.9}$$

which can be rewritten in terms of the mass quality x,

$$R_G = \frac{x\rho_L}{x\rho_L + (1-x)\rho_G} \tag{3.10}$$

Equations Describing the Homogeneous Model

Balance equations. The balance equations of the homogeneous model
are derived from the simplified balance equations written for the
mixture.

(i) Mass balance

Putting,

$$\rho \triangleq R_G\rho_G + (1-R_G)\rho_L \tag{3.11}$$

Eq. (3.2) becomes,

$$\frac{\partial \rho}{\partial t} + \frac{\partial \rho w}{\partial z} = 0 \qquad (3.12)$$

In steady-state flow we obtain

$$\rho w \stackrel{\Delta}{=} G = \text{constant} \qquad (3.13)$$

(ii) Momentum balance

Eq. (3.4) reads,

$$\frac{\partial}{\partial t} \rho w + \frac{\partial}{\partial z} \rho w^2 = -\frac{\partial p}{\partial z} + \rho F_z - \frac{P}{A} \tau_w \qquad (3.14)$$

where P is the wetted perimeter of the cross section and τ_w the averaged wall shear stress. For a tube of circular cross section of diameter D we get,

$$\frac{P}{A} = \frac{4}{D} \qquad (3.15)$$

Taking into account the mass balance (Eq. 3.12) we obtain for the momentum balance,

$$\rho \frac{\partial w}{\partial t} + \rho w \frac{\partial w}{\partial z} = -\frac{\partial p}{\partial z} + \rho F_z - \frac{P}{A} \tau_w \qquad (3.16)$$

In steady-state flow, this equation can be rewritten as,

$$\frac{dp}{dz} = (\frac{dp}{dz})_A + (\frac{dp}{dz})_G + (\frac{dp}{dz})_F \qquad (3.18)$$

(iii) Energy balance

Assuming that the heat flux at the wall is applied on the whole perimeter of the cross section and neglecting the kinetic and potential energies, Eq. (3.6) reads,

$$\frac{\partial}{\partial t} \rho u + \frac{\partial}{\partial z} \rho w i = \frac{P}{A} \psi \qquad (3.19)$$

where ψ is the heat flux density at the wall and where the internal energy and enthalpy are defined by,

$$u \stackrel{\Delta}{=} x u_G + (1-x) u_L \qquad (3.20)$$

$$i \stackrel{\Delta}{=} x i_G + (1-x) i_L \qquad (3.21)$$

In steady state flow we get,

$$\frac{d}{dz} \rho w i = \frac{P}{A} \psi \qquad (3.22)$$

Specified evolutions. In the homogeneous model we assume the following flow behavior,

(i) Mechanical equilibrium

The equality of the local velocities along with the assumption of flat profiles lead to Eq. (3.9),

$$R_G = \frac{x\rho_L}{x\rho_L + (1-x)\rho_G} \qquad (3.23)$$

This equation can be considered as the *void correlation* for the homogeneous model.

(ii) Thermal equilibria

Eq. (3.8) leads to,

$$i_G \equiv i_G^{sat}(p) \qquad (3.24)$$

$$i_L \equiv i_L^{sat}(p) \qquad (3.25)$$

Final set of equations. The *three* constraints on the solution (Eqs 3.23 to 3.25) which specify the flow evolution given by Eqs (3.7) and (3.8), added to the *three* mixture balance equations (3.13, 17 and 22), form a set of *six* equations corresponding to the general six equation set of the two-fluid model.

In an *adiabatic* flow, the quality is constant and is considered as a parameter whereas in a *diabatic* flow the quality is directly determined by means of the energy equation (3.22).

The set of equations (3.13, 17 and 22 to 25) is entirely closed if the wall shear stress τ_w is known. This will be dealt with in the following Section.

Frictional Pressure Drop Modelling

A friction factor C_f can be defined in the usual way,

$$C_f = \frac{\tau_w}{\frac{1}{2}\rho w^2} \qquad (3.26)$$

The frictional pressure drop thus reads,

$$\left(\frac{dp}{dz}\right)_F = \frac{4\tau_w}{D} = \frac{2C_f \rho w^2}{D} \qquad (3.27)$$

Constant friction factor. A frictional pressure drop estimate can
be obtained by taking a constant friction factor. For high
pressure boilers $C_f = 0.005$ is satisfactory for dispersed annular
flow when the film is not too thick, i.e. when the liquid is
practically entirely in the form of droplets. For flashing flow
of water at low pressure $C_f = 0.003$ seems adequate.

Single phase friction factor. In a *low quality* two-phase flow,
the friction factor can be calculated as in a *liquid* single phase
flow of mass flowrate equal to the total mass flowrate of the two-
phase flow.

In a *high quality* two-phase flow, the friction factor can be
calculated as in a *gas* single phase flow of mass flowrate equal
to the total mass flowrate of the two-phase flow.

Two-phase friction factor. A turbulent friction law for single
phase flow such as the Blasius equation, can be used with a
Reynolds number based on an equivalent viscosity μ and the total
mass velocity G of the actual two-phase flow,

$$C_f = f(Re) \qquad (3.28)$$

$$Re \triangleq \frac{GD}{\mu} \qquad (3.29)$$

(i)   In his study of steam-water mixtures vaporizing in a pipe,
      Owens (1963) took an equivalent viscosity equal to the liquid
      viscosity,

$$\mu = \mu_L \qquad (3.30)$$

   and a friction law identical to the one found in the liquid
   phase upstream of the boiling boundary. The result is
   equivalent to the use of a single phase liquid friction
   factor.

(ii) In their Case I (No slip) correlation, Dukler et al. (1964)
      used an equivalent viscosity based on the viscosities of each
      phase weighted by their respective volumetric quality,

$$\mu = \beta \mu_G + (1-\beta) \mu_L \qquad (3.31)$$

   along with the following friction law,

$$C_f = 0.0014 + 0.125 \, Re^{-0.32} \qquad (3.32)$$

Remark on the equivalent viscosity of a two-phase mixture. The
only justification of Eq. (3.31) is that the equivalent viscosity
is equal to the gas or liquid viscosity when β is equal to 1 or 0.
However for a static two-phase mixture this equation fails to
represent the reality. This is why a formula based on a static

concentration should be preferred. Ishii and Zuber (1979) proposed the following equation,

$$\frac{\mu}{\mu_c} = (1 - \frac{\alpha_d}{\alpha_{dm}})^{-2.5\alpha_{dm}(\mu_d + 0.4\mu_c)/(\mu_d + \mu_c)} \tag{3.33}$$

where the indices c and d refer to the continuous and dispersed phases, α is a volumetric fraction and where the index m denotes the maximum value of α corresponding to the densest packing,

$$\alpha_{dm} = 0.62 \text{ for } solid \text{ particles}$$

$$\alpha_{dm} = 1.0 \text{ for } fluid \text{ particles (gas or liquid).}$$

It can be easily seen that the wellknown Einstein and Taylor formulae can be recovered from Eq. (3.33).

Applicability of the Homogeneous Model

In the homogeneous model the underlying idea is to replace the two-phase fluid by an equivalent compressible single phase fluid. If one of the phases is finely dispersed, momentum and energy transfers will be sufficiently rapid for the local time-averaged velocities and temperatures of the two phases to be equal. As a result, the homogeneous model is only applicable if there is no rapid variation of flow parameters and if thermal nonequilibria have no great influence. Other rather obvious cases must also be excluded. For instance, counter-current flows cannot be represented by an equivalent single phase fluid flow. The homogeneous model has been frequently used to study problems in oil extraction, steam generation and refrigeration. The higher the pressures and velocities, the more realistic the homogeneous model.

A good example of the homogenous model effectiveness was given by Whitcutt and Chojnowski (1973) who carried out pressure drop measurements in high pressure steam generator tubes. A steam-water mixture, flowing upward in a vertical cylindrical tube, 7.62 m long and 32 mm in diameter, was uniformly heated with a wall heat flux density ranging from 20 to 140 W cm^{-2}. The pressure varied from 110 to 205 bar, the flowrate from 400 to 3500 kg m^{-2}s^{-1} and the exit quality from 0 to 0.80. The experimental results were compared with the total pressure drops determined by means of the homogeneous model using a single phase friction factor based on the liquid viscosity and the Moody chart with a relative roughness of 5.10^{-5}. The measured pressure drops ranged from 0.01 to 0.80 bar. The homogeneous model predicted 95% of total pressure drops measured with an error lower than $\pm$ 0.035 bar, i.e. an accuracy of $\pm$ 9% on a typical pressure drop of 0.40 bar.

3.4 Two-Fluid Models with Specified Evolutions

In this approach the two-fluid model equations are written in terms of three simplified balance equations for the mixture and three specified evolutions, two of these expressing the thermal equilibria and the third one being a void correlation.

Equations Describing the Two-Fluid Models with Specified Evolutions

Balance equations. The simplified balance equations for the mixture (Eqs. 3.2, 4 and 6) are written below for a steady state flow.

(i) Mass balance

$$\frac{d}{dz} \left[R_G \rho_G w_G + (1-R_G) \rho_L w_L \right] = 0 \tag{3.34}$$

which leads to,

$$R_G \rho_G w_G + (1-R_G) \rho_L w_L \triangleq G = \text{constant} \tag{3.35}$$

(ii) Momentum balance

$$\frac{d}{dz} \left[R_G \rho_G w_G^2 + (1-R_G) \rho_L w_L^2 \right] - \left[R_G \rho_G + (1-R_G) \rho_L \right] F_z$$

$$+ \frac{dp}{dz} + \frac{4\tau_w}{D} = 0 \tag{3.36}$$

which can be rewritten as,

$$G^2 \frac{d}{dz} \left[\frac{x^2}{R_G \rho_G} + \frac{(1-x)^2}{(1-R_G) \rho_L} \right] - \left[R_G \rho_G + (1-R_G) \rho_L \right] F_z$$

$$+ \frac{dp}{dz} - \left(\frac{dp}{dz}\right)_F = 0 \tag{3.37}$$

(iii) Energy balance

$$\frac{d}{dz} \left[R_G \rho_G i_G w_G + (1-R_G) \rho_L i_L w_L \right] = \frac{4\psi}{D} \tag{3.38}$$

or

$$G \frac{d}{dz} \left[x i_G + (1-x) i_L \right] = \frac{4\psi}{D} \tag{3.39}$$

Specified evolutions. In many practical cases the fluids flow at saturation conditions and the mechanical nonequilibrium is given by an experimental void correlation.

(i)   Mechanical nonequilibrium

$$R_G = f (x, p, \ldots) \tag{3.40}$$

(ii) Thermal equilibria

$$i_G \equiv i_G^{sat} (p) \tag{3.41}$$

$$i_L \equiv i_L^{sat} (p) \tag{3.42}$$

Final set of equations. The *three* constraints on the solution (Eqs. 3.40 to 3.42) added to the *three* mixture balance equations (3.35, 37 and 39) form a set of *six* equations corresponding to the general six equation set of the two fluid model.

In an *adiabatic* flow, the quality is constant and is considered as a parameter whereas in a *diabatic* flow the quality is directly determined by means of the energy equation (3.39).

The set of equations (3.35, 37 and 39 to 42) is entirely closed if the frictional pressure gradient $(dp/dz)_F$ is known. This will be dealt with in the following Sections.

Lockhart-Martinelli Correlations

Lockhart and Martinelli (1949) carried out experiments with air and different liquids flowing in *horizontal* pipes at pressures ranging from 1.1 to 3.6 bar. The diameter of the pipes used varied from 1.5 mm to 25.8 mm. The void fraction was measured by means of quick closing valves.

For two-component, two-phase flow in horizontal pipes the acceleration pressure drop can be neglected whereas the gravity pressure drop is zero. Consequently the measured pressure drop is equal to the frictional pressure drop.

Lockhart-Martinelli parameters. Consider a set of three experiments in the *same* horizontal pipe,

(i)   A *two-phase* mixture flows in the pipe with a total mass flowrate M equal to the sum of the gas mass flowrate M_G and the liquid mass flowrate M_L. The frictional pressure drop is then denoted by $(dp/dz)_F$.

(ii) The *gas* phase flows alone in the same pipe with a mass flowrate equal to M_G. The frictional pressure drop is then denoted by $(dp/dz)_{FG}$.

(iii) The *liquid* phase flows alone in the same pipe with a mass flowrate equal to M_L. The frictional pressure drop is then denoted by $(dp/dz)_{FL}$.

Lockhart and Martinelli (1949) introduced the following parameters,

$$\phi_G^2 \overset{\Delta}{=} \frac{(dp/dz)_F}{(dp/dz)_{FG}} \qquad (3.43)$$

$$\phi_L^2 \overset{\Delta}{=} \frac{(dp/dz)_F}{(dp/dz)_{FL}} \qquad (3.44)$$

$$X^2 \overset{\Delta}{=} \frac{(dp/dz)_{FL}}{(dp/dz)_{FG}} \qquad (3.45)$$

As the pressure drops in single phase flow can be calculated for both turbulent and laminar regimes, there are four possible combinations for calculating X^2: turbulent liquid and turbulent gas (tt), laminar liquid and turbulent gas (vt), turbulent liquid and laminar gas (tv), laminar liquid and laminar gas (vv). Lockhart and Martinelli assumed the following friction law to be used in Eq. (3.45):

$$C_f = A \, Re^{-n} \qquad (3.46)$$

where for *laminar* flow (Re < 1000) A = 16 and n = 1 whereas for *turbulent* flow (Re > 2000) A = 0.046 and n = 0.20.

For turbulent liquid and turbulent gas (tt) we thus obtain,

$$X_{tt} = \left(\frac{\mu_L}{\mu_G}\right)^{0.1} \left(\frac{1-x}{x}\right)^{0.1} \left(\frac{\rho_G}{\rho_L}\right)^{0.5} \qquad (3.47)$$

For viscous liquid and viscous gas (vv) we have,

$$X_{vv} = \left(\frac{\mu_L}{\mu_G}\right)^{0.5} \left(\frac{1-x}{x}\right)^{0.5} \left(\frac{\rho_G}{\rho_L}\right)^{0.5} \qquad (3.48)$$

The Lockhart and Martinelli method consists in plotting R_G and ϕ_G or ϕ_L versus the parameter X.

Void correlation. Figure 3.1 represents the variations of R_G and $(1-R_G)$ as a function of X. These curves represent the mechanical nonequilibrium and are equivalent to Eq. (3.40). They can be approximated by the following equation,

$$1-R_G = \frac{X}{\sqrt{X^2 + 20X + 1}} \qquad (3.49)$$

Frictional pressure drop correlation. Figure 3.2 shows the variations of ϕ_G and ϕ_L versus X for the different combinations of flow regimes. The curves are well represented by the following relationships,

$$\phi_L^2 = 1 + \frac{C}{X} + \frac{1}{X^2} \qquad (3.50)$$

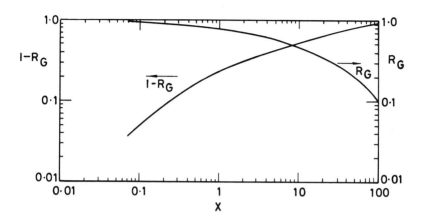

Figure 3.1: Lockhart-Martinelli void correlation.

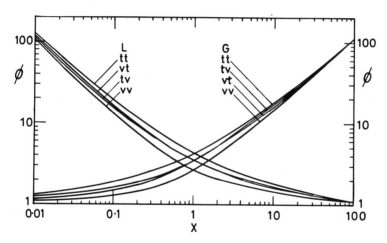

Figure 3.2: Lockhart-Martinelli frictional pressure drop correlation.

$$\emptyset_G^2 = 1 + CX + X^2 \tag{3.51}$$

where C is given by Table 3.1.

Comments on the method of Lockhart and Martinelli. In the discussion of the original paper Gazley and Bergelin brought up the following points:

(i) The choice of a representation involving the square root of the frictional pressure drop leads to an artificial reduction of the dispersion of the experimental results.

(ii) Parameters $\emptyset_G$ and $\emptyset_L$ depends not only on X but also on the liquid mass flowrate. This experimental fact was confirmed

Table 3.1

Values of C in Eqs (3.50) and (3.51)

Liquid	$Re_L \triangleq G(1-x)D/\mu_L$	Gas	$Re_G \triangleq GxD/\mu_G$	Index	C
Turbulent	>2000	Turbulent	>2000	tt	20
Laminar	<1000	Turbulent	>2000	vt	12
Turbulent	>2000	Laminar	<1000	tv	10
Laminar	<1000	Laminar	<1000	vv	5

theoretically for laminar-laminar two-phase flow by Delhaye (1966). However, Taitel and Dukler (1976) did not find any theoretical flowrate dependence for stratified flow but their calculation is not convincing. The effect of mass velocity will be taken into account by the method due to Baroczy (1966) which will be presented in a later section.

(iii) The experimental curves $\emptyset_G$ or $\emptyset_L$ versus X display several changes of slope indicating definite changes in the two-phase flow patterns.

Finally we would like to recall that the Lockhart and Martinelli method is based on low pressure experimental data. The pressure effect will be taken into account by Martinelli and Nelson (1948), whose method is presented in the next section.

Martinelli-Nelson Correlations

Martinelli and Nelson (1948) used experimental results on steam-water flow in boiler coiled tubes at a pressure ranging from 34.5 to 207 bar. They neglected the gravity pressure drop and compared the measured total pressure drop to the sum of acceleration and frictional pressure drops determined by the technique which is presented below.

Martinelli-Nelson parameters. Consider a set of two experiments in the *same adiabatic* pipe

(i)   A steam-water mixture flows in the pipe with a total mass flowrate M. The frictional pressure drop is then denoted by $(dp/dz)_F$.

(ii) Water flows alone in the same pipe with the same mass flow-rate M. The frictional pressure drop is then denoted by $(dp/dz)_{FLO}$.

Martinelli and Nelson (1948) introduced the following parameter,

$$\emptyset_{Lo}^2 \triangleq \frac{(dp/dz)_F}{(dp/dz)_{FLO}} \tag{3.52}$$

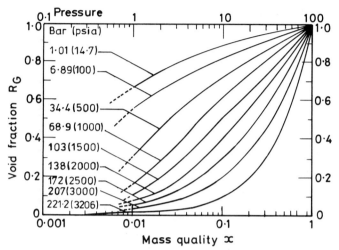

Figure 3.3: Martinelli-Nelson void correlation.

The Martinelli-Nelson method consists in representing R_G and $\emptyset_{LO}^2$ as a function of the quality x for different pressures.

Void correlation. Figure 3.3 represents the variation of R_G as a function of x for different pressures. At the critical pressure (221.2 bar) the densities of the phases are equal and the mixture can be considered as homogeneous. We thus have,

$$R_G = \beta \equiv x \qquad\qquad (3.53)$$

The curve $R_G(x)$ at atmospheric pressure is obtained from the Lockhart-Martinelli correlation (Fig. 3.1). Curves for inter- mediate pressures were obtained by interpolation. They represent the mechanical nonequilibrium and are equivalent to Eq. (3.40).

Frictional pressure drop correlation. Figure 3.4 shows the variations of $\emptyset_{LO}^2$ versus x for different pressures. At the critical pressure (221.2 bar) the phases cannot be distinguished so that

$$\emptyset_{LO}^2 \equiv 1 \qquad\qquad (3.54)$$

Total pressure drop calculation. Consider a steady-state steam- water flow in a tube of constant circular cross section, diameter D, inclined at an angle θ on the vertical. The length of the tube is denoted by L and the wall heat flux density by ψ. The fluid is at saturation conditions at the inlet of the pipe.

(i)   Mass balance

Eq. (3.35) reads,

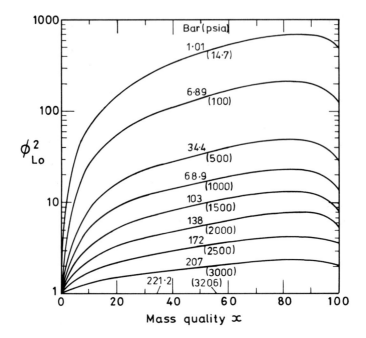

Figure 3.4: Martinelli-Nelson frictional pressure drop correlation.

$$G = \text{constant} \qquad\qquad (3.55)$$

(ii) Energy balance

The thermal equilibrium assumption (Eqs. 3.41 and 42) enables the energy balance (3.39) to be written as,

$$G \frac{d}{dz} \left[i_L^{sat}(p) + xL(\hat{p}) \right] = \frac{4\psi}{D} \qquad\qquad (3.56)$$

where L is the heat of vaporization given by

$$L(p) = i_G^{sat}(p) - i_L^{sat}(p) \qquad\qquad (3.57)$$

Assuming that the liquid saturation enthalpy and the heat of vaporization do not vary with the pressure significantly, we obtain,

$$\frac{dx}{dz} = \frac{4\psi}{DGL} \qquad\qquad (3.58)$$

As the fluid enters the pipe at saturation conditions we have,

$$x = 0 \quad \text{at} \quad z = 0 \qquad (3.59)$$

Consequently the quality is given by,

$$x = \frac{4\psi}{DGL} z \qquad (3.60)$$

At the pipe exit we thus have,

$$x_e = \frac{4\psi}{DGL} L \qquad (3.61)$$

(iii) Momentum balance

Eq. (3.37) reads,

$$G^2 \frac{d}{dz}\left[\frac{x^2}{R_G \rho_G} + \frac{(1-x)^2}{(1-R_G)\rho_L}\right] + \left[R_G \rho_G + (1-R_G)\rho_L\right] g \cos\theta$$

$$+ \frac{dp}{dz} - \left(\frac{dp}{dz}\right)_F = 0 \qquad (3.62)$$

Strictly speaking, the Martinelli-Nelson correlation is valid for horizontal pipes only. Nevertheless we will assume that it is still applicable in vertical or inclined pipes, despite the changes in the flow patterns. In these cases however one must account for the gravity component of the pressure drop. The frictional pressure drop is calculated as follows,

$$\left(\frac{dp}{dz}\right)_F = \emptyset_{Lo}^2 \left(\frac{dp}{dz}\right)_{Flo} \qquad (3.63)$$

$$= \emptyset_{Lo}^2 \frac{4C_{fLo}}{D} \frac{1}{2} \frac{G^2}{\rho_L} \qquad (3.64)$$

where C_{fLo} is the friction coefficient of the liquid flowing alone in the pipe with the total mass flowrate.

Integrating Eq. (3.62) from the inlet to the outlet of the pipe,

$$P_{z=0} - P_{z=L} = G^2 \left[\frac{x^2}{R_G \rho_G} + \frac{(1-x)^2}{(1-R_G)\rho_L}\right]_{z=0}^{z=L}$$

$$+ g\cos\theta \int_0^L \left[R_G \rho_G + (1-R_G)\rho_L\right] dz + \frac{2C_{fLo}}{D\rho_L} \int_0^L \emptyset_{Lo}^2 \, dz \qquad (3.65)$$

R_G and $\varnothing_{LO}^2$ being given in terms of x, it is necessary to change variables from z to x. We thus obtain,

$$P_{z=0} - P_{z=L} = G^2 \left\{ \frac{x_e^2}{R_{Ge}\rho_G} + \frac{1}{\rho_L} \left[\frac{(1-x_e)^2}{1-R_{Ge}} - 1 \right] \right\}$$

$$+ Lg\cos\theta \frac{1}{x_e} \int_0^{x_e} \left[\bar{R}_G\rho_G + (1-R_G)\rho_L \right] dx$$

$$+ \frac{2C_{fLO}G^2L}{D\rho_L} \frac{1}{x_e} \int_0^{x_e} \varnothing_{LO}^2 \, dx \qquad (3.66)$$

In order to simplify the calculations the following two-phase multipliers are introduced,

$$r_2 \triangleq \frac{x_e^2\rho_L}{R_{Ge}\rho_L} + \frac{(1-x_e)^2}{1-R_{Ge}} - 1 = r_2 \ (x_e, \ p) \qquad (3.67)$$

$$r_3 \triangleq \frac{1}{x_e} \int_0^{x_e} \varnothing_{LO}^2 \, dx = r_3 \ (x_e, \ p) \qquad (3.68)$$

$$r_4 \triangleq \frac{1}{x_e} \int_0^{x_e} \left[1-R_G\left(1 - \frac{\rho_G}{\rho_L}\right) \right] dx = r_4 \ (x_e, \ p) \qquad (3.69)$$

As a result, the total pressure drop reads,

$$P_{z=0} - P_{z=L} = \frac{G^2}{\rho_L} r_2 + \frac{2C_{fLO}G^2L}{D\rho_L} r_3 + Lg\rho_L\cos\theta \ r_4 \qquad (3.70)$$

In their paper, Martinelli and Nelson (1948) gave the curves of Figures 3.5 and 3.6 to determine the acceleration and friction multipliers for *horizontal* flows. Thom (1964) gave similar curves for the three multipliers to be used in *vertical* flow of steam-water mixtures. His results are based on experimental data obtained in a tube, 25.4 mm in diameter, at pressures ranging from 1 to 207 bar and for exit qualities from 0.03 to 1.00.

Comments on the method of Martinelli and Nelson. The Martinelli-Nelson correlations give more correct results than the homogeneous model for low mass velocities (G < 1300 kg m^{-2}s^{-1}). On the contrary the homogeneous model gives better results for high mass velocities.

The experimental curves $\varnothing_{LO}(x)$ show slope breaks due to changes in the flow pattern.

Surface tension is not taken into account although at high pressure, near the critical point, interfacial phenomena may have a drastic influence.

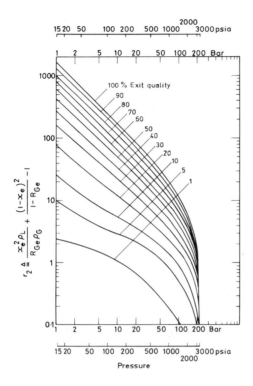

Figure 3.5: Acceleration pressure drop multiplier.

Finally, the Martinelli-Nelson representation does not take into account the mass velocity although its effect has been shown experimentally. This effect will be dealt with by the Baroczy correlation.

ARMAND-TRESHCHEV Correlations

Armand and Treshchev (1959) correlated the void fraction and the frictional pressure drop for steam-water mixtures flowing in rough pipes, 25.5 and 56 mm in diameter, at pressures ranging from 10 to 180 bar.

Void correlation.

$$R_G = (0.833 + 0.05 \ln p)\beta \qquad (3.71)$$

with p in bar.

Note that this correlation has the same form as the Bankoff void equation (Chapter 2, Eq. 2.180).

Frictional pressure drop correlation.

(i) $\beta < 0.9$

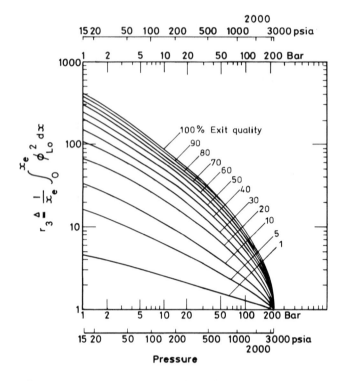

Figure 3.6: Frictional pressure drop multiplier.

$$R_G < 0.5; \quad \phi_{LO}^2 = \frac{(1-x)^{1.75}}{(1-R_G)^{1.2}} \qquad (3.72)$$

$$R_G > 0.5; \quad \phi_{LO}^2 = \frac{0.48(1-x)^{1.75}}{(1-R_G)^n} \qquad (3.73)$$

$$\text{with } n = 1.9 + 1.48 \ 10^{-3}p \ \text{(p in bar)} \qquad (3.74)$$

(ii) $\beta > 0.9$

$$\phi_{LO}^2 = \frac{0.0025 \ p + 0.055}{(1-\beta)^{1.75}} \ (1-x)^{1.75} \qquad (3.75$$

BAROCZY Correlation for Frictional Pressure Drop

The Lockhart and Martinelli method cannot represent the effects of pressure and mass velocity. Moreover it is based on a limited number of fluids. On the other hand the Martinelli-Nelson method takes account of the pressure but strictly applies to steam-water flow only. Again no mass velocity influence can be accounted for.

Baroczy (1966) succeeded in correlating the frictional pressure drop without ignoring the effects of pressure, mass velocity and nature of fluids. He introduced a *property index* involving the ratios of viscosities and densities of each phase,

$$\text{Property index} \triangleq \left(\frac{\mu_L}{\mu_G}\right)^{0.2} \left(\frac{\rho_G}{\rho_L}\right) \tag{3.76}$$

If a friction law in $Re^{-0.20}$ is adopted for each phase flowing alone in the pipe, we have,

$$\left(\frac{\mu_L}{\mu_G}\right)^{0.2} \left(\frac{\rho_G}{\rho_L}\right) = \frac{(dp/dz)_{FLO}}{(dp/dz)_{FGO}} \tag{3.77}$$

Baroczy (1966) used experimental data in adiabatic conditions where the acceleration pressure drop is zero and where the gravity pressure drop is either zero (horizontal flow) or could be neglected. He ended up with curves of $\varnothing_{LO}^2$ versus the property index for different qualities but for a given mass velocity (1360 kg m^{-2}s^{-1}). Figure 3.7 shows the Baroczy curves and the values of the property index for different fluids at saturation conditions.

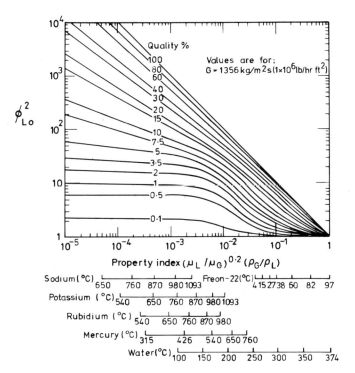

Figure 3.7: Baroczy frictional pressure drop correlation.

When the tube is full of liquid we have,

$$x = 0 \qquad \emptyset_{LO}^2 = 1 \tag{3.78}$$

whereas when the tube is full of gas,

$$x = 1 \qquad \emptyset_{LO}^2 = \frac{(dp/dz)_{FGO}}{(dp/dz)_{FLO}} = \frac{1}{\left(\dfrac{\mu_L}{\mu_G}\right)^{0.2} \left(\dfrac{\rho_G}{\rho_L}\right)} \tag{3.79}$$

At the critical point the property index and $\emptyset_{LO}^2$ are equal to one. The exponent 0.2 in the property index has no physical significance but allows the boundaries $x = 0$ and $x = 1$ to be represented in a simple way on Figure 3.7.

For mass velocities different from 1360 kg m^{-2}s^{-1} the value of $\emptyset_{LO}^2$ must be multiplied by a correcting factor Ω given in Figure

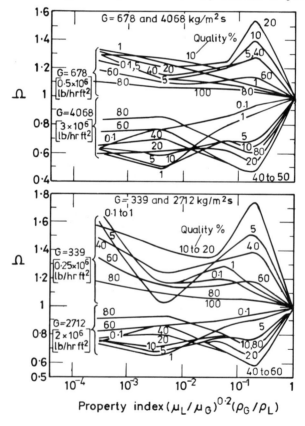

Figure 3.8: Baroczy correction factor for mass velocity effects.

3.8. These diagrams must be used by linearly interpolating between the qualities and mass velocities which surround the actual values.

3.5 Comparative Studies and Recommendations

Several extensive comparative studies have appeared in the literature for a number of years. However several of them are of lesser interest due generally to a lack of information on the way the data were processed and the correlations used. Among all the comparative studies available we have selected two of them. They cover two different industrial domains: petroleum engineering and nuclear engineering.

Gas-Liquid Flow in Horizontal Pipes

Mandhane et al. (1977) gathered 10,000 data points and compared sixteen frictional pressure drop correlations. When needed, the gravity pressure drop was determined by means of various void models whereas the acceleration pressure drop was neglected. They concluded that the best prediction technique required a determination of the flow pattern and the use of specific void and frictional pressure drop correlations for each flow pattern. The full set of recommended correlations is given in Table 3 of the original paper.

For a quick evaluation of the pressure drop, Mandhane et al. (1977) recommend the Dukler et al. (Case I - No slip) correlation.

Steam-Water Flow in Boiling Water Reactors

Idsinga et al. (1977) tested 3460 experimental steam-water pressure drop measurements against 18 frictional pressure drop correlations. The operating parameters had the following ranges: pressure, 17 to 103 bar; mass velocity, 270 to 4340 $kg\ m^{-2}s^{-1}$; steam quality: subcooled to 1.00; geometry: tube, annulus, rectangular channel and rod array; hydraulic diameter, 2.3 to 33.0 mm; flow directions: vertical upward or downward, horizontal.

The authors found that the best performing correlation was the homogeneous model using the liquid viscosity and the smooth tube friction law. However, for a restricted set of data corresponding to the conditions of BWR operation (pressure: 62 to 103 bar; mass velocity: 0 to 2700 $kg\ m^{-2}s^{-1}$; quality: 0 to 0.6), they recommend the Armand-Treshchev correlation for qualities lower than 0.3 and the Baroczy correlation for higher qualities.

Tentative Selection Table

The variety of geometries where two-phase pressure drop calculation is of interest is extremely large. As a result it is impossible to give recommendations for all the industrial applications involving two-phase flows. Table 3.2 is the embryo of a selection chart which will have to be improved and complemented continuously.

J M Delhaye

Table 3.2

Tentative Selection Table for Two-Phase Flow Pressure Drop Calculation

Domain	Use		See
Petroleum engineering: . Horizontal or inclined pipes . Pressure lower than 15 bar . Two-component flow	Detailed calculation: Table 3 of Mandhane et al. (1977)		Reference
	Quick calculation: Case I, Dukler et al. (1964)		Section 3.3
Nuclear engineering: . Vertical pipes . Pressure: 17 to 103 bar . Steam-water flow	Overall recommendation: Homogeneous model with $\mu = \mu_L$		Section 3.3
	BWR 8x8 rod bundle	x<0.3: Armand-Treshchev (1959)	Section 3.4
		x>0.3: Baroczy (1966)	Sect. 3.4
Steam generators: . Vertical pipes . Pressure: 110 to 205 bar . Steam-water flow	G<1000 kg m^{-2}s^{-1}: Thom (1964)		Reference and Sect. 3.4
	G>1000 kg m^{-2}s^{-1}: Homogeneous model with $\mu = \mu_L$		Section 3.3
Rod bundles	Grand (1981)		Reference
Tube banks	Grant et al. (1974)		Reference

Nomenclature

A	cross section area
A	space correlation coefficient, constant
B	space correlation coefficient
C	intersection between the interface and the cross section plane
C_f	friction factor (Eq. 3.26)
C	constant (Table 3.1)
D	pipe diameter
F	external force per unit of mass
F_z	z-component of the external force per unit of mass
f	arbitrary function

G	mass velocity
g	arbitrary function
i	specific enthalpy
L	pipe length
L	heat of vaporization
$\dot{m}$	mass transfer per unit area and per unit of time
n	normal unit vector
n	exponent (Eqs. 3.73, 74)
P	wetted perimeter
p	pressure
q	heat flux density vector
R	area fraction (Eq. 2.153)
Re	Reynolds number (Eq. 3.29)
r	two-phase multiplier (Eqs. 3.67, 68, 69)
T	temperature
t	time
u	specific internal energy
v	velocity vector
w	z-component of the velocity vector
X	Lockhart-Martinelli parameter (Eq. 3.45)
x	mass quality
z	coordinate along the axis of the pipe
α	volumetric fraction
β	volumetric quality (Eq. 2.177)
θ	angle of inclination on the vertical
μ	viscosity
ρ	density
τ	viscous stress tensor
$\emptyset$	Martinelli parameters (Eqs. 3.43, 44, 52)
ψ	heat flux density

Subscripts

A	acceleration
C	intersection of the interface with the cross section plane
c	continuous
F	friction
G	gas; gravity
k	phase index
L	liquid
Lo	liquid flowing alone with the total mass flowrate

m   maximum

sat   saturation

t   turbulent

v   viscous

W   wall

Operators

$\{\ \}$ area averaging operator over the total cross section area
          (Eq. 2.149)

$\overline{\quad}^{x}$ X-weighted time-averaging operator (Eq. 2.135)

$\overline{\quad}$ time averaging operator (Eq. 2.139)

References

Armand, A A, and Treshchev, G G, (1959) "Investigation of the resistance during the movement of steam-water mixtures in a heated boiler pipe at high pressures." AERE Lib./Trans. 816.

Baroczy, C J, (1966) "A systematic correlation for two-phase pressure drop." Heat Transfer - Los Angeles, *Chem. Engng. Progress, Symposium Series 64*, Knudsen, J G, Ed. **62**, 232-249.

Delhaye, J M, (1966) "Etude theorique des ecoulements diphasiques de type annulaire en regime laminaire-laminaire." CRAS, t. 263, Serie A, 324-327.

Dukler, A E, Wicks, M, and Cleveland, R G, (1964) "Frictional pressure drop in two-phase flow: B. An approach through similarity analysis." AIChEJ, 44-51.

Grand, D, (1981) "Pressure drops in rod bundles." *Thermohydraulics of Two-Phase Systems Applied in Industrial Design and Nuclear Engineering*, Delhaye, J M, Giot, M, and Riethmuller, M L, Eds, Hemisphere/McGraw-Hill (in the press).

Grant, I D R, Finlay, I C, and Harris, D, (1974) "Flow and pressure drop during vertically upward two-phase flow past a tube bundle with and without bypass leakage." *Multiphase Flow Systems*, **II**, I. Chem. E. Symposium Series No. 38, Paper I 1.

Idsinga, W, Todreas, N, and Bowring, R, (1977) "An assessment of two-phase pressure drop correlations for steam-water systems." *Int. J. Multiphase Flow*, **3**, 401-413.

Ishii, M and Zuber, N, (1979) "Drag coefficient and relative velocity in bubbly, droplet or particulate flows." *AIChEJ*, **25**, No. 5, 843-855.

Lockhart, R W, and Martinelli, R C, (1949) "Proposed correlation of data for isothermal two-phase, two-component flow in pipes." *Chem. Engng. Progress*, **45**, No. 1, 39-48.

Mandhane, J M, Gregory, G A, and Aziz, K, (1977) "Critical evaluation of friction pressure-drop prediction methods for gas-liquid flow in horizontal pipes." *J. Petroleum Technology*, 1348-1358.

Martinelli, R C, and Nelson, D B, (1948) "Prediction of pressure drop during forced-circulation boiling of water." *Transactions of the ASME*, 695-702.

Owens, W L, (1963) "Two-phase pressure gradient." *International Development in Heat Transfer*, ASME, 363-368.

Taitel, Y, and Dukler, A E, (1976) "A theoretical approach to the Lockhart-Martinelli correlation for stratified flow." *Int. J. Multiphase Flow*, **2**, 591-595.

Thom, J R S, (1964) "Prediction of pressure drop during forced circulation boiling of water." *Int. J. Heat Mass Transfer*, **7**, 709-724.

Whitcutt, R D B, and Chojnowski, B, (1973) "Two-phase pressure drop in high pressure steam generating tubes." *European Two-Phase Flow Group Meeting*, Brussels, paper A3.

Chapter 4

Singular Pressure Drops

J. M. DELHAYE

4.1 Introduction

It is worthwhile to notice that in many industrial piping
systems straight pipes are less encountered than singularities such
as bends, junctions, sudden area changes, manifolds, etc...
Actually, in the primary circuit of a nuclear reactor straight
pipes could be named singularities. However very few systematic
studies have been devoted so far to the behavior of two-phase flows
in singularities, even in steady-state conditions. Considering
the importance of the pressure drop evaluation in natural circulation loops it was judged necessary to complement Chapter 3 on
frictional pressure drops in straight channels by a tutorial
summary of the main singular pressure drop models which can be
recommended.

We will systematically examine the behavior of a single-phase
fluid flowing through a singularity before analyzing the two-phase
flow case. Due to the paucity of studies on nonstationary flow we
will restrict ourselves to *steady-state flows* through the following
elements: sudden expansion and contraction, inserts and grids, tees
and wyes, and finally manifolds. The sudden expansion will be
treated in detail to show how the other components could be
regarded. To conclude the Chapter we will give recommendations and
a tentative selection table.

4.2 Sudden Expansion

Single Phase Flow

In our opinion the best elementary discussion on sudden
expansion and contraction is given by Whitaker (1968). Consider
the steady-state, turbulent flow of an incompressible fluid across
the horizontal sudden expansion represented in Figure 4.1. The
areas of the upstream and downstream pipe cross sections are
respectively A_1 and A_2 and $\mathbf{k}$ is the unit vector of the pipe axis.
The mass and momentum balances and the mechanical energy equation
will be written for the fixed control volume V limited by cross
sections A_1 and A_2 where the flow can be considered as fully
developed in each pipe and by the pipe wall. We have used the
same notations A_1 and A_2 for the surface limiting the control
volume and the areas of cross sections of pipes 1 and 2. The wall

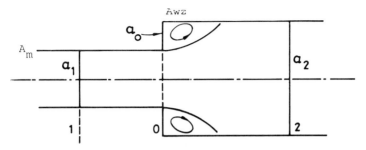

Figure 4.1: Control volume for the sudden expansion.

surface A_w can be divided into three components: A_{w1} and A_{w2} pertaining to the upstream and downstream pipes and A_o the area of the flange perpendicular to the pipe axis. We will note the total boundary of the control volume V,

$$A = A_1 + A_2 + A_w \qquad (4.1)$$

$$= A_1 + A_2 + A_{w1} + A_{w2} + A_o \qquad (4.2)$$

Mass balance: The *instantaneous* mass balance for volume V reads,

$$\frac{d}{dt} \int_V \rho \, dV = - \oint_A \rho \mathbf{v} . \mathbf{n} dA \qquad (4.3)$$

where $\mathbf{n}$ is the unit vector normal to A and outwardly directed. The *time-averaged* mass balance is then written,

$$\overline{\frac{d}{dt} \int_V \rho \, dV} = - \overline{\oint_A \rho \mathbf{v} . \mathbf{n} dA} \qquad (4.4)$$

As the control volume is *fixed* we have,

$$\frac{d}{dt} \int_V \overline{\rho} \, dV = - \oint_A \overline{\rho \mathbf{v} . \mathbf{n}} dA \qquad (4.5)$$

In case of a *steady-state* flow we find,

$$\frac{d}{dt} \int_V \overline{\rho} \, dV \equiv 0 \qquad (4.6)$$

The pipe wall being *impermeable* and the fluid *incompressible*, Eq. (4.5) becomes,

$$\langle \overline{w} \rangle_1 A_1 = \langle \overline{w} \rangle_2 A_2 \qquad (4.7)$$

where $\langle \overline{w} \rangle_i$ is the area-averaged of the time-averaged velocity along the pipe axis $(i = 1,2)$,

$$\langle \overline{w} \rangle_i \triangleq \frac{1}{A_i} \int_{A_i} \overline{w} \, dA \qquad (4.8)$$

Introducing the area ratio

$$\sigma \triangleq \frac{A_1}{A_2} \qquad (4.9)$$

we obtain for the mass balance

$$\sigma \langle \overline{w} \rangle_1 = \langle \overline{w} \rangle_2 \qquad (4.10)$$

Momentum balance. The *instantaneous* momentum balance for volume V reads,

$$\frac{d}{dt} \int_V \rho \mathbf{v} dV = - \oint_A \rho \mathbf{v} (\mathbf{v.n}) dA + \int_V \rho \mathbf{F} dV + \oint_A \mathbf{n.T} dA \qquad (4.11)$$

Introducing the viscous stress tensor by,

$$\mathbf{T} = - p \, \mathbf{U} + \tau \qquad (4.12)$$

the *time-averaged* momentum balance is written,

$$\overline{\frac{d}{dt} \int_V \rho \mathbf{v} dV} = - \overline{\oint_A \rho \mathbf{v} (\mathbf{v.n}) dA} + \overline{\int_V \rho \mathbf{F} dV}$$

$$- \overline{\oint_A p \mathbf{n} dA} + \overline{\oint_A \mathbf{n.\tau} dA} \qquad (4.13)$$

As the control volume is *fixed* we have,

$$\frac{d}{dt} \overline{\int_V \rho \mathbf{v} dV} = - \oint_A \overline{\rho \mathbf{v} (\mathbf{v.n})} dA + \int_V \overline{\rho \mathbf{F}} dV$$

$$- \oint_A \overline{p} \, \mathbf{n} dA + \oint_A \mathbf{n.\overline{\tau}} dA \qquad (4.14)$$

In case of a *steady-state* flow we obtain,

$$\frac{d}{dt}\int_V \overline{\rho \mathbf{v} dV} \equiv 0 \qquad (4.15)$$

For a *horizontal* flow and an *incompressible* fluid, the projection of Eq. (4.14) along the pipe axis thus reads,

$$\rho<\overline{w^2}>_2 A_2 + \rho<\overline{w'^2}>_2 A_2 - \rho<\overline{w^2}>_1 A_1 - \rho<\overline{w'^2}>_1 A_1$$

$$= <\overline{p}>_1 A_1 - <\overline{p}>_2 A_2 + <\overline{p}>_0 (A_2 - A_1)$$

$$-<(\mathbf{k}.\overline{\tau}).\mathbf{k}>_1 A_1 + <(\mathbf{k}.\overline{\tau}).\mathbf{k}>_2 A_2 - <(\mathbf{k}.\overline{\tau}).\mathbf{k}>_0 (A_2 - A_1)$$

$$+ \int_{A_{w1}+A_{w2}} (\mathbf{n}.\overline{\tau}).\mathbf{k} dA \qquad (4.16)$$

(i) Turbulent terms:

According to Hinze (1975) we have in fully developed flows,

$$\text{Max} \frac{\sqrt{\overline{w'^2}}}{\overline{w}} \simeq 0.08 \text{ (in a region close to the wall)}$$

As a result,

$$\text{Max} \frac{\overline{w'^2}}{\overline{w}} \simeq 0.006$$

and the terms in $\rho<\overline{w'^2}>$ are thus negligible with respect to the terms in $\rho<\overline{w^2}>$.

(ii) Viscous terms on A_1, A_2 and A_0, surfaces perpendicular to the pipe axis:

If τ_{ij} is the element of the viscous stress tensor τ, we have,

$$(\mathbf{k}.\overline{\tau}).\mathbf{k} = \overline{\tau}_{33} \qquad (4.17)$$

For an incompressible fluid we find in cylindrical coordinates,

$$\overline{\tau}_{33} = 2\mu \frac{\partial \overline{w}}{\partial z} \qquad (4.18)$$

Recalling that in sections 1 and 2 the flow is fully developed, we obtain on A_1 and A_2

$$\overline{\tau}_{33} \equiv 0 \tag{4.19}$$

To evaluate $\overline{\tau}_{33}$ on the flange A_0 we use the continuity equation in cylindrical coordinates,

$$\frac{\partial \overline{w}}{\partial z} = - \frac{1}{r} \frac{\partial}{\partial r} (r \, \overline{v}_r) \tag{4.20}$$

where we assume that the flow is *axisymmetric*.

But on the flange we have

$$\overline{v}_r \equiv 0 \tag{4.21}$$

so that,

$$\frac{\partial \overline{w}}{\partial z} \equiv 0 \tag{4.22}$$

Eq. (4.18) shows that on the flange we also have

$$\overline{\tau}_3 \equiv 0 \tag{4.23}$$

To conclude, all the viscous terms on surfaces perpendicular to the pipe axis (i.e. on A_1, A_2 and A_0) are strictly zero.

(iii) Viscous terms on A_{w1} and A_{w2}, surfaces parallel to the pipe axis:

We have,

$$(\boldsymbol{n}.\overline{\tau}).\boldsymbol{k} = \tau_{13} \tag{4.24}$$

and for an incompressible fluid in cylindrical coordinates,

$$\overline{\tau}_{13} = \mu (\frac{\partial w}{\partial r} + \frac{\partial v_r}{\partial z}) \tag{4.25}$$

where v_r is the radial component of the velocity vector.

In the momentum balance equation (4.16) the term,

$$\int_{A_{w1}+A_{w2}} (\boldsymbol{n}.\overline{\tau}).\boldsymbol{k} \, dA = \int_{A_{w1}+A_{w2}} \mu (\frac{\partial w}{\partial r} + \frac{\partial v_r}{\partial z}) \tag{4.26}$$

represents the friction on the pipe wall. *We will assume that this term is negligible.*

(iv) Final form of the momentum balance equation:

The momentum balance equation (4.16) finally reads,

$$\rho <\overline{w}^2>_2 A_2 - \rho <\overline{w}^2>_1 A_1$$

$$= <\overline{p}>_1 A_1 - <\overline{p}>_2 A_2 + <\overline{p}>_0 (A_2 - A_1) \tag{4.27}$$

Being interested in the pressure difference between sections A_1 and A_2 we have to express $<\overline{p}>_0$ in terms of the upstream or downstream pressures. *We will assume,*

$$<\overline{p}>_0 = <\overline{p}>_1 \tag{4.28}$$

Moreover we know that for fully-developed *turbulent* flow we have (Whitaker, 1968),

$$\text{Max } <\overline{w}^2>/<\overline{w}>^2 \simeq 1.03 \tag{4.29}$$

whereas for *laminar* flow,

$$<\overline{w}^2>/<\overline{w}>^2 = 3.20 \tag{4.30}$$

Consequently we will write for *turbulent* flows,

$$<\overline{w}^2> \equiv <\overline{w}>^2 \tag{4.31}$$

Finally the momentum balance equation (4.27) becomes,

$$\rho <\overline{w}>_2^2 A_2 - \rho <\overline{w}>_1^2 A_1 = (<\overline{p}>_1 - <\overline{p}>_2) A_2 \tag{4.32}$$

Putting for the sake of simplicity,

$$\left\{ \begin{array}{l} p_1 \triangleq <\overline{p}>_1 \\ p_2 \triangleq <p>_2 \end{array} \right. \tag{4.33}$$
$$\tag{4.34}$$

and recalling the definition (4.9) of the area ratio we obtain,

$$p_2 - p_1 = \rho <\overline{w}>_1^2 \sigma - \rho <\overline{w}>_2^2 \tag{4.35}$$

(v) Expression for the pressure difference

Combining the mass and momentum balances (Eqs 4.10 and 35) yields,

$$\frac{P_2 - P_1}{\frac{1}{2}\rho <\overline{w}>_1^2} = 2\sigma \ (1-\sigma) \tag{4.36}$$

Note that this equation has been obtained with the following major hypotheses,

- negligible frictional effects on the *pipe wall*

- $<\overline{p}>_0 = <\overline{p}>_1$

Mechanical energy equation. The integral mechanical energy equation is *not* a balance equation in the sense that it cannot be deduced directly from a combination of the instantaneous mass and momentum balances written for a control volume. Instead we must start with the *local* instantaneous mechanical energy equation and integrate it over the control volume.

The *local instantaneous* mechanical energy equation reads,

$$\frac{\partial}{\partial t}(\tfrac{1}{2}\rho v^2) + \nabla.(\tfrac{1}{2}\rho v^2 \boldsymbol{v}) - \rho \boldsymbol{F}.\boldsymbol{v} - \nabla.(\boldsymbol{T}.\boldsymbol{v}) + \boldsymbol{T}:\nabla \boldsymbol{v} = 0 \tag{4.37}$$

If the fluid is *incompressible* and if the body force $\boldsymbol{F}$ is such that

$$\boldsymbol{F} = -\nabla \Omega \tag{4.38}$$

with

$$\frac{\partial \Omega}{\partial t} = 0 \tag{4.39}$$

then Eq. (4.37) becomes,

$$\frac{\partial}{\partial t}(\tfrac{1}{2}\rho v^2) + \nabla.(\tfrac{1}{2}\rho v^2 \boldsymbol{v}) + \nabla.\rho \Omega \boldsymbol{v} - \nabla.(\boldsymbol{T}.\upsilon) + \tau:\nabla \boldsymbol{v} = 0 \tag{4.40}$$

Integrating over control volume V,

$$\int_V \frac{\partial}{\partial t}(\tfrac{1}{2}\rho v^2)dV + \int_V \nabla.(\tfrac{1}{2}\rho v^2 \boldsymbol{v})dV + \int_V \nabla.\rho \Omega \boldsymbol{v}dV$$

$$- \int_V \nabla.(\boldsymbol{T}.\boldsymbol{v})dV + \int_V \tau:\nabla \boldsymbol{v} \ dV = 0 \tag{4.41}$$

Transforming this equation by means of the Leibniz rule and the Gauss theorem we obtain the *integral instantaneous* mechanical energy equation,

$$\frac{d}{dt} \int_V \tfrac{1}{2}\rho v^2 dV + \oint_A \tfrac{1}{2}\rho v^2 \mathbf{v} \cdot \mathbf{n} dA + \oint_A \rho \Omega \mathbf{v} \cdot \mathbf{n} dA$$

$$- \oint_A (\mathbf{n} \cdot \mathbf{T}) \cdot \mathbf{v} dA + \int_V \tau : \nabla \mathbf{v} dV = 0 \qquad (4.42)$$

The *time-averaged* integral mechanical energy equation then reads,

$$\overline{\frac{d}{dt} \int_V \tfrac{1}{2}\rho v^2 dV} + \overline{\oint_A \tfrac{1}{2}\rho v^2 \mathbf{v} \cdot \mathbf{n} dA} + \overline{\oint_A \rho \Omega \mathbf{v} \cdot \mathbf{n} dA}$$

$$- \overline{\oint_A (\mathbf{n} \cdot \mathbf{T}) \cdot \mathbf{v} dA} + \overline{\int_V \tau : \nabla \mathbf{v} dV} = 0 \qquad (4.43)$$

As the control volume is *fixed* we have,

$$\frac{d}{dt} \int_V \overline{\tfrac{1}{2}\rho v^2} dV + \oint_A \overline{\tfrac{1}{2}\rho v^2 \mathbf{v} \cdot \mathbf{n}} dA + \oint_A \overline{\rho \Omega \mathbf{v} \cdot \mathbf{n}} dA$$

$$- \int_A \overline{(\mathbf{n} \cdot \mathbf{T}) \cdot \mathbf{v}} dA + \int_V \overline{\tau : \nabla \mathbf{v}} dV = 0 \qquad (4.44)$$

In case of a *steady-state* flow we get,

$$\frac{d}{dt} \int_V \overline{\tfrac{1}{2}\rho v^2} dV \equiv 0 \qquad (4.45)$$

For an incompressible fluid the mechanical energy equation (4.44) thus reads,

$$\tfrac{1}{2}\rho <\overline{w}^3>_2 A_2 + \tfrac{1}{2}\rho <\overline{w'^3}>_2 A_2 + \tfrac{1}{2}\rho <3\overline{ww'^2}>_2 A_2$$

$$- \tfrac{1}{2}\rho <\overline{w}^3>_1 A_1 - \tfrac{1}{2}\rho <\overline{w'^3}>_1 A_1 - \tfrac{1}{2}\rho <3\overline{ww'^2}>_1 A_1$$

$$+ \rho \Omega (<\overline{w}>_2 A_2 - <\overline{w}>_1 A_1) + \int_V \overline{\tau : \nabla \mathbf{v}} dV$$

$$+ <\overline{(\mathbf{k} \cdot \mathbf{T}) \cdot \mathbf{v}}>_1 A_1 - <\overline{(\mathbf{k} \cdot \mathbf{T}) \cdot \mathbf{v}}>_2 A_2 = 0 \qquad (4.46)$$

(i) Body force term

From the mass balance (4.7) we obtain,

$$\rho\Omega(<\overline{w}>_2 A_2 - <\overline{w}>_1 A_1) \equiv 0 \tag{4.47}$$

(ii) Stress terms on A_1 and A_2, surfaces perpendicular to the pipe axis:

If T_{ij} is the element of the total stress tensor T, we have,

$$\overline{(\boldsymbol{k}.T).\boldsymbol{v}} = \overline{w\, T_{33}} = -\,\overline{pw} + \overline{\tau_{33}w} \tag{4.48}$$

For fully developed flow we obtain (see Eqs 4.18, 19),

$$\overline{(\boldsymbol{k}.T).\boldsymbol{v}} = -\,\overline{pw} + \overline{\tau'_{33}w'} = -\,\overline{pw} + \overline{\tau'_{33}w'} \tag{4.49}$$

(iii) Turbulent terms:

We will assume that all the terms involving velocity and pressure turbulent fluctuations can be neglected.

(iv) Dissipation term

We will assume that the viscous dissipation term is negligible

(v) Final form of the mechanical energy equation:

The mechanical energy equation (4.46) finally reads,

$$\tfrac{1}{2}\rho<\overline{w^3}>_2 A_2 - \tfrac{1}{2}\rho<\overline{w^3}>_1 A_1$$

$$+ <\overline{p}\;\overline{w}>_2 A_2 - <\overline{p}\;\overline{w}>_1 A_1 = 0 \tag{4.50}$$

We know that for fully-developed *turbulent* flow we have (Whitaker, 1968)

$$\text{Max } <\overline{w^3}>/<\overline{w}>^3 \simeq 1.08 \tag{4.51}$$

whereas for *laminar* flow,

$$<\overline{w^3}>/<\overline{w}>^3 = 7.32 \tag{4.52}$$

Consequently we will write for *turbulent* flows

$$<\overline{w^3}> \equiv <\overline{w}>^3 \tag{4.53}$$

Moreover we will assume that

$$\langle \overline{pw} \rangle \equiv \langle \overline{p} \rangle \langle \overline{w} \rangle \tag{4.54}$$

Finally the mechanical energy equation (4.5) becomes,

$$\tfrac{1}{2}\rho \langle \overline{w} \rangle_2^3 A_2 - \tfrac{1}{2}\rho \langle \overline{w} \rangle_1^3 A_1$$

$$+ \langle \overline{p} \rangle_2 \langle \overline{w} \rangle_2 A_2 - \langle \overline{p} \rangle_1 \langle \overline{w} \rangle_1 A_1 = 0 \tag{4.55}$$

or recalling Eqs (4.9, 33 and 34),

$$\tfrac{1}{2}\rho \langle \overline{w} \rangle_2^3 - \tfrac{1}{2}\rho \langle \overline{w} \rangle_1^3$$

$$+ p_2 \langle \overline{w} \rangle_2 - p_1 \langle \overline{w} \rangle_1 \sigma = 0 \tag{4.56}$$

(vi) Expression for the pressure difference

Combining the mechanical energy equation (4.56) and the mass balance (4.1) yields,

$$\frac{p_2 - p_1}{\tfrac{1}{2}\rho \langle \overline{w} \rangle_1^2} = 1 - \sigma^2 \tag{4.57}$$

Note that this equation has been obtained with the following major hypothesis,

- negligible viscous dissipation within the control *volume*

Comparisons with experimental results. Figure 4.2 shows the experimental results obtained by Archer (1913) and the nondimensional pressure increase determined from the momentum equation (4.36) and the mechanical energy equation (4.57) as a function of the diameter ratio $\sqrt{\sigma}$. We recall that,

(i) from the momentum equation

$$\frac{p_2 - p_1}{\tfrac{1}{2}\rho \langle \overline{w} \rangle_1^2} = 2\sigma(1 - \sigma) \tag{4.58}$$

(ii) from the mechanical energy equation without dissipation

$$\frac{p_2 - p_1}{\tfrac{1}{2}\rho \langle \overline{w} \rangle_1^2} = 1 - \sigma^2 \tag{4.59}$$

According to Figure 4.2 the pressure *increase* given by the momentum equation (4.58) corresponds fairly well to the experimental data. As a result it can be considered as the sum of a *reversible*

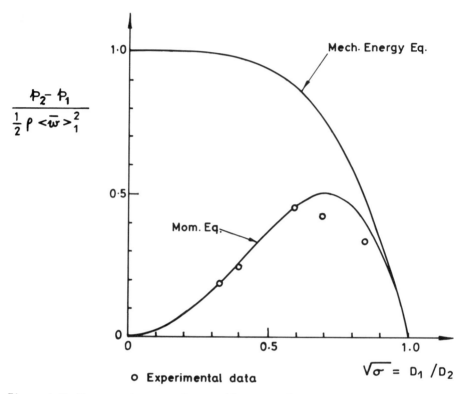

o **Experimental data**

Figure 4.2: Pressure increase in a sudden expansion.

pressure increase $(p_2-p_1)_R$ due to the area change only, and of an *irreversible pressure drop* $(p_2-p_1)_I$ due to the viscous dissipation within the control volume. The pressure increase given by the mechanical energy equation (4.59) without viscous dissipation (Eq. 4.59) corresponds to the reversible pressure increase,

$$\frac{(p_2-p_1)_R}{\frac{1}{2}\rho<\bar{w}>_1^2} = 1 - \sigma^2 \tag{4.60}$$

The irreversible pressure drop is,

$$(p_2-p_1)_I = p_2-p_1 - (p_2-p_1)_R \tag{4.61}$$

Consequently, we obtain,

$$\frac{(p_2-p_1)_I}{\frac{1}{2}\rho<\bar{w}>_1^2} = 2\sigma(1 - \sigma)-(1 - \sigma^2) \tag{4.62}$$

or,

$$\frac{(p_2 - p_1)_I}{\frac{1}{2}\rho \langle w \rangle_1^2} = -(1 - \sigma)^2 \tag{4.63}$$

Two-Phase Flow

Consider the steady-state, turbulent flow of a mixture of two incompressible fluids across the horizontal sudden expansion represented in Figure 4.1.

We will not present the full derivation of the mass, momentum and mechanical energy equations as was done for a single-phase flow. However, the same method must be used so that all the hypotheses can appear clearly. For the sake of simplicity all the averaging operator symbols have been removed from the equations.

Mass balance for the mixture. For a steady-state flow the mixture mass flowrate is constant,

$$M_1 \equiv M_2 \tag{4.64}$$

The mass velocity is defined by

$$G \triangleq \frac{M}{A} \tag{4.65}$$

As a result, the mass balance (4.64) becomes

$$G_1 A_1 = G_2 A_2 \tag{4.66}$$

or in terms of the area ratio (Eq. 4.9)

$$G_2 = \sigma G_1 \tag{4.67}$$

Momentum balance for the mixture. The momentum balance for the mixture within control volume V reads,

$$p_2 A_2 - p_1 A_1 = \rho_L w_{L1}^2 A_{L1} + \rho_G w_{G1}^2 A_{G1}$$

$$- \rho_L w_{L2}^2 A_{L2} - \rho_G w_{G2}^2 A_{G2} \tag{4.68}$$

The following quantities are now introduced,

- the area ratio σ (Eq. 4.9)
- the void fraction,

$$R_G \triangleq \frac{A_G}{A} \tag{4.69}$$

● the mass quality,

$$x \triangleq \frac{M_G}{M} \tag{4.70}$$

● the mass velocity G (Eq. 4.65)

Eq. (4.68) becomes,

$$P_2 - P_1 = \sigma \left[\frac{(1-x_1)^2 G_1^2}{(1-R_{G1}) \rho_L} + \frac{x_1^2 G_1^2}{R_{G1} \rho_G} \right]$$

$$- \left[\frac{(1-x_2)^2 G_2^2}{(1-R_{G2}) \rho_L} + \frac{x_2^2 G_2^2}{R_{G2} \rho_G} \right] \tag{4.71}$$

Combining the mass and momentum balances (Eqs. 4.67 and 71) yields,

$$P_2 - P_1 = \sigma G_1^2 \left\{ \left[\frac{(1-x_1)^2}{(1-R_{G1}) \rho_L} + \frac{x_1^2}{R_{G1} \rho_G} \right] - \sigma \left[\frac{(1-x_2)^2}{(1-R_{G2}) \rho_L} + \frac{x_2^2}{R_{G2} \rho_G} \right] \right\} \tag{4.72}$$

According to Lottes (1961) this equation was first derived by Romie in 1958.

If there is no phase change between sections 1 and 2, then we have,

$$x_1 \equiv x_2 \triangleq x \tag{4.73}$$

For vertical air-water flow at atmospheric pressure, Petrick and Swanson (1959) proposed an empirical relationship between R_{G1}, R_{G2}, P_1, P_2 and the area ratio σ. However, Richardson (1958) for horizontal air-water flows at atmospheric pressure and Velasco (1975) for vertical flows in the same conditions showed that, to a good approximation,

$$R_{G1} \simeq R_{G2} \triangleq R_G \tag{4.74}$$

If equations (4.73) and (4.74) are adopted, then the momentum balance (4.72) becomes,

$$P_2 - P_1 = \sigma (1-\sigma) G_1^2 \left[\frac{(1-x)^2}{(1-R_G) \rho_L} + \frac{x^2}{R_G \rho_G} \right] \tag{4.75}$$

Note that, as in single phase flows, this equation has been obtained with the following major hypotheses,

- negligible frictional effects on the *pipe wall*

- $<\bar{p}>_o = <\bar{p}>_1$ (4.76)

- flat profiles

In *single phase flow* the momentum balance (4.36) also reads,

$$P_2 - P_1 = \sigma(1-\sigma)G_1^2 \frac{1}{\rho}$$ (4.77)

In *two-phase flow*, Eq. (4.75) allows a relative velocity between the phases. If one considers a homogeneous model the momentum equation would be,

$$P_2 - P_1 = \sigma(1-\sigma)G_1^2 \left(\frac{x}{\rho_G} + \frac{1-x}{\rho_L}\right)$$ (4.78)

Mechanical energy equation for the mixture. *Neglecting the viscous dissipation*, the mechanical energy equation obtained by space and time integration of the local instantaneous mechanical energy equations for each phase reads,

$$\tfrac{1}{2}\rho_G w_{G2}^3 A_{G2} + \tfrac{1}{2}\rho_L w_{L2}^3 A_{L2} - \tfrac{1}{2}\rho_G w_{G1}^3 A_{G1} - \tfrac{1}{2}\rho_L w_{L1}^3 A_{L1}$$

$$+ P_2 w_{G2} A_{G2} + P_2 w_{L2} A_{L2} - P_1 w_{G1} A_{G1} - P_1 w_{L1} A_{L1} = 0$$ (4.79)

As was done for the momentum equation, we will introduce in this equation the area ratio σ (Eq. 4.9), the void fraction R_G (Eq. 4.69), the mass quality x (Eq. 4.70) and the mass velocity G (Eq. 4.65). Moreover we will assume that Eqs. (4.73) and (4.74) still hold. Finally, taking account of the mass balance for the mixture (Eq. 4.67), the mechanical energy equation (4.79) becomes,

$$P_2 - P_1 = \tfrac{1}{2}(1-\sigma^2)G_1^2 \left[\frac{x^3}{R_G^2 \rho_G^2} + \frac{(1-x)^3}{(1-R_G)^2 \rho_L^2}\right] \left(\frac{x}{\rho_G} + \frac{1-x}{\rho_1}\right)^{-1}$$ (4.80)

Note that, as in single phase flow, this equation has been obtained with the following major hypotheses,

- negligible viscous dissipation within the control volume

- flat profiles

In *single phase flow*, the mechanical energy equation (4.57) also reads,

$$P_2 - P_1 = \tfrac{1}{2}(1-\sigma^2)G_1^2 \frac{1}{\rho}$$ (4.81)

In *two-phase flow*, Eq. (4.80) allows a relative velocity
between the phases. If one considers a homogeneous model the
mechanical energy equation would be,

$$P_2 - P_1 = \tfrac{1}{2}(1-\sigma^2)G_1^2 \; (\frac{x}{\rho_G} + \frac{1-x}{\rho_L}) \tag{4.82}$$

Comparisons with experimental results. Already recommended by
Lottes (1961), Romie's equation (Eq. 4.72) or its particular form
(Eq. 4.75) were successfully checked in steam-water flow by
Ferrel and McGee (1966) and in freon-freon vapor flow by Weisman
et al. (1978).

Ferrel and McGee (1966) carried out experiments with *steam-
water* mixtures flowing adiabatically through sudden expansions of
area ratios 0.332, 0.546 and 0.608. The pressure ranged from 4.1
to 16.5 bar and the mass velocity from 330 to 2470 kg $m^{-2}s^{-1}$.
The void fractions entering Romie's equation was the measured void
fractions. When the particular form was used (Eq. 4.75), the void
fraction was put equal to the average of the measured upstream
and downstream void fractions. The agreement with the Romie's
equation and its particular case (Eqs. 4.72 and 75) was of the
order of $\pm$ 40%.

Romie's equation (4.72) was also successfully checked by
Weisman et al. (1978) who performed experiments with *freon-freon
vapor* mixtures flowing through sudden expansions of area ratios
0.25 and 0.56. The mass velocities ranged from 1500 to 6200 kg
$m^{-2}s^{-1}$. The void fractions entering Romie's equation were
calculated with the Hughmark correlation. For mass velocities
above 2700 kg $m^{-2}s^{-1}$ Weisman et al. (1978) found a good agreement
with the pressure rise given by the homogeneous model (Eq. 4.78).

To calculate the irreversible pressure drop we recall that,

(i) from the momentum equation

$$P_2 - P_1 = \sigma(1-\sigma)G_1^2 \left[\frac{(1-x)^2}{(1-R_G)\rho_L} + \frac{x^2}{R_G\rho_G} \right] \tag{4.83}$$

(ii) from the mechanical energy equation without dissipation

$$P_2 - P_1 = \tfrac{1}{2}(1-\sigma^2)G_1^2 \left[\frac{x^3}{R_G^2\rho_G^2} + \frac{(1-x)^3}{(1-R_G)^2\rho_L^2} \right] \left(\frac{x}{\rho_G} + \frac{1-x}{\rho_L} \right)^{-1} \tag{4.84}$$

As was shown above, the pressure *increase* given by the momentum
equation (4.83) corresponds fairly well to the experimental data.
As a result it can be considered as the sum of a *reversible
pressure increase* $(p_2-p_1)_R$ due to the area change only, and of an
irreversible pressure drop $(p_2-p_1)_I$ due to the viscous dissipation
within the control volume. The pressure increase given by the
mechanical energy equation without viscous dissipation (Eq. 4.84)
corresponds to the reversible pressure increase,

$$(P_2 - P_1)_R = \frac{1}{2}(1-\sigma^2)G_1^2 \left[\frac{x^3}{R_G^2\rho_G^2} + \frac{(1-x)^3}{(1-R_G)^2\rho_L^2}\right]\left(\frac{x}{\rho_G} + \frac{1-x}{\rho_L}\right)^{-1} \quad (4.85)$$

The irreversible pressure drop is,

$$(P_2 - P_1)_I = P_2 - P_1 - (P_2 - P_1)_R \quad (4.86)$$

Consequently we obtain,

$$(P_2 - P_1)_I = (1-\sigma)G_1^2 \left\{\sigma\left[\frac{(1-x)^2}{(1-R_G)\rho_L} + \frac{x^2}{R_G\rho_G}\right]\right.$$

$$\left. - \frac{1+\sigma}{2}\left[\frac{x^3}{R_G^2\rho_G^2} + \frac{(1-x)^3}{(1-R_G)^2\rho_L^2}\right]\left(\frac{x}{\rho_G} + \frac{1-x}{\rho_L}\right)^{-1}\right\} \quad (4.87)$$

The homogeneous model would have given the following result,

$$(P_2 - P_1)_I = -\frac{1}{2}(1-\sigma)^2 G_1^2 \left(\frac{x}{\rho_G} + \frac{1-x}{\rho_L}\right) \quad (4.88)$$

Another formula for the pressure rise in a sudden expansion was proposed by Chisholm (1969-70). It reads,

$$P_2 - P_1 = \sigma(1-\sigma)G_1^2(1-x)^2\left(1 + \frac{C}{X} + \frac{1}{X^2}\right)\frac{1}{\rho_L} \quad (4.89)$$

For *rough tubes* we have,

$$X = \frac{1-x}{x}\left(\frac{\rho_G}{\rho_L}\right)^{0.5} \quad (4.90)$$

and,

$$C = \left[1 + 0.5\left(\frac{\rho_L-\rho_G}{\rho_L}\right)^{0.5}\right]\left[\left(\frac{\rho_L}{\rho_G}\right)^{0.5} + \left(\frac{\rho_G}{\rho_L}\right)^{0.5}\right] \quad (4.91)$$

4.3 Sudden Contraction

Single Phase Flow

Consider the steady state, turbulent flow of an incompressible fluid across the horizontal sudden contraction represented in Figure 4.3. The areas of the upstream and downstream pipe cross sections are respectively A_1 and A_2 and **k** is the unit vector of the pipe axis. As was done in Section 4.2 for the sudden

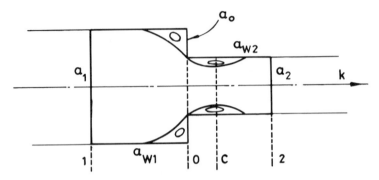

Figure 4.3: Control volume for the sudden contraction.

expansion, the mass and momentum balances and the mechanical
energy equation are written for the fixed control volume V limited
by cross sections A_1 and A_2 where the flow can be considered as
fully developed in each pipe and by the pipe wall A_w. The wall
surface A_w can be divided into three components: A_{w1} and A_{w2}
pertaining to the upstream and downstream pipes and A_o the area
of the flange perpendicular to the pipe axis. We will define A
the total boundary of the control volume V,

$$A = A_1 + A_2 + A_w \tag{4.92}$$

$$= A_1 + A_2 + A_{w1} + A_{w2} + A_o \tag{4.93}$$

Momentum balance. The same method as in Section 4.2 leads to the
following equation for the pressure drop between sections 1 and 2,

$$\frac{p_1 - p_2}{\frac{1}{2}\rho <w>_2^2} = 2(1 - \frac{1}{\sigma}) \tag{4.94}$$

This equation is obtained with the following major hypotheses,

- negligible frictional effects on the *pipe wall*

- $<\bar{p}>_o = <\bar{p}>_1$ $\tag{4.95}$

Note that in Eq. (4.94) the pressure drop is scaled with the
kinetic energy of the downstream pipe whereas in a sudden expansion
the pressure rise is scaled with the kinetic energy of the upstream
pipe. Note also that there is no obvious reason to support Eq.
(4.95).

Mechanical energy equation. We would obtain the following equation,

$$\frac{p_1 - p_2}{\frac{1}{2}\rho <\bar{w}>_2} = 1 - \frac{1}{\sigma^2} \tag{4.96}$$

providing the following major hypothesis is made,

- negligible viscous dissipation within the control *volume*.

Comparisons with experimental results. Figure 4.4 shows the experimental results obtained by Archer (1913) and the nondimensional pressure drop determined from the momentum equation (4.94) and the mechanical energy equation (4.96) as a function of the diameter ratio $1/\sqrt{\sigma}$. According to Figure 4.4 the experimental results lie between the two curves. Looking at the *momentum balance,* we could think of adding a friction term. However the calculated pressure drop would be larger and the only way to lower the calculated pressure drop would be to say that,

$$<\bar{p}>_1 \; < \; <\bar{p}>_0 \tag{4.97}$$

On the other hand, accounting for viscous dissipation in the *mechanical energy equation* would improve the agreement with the experimental data. For that we assume that the *irreversible pressure drop* of the sudden contraction is localized between the vena contracta (section c on Figure 4.3) and the downstream boundary of the control volume (section 2 on Figure 4.3). Moreover we assume that this irreversible pressure drop corresponds to a

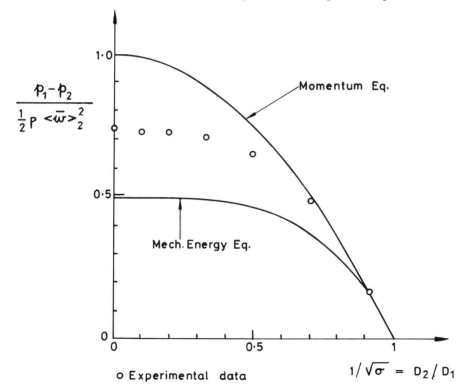

Figure 4.4: Pressure drop in a sudden contraction.

sudden expansion between sections c and 2. In consequence we have from Eq. (4.63),

$$\frac{(p_1 - p_2)_I}{\frac{1}{2}\rho \overline{\langle w \rangle}_c^2} = (1 - \sigma_c)^2 \qquad (4.98)$$

where σ_c is the *contraction coefficient* defined by,

$$\sigma_c \triangleq \frac{A_c}{A_2} \qquad (4.99)$$

Taking into account the mass balance, Eq. (4.98) becomes,

$$\frac{(p_1 - p_2)_I}{\frac{1}{2}\rho \overline{\langle w \rangle}_2^2} = \left(\frac{1}{\sigma_c} - 1\right)^2 \qquad (4.100)$$

The *reversible pressure drop* due to the area change only is given by the mechanical energy equation (4.96),

$$\frac{(p_1 - p_2)_R}{\frac{1}{2}\rho \overline{\langle w \rangle}_2^2} = 1 - \frac{1}{\sigma^2} \qquad (4.101)$$

Finally, the actual pressure drop is given by,

$$\frac{p_1 - p_2}{\frac{1}{2}\rho \overline{\langle w \rangle}_2^2} = 1 - \frac{1}{\sigma^2} + \left(\frac{1}{\sigma_c} - 1\right)^2 \qquad (4.102)$$

Figure 4.4 enables the contraction coefficient σ_c to be calculated as a function of the area ratio σ (Table 4.1).

Table 4.1

Weisbach value of the contraction coefficient

$\frac{1}{\sigma}$	0	0.2	0.4	0.6	0.8	1.0
σ_c	0.586	0.598	0.625	0.686	0.790	1.0

Two-Phase Flow

As was done in single phase flow one again assumes that the *irreversible pressure* drop of the sudden contraction takes place between the vena contracta, provided it exists in two-phase flow (!), and the downstream boundary of the control volume (Fig. 4.3). Moreover the assumption is made that this irreversible pressure drop corresponds to a sudden expansion between sections c and 2. In consequence, we obtain if there is *no phase change* and if the *homogeneous* model holds,

$$(p_1-p_2)_I = \frac{1}{2}\left(\frac{1}{\sigma_c} - 1\right)^2 G_2^2 \left(\frac{x}{\rho_G} + \frac{1-x}{\rho_L}\right) \qquad (4.103)$$

and

$$p_1-p_2 = \frac{1}{2}\left[1 - \frac{1}{\sigma^2} + \left(\frac{1}{\sigma_c} - 1\right)^2\right]G_2^2 \left(\frac{x}{\rho_G} + \frac{1-x}{\rho_L}\right) \qquad (4.104)$$

In the preceding equations the contraction coefficient σ_c is the same as in single phase flow and given by Table 4.1.

The use of the homogeneous model is recommended by several authors. Ferrel and McGee (1966) carried out experiments with adiabatic *steam-water* flows with contraction of area ratios 1.831 and 1.645. The pressures ranged from 4.1 to 16.5 bar and the mass velocity from 330 to 2470 kg $m^{-2}s^{-1}$. The homogeneous model agreed within $\pm$ 40% with the experimental data despite the fact that the measured void fraction was different from the calculated value. Geiger and Rohner (1966) performed similar tests with the following area ratios: 6.944, 3.953 and 2.513. The pressure varied from 13.8 to 34.5 bar and the mass velocity from 700 to 6535 kg $m^{-2}s^{-1}$. Weisman et al. (1978) used freon for contractions of area ratios 4 and 1.786 with mass velocities from 1500 to 6200 kg $m^{-2}s^{-1}$. All of these authors as well as Chisholm and Sutherland (1969-70) consider the homogeneous model (Eq. 4.104) as a very convenient design tool.

4.4 Inserts and Grids

No definite method can be recommended to calculate the pressure drops due to restrictions located in a channel. Very few articles dealing with this subject have been published so far.

Single Hole Restriction

Weisman et al. (1978) studied the pressure drop across the single hole restriction represented in Figure 4.5. Freon and its vapor was the circulating two-phase flow. The geometric characteristics of the restrictions used are given in Table 4.2.

The reader is referred to the original article for the pressure drop correlation to be used. Essentially, the type of correlation depends whether the vena contracta following the sudden contraction is located within or outside the restriction.

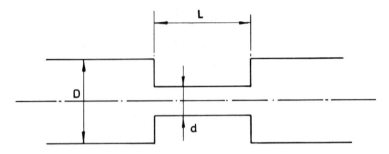

Figure 4.5: Single hole restriction.

Table 4.2

Single hole restriction characteristics in the Weisman et al. experiments (1978)

d (mm)	D (mm)	L (mm)	Area ratio
19.3	25.2	50.8	0.590
19.4	25.2	12.8	0.595
19.7	25.2	5.5	0.615
12.6	25.2	13.2	0.249

Multiple Hole Restriction

The same conclusions were obtained for multiple hole restrictions. Weisman et al. (1978) performed tests with the restrictions the characteristics of which are given in Table 4.3. In their experiments the freon mass velocities ranged from 1500 to 6200 kg $m^{-2}s^{-1}$. Similar studies were carried out by Al'ferov and Shul'zhenko (1977) with steam-water flow (pressures from 10 to 65 bar, quality from 0 t 1) and with air water flows (pressure: 5 bar, quality from 0 to 0.5). Table 4.4 gives the characteristics of the restriction.

Table 4.3

Multiple hole restriction characteristics in the Weisman et al. experiments (1978)

Number of hole	Hole diameter (mm)	D (mm)	L (mm)	Area ratio
4	8.9	25.2	4.8	0.505
4	9.2	25.2	15.7	0.934

Table 4.4

*Multiple hole restriction characteristics in the
Al'ferov and Shul'zhenko experiments (1977)*

Number of hole	Hole diameter (mm)	D (mm)	L (mm)	Pitch (mm)
7	2.4	10	2	3.1

Grids

Borishanskiy et al. (1977) studied the pressure drops
produced by spacing grids of various designs used in nuclear
engineering. Wet steam and air-water mixtures flowed through the
grids at pressures ranging from 20 to 130 bar and with mass
velocities from 500 to 3000 kg m^{-2}s^{-1}.

4.5 Tees and Wyes

Single Phase Flow

Fluid mechanics in tees and wyes is discussed in detail by
Ginzburg (1963) and by Gardel and Rechsteiner (1970). In this
last paper the authors present an extensive set of data.
Approximately 6000 runs were analyzed of water flowing in straight
pipes of circular cross sections. The experimental parameters
were the following, (i) the ratios of the secondary pipe diameter
to the main pipe diameter (corresponding area ratios: 0.44 and
0.69), (ii) junction angles ranging from 45 to 135°, (iii) rounding
radius: 0, 0.1 D, 0.2 D where D is the main pipe diameter. Figure
4.6 shows the four modes which were studied.

Two-Phase Flow

Very few studies have been devoted to two-phase flow in tees
and wyes. Idel'chick (1976) proposed a theory of irreversible
pressure losses in a tee for dispersed flows of solid particles,
droplets or bubbles. However no comparisons with experimental
results are given. Some experimental data obtained with gas-liquid
flows can be found in Fouda's thesis (1975). Finally Kubie and
Gardner (1978) examined the behavior of a two-phase mixture in a

Figure 4.6: Types of junctions studied in single phase flow by Gardel and
Rechsteiner (1970).

vertical wye where the gas phase was freely admitted in the sloping
limbs.

4.6 **Manifolds**

Single Phase Flow

In Bajura and Jones (1976) the flow distribution in the
lateral branches of discharge, collecting, reverse and parallel
flow manifolds (Figure 4.7) is studied both analytically and
experimentally.

This article contains a good review of existing works but
some statements regarding the first law of thermodynamics seem
rather unjustified.

Two-Phase Flow

Apart from the unpublished work done at Harwell very few
systematic studies were devoted to two-phase flow pressure drops
in manifolds.

Mochan (1969) studied the resistance of headers with multiple
delivery and discharge tubes. His work seems very extensive and
directly related to industrial problems. However the parameters
used in the data processing are not clearly defined which makes
this article rather difficult to understand.

Kubo and Ueda (1973) determined the flow distribution in
branch pipes for air-water mixtures in a horizontal piping system
with a collecting header.

Finally interesting experimental data can be found in Fouda's
dissertation (1975).

4.7 **Bends**

90^0 Bends

Chisholm (1967) proposed the following correlation,

$$\Delta p_F = (1-x)^2 \; \Delta p_{FLO} \; (1 + \frac{C}{X} + \frac{1}{X^2}) \qquad (4.105)$$

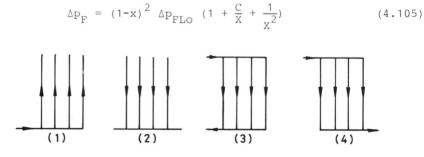

Figure 4.7: Different types of manifolds, (1) discharge, (2) collecting,
(3) reverse, (4) parallel.

with

$$C = \left[1 + (C_3-1)\left(\frac{\rho_L-\rho_G}{\rho_L}\right)^{0.5}\right]\left[\left(\frac{\rho_L}{\rho_G}\right)^{0.5} + \left(\frac{\rho_G}{\rho_L}\right)^{0.5}\right] \qquad (4.106)$$

For steam-water mixtures, constant C_3 is given in Figure 4.8 in terms of the ratio of the bend radius to the pipe diameter. In this figure the upstream disturbance refers to a contraction located at 56 D upstream of the bend. Both experimental and predicted losses are as much as 2.5 times the losses calculated using the homogeneous model ($C_3 = 1$).

Other experimental data can be found in Mochan (1969) for resistance of bends in discharge lines.

180° Bends

Giri (1977) performed some tests with air-water mixtures flowing at atmospheric pressure in 180° horizontal bends. The frictional pressure drop was well correlated by the Lockhart-Martinelli correlation for turbulent-turbulent flow where the X parameter is calculated from the experimental pressure gradients of each phase.

4.8 Recommendations and Selection Table

To conclude this Chapter, we will first emphasize the need for more work on steady-state flow in tees and wyes, manifolds and bends. As far as nonstationary flow in singularities is concerned everything remains to be done.

Although this overview concludes with a rather pessimistic note, the reader is referred to Table 4.5 for some recommendations.

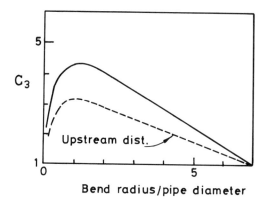

Figure 4.8: Coefficient C_3 in Eq. (4.106).

Table 4.5

Selection table for steady-state singular pressure drop

Singularity	Recommendation
Sudden expansion	Romie's equation (4.72 or 75)
Sudden contraction	Homogeneous model (Eq. 4.102 and Table 4.1)
Inserts and grids	None
Tees and wyes	None
Manifolds	None
Bends	Chisholm equations (Eqs. 4.105, and 106 and Fig. 4.8)

Nomenclature

A - area
C - Chisholm parameter (Eqs. 4.89, 91)
D - diameter
d - diameter
F - external force per unit of mass
G - mass velocity (Eq. 4.65)
k - unit vector of the pipe axis
L - length
M - mass flowrate
n - normal unit vector
p - pressure
R_G - void fraction (Eq. 4.69)
r - radial coordinate
T - total stress tensor
T - element of the total stress tensor
t - time
V - control volume
v - velocity vector
v_r - radial component of the velocity vector
w - component of the velocity vector along the pipe axis
w' - fluctuation of the velocity component along the pipe axis

X   - Lockhart-Martinelli parameter (Eq. 4.70)

x   - mass quality (Eq. 4.70)

z   - coordinate along the pipe axis

μ - viscosity

ρ - density

σ - area ratio (Eq. 4.9)

σ_c - contraction coefficient (Eq. 4.99)

τ - viscous stress tensor

τ - element of the viscous stress tensor

Ω - external force potential (Eq. 4.38)

Subscripts

c   - vena contracta

F   - frictional

G   - gas

I   - irreversible

L   - liquid

Lo - liquid flowing alone in the pipe with a mass flowrate equal
      to the total mass flowrate of the two-phase mixture

R   - reversible

W   - wall

Operator

──

     time-averaging operator

References

Al'Ferov, N S, and Shul'zhenko, Ye, N, (1977) "Pressure drops in two-phase flows through local resistances." *Fluid Mechanics-Soviet Research,* **6**, No 6, 20-34.

Archer, W H, (1913) "Experimental determination of loss of head due to sudden enlargement in circular pipes." *Trans. Americ. Soc. Civil Engineers,* **76**, 999-1026.

Bajura, R A, and Jones, E H, (1976) "Flow distribution manifolds." J. *Fluids Engng.*, 654-666.

Borishanskiy, V M, Andreyevskiy, A A, Bykov, G S, Volukhova, T G, Shleyfer, V A, and Belen'kiy, M, Ya, (1977) "Calculation of the hydraulics of local resistances in two-phase flow." *Fluid Mechanics-Soviet Research,* **6**, No 6, 35-50.

Chisholm, D, "Calculate pressure losses in bends and tees during steam/water flow." *Engng. and Boiler House Review,* **82**, No 8, 235-237.

Chisholm, D, and Sutherland, L A, (1969-70) "Prediction of pressure gradients in pipeline systems during two-phase flow." *Proc. Instn. Mech. Engrs.*, **184**, Pt 3 C, 24-32.

Ferrel, J K, and McGee, J W, (1966) "Two-phase flow through abrupt expansions and contractions." TID-23394, **3**.

Fouda, A E, (1975) "Two-phase flow behaviour in manifolds and networks." Ph.D. Thesis, Univ. of Waterloo (Canada), Dept of Chem. Engng.

Gardel, A, and Rechsteiner, G F, (1970) "Les pertes de charge dans les branchements en té des conduites de section circulaire." *Bulletin Technique de la Suisse Romande*, No 25, 363-391.

Geiger, G E, and Rohner, W N, (1966) "Sudden contraction losses in two-phase flow." *J. Heat Transfer*, 1-9.

Ginzburg, I P, (1963) *Applied Fluid Dynamics*, Israel Program for Scientific Translations.

Giri, N K, (1977) "Pressure drop in adiabatic two-phase, two-component flow in 180° bends." *Indian J. of Technology*, **15**, 1-7.

Hinze, J O, (1975) *Turbulence*, Sec. Ed., McGraw-Hill.

Idel'chik, I, Ye, (1976) "Determination of the hydraulic resistance of channels in which separation or merging of two-phase (or multiphase) incompressible flows take place." *Fluid Mechanics-Soviet Research*, **5**, No 3, 50-59.

Kubite, J, and Gardner, G C, (1978) "Two-phase gas-liquid flow through Y-junctions." *Chem. Engng. Science*, **33**, 319-329.

Kubo, T, and Ueda, T, (1973) "On the characteristics of confluent flow of gas-liquid mixtures in headers." *Bull. JSME*, **16**, No 99, 1376-1384.

Lottes, P A, (1961), "Expansion losses in two-phase flow." *Nucl. Science and Engng.* **9**, 26-31.

Mochan, S I, (1969) "Local resistances in the flow of two-phase mixtures." *Problems of Heat Transfer and Hydraulics of Two-Phase Media*, Kutateladze, S S, Ed, Pergamon, 327-384.

Petrick, M, and Swanson, B S, (1959) "Expansion and contraction of an air-water mixture in vertical flow." *AIChEJ*, **5**, No 4, 440-445.

Richardson, B E, (1958) "Some problems in horizontal two-phase two-component flow." ANL 5949.

Velasco, I, (1975) L'écoulement diphasique à travers un élargissement brusque, Thèse de Maîtrise en Sciences Appliquées, Université Catholique de Louvain, Faculté des Sciences Appliquées, Unité de Thermodynamique et Turbomachines.

Weisman, J, Husain, A, and Harshe, B, (1978) "Two-phase pressure drop across abrupt area changes and restriction." *Two-Phase Transport and Reactor Safety*, **IV**, Veziroglu, T N, and Kakac, S, Eds, Hemisphere, 1281-1316.

Whitaker, S, (1968) *Introduction of Fluid Mechanics*, Prentice-Hall.

Chapter 5

Annular Flow

G. F. HEWITT

5.1 Introduction

Annular two-phase flow is arguably the most important flow regime. It occurs widely in nuclear, chemical and process equipment. Annular flow is complicated by the existence of a complex pattern of interfacial waves and by the entrainment of liquid droplets off these waves. Often, a distinction is drawn between "annular flow" and "annular mist flow". In my view, this distinction is rather artificial since conditions in which there is no droplet entrainment are rather rare. Furthermore, conditions in which there is complete entrainment are difficult to achieve in adiabatic systems. Thus, I prefer the generic name "annular flow" to cover all the cases.

In analysing annular flow systems, one must make a careful distinction between those variables which are *independent* and those variables which are *dependent*. Taking the case of annular upwards flow in a round tube, the main independent variables are as follows:

(1) The tube diameter, d_o, or radius, r_o.

(2) The flow rates of the gas and liquid phases, W_G and W_L.

(3) The physical properties of the two fluids.

(4) The flow geometry upstream of the point being considered; this includes particularly the mode by which the phases were introduced into the channel. In other words, the flow is not completely specified in terms of the channel diameter, phase flow rates and phase physical properties. The influence of upstream effects is much more severe than in the case of single phase flow.

The main dependent variables in the system are as follows:

(1) The mean liquid film thickness, m, and the interfacial structure, m (t) or m (z).

(2) The axial pressure gradient, (dp/dz).

(3) The liquid film flow rate W_{LF} (or, if the total liquid flow rate W_L is known, the entrained liquid flow rate W_{LE}).

(4)   The rate of droplet deposition onto the film, per unit channel
       surface area (D) and the rate of entrainment from the film per
       unit area (E).

(5)   The distribution of dropsize and drop velocity within the core
       of the flow.

       An annular flow may be said to be in "hydrodynamic equilibrium"
if the rate of deposition of droplets onto the film surface is
equal and opposite to the rate of entrainment of droplets from the
film (ie $D = E$).

       In the first part of the chapter I shall describe briefly
the measurement techniques which have been developed for the
determination of the various parameters of annular flow (ie film
thickness, pressure drop, film flow rate, deposition rate,
entrainment rate and dropsize). I will also describe briefly
methods which have been developed for photographing the interfacial structure and detailed behaviour of annular flows. In the
second part of the chapter I will describe analytical methods for
annular flow, and their application. This second part starts
with a description of the so-called triangular relationship.
I shall then discuss correlations for deposition and entrainment
in annular flow and, finally, briefly describe some applications
of annular flow modelling.

5.2 **Measurement of Annular Flow Parameters**

       For the purposes of this chapter, I will give only a brief
description of the principal measurement techniques. Fuller
reviews are available elsewhere (Hewitt and Hall-Taylor (1970),
Hewitt (1978a) and Hewitt (1978b)).

Film Thickness and Interfacial Structure

       The most commonly used devices for the measurement of film
thickness are conductance probes. A conductance probe pair
consists of two metal elements, placed a little distance apart,
and mounted flush with the inside of the (nonconducting) wall of
the channel. The conductance, through the film, between the
probes is a measure of the film thickness.

       One problem with conductance probes is that they are inherently
unsuitable for obtaining highly localised measurements. As the
probes are brought nearer together, then the lines of conductance
through the film become curved and, if the probes are close enough,
then they become insensitive to changes in the film thickness. A
more suitable method for point measurements is the fluorescence
technique (Hewitt et al (1964)) which is illustrated in Figure 5.1.
Blue light from a mercury vapour lamp excites a fluorescence from
a dyestuff which has been added to the circulating liquid and the
intensity of this fluorescence is an indication of the thickness
of the film.

       The output from film thickness measurement transducers can be
analysed in various ways. Suppose that there are two transducers
in the tube, giving an output m_1 (t) at position z_1 and m_2 (t) at
position z_2. The signals may be processed as follows:

A Mercury vapour lamp.   G Liquid film
B Filters.   H Barfitt spectrometer
C Prisms.   I Photomultiplier
D Half-silvered mirror.   J Amplifying microammeter
E Objective lens.   K Recording oscilloscope
F Tube wall.   L Multichannel switch

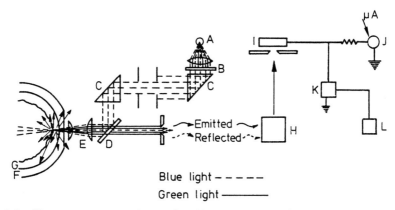

Blue light ─ ─ ─ ─ ─
Green light ───────

Figure 5.1: Fluorescence techniques for liquid film thickness measurement.

(1)   The average film thickness may be determined for each position.

(2)   The output traces may be examined in detail to give velocities and shapes of specific waves which can be identified in the traces.

(3)   The amplitude characteristics of the traces (ie the percentage of total time over which the film thickness exceeds a given value) can be determined electronically.

(4)   The auto-correlation function:

$$R_{m_1 m_1}(\tau) = \lim_{t \to \infty} \frac{1}{t} \int_0^t m_1(t) m_1(t+\tau)\, dt \tag{5.1}$$

can be calculated from either signal and the power spectral density can be determined by a Fourier transform of the auto-correlation:

$$G_{m_1 m_1}(f) = 4 \int_0^\infty R_{m_1 m_1}(\tau) \cos 2\pi f \tau\, d\tau \tag{5.2}$$

$$= 1 \int_{-\infty}^\infty R_{m_1 m_1}(\tau) \exp(-i\, 2\pi f \tau)\, d\tau$$

where f is the frequency.

(4)   The wave transit time (and hence wave velocity) between
       positions z_1 and z_2 can be determined from the cross correla-
       tion function:

$$R_{m_1 m_2}(\tau) = \lim_{t \to \infty} \frac{1}{t} \int_0^t m_1(t) m_2(t+\tau) dt \qquad (5.3)$$

Pressure Drop Measurement

     Measurement of two phase pressure drop presents a number of
special problems and I have reviewed these in some detail elsewhere
(Hewitt (1978c)). In particular, it is important to avoid
ambiguity about the nature of the contents of the lines joining the
tappings to the manometric system. We find it best to use liquid-
purge systems since this reduces the compressibility of the line
fluids and prevents gas ingress into the lines. A typical liquid
purge pressure drop system is illustrated in Figure 5.2.

Liquid Film Flow Rate/Entrained Liquid Flow Rate

     The most common method for measurement of liquid flow rate is
simply to suck off the film through a porous extraction device; a

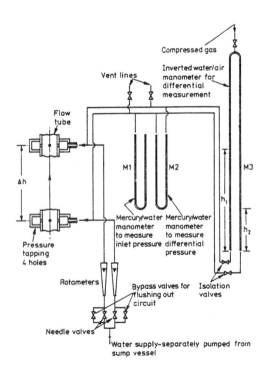

Figure 5.2: Liquid purge system for pressure drop mechanism in annular two phase
          flow.

typical device is illustrated in Figure 5.3. The film flow rate
fluctuates with time due to interfacial waves so that a variable
proportion of the extraction sinter is wetted by the liquid. Gas
is extracted through the unwetted portion, but the liquid droplets
entrained in the core have sufficient momentum not to be diverted
with the gas. A plot of film extraction rate versus gas extract
rate shows a plateau (sometimes not very distinct!) which
corresponds to the film flow rate. With single-component flows,
the extract may be condensed and the vapour content in the extract
determined by a heat balance.

An alternative to the suction method for film flow rate
measurement is to use traversable liquid collection probes. These
are of two main types:

(a)   Sampling probes; these consist simply of a bent tube with its
      open end facing into the flow. The rate of liquid drop
      collection through the probe is relatively insensitive to the
      rate of gas collection.

(b)   Isokinetic probes; here, both the gas and liquid are extracted
      together under "isokinetic conditions" by controlling the
      pressure in the probe mouth to equal that read from a static
      pressure tapping on the tube wall.

Using such probes, it is possible to integrate the mass flow
distribution of the liquid (droplet) phase across the gas core. A
difficulty, however, is in identifying precisely the boundary
between the core and the film which is uncertain because of the
presence of interfacial waves.

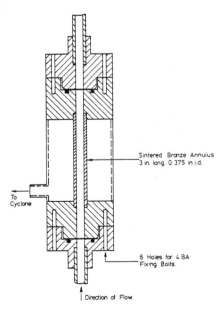

Figure 5.3: Film extraction device for film flow measurement.

Measurement of Droplet Deposition Rate

It is possible to measure the uni-directional deposition rate by a film extraction technique as illustrated in Figure 5.4. The liquid film is removed through a porous wall section or through a slot and the rate of build up of a new film is observed. The latter is done by either extracting the newly deposited film (Figure 5.4a and Figure 5.4b), by measuring the local film velocity in the newly deposited film (Figure 5.4c) or by measuring the remaining droplet mass flux (Figure 5.4d).

The deposition rate can also be infered from measurement of burnout heat flux in a non-uniformally heated channel. The principle is illustrated in Figure 5.5(a). The film flow rate decreases with increasing power input and, at the burnout point, reaches a value of zero upstream of the end of the test section. At this point, the rate of deposition of droplets is equal and opposite to the evaporation rate (ie $D = \phi/\lambda$ where ϕ is the heat flux and λ the latent heat). It is often rather difficult to identify precisely the point of dryout and more accurate data can be obtained by the special thermal deposition experiment shown in Figure 5.5(b). Conditions are adjusted so that burnout occurs simultaneously at lengths z_1 and z_2. In the zone between z_1 and z_2, the liquid deposition rate is balanced by the evaporation rate.

A difficulty with uni-directional deposition experiments is that the deposition rate may not be representative of fully developed annular flow; the presence of interfacial waves may change the characteristics of the gas core in such a way as to change the deposition rate.

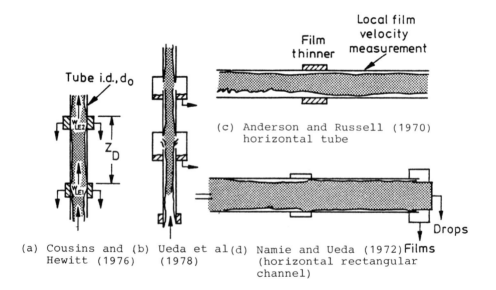

(a) Cousins and Hewitt (1976) (b) Ueda et al (1978) (c) Anderson and Russell (1970) horizontal tube (d) Namie and Ueda (1972) (horizontal rectangular channel)

Figure 5.4: Film extraction/deposition method for measuring deposition rate.

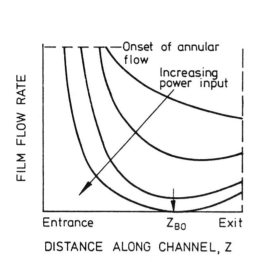

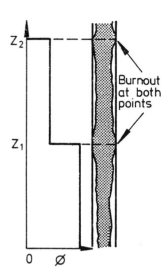

(a) Depostion rate from non-uniform
heat flux upstream burnout data

(b) Special thermal
deposition experiment

Figure 5.5: Thermal methods for determination of deposition rate in evaporating
channels.

Entrainment Rate Measurement

If the rate of deposition (D) is known for a hydrodynamic
equilibrium condition, then the entrainment rate (E) is also known
by definition. However, there are difficulties in estimating the
deposition rate (as explained above) and separate experiments for
entrainment rate are desirable. Figure 5.6 illustrates the use
of tracer techniques for the measurement of interchange rate. A
dye or salt solution is added to the liquid film and the rate of
interchange is determined by examining the variation of the
concentration of the dyestuff in the film as a function of length
Simply sampling the film tends to give incorrect data since mixing
of the dyestuff within the film itself (particularly in the laminar
region near the wall) is rather slow. Sampling of the whole film
(Figure 5.6(b)) gives better results but there is still the
problem of estimating the mixing within the gas core. This can
be studied by coupling the use of the dye tracer technique with
the use of isokinetic sampling as illustrated in Figure 5.6(c).
The dye concentration varies with radial position on the distance
as shown in Figure 5.6(d) and the interchange rate can be deduced
from this data. One may tentatively conclude that the mass
transfer coefficients for interchange (defined as $k = D/C$, where C
is the homogeneous mean concentration of droplets in the gas core)
are approximately the same as those for the uni-directional
deposition coefficients. However, there is a need for much more
data in this area and for the development of better techniques for
studying the interchange process.

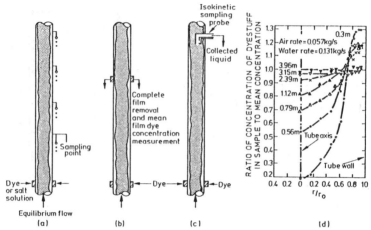

Figure 5.6: Dye tracer methods for measuring droplet-film interchange rates.

Dropsize Measurement

A variety of techniques are available for the measurement of dropsize distribution in annular flow. These are reviewed in detail by Hewitt (1978a) (1978b). One of the techniques which have been used recently is illustrated in Figure 5.7. Laser light is scattered by the droplets, the scattering angle depending on droplet size. The scattered light is collected on a detector which has the form of a number of annular rings and the distribution of intensity on this detector can be transformed by a mini-computer to give a droplet size distribution.

Photographic Techniques for Studying Interfacial Behaviour

There has been continued development of photographic methods for studying the nature of the interfacial waves in annular flow. Figure 5.8 shows an axial view device which allows the photography of a cross section of the channel at a particular axial position. Recent developments in axial view photography include the use of parallel (laser) light beams. Here, the whole of the channel is seened simultaneously and the motion of the droplets from the point of their inception, to the point of their deposition may be clearly observed.

5.3   Analytical Methods for Annular Flow

The experimental studies of annular flow reveal the following main features:

(1)   The flow takes a long time to reach equilibrium, with the independent variables still changing after several hundred diameters. Thus, in nearly all practical cases, the flow is essentially non-equilibrium and any analytical model which is to have practical utility must deal with this case.

(2)   The interface in annular flow is highly complex and is

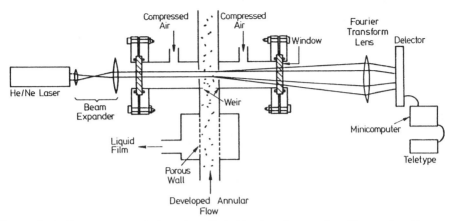

Figure 5.7: Measurement of dropsize distribution in annular flow using the light scattering method (Azzopardi et al (1978)).

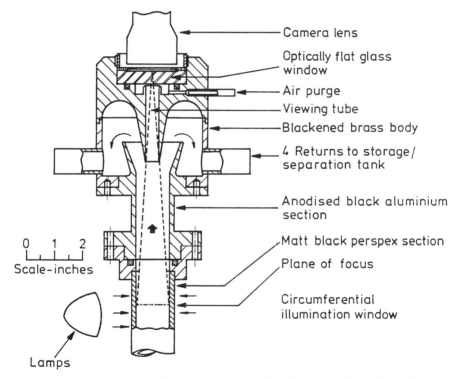

Figure 5.8: Apparatus for axial view photography of annular two phase flow.

traversed by large "disturbance" or "roll" waves which are a necessary condition for liquid entrainment.

(3) Extensive entrainment of the liquid phase is the norm and its variation along the channel of critical importance.

(4)   The pressure drop in annular flow is high compared to that
      for a single phase gas flow, mainly due to the influence of
      the interface in providing a rough surface over which the gas
      flows.

      Suppose that the film flow rate W_{LF} is known at a point z
along the channel. The objectives of analytical modelling of
annular flow are two-fold:

(1)   To predict the values of the other dependent variables (and
      in particular the pressure gradient dp/dz and the film thick-
      ness m) from the known value of W_{LF}.

(2)   To calculate the change δW_{LF} in film flow rate occurring over
      a distance δz and to repeat the first stage of the calculation
      and calculate new values of pressure gradient and film
      thickness. In this way, the development of the annular flow
      along the channel may be calculated.

      The first objective of the calculation is achieved by
solving simultaneously and iteratively a relationship linking
film flow rate, film thickness and pressure gradient (the
"triangular" relationship) and also a relationship linking pressure
gradient to film thickness (the interfacial roughness relationship).
The second objective of the calculation is achieved by estimating
the local values of D and E and estimating the change in film
flow rate from the relationship:

$$\delta W_{LF} = \delta z \ S \ (D-E) \qquad\qquad\qquad (5.4)$$

where S is the tube periphery. As we shall see below, D is
calculated from the equation:

$$D = kC \qquad\qquad\qquad (5.5)$$

where k is the deposition of mass transfer coefficient and C the
homogeneous mean droplet concentration in the gas core, which can
be calculated provided that W_{LF} (and hence, W_{LE}) is known. The
entrainment rate, E is calculated from a correlation which may
involve a knowledge of interfacial shear stress (calculated from
pressure gradient) and film thickness. It will thus be seen that
the values of the dependent variables are intimately linked
together and iterative solutions are required.

      In what follows, each of the main constituents of the
analytical model described in turn, followed by some examples of
its application.

The Triangular Relationship

      This relationship is best understood by discussing the
calculation of W_{LF} from a knowledge of dp/dz and m. The steps
are as follows:

(1)   The interfacial shear stress is calculated from the known
      value of pressure gradient and film thickness. Simplifying

the expression by ignoring acceleration and the effect of droplets on the core density, we have:

$$\tau_i = \frac{r_o^{-m}}{2} \left[-\frac{dp}{dz} - \rho_G g \right] \tag{5.6}$$

where ρ_G is the gas phase density.

(2)   Calculate the shear stress distribution in the liquid film. This is done by carrying out a force balance on an annular ring of outer radius r and inner radius r_i, the interfacial radius. This gives:

$$2\pi r\tau\delta z = 2\pi r_i \tau_i \delta z + \rho_L g \pi \delta z (r_i^2 - r^2)$$

$$+ \pi \left[p - (p + \delta z \frac{dp}{dz}) \right] \left[r^2 - r_i^2 \right] \tag{5.7}$$

where ρ_L is the liquid density and τ the shear stress radius r. This simplifies to:

$$\tau = \tau_i (\frac{r_i}{r}) + \frac{1}{2} (\rho_L g + \frac{dp}{dz}) (\frac{r_i^2 - r^2}{r}) \tag{5.8}$$

(3)   Calculate the velocity profiles from the expression:

$$\tau = \mu_E \frac{du}{dy} \tag{5.9}$$

where u is the local velocity within the liquid film at distance y from the wall and μ_E is the effective viscosity. For laminar flow, the effective viscosity is equal to the liquid viscosity (μ_L) and integrating equation 5.9 with a boundary condition u = 0 and y = 0, we have:

$$u = \frac{1}{\mu_L} \left\{ \left[\tau_i r_i + \frac{1}{2} (\rho_L g + \frac{dp}{dz}) r_i^2 \right] \ln \frac{r_o}{r} \right.$$

$$\left. - \frac{1}{4} \left[(\rho_L g + \frac{dp}{dz}) (r_o^2 - r_i^2) \right] \right\} \tag{5.10}$$

For turbulent flow, the effective viscosity is a function of the wall shear stress and of the distance from the wall. Commonly, this is calculated from the "eddy viscosity model" where:

$$\mu_E = \mu_L + \rho_L \epsilon \tag{5.11}$$

where ϵ is the eddy diffusivity. Single phase flow correlations for eddy diffusivity are often applied in two

phase flow. An example of such a model is that of Dukler
(1960) for downwards flow which was adopted by Hewitt (1961)
for upward flow. Here, the choice of expression is governed
by the value of y^+ where:

$$y^+ = u*y \, \rho_L/\mu_L \qquad (5.12)$$

where u* is the "friction velocity" defined by:

$$u* = \sqrt{\tau_o/\rho_L} \qquad (5.13)$$

Thus, for $y^+ < 20$, the eddy diffusivity may be calculated
from the Deissler expression:

$$\epsilon = n^2 uy \left[1-\exp\,(-\rho n^2 uy/\mu)\right] \qquad (5.14)$$

and for $y^+ > 20$, the von Karman expression may be used:

$$\epsilon = \frac{k(du/dy)^3}{(d^2u/dy^2)^2} \qquad (5.15)$$

where n and k are constants. Introducing the eddy diffusivity
expressions (equations (5.14) and (5.15)) into the expression
for the effective viscosity (equation (5.11)) it is possible
to numerically integrate equation (5.9) with the boundary
condition u = 0 at y = 0.

(4)   Calculate film flow rate by the integral:

$$W_{LF} = \int_o^m \rho_L u \, dy \qquad (5.16)$$

For laminar flow, the result is as follows:

$$W_{LF} = \frac{2\pi\rho_L}{\mu_L} \left\{ \left[\tau_i r_i + \tfrac{1}{2}\,(\rho_L g + \frac{dp}{dz})\,r_i^2\right]\left[\tfrac{1}{4}(r_o^2 - r_i^2) - \tfrac{1}{2}r_i^2 \ln \frac{r_o}{r_i}\right.\right.$$

$$\left.\left. - \tfrac{1}{16}\left[(\rho_L g + \frac{dp}{dz})\,(r_o^2 - r_i^2)^2\right]\right\}\right. \qquad (5.17)$$

and for turbulent flow, a numerical integration may again be
carried out of the velocity profile determined from the
numerical integration of the previous step.

The triangular relationship gives quite good predictions of
film flow rate from measured values of film thickness and pressure
gradient as is illustrated in Figure 5.9. However, it does not

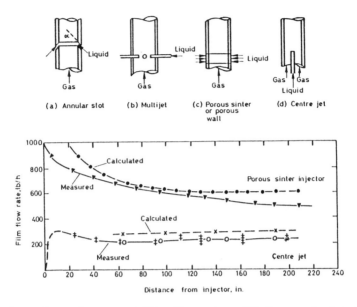

Figure 5.9: Comparison of measured film flow rates with those calculated from
the triangular relationship from measured values of film thickness
and pressure gradient.

provide a complete solution of the system as Figure 5.9 also
illustrates; for the same total flow rates of the two phases, and
for the same physical properties, the actual values of film flow
rate (and, thus, of film thickness and pressure gradient) are
quite different depending on the method of introducing the liquid
phase, even at considerable distances along the channel.

Interfacial Roughness Relationships

An interesting empirical observation about annular flow is
that the wave configuration on the interface seems to be approximately the same, for a given average film thickness, irrespective
of the phase flow rates which produce that film thickness. This
was confirmed, for instance, in detailed measurements on waves
reported by Hewitt (1969) who showed that the disturbance waves
typically had a magnitude which was five times the film thickness
and a spacing which also depended, to a first approximation, only
on film thickness. This "geometrical similarity" of the interface
is the underlying reason why plots of the ratio of "equivalent
sand roughness" of the interface to the tube diameter, against the
ratio of film thickness to tube diameter (Gill et al (1963)) and
plots of interfacial friction factor against m/d_O (Hartley and
Roberts (1961) and Wallis (1970)), tend to correlate a rather
wide range of experimental data for pressure drop. The most
commonly used form of the interfacial roughness correlation is
that of Wallis (1970):

$$f_{gsci} = \tau_i / \left[\tfrac{1}{2} \, \rho_{gc} V^2_{gc} \right]$$

$$f_{gsci} = f_{sgc} \ (1 + 360 \ m/d_o) \tag{5.18}$$

where f_{sgc} is the smooth pipe single phase friction factor from:

$$f_{sgc} = fn \ (Re_{sgc}) = fn \ (V_{gc}\rho_{gc} \ d_o/\mu_G) \tag{5.19}$$

where the gas core density ρ_{gc} is calculated on the basis of a homogeneous mixture in the gas core and is thus approximated by:

$$\rho_{gc} \sim \rho_G \ (\frac{W_G + W_{LE}}{W_G}) \tag{5.20}$$

and the gas core superficial velocity V_{gc} is approximated by:

$$V_{gc} \sim \frac{4W_G}{\pi \rho_G d_o{}^2} \tag{5.21}$$

Correlations for Deposition Rate, D

The deposition rate is caluclated from equation 5.5 provided k is known. Measurements of k exist for some situations, but some form of general correlation is required. In the annular flow modelling work at Harwell, the relationship illustrated in Figure 5.10 has been employed; however, it is unlikely that a simple relationship exists between deposition coefficient and surface tension as illustrated and this has led to the search for a more general correlation. McCoy and Hanratty (1977) suggest the dimensionless relationship:

$$\frac{k}{u^*} = 4390 \ \sqrt{\mu^2{}_G/d_o\sigma\rho_L} \tag{5.22}$$

but this does not fit data for high pressure as illustrated in Figure 5.11. A relationship (without direct physical justification)

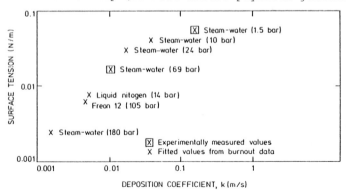

Figure 5.10: Tentative relationship between deposition coefficient and surface tension.

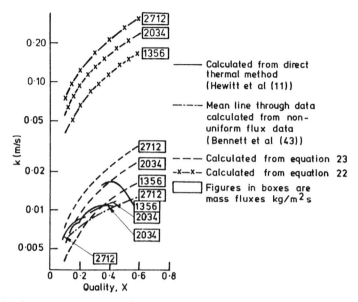

Figure 5.11: Comparison of correlations for k with experimental data obtained by the thermal method for steam-water flow at 70 bar.

which fits both high pressure and low pressure data is as follows:

$$\frac{k}{u^*} = 87 \sqrt{\frac{\mu_L^2}{d_o \sigma \rho_L}}$$ (5.23)

In equation (5.22) u^* is defined as $\sqrt{\tau_i/\rho_G}$ and in equation (5.23) as $\sqrt{\tau_i/\rho_{gc}}$.

Correlations for Entrainment Rate, E

As was mentioned above, there is a lack of satisfactory techniques for the measurement of entrainment rate and the normal method is to deduce it from the data for entrainment flow rate at hydrodynamic equilibrium. Here, we have:

$$E = D = k\,C_E$$ (5.24)

where C_E is the equilibrium homogeneous concentration of droplets in the gas core. Hutchinson and Whalley (1972) suggested correlating C_E as a function of the dimensionless group $(\tau_i m/\sigma)$. Their correlation is illustrated in Figure 5.12. Though this correlation has been useful, it has the disadvantage of being in dimensional form and also that, when converted to a correlation of entrainment rate, it gives a separation of the points. Recent attempts to produce an improved correlation are described by Whalley and Hewitt (1978) and a tentative correlation form is illustrated in Figure 5.13. Although this looks inferior to that

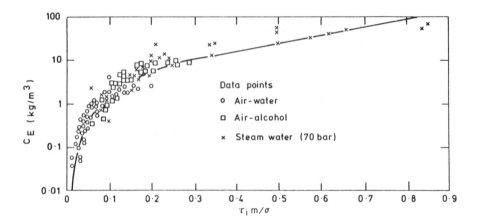

Figure 5.12: Hutchinson and Whalley (1972) correlation for equilibrium
entrained droplet concentration.

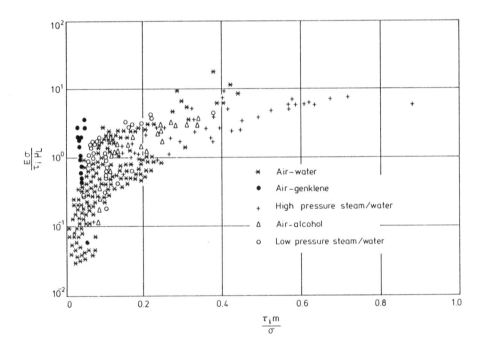

Figure 5.13: Dimensionless correlation of entrainment rate (Whalley and Hewitt
(1978)).

shown in Figure 5.12, the overall calculation accuracy for annular
flow is about the same, and the correlation in Figure 5.13 has
the advantage of being in dimensionless form. Clearly, there is
a need for improved correlations in this area.

Applications of Annular Flow Analysis

Figure 5.14 shows predictions of entrained liquid mass flow
as a function of quality in both hydrodynamic equilibrium
conditions and in an evaporating flow. Reasonable quantitative
agreement, and good qualitative agreement, is obtained. Note the
influence of "cold patches", ie zones of the channel in which
there is no heat input. Here, there is a shift in entrained flow
at constant quality. Figure 5.15 shows calculations of pressure
gradient in evaporating flow at high pressure; note the peak in
pressure gradient and the differences between the evaporating and
adiabatic equilibrium flows. Figure 5.16 shows an attempt to
predict, from the closed form model, the results illustrated in
Figure 5.9. Too rapid an approach to equilibrium is predicted and
some allowance must be made for the finite time required for interfacial development. This point is illustrated in Figure 5.17(a).
Considering the case where the liquid is injected smoothly into
the channel (porous sinter injector) it will be seen that the film
thickness decreases with increasing length. This would lead to a
prediction of an initially high entrainment rate decreasing to an
asymptotic value. However, since the disturbance waves required for
entrainment do not exist initially, the actual curve for entrainment rate may be different as illustrated. Figure 5.7(b) indicates
the opposite case where a liquid film is being evaporated away
and waves existing further upstream take a finite time to decay
and this makes the entrainment rate higher than that predicted for
the local equilibrium value of film thickness.

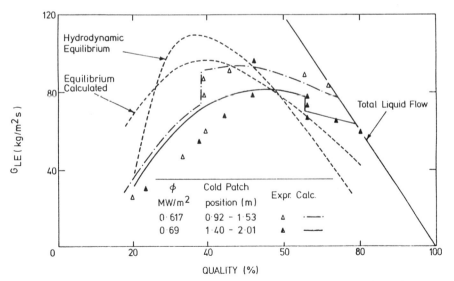

Figure 5.14: Comparison of predicted and experimental entrained liquid mass
flux in an evaporating steam-water flow at low pressure.

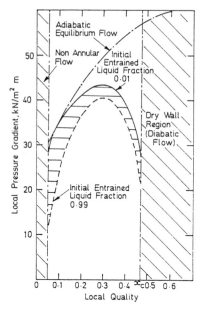

Figure 5.15: Predicted local pressure gradients in evaporation of water at 9.8MN/m² pressure with a mass flux of 1500kg/m²s.

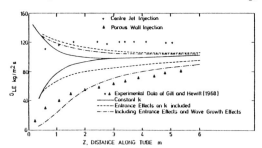

Figure 5.16: Comparison of predicted entrained liquid mass fluxes with those observed in air-water flow with two different injectors (compare Figure 5.9).

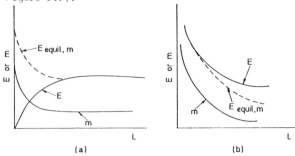

Figure 5.17: Departures from equilibrium entrainment rate due to interfacial development.

The results of prediction methods in annular flow are very encouraging but there is much further work to be done on the development of the interfacial characteristics and on providing better and more general correlations for the deposition and entrainment rates.

Nomenclature

C	– Droplet concentration in gas core	(kg/m^3)
C_E	– Gas core droplet concentration for conditions of hydrodynamic equilibrium (where $D = E$)	(kg/m^3)
d_o	– Tube diameter	(m)
D	– Droplet deposition rate per unit tube surface area	(kg/m^2s)
E	– Droplet entrainment rate per unit tube surface area	(kg/m^2s)
f	– Frequency	$(1/s)$
f_{sgc}	– Smooth pipe friction factor corresponding to gas core Reynolds number Re_{sgc} (equation (5.19)	$(-)$
f_{gsci}	– Interfacial friction factor defined by and calculated from equation 5.18	$(-)$
g	– Acceleration due to gravity	(m/s^2)
$G_{m_1m_1}$	– Power spectral density functions for film thickness (equation 5.2)	(m^2s)
k	– von Karman constant	$(-)$
	– Mass transfer coefficient	(m/s)
m	– Liquid film thickness	(m)
n	– Constant in equation 5.14	$(-)$
p	– Pressure	(N/m^2)
r	– Radial position	(m)
r_i	– Interfacial radius	(m)
r_o	– Tube radius	(m)
$R_{m_1m_1}$	– Autocorrelation function for film thickness (equation 5.1)	(m^2)
$R_{m_1m_2}$	– Cross-correlation function for film thickness (equation 5.3)	(m^2)
Re_{sgc}	– Superficial gas core Reynolds number defined by equation 5.19	$(-)$
S	– Channel periphery	(m)
t	– Time	(s)
T	– Time over which $R_{m_1m_2}$ or $R_{m_1m_1}$ is determined	(s)
u	– Local velocity in liquid film	(m/s)
$u*$	– Friction velocity in liquid film $(=\sqrt{\tau_o/\rho_L})$ in equation 5.13	(m/s)
	Friction velocity in gas core $(=\sqrt{\tau_i/\rho_G}$ in equations 5.22 and $\sqrt{\tau_i/\rho_{gc}}$ in equation 5.23	(m/s)

V_{gc}	– Superficial velocity of gas core (equation 5.21)	(m/s)
W_G	– Gas mass rate of flow	(kg/s)
W_L	– Liquid mass rate of flow	(kg/s)
W_{LE}	– Entrained liquid mass rate of flow	(kg/s)
W_{LF}	– Film liquid mass rate of flow	(kg/s)
y	– Distance from channel wall	(m)
y^+	– Dimensionless distance parameter (equation 5.11)	(–)
z	– Axial distance	(m)

Greek

ϵ	– Eddy diffusivity	(m^2/s)
λ	– Latent heat	(J/kg)
ϕ	– Heat flux	(W/m^2)
μ_E	– Effective viscosity	(kg/ms)
μ_G	– Gas phase viscosity	(kg/ms)
μ_L	– Liquid phase viscosity	(kg/ms)
ρ_G	– Gas phase density	(kg/m^3)
ρ_L	– Liquid phase density	(kg/m^3)
ρ_{gc}	– Gas core density (equation 5.20)	(kg/m^3)
σ	– Surface tension	(m/s^2)
τ	– Time interval (equations 5.1-5.3)	(s)
	– Shear stress	(N/m^2)
τ_i	– Interfacial shear stress	(N/m^2)
τ_o	– Wall shear stress	(N/m^2)

References

Anderson, R J and Russell, T W F, (1970) "Film formation in two-phase annular flow". *A.I.Ch.E.J.* **16**, 626.

Azzopardi, B J, Freeman, G and Whalley, P B, (1978) "Drop sizes in annular two-phase flow". *AERE-R9074*.

Cousins, L B, Denton, W H and Hewitt, G F, (1976) "Liquid mass transfer in annular two-phase flow". *Symp. on Two Phase Flow, Exeter, Proceedings* II, C401.

Cousins, L B and Hewitt, G F, (1968) "Liquid phase mass transfer in annular two-phase flow: Droplet deposition and liquid entrainment". *AERE-R5657*.

Cousins, L B and Hewitt, G F, (1968) "Liquid phase mass transfer in annular two-phase flow: Radial liquid mixing". *AERE-R5693*.

Dukler, A E, (1960) "Fluid mechanics and heat transfer in falling film systems". *Chem. Eng. Prog. Symp. Ser.* **56** (30), 1.

Gill, L E, Hewitt, G F and Lacey, P M C, (1964) "Sampling probe studies of the gas core in two phase flow: II Studies of the effect of phase flow rates on phase and velocity distribution". *Chem. Eng. Sci.* **19**, 665. *(Vol-4) 66*

Hewitt, G F, (1961) "Analysis of annular two-phase flow". AERE-R3680

Hewitt, G F, Lovegrove, P C and Nichols, B, (1964) "Film thickness measurement using a fluorescence technique. Part I: Description of the method". AERE-R4478.

Hewitt, G F, (1969-70) "Disturbance waves in annular two-phase flow". *Proc. Inst. Mech. Eng.* **184**, 142.

Hewitt, G F and Hall-Taylor, N S, (1970) "Annular two-phase flow". *Pergamon Press*, Oxford.

Hewitt, G F, (1978a) "Observation and quantity measurement in annular two phase flow". Lecture presented at the *Fluid Dynamics Institute (FDI) Course on Two Phase Flow Measurements, Dartmouth College*.

Hewitt, G F, (1978b) "Measurement of two phase flow parameters". *Academic Press*.

Hewitt, G F, (1978c) "Liquid mass transport in annular two-phase flow". Paper presented at *1978 Seminar of the International Centre for Heat and Mass Transfer, Dubrovnik*.

Hutchinson, P and Whalley, P B, (1972) "A possible characterisation of entrainment in annular flow". AERE-R7126.

McCoy, D D and Hanratty, T J, (1977) "Rate of deposition of droplets in annular two-phase flow". *Int. J. Multiphase Flow*, **3**, 319.

Namie, S and Ueda, T, (1972) "Droplet transfer in two-phase annular mist flow". *Bull. Japanese Soc. Mech. Engrs.*, **15**, 1568-1580.

Ueda, T, Tanaka, H and Koizumi, Y, (1978) "Dryout of liquid films in high quality R-113 upflow in a heated tube". *Sixth International Heat Transfer Conference, Toronto*.

Whalley, P B and Hewitt, G F, (1978) "The correlation of liquid entrainment fraction and entrainment rate in annular two-phase flow". AERE-R9187.

Chapter 6

Introduction to Two-Phase Heat Transfer

G. F. HEWITT

6.1 Introduction

This chapter is aimed at providing a general background to the succeeding chapters on boiling and condensation. Here, only brief descriptions are given with the objective of setting a framework into which the later more detailed material can be fitted.

First, general aspects of vapour-liquid equilibrium as they affect phase change and, more specifically, the equilibrium relationships for bubbles and droplets, are presented. Next, the mechanisms of nucleation are reviewed, including homogeneous nucleation and heterogeneous nucleation, the latter including a discussion of bubble growth from an idealised cavity.

Various aspects of heat transfer in two-phase flows are then discussed. These include typical temperature profiles, interfacial temperature differences, gas phase resistance in two-phase heat transfer, definitions of the modes of evaporation and condensation, and various types of departure from thermodynamic equilibrium in two-phase systems.

Finally, examples of two-phase heat transfer are discussed as follows:

(1) Condensation in a horizontal tube. The variations of flow regime along the channel, resulting from the change in vapour flow rate, are discussed.

(2) Pool boiling heat transfer. The behaviour of this system for both heat flux-controlled and wall temperature-controlled conditions is reviewed.

(3) Regions of boiling and flow in evaporation in a vertical tube. Here, the link between flow patterns and heat transfer regimes is discussed, and the differences between the flow situation and that occurring in pool boiling are emphasized.

6.2 Vapour-Liquid Equilibrium

It is convenient to discuss vapour-liquid equilibrium in terms of the relationship between system pressure p and fluid

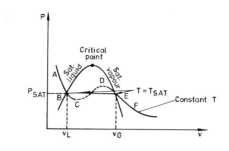

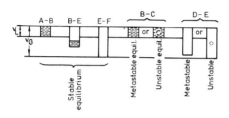

Figure 6.1: States of vapour-liquid equilibrium and their relationship to the
p-v relationship (from W B Hall).

specific volume v. The p-v curve is illustrated in Figure 6.1,
together with the various possible fluid states (this diagram is
derived from Professor W B Hall, University of Manchester,
Manchester, England). At constant temperature the "equilibrium"
p-v curve consists of liquid (AB), two-phase mixture (BE), and
vapour (EF). However, in the absence of the vapour nucleation it
is possible to increase the specific volume of the liquid along
line BC or to decrease the specific volume of the vapour along
line ED, and the liquid and vapour phases, under these conditions
are said to be in "metastable equilibrium". As will be shown
below, there is a critical size of bubble or droplet that is in
equilibrium with the liquid or vapour along lines BC and ED,
respectively. However, a slight decrease in the bubble or droplet
diameter will cause it to spontaneously collapse, and a slight
increase will cause it to spontaneously grow. This situation is
therefore called "unstable equilibrium". In many instances line
EF can be approximated by the ideal gas law:

$$pMv_G = RT \tag{6.1}$$

where M is the molecular weight of the vapour, v_G the vapour
specific volume, R the gas constant, and T the absolute temperature.
The curve ABCDEF can be represented approximately by the van der
Waals equation:

$$\left[p + \frac{a}{(Mv)^2}\right]\left[(Mv) - b\right] = RT \tag{6.2}$$

Figure 6.2: Vapour-liquid equilibrium at planar and curved surfaces.

where a and b are constants. A further useful relationship in what
follows is the Clapeyron-Clausius equation:

$$\frac{dp}{dT} = \frac{\lambda}{T(v_G - v_L)} \qquad (6.3)$$

which applies along the saturation curve where λ is the latent
heat o evaporation, T the saturation temperature, and v_G and v_L
the vapour and liquid specific volumes respectively. For $v_G \gg v_L$
we have

$$\frac{dp}{dT} \approx \frac{\lambda}{v_G T} \approx \frac{\lambda M p}{RT^2} \qquad (6.4)$$

It should be stressed that the approximate form of the Clapeyron-
Clausius equation (equation 6.4) is often significantly in error,
particularly at high pressures.

   Figure 6.2 shows the situation for equilibrium between liquid
and vapour at planar and curved interfaces. For the planar interface we have

$$p_L = p_g \qquad (6.5)$$

$$T_L = T_G = T_{SAT} \qquad (6.6)$$

where p_L and p_G are the pressure of the liquid and gas phases, T_L
and T_G their temperatures, and T_{SAT} the situation temperature.
For the case of the curved interface (Figure 6.2b) we have

$$p_L < p_G \qquad (6.7)$$

$$T_L = T_G > T_{SAT} \qquad (6.8)$$

Thus to maintain equilibrium with a curved interface, the temperature must be higher than that required for equilibrium with a
planar interface. The difference between the vapour and liquid
pressures for a spherical vapour bubble of radius r is given by

$$p_G - p_L = \frac{2\sigma}{r} \qquad (6.9)$$

where σ is the interfacial tension.

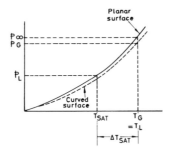

Figure 6.3: Relationship between saturation pressure and saturation temperature for planar and curved interfaces.

In Figure 6.3 the relationship between vapour pressure and temperature for a planar interface is shown by the solid line. The equilibrium saturation temperature for a planar interface, at the liquid (system) pressure p_L, is defined as T_{SAT}, as illustrated. Suppose the system is at a temperature $T_G (= T_L)$; it is important to calculate the temperature difference (or "superheat") $\Delta T_{SAT} = T_G - T_{SAT}$. The relationship between saturation pressure and saturation temperature for a curved interface is slightly different from that for a planar interface and is shown by the dashed line in Figure 6.3. By defining p_∞ as the vapour pressure corresponding to T_G for a planar interface it can be shown that

$$p_G = p_\infty \exp\left(-\frac{2\sigma v_L M}{rRT}\right) \approx p_\infty \left(1 - \frac{2\sigma v_L}{p_\infty r v_G}\right) \tag{6.10}$$

and thus

$$p_\infty - p_G = \frac{2\sigma v_L}{r v_G} \tag{6.11}$$

Combining equations (6.9) and (6.11), we have

$$p_\infty - p_L = \frac{2\sigma}{r}\left(1 + \frac{v_L}{v_G}\right) \tag{6.12}$$

To convert this relationship into a relationship for temperature difference, we integrate the Clapeyron-Clausius relationship (equation 6.4) over the range $p = p_\infty$ at $T = T_G$ and $p = p_L$ at $T = T_{SAT}$, giving

$$\ln \frac{p_\infty}{p_L} = \frac{\lambda M}{R}\left(\frac{1}{T_{SAT}} - \frac{1}{T_G}\right) = \frac{\lambda M}{R}\frac{\Delta T_{SAT}}{T_G T_{SAT}} \tag{6.13}$$

and thus

$$\Delta T_{SAT} = \frac{RT_G T_{SAT}}{\lambda M}\ln\left(1 + \frac{p_\infty - p_L}{p_L}\right) \approx \frac{RT_G}{\lambda M p_L}T_{SAT}(p_\infty - p_L)$$

$$\approx \frac{T_{SAT} v_G}{\lambda} (p_\infty - p_L) \qquad (6.14)$$

For $v_L/v_G \ll 1$, combining equations (6.12) and (6.14), we have

$$\Delta T_{SAT} \approx \frac{2\sigma T_{SAT} v_G}{\lambda r} \qquad (6.15)$$

Thus the ratius r^* of a bubble in equilibrium with a liquid whose superheat is ΔT_{SAT} is given by

$$r^* = \frac{2\sigma T_{SAT} v_G}{\lambda \Delta T_{SAT}} \qquad (6.16)$$

The case of a droplet of radius r, in equilibrium with a vapour temperature T_G, can be treated in a similar way. Here, we have

$$p_G - p_\infty \approx \frac{2\sigma}{r} \frac{v_L}{v_G} \qquad (6.17)$$

$$\Delta T_{SUB} = T_G - T_{SAT} \approx \frac{2\sigma v_L}{r\lambda} \frac{T_G}{} \qquad (6.18)$$

where p_∞ is the vapour pressure for a planar interface at temperature T_G and T_{SAT} is the saturation temperature for a planar interface at system pressure p_G.

The free energy ($\Delta G(r)$) of a bubble of radius r is given by

$$\Delta G(r) = -\frac{4\pi r^3}{3} (p'_\infty - p_L) + 4\pi r^2 \sigma = 4\pi \sigma (-\frac{2}{3} \frac{r^3}{r^*} + r^2) \quad (6.19)$$

where p'_∞ is the vapour pressure inside a nucleus of critical size r^*. The relationship between $\Delta G(r)$ and r is sketched in Figure 6.4. With $r = r^*$, both increasing and decreasing the bubble radius gives rise to a decrease in free energy, and thus the system is unstable. However, once the bubble exceeds the critical radius, then it will continue to grow.

6.3 Mechanism of Nucleation

We shall now consider the formation of nuclei in a super-heated liquid. One way in which such nuclei could be formed would

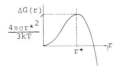

Figure 6.4: Relationship between free energy and bubble formation as a function of bubble radius.

be by "homogeneous nucleation"; that is, a vapour embryo of the
critical radius could be formed by the normal statistical
fluctuations of the liquid. From statistical theory the number
of embryos of radius r in a system with C molecules is given by

$$N_r = C \exp\left(- \frac{\Delta G(r)}{kT}\right) \qquad (6.20)$$

where $\Delta G(r)$ is the free energy of formation of a bubble embryo of
radius r and k is the Boltzmann constant. The number of bubbles
of the critical size r* for growth is given (see equation 6.19)
by

$$N_{r*} = C \exp\left(- \frac{4\pi\sigma r*^2}{3kT}\right) \qquad (6.21)$$

At first sight it would seem to be necessary to have only one
critical embryo present; in practice, to obtain "significant"
nucleation, a very large number of critically sized embryos are
required, though the calculation of superheat is not too sensitive
to this number. Calculation of teh temperature for homogenous
nucleation in water and benzene at atmospheric pressure indicates
that it will occur at $320^\circ C$ and $225^\circ C$, respectively, compared
to saturation temperatures of $100^\circ C$ and $80^\circ C$. The superheat
required for homogeneous nucleation is very large in comparison to
the actual superheats observed in practice, that is, of the order
of $10^\circ C$.

A similar calculation can be carried out for homogeneous
nucleation of droplets from a subcooled vapour, and the value of
ΔT_{SUB} is calculated, for water, to be of the order of $10^\circ C$.
Again, this is much higher than the values observed in practical
condensing systems, that is, less than $1^\circ C$.

The discrepancy between the superheat required for homogeneous bubble nucleation and that required for nucleation in practical systems could be a result of the presence of the containing
surface. Bankoff (1957) considered a bubble forming on a solid
surface with a contact angle (see Figure 6.5) of θ. For this
case he showed that the free energy of bubble formation of a
critically sized bubble $\Delta G(r*)$, calculated from equation (6.19)
should be multiplied by a factor Z, calculated as follows:

$$Z = \frac{(2 - \cos\theta)(1 + \cos\theta)^2}{4} \qquad (6.22)$$

This result implied a zero superheat for $\theta = 180^\circ$. However,
there is no significant change in the superheat required for
values of θ commonly found in practice (i.e., $< 140^\circ$).

Kottowski et al. (1974) considered the case of nucleation
from a conical cavity with included angle β (see Figure 6.6),
which is initially liquid filled. In this case $\Delta G(r*)$ should be
multiplied by a factor H, which is calculated from the equation

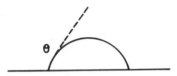

Figure 6.5: Bubble formation at solid surface.

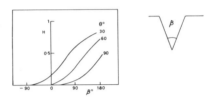

Figure 6.6: Bubble formation cavities that are initially liquid filled.

$$H = \frac{1}{4} \left[2 - 3 \sin \left(\frac{\Theta - \beta}{2} \right) + \sin^3 \left(\frac{\Theta - \beta}{2} \right) \right] \qquad (6.23)$$

The relationship between H and β, with Θ as a parameter, is illustrated in Figure 6.6. Though the superheat requirements are less for this case, they are still far too large in comparison with the practical values, and other mechanisms of heterogeneous nucleation must be postulated. These could include the following:

(1) The presence of preexisting vapour or inert gas pockets within nucleation cavities. The preexistence of a vapour-liquid interface would allow nucleation at low superheats, and the cavity may remain active providing it is not extinguished in the bubble growth/release cycle.

(2) Reentrant cavities. It is possible to postulate a shape of reentrant cavity that the liquid phase, on wetting the containing surface, would be unable to penetrate because of capillarity. Such reentrant cavities can be obtained by coating the surface, for instance, with a layer of porous material, and this method is now used commercially, though there are difficulties in its application where fouling of the surface is possible.

(3) As boiling develops, new cavities can be activated during the boiling process from adjacent cavities. This is probably a very significant mechanism in spreading nucleate boiling along the surface, though an initial nucleation site is still required.

Nucleate boiling is often considered in terms of vapour growth from an idealised conical cavity, as illustrated in Figure 6.7. As the bubble grows out of the cavity, penetrating to a distance b into the fluid, the bubble radius passes through a minimum, corresponding to a hemispherical bubble over the cavity, as illustrated in Figure 6.8. The superheat required for growth ΔT_{SAT} reaches a maximum corresponding to this hemispherical condition, as is also shown in Figure 6.8. This concept, and its implications will be discussed further in Chapter

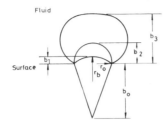

Figure 6.7: Stages in bubble growth from an idealised cavity.

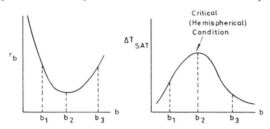

Figure 6.8: Vibration bubble radius at equilibrium superheat during bubble growth from an idealised conical cavity.

6.4 Aspects of Heat Transfer in Two-Phase Flows

Temperature Profiles

Consider the idealised two-phase heat transfer situation illustrated in Figure 6.9 in which a liquid film is being evaporated (Figure 6.9a) or in which vapour is being condensed onto a similar film. The total temperature difference ΔT between the wall temperature T_W and the gas temperature T_G can be divided into a number of parts:

(1) The film temperature difference ΔT_{film} due to conduction and convection in the liquid film.

(2) An interface temperature difference ΔT_{int}, which arises because of the presence of *net* evaporation or condensation from or to the film.

(3) A temperature difference ΔT_{gas} in the gas core.

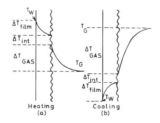

Figure 6.9: Typical temperature profiles in idealised two-phase heat transfer.

In the evaporation or condensation of pure fluids the vapour phase is normally quite close to the saturation temperature, and since ΔT_{int} is often small, the situation is governed by the film resistance. However, when more than one component is present in a system, then the temperature difference in the gas phase can be important. An example here is the condensation of vapours in the presence of incondensable gases.

The subject of interfacial temperature drops is disucssed in the following section, and the case of condensation in the presence of an incondensable gas is discussed in the section "Gas phase resistance" below.

Interfacial Temperature Drops

For an interface in thermodynamic equilibrium, the flux of vapour molecules to the interface is equal and opposite to the flux of molecules away from it. For net evaporation or condensation the flux in one direction or the other is greater. However, the proportionate change in molecular flux due to normal rates of condensation and evaporation is very small. Thus only a small change in interface temperature away from the saturation temperature is required to accommodate net condensation or evaporation.

A detailed discussion of interfacial temperature drops is given by Collier (1972); a convenient way of expressing such temperature drops is to use an equivalent interfacial heat transfer coefficient h_i, which is defined and obtained from the approximate relationship

$$h_i = \frac{G_c \lambda}{\Delta T_{int}} = \frac{2\sigma_a}{2 - \sigma_a} \ (\frac{M}{2\pi RT})^{1/2} \frac{\lambda^2 \ pM}{RT^2} \qquad (6.24)$$

where G_c is the mass rate of condensation per unit area of interface, λ the latent heat, M the molecular weight, R the gas constant, T the absolute temperature, and p the pressure. The term σ_a is known as the "accommodation coefficient" or "condensation coefficient"; this term represents the fraction of molecules impinging on the interface that are adsorbed by the interface. Various values have been ascribed to this term, though the present consensus is that it should be close to unity. For the condensation of water, with $\sigma_a = 1$, the following values of h_i can be calculated from equation 6.24.

p(atm)	h_i (W/m^2-K)
0.05	1.47×10^6
0.1	2.57×10^6
1.0	1.57×10^7

The values of h_i are several orders of magnitude greater than the heat transfer coefficients for conduction and convection through the liquid film (which lie typically in the range 0.5×10^3

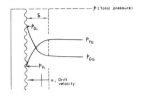

Figure 6.10: Gas phase resistance to heat transfer due to the presence of incondensible gases.

to 1×10^4 W/m²-K). Thus the resistance at the film-vapour interface is usually small in comparison to the resistance in the liquid film and can usually be ignored in engineering calculations.

Gas Phase Resistance

Consider the situation illustrated in Figure 6.10. The vapour is condensing at a mass rate G_C per unit of interface in the presence of an inert gas. The bulk partial pressure of the vapour is p_{vb}, and the partial pressure of the incondensible gas in the bulk is p_{Gb}. At the interface the concentrations are p_{vi} and p_{Gi}, respectively. Since there is a drift of vapour toward the interface the inert gas is swept in the direction of the interface and has to diffuse back. There is a net drift velocity u toward the interface at any point. The situation can be represented in a simple form by considering that the vapour-gas mixture has a uniform concentration apart from a layer of thickness δ adjacent to the interface. Within this layer one can write

$$G_c = u \frac{p_v}{p} \rho_m + \frac{M_v D}{RT} \frac{dp_v}{dy} = G_c \frac{p_v}{p} + \frac{M_v D}{RT} \frac{dp_v}{dy} \qquad (6.25)$$

where ρ_m is the mean density of the mixture, M_v the molecular weight of the vapour, p_v the local vapour pressure, p the total pressure, y the distance from the interface, and D the diffusion coefficient. Integrating equation 6.25 over the equivalent boundary layer thickness δ, we have

$$G_c = \frac{D}{\delta} \frac{pM_v}{RT} \ln \frac{p_{Gi}}{p_{Gb}} = k \frac{pM_v}{RT} (p_{Gi} - p_{Gb}) = k \frac{pM_v}{RT} (p_{vb} - p_{vi}) \qquad (6.26)$$

where k is a mass transfer coefficient defined by equations 6.26 and equivalent to

$$k \equiv \frac{D}{\delta (p_G)_{lm}} \qquad (6.27)$$

where $(p_G)_{lm}$ is defined by

$$(p_G)_{lm} = \frac{p_{Gi} - p_{Gb}}{\ln p_{Gi}/p_{Gb}} \qquad (6.28)$$

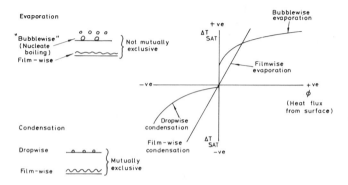

Figure 6.11: Modes of condensation and evaporation.

Since the interfacial vapour pressure, p_{vi} is considerably lower
than the bulk vapour pressure p_{vb}, the equilibrium temperature at
the interface is also lower. This temperature difference between
the interface and the bulk gas phase leads to heat transfer
simultaneous to the mass transfer due to condensation. The
classical treatment of this situation is that due to Colburn and
Hougen (1934); a more detailed discussion of mass transfer in
condensation is given by Collier (1972) and in Chapter 18.

Types of Evaporation and Condensation

Figure 6.11 illustrates the main modes of condensation and
evaporation and also shows schematically the relationship between
ϕ, the heat flux from the surface, and ΔT_{SAT}, the difference
between the surface temperature and the saturation temperature.
In "filmwise" condensation and evaporation, for a given liquid
film thickness and flow characteristics (giving a constant heat
transfer coefficient), the heat flux varies linearly with ΔT_{SAT} as
is shown. In the case of "bubblewise" evaporation in which bubbles
are formed at the heated surface, a finite value of ΔT_{SAT} is
required before heat is transferred by this mechanism. However,
in bubblewise evaporation (more commonly known as "nucleate
boiling") the heat flux increases more rapidly with ΔT_{SAT} than for
most situations of filmwise evaporation. The same is true for
"dropwise" condensation, in which droplets are nucleated on the
surface. Note that for a given location on a surface, dropwise
and filmwise condensation are mutually exclusive, whereas nucleate
boiling and filmwise evaporation are not mutually exclusive and
can occur simultaneously. A more extensive discussion of the
properties of nucleate boiling is given in Chapter 6.7.

Departures from Thermodynamic Equilibrium in Two-Phase Systems

In two-phase flow one may define the vapour flow mass
fraction or "quality" x as follows:

$$x = \frac{W_G}{W_G + W_L} \qquad\qquad (6.29)$$

where W_G is the rate of gas or vapour flow and W_L the mass rate of liquid flow. Usually, in a situation with boiling or condensation the sum of W_G and W_L is constant along the channel and is equal to GA, where G is the mass flux and A the cross-sectional area. However, W_G is usually not known independently, though the average enthalpy of the flow i_{TP} can often be estimated from a heat balance. If the two phases are in thermodynamic equilibrium, it follows that

$$i_{TP} = \frac{W_G i_G + W_L i_L}{W_G + W_L} = \frac{GA(xi_G + (1-x)i_L)}{GA} = xi_G + (1-x)i_L \qquad (6.30)$$

where i_G and i_L, are the saturated enthalpies of the vapour and the liquid phases, respectively. The quality x can be calculated from equation 6.30 and is as follows:

$$x = \frac{i_{TP} - i_L}{i_G - i_L} \qquad (6.31)$$

The quality defined by equation 6.31 is used widely in performing calculations, presenting data, etc., in boiling and condensation. Note that if $i_{TP} < i_L$, then the bulk fluid temperature at equilibrium is less than the saturation temperature and x is, by definition, negative. Furthermore, if $i_{TP} > i_G$, x is greater than unity. In this case the equilibrium condition corresponds to superheated vapour. However, it is possible to have two-phase (vapour-liquid) conditions when x, defined by equation 6.31, is either greater than unity or less than zero. Similarly, it is possible to have single-phase conditions of vapour or liquid within the quality range 0 < x < 1. This happens because of the departures from thermodynamic equilibrium that can be broadly classified as follows:

(1) "Continuum" effects. Here, a departure from thermodynamic equilibrium occurs owing to the failure to nucleate the second phase within the first. Under this heading we have subcooled vapour without droplet formation and superheated liquid without bubble formation.

(2) "Transient" effects. Here, the system is such that there has been insufficient time to reach equilibrium. Examples here would be superheated vapour with droplets and subcooled liquid with vapour bubbles.

(3) "Local" effects. Here, though the total mixed flow in a channel would be single-phase liquid or vapour, temperature differences across the channel cause the local formation of the second phase. An example is subcooled boiling. Here, though the average enthalpy of the fluid is less than that of the saturated liquid, the flux at the channel wall gives rise to a temperature there that exceeds the temperature required for the onset of boiling, and vapour bubbles are formed that condense as they mix with the fluid in the centre of the channel. A second example is condensation of superheated vapour. If the wall temperature is low enough, then condensation occurs even though the bulk enthalpy is greater than that of saturated steam.

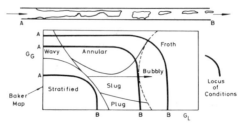

Figure 6.12: Flow regimes in condensation in a horizontal tube and their relationship to the Baker (1954) flow pattern map.

6.5 Examples of Two-Phase Heat Transfer

Condensation in Horizontal Tubes

This case is illustrated schematically in Figure 6.12. Initially, condensation occurs at the tube surface, giving rise to an annular flow. Film drainage leads to a thickening of the film at the bottom of the tube, and as the vapour flow rate decreases and liquid flow rate increases, the flow regime changes from annular flow through the horizontal flow regimes in a manner depending on the total flow rate. In Figure 6.12 the locus of condensation for a variety of fixed mass flows is plotted on the Baker flow pattern diagram (see Chapter 1). In considering flow patterns with condensation, it should be borne in mind that the flow pattern very much depends on the entrance conditions and the way the flow develops. The applicability of the adiabatic flow pattern diagrams to the case of condensation should therefore be regarded as somewhat uncertain.

Pool Boiling

The stages in boiling from the heated horizontal cylinder in a static pool of liquid are illustrated in Figure 6.13. At low heat fluxes the cylinder is cooled by natural convection, and there is no bubble formation. As the heat flux is increased, bubbles grow but do not detach from the surface (Figure 6.13b). With a further increase in heat flux, bubbles increase in number, and some detach and, if the liquid is subcooled, condense in the bulk fluid. Ultimately, as heat flux increases further, the bubbles on the surface become so numerous that they coalesce to form a continuous vapour layer.

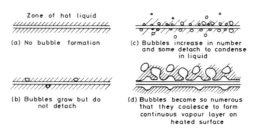

Figure 6.13: Modes of heat transfer in pool boiling.

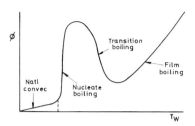

Figure 6.14: Pool boiling curve for a surface whose temperature is controlled.

Figure 6.14 shows how the heat flux from the surface varies with the surface temperature. After an initial region of natural convection the heat flux rises rapidly with surface temperature in the region of nucleate boiling. As the surface becomes crowded with vapour, maximum heat flux is reached after which the heat flux drops with increasing wall temperature. Finally, with a high enough temperature the flux again increases with wall temperature in the film boiling region. The curve illustrated in Figure 6.14 is obtained for surfaces whose temperature is controlled at the various levels. This might be achieved by heating with a vapour at various pressures. If, on the other hand, the heat flux is controlled (by, say, electrical heating) then the wall temperature is the dependent variable and varies as shown in Figure 6.15. Initially, in the natural convection and nucleate boiling region the curve is similar to that for the temperature-controlled surface. When the maximum heat flux is reached, however, there is an excursion of temperature, and the film boiling region is entered without an intermediate transition boiling region. On reducing the heat flux there is a hysteresis, as is illustrated in Figure 6.15. Film boiling is maintained until the minimum heat flux is reached, corresponding to the entry into the transition boiling region. At this point, an excursion back to nucleate boiling occurs as shown. This hysteresis is characteristic of electrically heated systems, and of course, the temperature rise due to the excursion into film boiling can often be sufficient to melt the heated surface.

Evaporation in a Vertical Tube

Figure 6.16 shows schematically the various stages of two-phase flow and boiling during the evaporation of a liquid in a vertical tube as the heat flux is increased. In tube A the heat

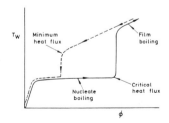

Figure 6.15: Pool boiling curve for a surface where the heat flux is controlled, showing hysteresis in re-entering the nuclear boiling region.

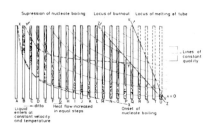

Figure 6.16: Regimes of boiling and flow in a vertical tube.

flux is just sufficient to bring the fluid up to saturation
temperature at the outlet. Tube B has twice this heat input,
tube C three times, and so on. The main features shown on Figure
6.16 are as follows:

(1)   Lines of constant quality (x) are shown, and obviously, the
      range of quality covered as the flow passes through the tube
      is increased as the heat input is increased. Note particu-
      larly the position of the line for x = 0.

(2)   The line XX denotes the onset of nucleate boiling. At low
      heat inputs (tube B), nucleation does not start until the
      bulk fluid has become superheated. At high heat fluxes,
      nucleation occurs near the wall even when x < 0, corresponding
      to the onset of subcooled boiling.

(3)   It will be seen that the flow patterns develop progressively
      as the heat input increases. Initially, bubble flow occurs
      at the end of a tube, and as the vapour constant increases,
      this changes to slug flow, churn flow, and, ultimately,
      annular flow.

(4)   As the two-phase quality increases, the wall temperature
      decreases because of improved heat transfer. Ultimately,
      the wall temperature is insufficient to sustain nucleate
      boiling and this is suppressed along the line YY. Down-
      stream of this line, the evaporation is filmwise.

(5)   The liquid film thins along the channel by dint of evapora-
      tion and entrainment of droplets from it, and ultimately,
      the film dries up, giving rise to a region of poor heat
      transfer. This region is propagated upstream as the heat
      flux is increased and ultimately penetrates the region of
      subcooled boiling, where the transition is of a somewhat
      different nature. This transition, known as "burnout", is
      shown as line ZZ on Figure 6.16.

(6)   In the region beyond burnout the wall temperature can be
      very high, and locus VV might be representative of the melt-
      ing of the tube. Note that to obtain burnout at low

qualitites, it is necessary to use rather short tubes to prevent physical melting.

An alternative view of the region bounded by the lines YY and ZZ is presented by Mesler (1977). He postulates that the increase in heat transfer coefficients in this region is due to enhancement of nucleate boiling as the film thins, rather than the suppression of nucleate boiling and increased forced convective heat transfer. He bases this hypothesis on results for static films where earlier data suggested an increase in nucleate boiling coefficient with decreasing film thickness; however, more recent data by Tolubinsky (1978) shows evidence for nucleation suppression even with static films.

Figure 6.16 shows the sequence of events for *increasing* heat flux. The situation is not necessarily reversable when the heat flux is *decreased*. In particular:

(a)   There may be hysteresis in the nucleate boiling, higher coefficients being maintained in reducing the heat flux.

(b)   The film boiling regime may be maintained on reducing the flux. However, if burnout has occurred due to film dryout then the hysteresis is small.

Nomenclature

a	Constant in van der Waals equation (equation 6.2)	Nm^4
A	Channel cross-sectional area	m^2
b	Constant in van der Waals equation (equation 6.2)	$m^3/kmol$
b_0-b_3	Distances defined in Figure 6.7	m
C	Number of molecules in system (equation 6.20)	-
D	Diffusion coefficient	m^2/s
G	Mass flux	kg/m^2s
G_c	Condensation mass flux	kg/m^2s
$\Delta G(r)$	Free energy of formation of bubble of radius r	J
h_i	Interfacial heat transfer coefficient	W/m^2K
H	Factor defined by equation 6.23	-
i_G	Saturation enthalpy of vapour	J/kg
i_L	Saturation enthalpy of liquid	J/kg
i_{TP}	Enthalpy of two phase mixture	J/kg
k	Mass transfer coefficient	m/s.Pa
M	Molecules weight	kg/kmol
N_r	Number of nuclei of radius r	-
p	Pressure	Pa
p_{Gb}	Partial pressure of gas in bulk mixture	Pa

p_{Gi}	Partial pressure of gas at interface	Pa
p_{Vb}	Partial pressure of vapour in bulk mixture	Pa
p_{Vi}	Partial pressure of vapour at interface	Pa
p_L	Pressure in liquid	Pa
p_G	Pressure in vapour	Pa
$(p_G)_{lm}$	Log mean gas pressure difference (equation 6.28)	Pa
p_∞	Saturation pressure for a planar interface at the system temperature	Pa
r	Bubble radius	m
r_a	Cavity radius	m
r_b	Bubble radius in cavity (Figure 6.7)	m
$r*$	Critical bubble radius for growth	m
R	Gas constant	Nm/kmol
T	Temperature	K
T_G	Gas temperature	K
T_L	Liquid temperature	K
T_{SAT}	Saturation temperature	K
ΔT_{film}	Film temperature drop	K
ΔT_{int}	Interfacial temperature drop	K
ΔT_{SAT}	Difference between system temperature and saturation temperature	K
ΔT_{SUB}	Difference between T_{SAT} and system temperature	K
u	Drift velocity (Figure 6.10)	m/s
v_G	Gas (vapour) specific volume	m^3/kg
v_L	Liquid specific volume	m^3/kg
W_G	Gas mass rate of flow	kg/s
W_L	Liquid mass rate of flow	kg/s
x	Quality	-
y	Distance from wall	m
z	Parameter defined by equation 6.22	-
β	Cavity angle (Figure 6.6)	-
δ	Laminar film thickness	m
k	Boltzman constant	J/K
σ	Surface tension	N/m
σ_a	Accommodation coefficient	-
λ	Latent heat of vapourisation	J/kg
Θ	Contact angle	

References

Baker, O, (1954) "Simultaneous flow of oil and gas." *Oil Gas Journal*, **53**, p. 185.

Bankoff, S G, (1957) "Ebullition from solid surfaces in the absence of a preexisting gaseous phase." *Trans*. ASME, **79**, 735-740.

Colburn, A P, and Hougen, O A, (1934) "Design of cooler condensers for mixtures of vapours with noncondensing gases." *Ind. Eng. Chem.*, **26**, p. 1178.

Collier, J G, (1972) "Convective boiling and condensation." McGraw-Hill Book Company, New York.

Kottowski, H M, Grass, G, and Spiller, K H, (1974) "Contribution to the boiling nucleation mechanism on heating surfaces for liquid metals." *Proc. 5th Int. Heat Transfer Conf*. **4**, no. B7.2, 310-314.

Mesler, R B, (1977) "An alternate to the Dengler and Addoms convection concept of forced convection boiling heat transfer." *AIChE J.*, **23**, No. 4, 448-453.

Tolubinsky, V I, Antonenko, V A, and Ostrovsky, Ju N, (1978) "Heat transfer at liquid boiling in thin films." Int. Seminar Momentum Heat and Mass Transfer in Two-Phase Energy and Chemical Systems, Dubrovnik, Yugoslavia, 4-9 Sept. 1978, Spons. *ICHMT*.

Chapter 7

Pool Boiling

A. E. BERGLES

7.1 Introduction

Pool boiling represents the traditional starting point for a discussion of heat transfer in boiling systems. With pool boiling it is possible to minimize the number of variables which must be considered in an experimental apparatus or analytical formulation. Due to extensive research effort within the past thirty years, the mechanism of pool boiling is relatively well understood. However, it is still not possible to predict heat-transfer characteristics for this simplest of boiling systems with the precision associated with single-phase systems. This is due largely to the inherent difficulty of quantitatively describing all the important surface and fluid characteristics.

7.2 The Boiling Curve

The first complete characteristics of pool boiling were reported by Nukiyama in 1934. The popular version of his "boiling curve" is chosen here as the basis of discussion since it is still the most meaningful representation.

As shown in Figure 7.1, the boiling characteristics are generally represented as a log-log plot of heat flux versus wall superheat. The data represent saturated pool boiling of degassed liquid, which provides a suitable reference case since it involves the least number of system variables. The boiling curve can be traced out entirely by heaters with constant wall temperature (high temperature fluid or condensing vapor as the heat source) or partially with a constant heat flux heaters (electrical heating). The basic regions and points on the boiling curve are now described.

I. *Single Phase Natural Convection*. Conduction with convection occurs at the heated surface. There is then convective transport to the vicinity of the free liquid surface where the energy is transported by conduction with convection and evaporation (Heidrich (1931)).

(a) *Incipient boiling*. The first bubbles form at the heated surface depart, and rise to the free surface. In practice, the initial bubbles may consist largely of air trapped in surface cavities. It is also difficult to maintain a saturated pool at low heater

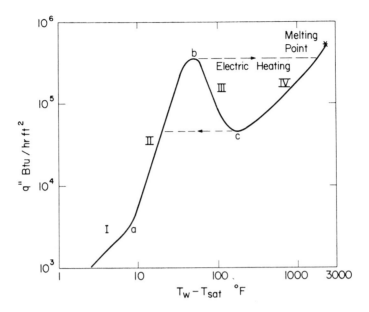

Figure 7.1: Typical boiling curve for saturated pool boiling of water at
atmospheric pressure.

power without the use of auxiliary heaters, thus, vapor bubbles
may condense before reaching the surface.

II. *Nucleate Boiling.* Bubbles form at many favored sites on the
surface. At lower heat flux, individual bubbles can be distinguished and there is little interaction between bubbles generated
from adjacent sites. At higher heat flux, the bubble generation
rate is so high that continuous columns of vapor appear, with
apparent interaction between these columns.

(b) *Peak nucleate boiling heat flux condition.* The vapor generation rate becomes so high that it restricts liquid flow to the
surface, whereupon the surface becomes blanketed with a vapor
film.

III. *Partial Film Boiling.* The vapor film is unstable, collapsing
and reforming under the influence of convection currents and surface
tension. Large vapor bubbles originate at the outer edge of the
film and at the fluctuating locations where liquid touches the
surface. As the surface temperature is increased, the average
wetted area of the surface decreases and a lower heat flux is
obtained.

(c) *Minimum film boiling condition.* A continuous film just covers
the heated surface at this condition, frequently referred to as
the Leidenfrost point.

IV. *Stable Film Boiling.* An orderly bubble discharge occurs from
the edge of the vapor film covering the surface; however, the
shape of the interface varies continuously. At higher surface

temperatures, radiation supplements film conduction.

All regions of the boiling curve can be exhibited with a
constant temperature heating system. For instance, condensing
steam at various pressures might be used to generate the curve of
Figure 7.1 up to the lower region of film boiling. The transient
calorimeter or quenching technique has also been used for some
systems (Merte and Clark (1964)) however, significant alterations
in the curve due to transient effects have been noted (Bergles
and Thompson (1970), Peyayopanakul and Westwater (1978)). With a
nearly constant heat flux system, such as electric heating, it is
not possible to operate in Region III. When the power is increased
to point b, a first-order instability ensues and the operating
point shifts rapidly to the film boiling region. For many systems,
this new operating point corresponds to a temperature greater than
the melting temperature; hence, point b is commonly referred to as
the burnout point. If operation in the film boiling region is
obtained, the power may be increased to actual burnout or reduced
to the Leidenfrost point, where the system reverts back to operation in the nucleate region. Successful attempts have been made
to develop electrically heated systems with appropriate feedback
control so that Region III can be studied, e.g. Peterson and
Zaalouk (1971) and Sakurai and Shiotsu (1974).

A summary of present concepts regarding the mechanism, analysis,
and prediction of the various regions of the boiling curve is now
presented. The discussion is oriented toward providing representative data and correlations which might be used for design. No
attempt is made to incorporate all of the available analytical
models or experimental observations.

7.3 Natural Convection

The initial portion of the boiling curve is predicted by
standard data for heat transfer with free convection. Correlations
are generally presented in the form

$$Nu = f(Gr.Pr) \tag{7.1}$$

where the properties are usually evaluated at the film temperature,
$(T_w + T_b)/2$. The functional relationship is such that the
characteristic length can be loosely specified; for example, "a
side" in the case of the horizontal plate. Figure 7.2 presents
curves based on available data for the common pool boiling
geometries: vertical plate, horizontal cylinder, and horizontal
plates with the heated side up. Numerous linear approximations to
the curves are usually presented for the laminar and turbulent
regions; however, it would appear that a single plot is more
useful, especially since there is no distinct break between laminar
and turbulent flow.

Once the heat-transfer coefficient is obtained, the curve for
Region I can be obtained since

$$q" = h(T_w - T_{sat}) \tag{7.2}$$

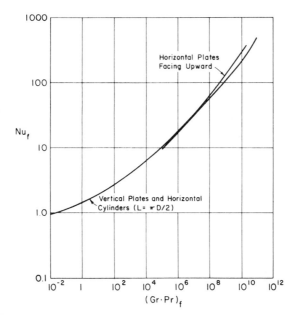

Figure 7.2: Correlation of natural convection data (McAdams (1954)).

The heat-transfer coefficient for natural convection thus governs the entry into boiling conditions.

7.4 Nucleation*

Metastable States

In nucleate boiling we observe two separate processes - the formation of bubbles (nucleation) and the subsequent growth and motion of these bubbles. In general, nucleation may be either of the homogenous or heterogenous variety; however, both types involve metastable states. Figure 7.3 depicts the superheated liquid state A where the liquid is heated above the saturated temperature T_1 for the constant system pressure p_1. This is referred to as $(T_2 - T_1)$ degrees of liquid superheat; or for the corresponding subcooled vapor state B, $(T_1 - T_3)$ degrees of vapor subcooling. Nucleation theory and experience are concerned with finding the superheat required to initiate vapor formation.

Homogeneous Versus Heterogeneous Nucleation

Several theories based on statistical mechanics have been proposed to account for homogeneous nucleation in a pure liquid. One approach using classical rate theory, e.g. Volmer (1939),

---

*Cole (1974) can be consulted for more details about many of the concepts presented in this section.

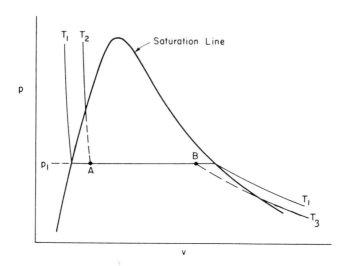

Figure 7.3: Illustration of metastable states for liquid and vapor.

presumes that numerous molecules have the activation energy
required for existence in the vapor phase. These energetic
molecules could combine through collisions to form a cluster,
which is then a vapor bubble. Theories of this type yield
extremely high liquid superheats for nucleation in a pure liquid,
for example, 90°F for water at 1 atm. Such high superheats are
contrary to experimental observations with real systems.

       In a real system, of course, the liquid contains foreign
particles and dissolved gas which could act as nuclei. The
predicted nucleation superheats would be considerably less in the
presence of a pre-existing gas phase. This form of homogeneous
nucleation implies that vapor formation would be noted at random
points where the nuclei happen to be located. In actual practice,
however, bubbles form at specific locations associated with the
heated surface, not the fluid. It has furthermore been found by
microscopic observation that these locations are small imperfections or cavities on the heated surface (Clark et al (1959)).
For practical boiling systems, then, one is forced to discard
homogeneous nucleation theories and concentrate upon heterogeneous
or cavity nucleation.*

Nucleation From Cavities

       Consider an idealized conical cavity with a pre-existing gas
phase, presumed to be pure vapor for the moment, as shown in
Figure 7.4. A wetting situation, $\beta < 90°$, is presumed. For a

_____

*An understanding of homogeneous nucleation is required for the
 analysis of "vapor explosions" which may occur when a hot liquid
 is brought into intimate contact with a cold liquid. The industrial safety aspects of this phenomenon are discussed by Porteous
 and Reid (1976) and a review of research is presented by Bankoff
 (1978).

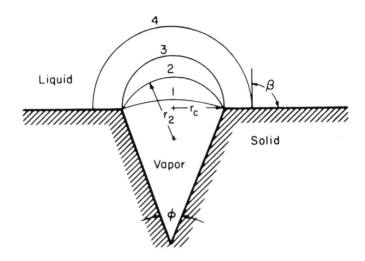

Figure 7.4: Vapor bubbles in a conical cavity with $\beta = 90°$

segment of a spherical bubble, mechanical equilibrium requires
that

$$P_{vr} - P_\ell = \frac{2\sigma}{r} \qquad (7.3)$$

where $P_{vr} = f(T,r)$.[1] For the radii of curvature of interest, it
can be shown that $P_{vr} = f(T)$ (Keenan (1941)). It can also be
demonstrated that thermodynamic equilibrium requires that $T_\ell = T_v$.
At equilibrium, then, it is evident that the liquid temperature
is superheated with respect to the liquid pressure.

    In order to arrive at a general expression for the superheat
required for nucleation, it is necessary to relate T and p along
the saturation line. It is most convenient to employ the Clapeyron
relation:

$$\frac{h_{fg}}{v_{fg}\,T} = \frac{dp}{dT} \qquad (7.4)$$

Assuming that the quantity $h_{fg}/v_{fg}\,T$ is constant, with $T \sim T_{sat}$,[2]

---

[1] The same relation holds for the usual, non-conical cavity; however,
2/r is to be interpreted as the sum of the inverse principal
radii of curvature.

[2] Many other approximations have been employed, such as introduction
of the perfect gas approximation or use of another p-T relation
for the saturation line. The present result is generally
reliable for moderate pressures.

this relation can be integrated to yield

$$P_v - P_\ell = \frac{h_{fg}}{v_{fg} T_{sat}} (T_v - T_{sat})$$
(7.5)

Substituting equation (7.5) in equation (7.3), the relation for the equilibrium superheat is obtained:

$$T_v - T_{sat} = \frac{T_{sat} v_{fg}}{h_{fg}} \frac{2\sigma}{r} = T_\ell - T_{sat}$$
(7.6)

which is identical to equation (6.15), which was developed using some of the alternate approximations noted above. This equation indicates that as the superheat is raised, the radius of curvature decreases, from (1) to (2) in Figure 7.4, for example. It is noted, however, that the minimum radius corresponds to the hemispherical state (3) with the radius equal to the cavity mouth radius. With further increases in superheat (4), the bubble is no longer stable since its radius must increase in violation of the requirement for mechanical equilibrium. The bubble "nucleates" and grows at a rate dependent upon the rate of heat transfer from the surrounding liquid, shortly becoming visible to the naked eye. It is particularly significant that the superheat required for incipient boiling is dependent on a single dimension of the cavity, the mouth radius. The validity of equation (7.6) has been established by Griffith and Wallis (1960), who manufactured cavities of a known size on a copper surface and produced nucleation by uniformly superheating the water pool.

It is evident that equation (7.6) will not be adequate for predicting the incipience of boiling in the usual pool with a submerged heater where a temperature gradient exists in the liquid adjacent to the heated surface. Nucleation cannot take place from a cavity unless the environment that the bubble sees is sufficiently high in temperature. Two basic assumptions must be made regarding the liquid temperature profile and the relation of the liquid temperature to the vapor temperature. The following illustrates how this problem might be handled.

Assume that the liquid temperature profile is linear:

$$T_\ell = T_w - q''y/k_\ell$$
(7.7)

Further assume that the liquid at all points adjacent to the bubble must be at a higher temperature than the required vapor temperature. Referring now to Figure 7.5, it is seen that the liquid and vapor temperature are related as follows:

$$T_\ell = T_v \text{ and } \frac{d T_\ell}{dy} = \frac{d T_v}{dr} \text{ at } y = r_c$$
(7.8)

Utilizing equations (7.6) and (7.7), the cavity size at incipient boiling is

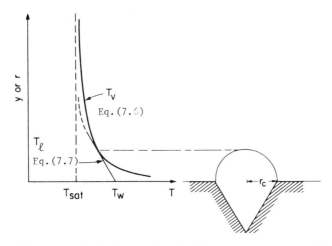

Figure 7.5: Conditions required for nucleation in a temperature gradient.

$$r = \sqrt{\frac{T_{sat} \; v_{fg} \; 2\sigma \; k}{h_{fg} \; q''}} \qquad\qquad (7.9)$$

and the heat flux - wall superheat relation is

$$q'' = \frac{h_{fg} \; k_\ell}{8 \; T_{sat} \; v_{fg} \; \sigma} \; (T_w - T_{sat})^2 \qquad\qquad (7.10)$$

To apply this criterion, it is convenient to plot equation (7.10) using the coordinates of Figure 7.1; the intersection of this equation with the convection relation, equation (7.2), defines the point of incipient boiling. In practice, however, this simple criterion is not accurate for pool boiling systems. The liquid temperature has a characteristic transient profile which is distorted by the presence of the bubble and circulation currents in the bulk liquid (Hsu (1962) and Han and Griffith (1965)). In addition, the large cavities called for may not be available as nucleation sites. Both of these factors help to explain why pool boiling incipient superheats for degassed systems are higher than those predicted from equation (7.10).

Inert gas is frequently present in boiling systems, either in solution or trapped in surface cavities. It is quite possible that homogeneous nucleation can occur with large dissolved gas concentration. The gas is essentially driven out of solution near the heated surface, presenting the appearance of normal boiling. The previous analysis can be modified to account for the presence of an inert gas. Equation (7.3) becomes

$$P_v - P_\ell = \frac{2\sigma}{r} - P_g \qquad\qquad (7.11)$$

where p_g is the partial pressure of the inert gas in the homogene-
ouse or heterogeneous nucleation site. Following a similar
development the equilibrium superheat is obtained:

$$T_v - T_{sat} = \frac{T_{sat}\ v_{fg}}{h_{fg}} \left(\frac{2\sigma}{r} - p_g\right) \qquad (7.12)$$

The inert gas thus reduces the superheat required for nucleation;
in fact, nucleation can occur for temperatures below saturation.
Unfortunately, it is virtually impossible to determine p_g, so
only a qualitative assessment can be made.

Cavity Stability

It is appropriate to examine under what conditions a surface
imperfection becomes a nucleation site. When the pool is filled,
air can be trapped in a cavity when the advancing liquid touches
the far side of the groove or pit before touching the base (Bankoff
(1959)). This criterion can be stated in terms of the contact
and cavity angles shown in Figure 7.4 as $\beta > \emptyset$. Nucleation
proceeds according to the previous analysis; however, the point of
incipient boiling shifts to higher wall superheat as the air
becomes depleted. Eventually a steady-state condition is reached
where there is only vapor trapped in the cavity.

Boiling systems are normally operated so that there are
repeated cycles of heating and cooling. If the air in a conical
cavity has been exhausted, and the temperature is reduced below
the saturation temperature, the requirements for equilibrium can
no longer be satisfied and the interface recedes steadily into
the cavity until there is no vapor left. The nucleation site
has then been deactivated or "snuffed out". Re-entrant cavities,
either natural or artificial can remain active for considerable
subcooling. As indicated in Figure 7.6, the square re-entrant
cavity retain vapor for a subcooling dependent on the inner mouth
radius in accordance with equation 7.6. It is noted, however,
that the degree of subcooling is reduced as the contact angle
becomes smaller. For the alkali liquid metals, where $\beta \sim 0$, the
cavity of Figure 7.6a will flood for any subcooling. The doubly
re-entrant cavity shown in Figure 7.6b has been suggested to
remedy the notorious nucleation instability of these fluids,
since the cavity will sustain subcooling corresponding to the

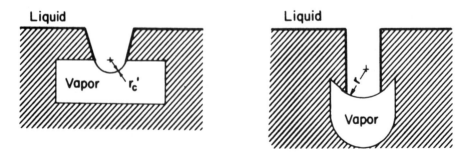

Figure 7.6: Re-entrant cavities used to improve boiling stability.

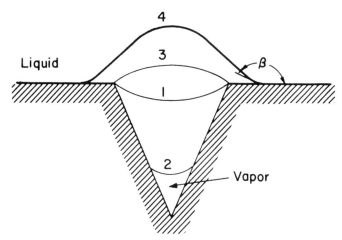

Figure 7.7: Vapor bubbles in a conical cavity with $\beta > 90^{\circ}$.

indicated radius of curvature. The stability of cavities
containing non-condensible gases is discussed by Mizukami (1979).

     Cavity stability is much enhanced if nonwetting conditions are
present. From Figure 7.7 it is seen that the cavity can retain
vapor for virtually any amount of subcooling (1), (2). It is
noted, however, that the cavity radius no longer defines the point
of incipient boiling, since the bubble becomes unstable before
reaching the hemispherical state (3). The boiling process is
generally erratic due to the tendency of the bubble to cover a
large portion of the surface (4). Such behavior is encountered
with mercury or when boiling from teflon or oily surfaces.

Determination of Active Nucleation Sites

     As noted in the following section, all theoretical models of
nucleate boiling need input data on the size distribution of
active cavities. Visual observation of the surface during
boiling is not reliable as the bubbles obscure the surface, except
at very low heat flux. Post boiling observations of deposits
surrounding nucleation sites, e.g. Heled and Orell (1967), are
likely to be suspect due to modification of the nucleation sites
by deposits. It is difficult to obtain quantitative information
by analyzing surface roughness parameters or by studying photographs such as those obtained by a scanning electron microscope
(SEM), as it is not clear which "pits" represent potential active
nucleation sites. The work by Nail et al (1974) is an example of
efforts to examine surfaces by SEM. Recently, Kopp (1976) claimed
success in relating surface roughness and SEM observation to
boiling performance.

     Brown (1967) suggested that population data for boiling
nucleation sites could be obtained by studying air/vapor bubble
nucleation in supersaturated gas solutions. The growth rate of
gas bubbles is so slow that it is possible to count nucleation
sites directly or from photographs. Lorenz et al (1975) modified
the technique to allow for thermodynamic non-equilibrium, and
found fair agreement between effective cavity radii derived from

boiling experiments, study of photomicrographs, and the gas
diffusion tests - all for the same copper surface in water.
Smaller effective radii are observed with organic fluids, possibly
due to penetration of the liquid into the cavity (Lorenz et al
(1974), and Singh et al (1976)). Eddington et al (1978 and 1979)
made further refinements and applied this technique to low velocity
flow boiling; however, the discrepancies between gas diffusion and
boiling tests were still large.

7.5 Saturated Nucleate Pool Boiling

Description of Heat Transfer in the Isolated Bubble Regime

Upon reaching the incipient boiling condition, further heat
transfer promotes bubble growth, due to the excess vapor pressure
that is no longer balanced by the surface tension forces. The
growth of bubbles is a dynamics problem coupled with a heat-transfer
problem. Considerable effort has been devoted to examining the
behavior of single bubbles. These isolated bubbles which are
amenable to analysis, are found at low heat flux.

In general, heat is transferred from the wall to the liquid
so as to establish a superheated layer. This layer represents
free convection heat transfer, perhaps augmented by "bubble
agitation", over that portion of the surface not directly affected
by bubbles. Transient free convection is observed prior to
nucleation or in the vicinity of the bubble after nucleation.
Directly under the bubble, there is heat transfer from the wall to
liquid or to vapor.

After nucleation, bubble growth is promoted by heat transfer
from the superheated layer and vaporization of a "microlayer" of
liquid between the bubble and the wall. There may also be
evaporation at or near the base of the bubble and condensation at
the top of the bubble ("latent heat transport"). Eventually, the
bubble grows to the point where it departs due to buoyancy, carrying with it a substantial portion of the superheated liquid layer
("bubble pumping" or "microconvection").

A miniature thermocouple (Moore and Mesler (1961)) or thin
film sensor (Cooper and Lloyd (1966)) installed near a nucleation
site records temperature fluctuations, $T_w(t)$, which would not be
noted by the usual instrumentation, which records merely $\overline{T}_w$.
These temperature fluctuations and associated bubble states are
shown schematically in Figure 7.8. We start with the condition
subsequent to incipient boiling where the bubble has grown rapidly
outside the cavity. The microlayer vaporizes rapidly, producing
a drop in surface temperature (1). When the layer has evaporated,
the surface rises in temperature (2) since heat transfer to the
vapor is poor. During this period the bubble is also growing due
to evaporation at the interface between the bubble and the superheated liquid layer. When the bubble departs, the surface is
quenched by cooler bulk liquid and there is a sharp drop in
temperature (3). The thermal boundary layer is then re-established
and the surface rises in temperature until conditions are again
suitable for the incipience of boiling (4). The temperature oscillations depend strongly upon the system; for instance a glass
heater would produce substantial fluctuations, while the

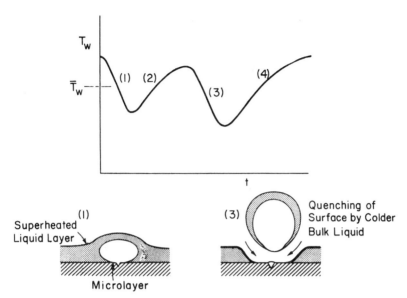

Figure 7.8: Surface temperature variation in vicinity of a nucleation site with corresponding bubble states.

fluctuations would be small for a metal heater.

The relative importance of these energy transport mechanisms varies according to the boiling system. Graham and Hendricks (1967), for example estimated that microlayer evaporation accounted for about 50% of the wall heat transfer in the case of methanol but only about 25% of the wall heat transfer in the case of water. Microlayer evaporation appears to be important in boiling of sodium (Deane and Rohsenow (1970)).

Bubble Growth and Departure

Numerous attempts have been made to formulate and solve reasonable mathematical models for the growth of vapor bubbles. The classic formulation of Rayleigh, originally applied to cavitation bubbles, has been utilized to examine the initial stage of bubble growth which is controlled by surface tension and inertia. During the latter stage of growth, the problem can be formulated relatively simply in terms of heat transfer since the growth rate is controlled by the rate of evaporation at the interface which, in turn, depends on heat conduction from the surrounding superheated liquid layer.

The following expression, derived by Mikic et al (1970), was found to satisfactorily describe experimental results for bubble growth in a uniformly superheated liquid:

$$R^+ = \tfrac{2}{3} \left[(t^+ + 1)^{\frac{3}{2}} - (t^+)^{\frac{3}{2}} - 1 \right] \tag{7.12}$$

where

$$R^+ = \frac{AR}{B^2}$$

$$t^+ = \frac{A^2 t}{B^2}$$

$$A = \left[b \frac{(T_\infty - T_{sat}) h_{fg} \rho_v}{T_{sat} \rho_\ell} \right]^{\frac{1}{2}}$$

$$B = \left[\frac{12}{\pi} Ja^2 (\frac{k}{\rho c_p})_\ell \right]^{\frac{1}{2}}$$

and $b = 2/3$ for spherical bubble growth in an infinite medium, while $b = \pi/7$ for a spherical bubble touching a plane surface. The following result is obtained for growth of a spherical bubble at a heated wall for $t^+ \gg 1$ (Mikic and Rohsenow (1969)):

$$R^+ = (t^+)^{\frac{1}{2}} \left\{ 1 - \frac{T_w - T_b}{T_w - T_{sat}} \right\} \left[(1 + \frac{t^+_w}{t^+})^{\frac{1}{2}} - (\frac{t^+_w}{t^+})^{\frac{1}{2}} \right] \qquad (7.13)$$

The bubble departure size is an important parameter in the study of pool boiling since it has a direct bearing on the heat-transfer characteristics. Fritz (1935), utilized the theory of capillarity to get an equilibrium bubble shape and formulated a differential equation representing a balance of gravity and surface-tension forces. An approximate solution to this equation yielded the bubble diameter at departure:

$$D_b = 0.0148\beta \left[\frac{2 g_c \sigma}{g(\rho_\ell - \rho_v)} \right]^{\frac{1}{2}} \qquad (7.14)$$

which has been verified for numerous systems including steam and hydrogen bubbles in water.

Prediction of the Boiling Performance of a Surface

With the preceding information, it is possible to combine the individual processes of bubble inception, growth, and departure to predict the heat-transfer performance of a boiling surface. It is instructive to outline the method of Han and Griffith (1965a, 1965b even though it will become apparent that the procedure is too involved to be used for engineering calculations.

The model is formulated for the isolated bubble regime as indicated in Figure 7.9. An idealized grid of nucleation sites is postulated which defines the so-called bulk convection area. The mechanism of heat removal from the surface consists of two parts: 1) the enthalpy transport represented by the repeated removal (bulk convection) of the superheated layer in the vicinity of the bubbles, and 2) continuous removal of heat by the usual free-

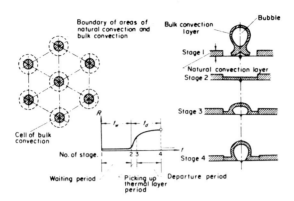

Figure 7.9: Model for nucleate boiling in the isolated bubble region (Han and Griffith (1965b)).

convective process in the area uninfluenced by the bubbles. Referring to the earlier discussion of the mechanism of nucleate pool boiling, it is seen that only Regions 3 and 4 of Figure 7.8 are considered.

At Stage 1 the bubble departs, carrying with it a section of the superheated transient thermal layer. Colder liquid from the bulk of the pool quenches the heated surface and the transient thermal layer is reformed. A waiting period is required before the layer is superheated sufficiently to activate the cavity (Stage 2). The bubble then grows (Stage 3) until the departure diameter is reached (Stage 4), at which point the cycle is repeated. An experiment indicated that the area from which the superheated liquid is pumped away corresponds to twice the bubble departure diameter. Assuming only pure conduction to the superheated liquid layer in the area of influence, the problem was modelled as conduction to a semi-infinite body with a step change in temperature at the surface. The superheated layer is replaced with a frequency corresponding to the frequency of bubble departures.

In somewhat simplified form, the final expression for the heat flux is

$$q'' = q''_{nc} + q''_{bc} \tag{7.15}$$

$$q''_{nc} = (1 - \frac{\pi}{4} n (2D_b)^2) h_{nc} (T_w - T_b) \tag{7.16}$$

$$q''_{bc} \simeq 2 [\pi (k\rho c)_\ell f]^{\frac{1}{2}} D_b^2 n (T_w - T_b) \tag{7.17}$$

where an average departure frequency is presumed to be valid for the bubbles issuing from the n active sites per unit area of the heating surface. To use this equation it is necessary to specify f, D_b, and $n(r)$. Han and Griffith (1965b) obtained the frequency from transient conduction calculations, in essence getting the time required for nucleation of a specified cavity (t_w) and the time required for the bubble to grow to departure size (t_d). The

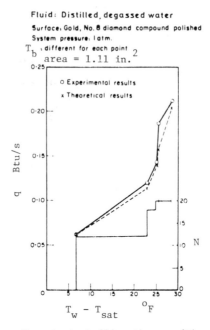

Figure 7.10: Comparison of nucleate boiling theory with experimental data (Han and Griffith (1965b)).

departure diameter was obtained from the Fritz relation, equation (7.14); however, the contact angle was taken as the measured dynamic value (from motion pictures) at the average bubble growth rate. The cavity size distribution was inferred from the wall heat flux and wall superheat at which incipient boiling was observed. The formulation thus represents a "two adjustable parameter" model. As shown in Figure 7.10, the agreement between measured and predicted heat transfer rates is quite satisfactory.

    Mikic and Rohsenow (1969) modified this procedure to take into account the generally observed functional form of the cavity size distribution, $n \sim 1/r^m$. Since the departure diameter and the frequency were taken from correlations not dependent on the particular boiling system, the only input to the model is the cavity size distribution. This "one adjustable parameter" model was tested against the boiling data of Gaertner and Westwater (1960) who had available the active cavities as a function of wall superheat (obtained with an electro-plating technique). As shown in Figure 7.11, the agreement is quite good. As can be seen from this Figure and Figure 7.10, the method works quite well for heat fluxes well beyond the isolated bubble region. Another test of the correlation was to deduce the cavity size distribution from the boiling curve at one pressure and then use this distribution to predict the data at other pressures. The Mikic-Rohsenow model was subsequently utilized by Lorenz et al (1974) who demonstrated the importance of geometry (cone angle) and contact angle in deter-mining the effective cavity radius for fluids with contact angle β > 30° (See section "Cavity Stability").

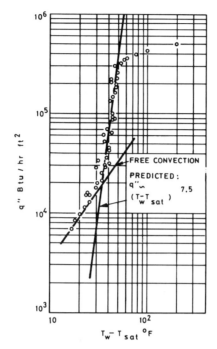

Figure 7.11: Comparison of measured (Gaertner and Westwater (1960)) and
             predicted (Mikic and Rohsenow (1969)) data for nucleate boiling
             of water.

Flow Regime Transition in Nucleate Boiling

    As the heat flux is increased, the number of active sites
increases and the bubble frequency is raised to the point where
continuous vapor columns are observed. Using basic hydrodynamic
information concerning bubble behavior, including equation (7.14),
Moissus and Berenson (1963) arrived at an expression for the
transition from isolated bubbles to vapor columns for horizontal
surfaces:

$$q''_{tr} = 0.11 \, \rho_v \, h_{fg} \beta^{\frac{1}{2}} \left[\frac{g\sigma}{\rho_\ell - \rho_v} \right]^{\frac{1}{4}} \qquad (7.18)$$

where the empirical constant appears to be valid for a fairly wide
range of fluid-surface combinations. In the column region, one
would suspect that the mechanism of heat removal is substantially
different than the pumping model proposed for the isolated bubble
region. As noted in the previous section, however, there is no
particular change in the boiling characteristics at this transition, which occurs at the lower end of the fully-developed portion
of the boiling curve.

Nucleate Boiling Trends

    The preceding discussion has given some idea of the variables
which have an effect on nucleate boiling. A more complete summary

is given below.

Fluid State - The fluid properties p, v_{fg}, h_{fg}, σ, k, ρ, c, and the fluid-surface property β are important for the analysis in the section *"Prediction of the boiling performance of a surface"*. In addition, the dissolved-gas concentration affects nucleation as noted in the section *"Homogeneous versus heterogeneous nucleation"*.

Surface Condition - The material properties k, ρ, c are important in certain cases. Mechanical properties have a bearing on how a particular finishing operation produces the characteristic cavity size distribution. The amount of gas trapped in the cavities also affects nucleation. Surface coating, oxidation, or fouling can markedly affect the wettability and effective cavity sizes.

Heater and Pool Geometry - These variables determine the pool convective conditions which have an important bearing on the natural convection heat-transfer coefficient as well as bubble motion.

Body Forces - Signficant differences in boiling behavior have been noted in fractional- and multi-gravity situations.

Method of Heating - The heating method is immaterial except for the case of a.c. resistance heating which may cause bubble generation in phase with the power supply.

History - Due to the peculiarities of nucleation, hysteresis behavior may be noted where the boiling curve is different when the heat flux is increased than when it is reduced. An ageing phenomenon can also occur as the surface becomes degassed or contaminated.

Several sets of data illustrate certain of these effects. Figure 7.12 demonstrates the effect of relatively large-scale surface roughness (5), oxidation (2), and ageing due to surface degassing (4, 6, 7). Additional roughness effects are illustrated in Figure 7.13, where it is seen that a four-fold increase in the boiling heat transfer coefficient can be obtained by choosing a lap over a mirror finish. Data which demonstrate hysteresis are presented in Figure 7.14. It is apparent that neither a precise description nor a universal correlation will be possible for nucleate boiling due to the number of variables, in particular, those which might be termed nuisance variables.

Fully Developed Nucleate Boiling Correlation

Numerous attempts have been made to correlate that portion of the boiling curve which is log-linear. The most successful models are semi-analytical and involve a form of dimensional analysis. The most widely used correlation, developed by Rohsenow (1952) is described here. Adopting the argument, similar to that discussed in the section *"Prediction of the boiling performance of a surface"* that the major portion of the heat is transferred directly from the surface to the liquid, a conventional combination of dimensionless groups is sought:

$$Nu = f(Re,\ Pr) \qquad\qquad (7.19)$$

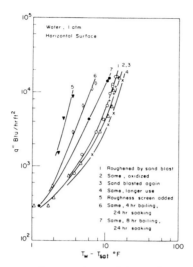

Figure 7.12: Boiling curves for surface with various surface treatment and operating hsitory (Jakob and Fritz (1931)).

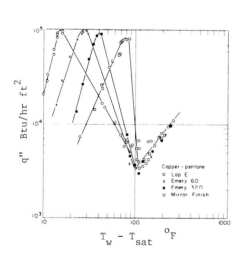

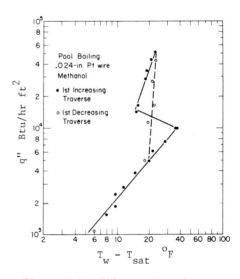

Figure 7.13: Effect of surface finish on boiling curves (Berenson (1962)).

Figure 7.14: Illustration of hysteresis in nucleate boiling (Bankoff et al (1958)).

The forced-convection turbulent flow groups are pertinent since the bubbles are pumping the liquid.

The Reynolds number represents the ratio of bubble inertia to frictional forces on the bubble:

$$Re_b = \frac{G_b \, D_b}{\mu_\ell} \tag{7.20}$$

Use of G_b is permitted since $G_b = G_\ell$, by continuity. The bubble superficial mass velocity is given by

$$G_b = \frac{\pi}{6} D_b^{\,3} \, \rho_v \, f \, n \tag{7.21}$$

A bubble Nusselt number is given by

$$Nu_b = \frac{q'' D_b}{(T_w - T_{sat}) k_\ell} \tag{7.22}$$

The latent heat transfer of the bubbles is represented as

$$q''_b = h_{fg} \frac{\pi}{6} D_b^{\,3} \, f \, \rho_v \, n \tag{7.23}$$

Experiments indicate that for a wide range of conditions,

$$q'' \sim n \tag{7.24}$$

Thus,

$$q'' = C_q \, q''_b \tag{7.25}$$

The Prandtl number of the liquid is appropriate:

$$Pr_\ell = \left(\frac{c\mu}{k}\right)_\ell \tag{7.26}$$

Introducing equation (7.14) for D_b, and formulating the functional relationship as

$$\frac{Re_b \, Pr_\ell}{Nu_b} = C \, Re_b^{\,n} \, Pr_\ell^{\,m} \tag{7.27}$$

the final correlation can be obtained as

$$\frac{c_\ell (T_w - T_{sat})}{h_{fg}} = \underbrace{C_q \left[\frac{\sqrt{2} \; 0.0148\beta}{C_q}\right]^n}_{C_{sf}} \left[\frac{q''}{h_{fg}\mu_\ell} \sqrt{\frac{\sigma}{g(\rho_\ell - \rho_v)}}\right]^n \left[\frac{c\mu}{k}\right]_\ell^m \tag{7.28}$$

The collection of constants preceding the slightly redefined Reynolds number function is designated as C_{sf}, a constant reflecting the condition of a particular fluid and surface combination. Since all properties are evaluated at the saturation temperature, this expression is analogous to the general approximation for fully established boiling:

$$q'' = f(p, \text{ fluid, surface}) (T_w - T_{sat})^{1/n} \qquad (7.29)$$

It was possible to correlate data for a wide range of fluids at different pressures utilizing $n = 0.33$ and $m = 1.7$,[*] where C_{sf} has a different value for each fluid-surface combination as listed in Table 7.1.

Table 7.1

Surface-Fluid Combination	C_{sf}
Water-nickel	0.006
Water-platinum	0.013
Water-copper	0.013
Water-brass	0.006
Carbon tetrachloride-copper	0.013
Benzene-chromium	0.010
n-Pentane-chromium	0.015
Ethyl alcohol-chromium	0.0027
Isopropyl alcohol-copper	0.0025
35% Potassium carbonate-copper	0.0054
50% Potassium carbonate-copper	0.0027
n-Butyl alcohol-copper	0.0030

This "one adjustable constant" correlation is useful in preliminary design. It is, of course, desirable to have some data for the actual system at hand to establish the appropriate C_{sf}. This testing might be done in a simple pool at 1 atm; the correlation then permits extrapolation to high pressure. Some caution is to be advised in performing this extrapolation if such reference experiments indicate that $n \neq 0.33$, since the property groups are affected by this exponent. In actual practice, a range of $n = 0.04$ to 1.0 has been reported.

At this point it is apparent that the designer is still at a loss to predict nucleate boiling without direct information on the fluid and surface of interest. A natural question is: In view of

---

[*]A subsequent re-evaluation of the data indicates that $m = 1.0$ is more accurate over the entire pressure range for *water* (Rohsenow (1973)). This is a consequence of the Prandtl number for water not being a monotonic function of pressure.

the vast amount of data accumulated on nucleate boiling, isn't it
possible to get a correlation by purely statistical means?
According to the recent work of Stephan and Abdelsalam (1980),
the answer is a qualified yes. They identified the physical
properties and variables characterizing the process and developed
a possible set of 13 independent dimensionless groups. A product
relation among these groups was postulated, and a linear regression
analysis was used to obtain the constant and exponents of the
most important groups. The fluids were divided into four categories: water, hydrocarbons, cryogenic fluids, and refrigerants;
about 5000 data points were considered. While only 8 of the
groups were ultimately utilized and only 3-5 groups appeared in
the correlation for a given category, the correlations are quite
involved. Accordingly, a simple set of heat transfer coefficients
was proposed as

$$h = C(q")^n \qquad\qquad (7.30)$$

where the exponent was fixed for each category and the constant is
given as a function of pressure for each fluid. For water, for
example, the expression is

$$h = C(q")^{0.673} \qquad\qquad (7.31)$$

or

$$q" = C(T_w - T_{sat})^{3.06} \qquad\qquad (7.32)$$

which is in excellent agreement with Rohsenow's recommendation of
an exponent of 3.03 for equation (7.29). It is important to note,
however, that the contact angle was fixed (45°) and the surface
roughness does not appear in the correlation even though these
parameters generally have a strong influence on the boiling curve.

7.6 Peak Nucleate Boiling Heat Flux

The peak heat flux is generally the most important piece of
information required for design of a nucleate boiling system. In
spite of intensive research effort, there is still disagreement
regarding the actual mechanism of the peak nucleate or critical
heat flux. Two models of the critical conditions are

Bubble Packing. The number of nucleation sites becomes so numerous that neighboring bubbles or vapor columns coalesce causing a
vapor blanketing of the surface (Rohsenow and Griffith (1956)).

Hydrodynamic Instability. At high heat flux, the number of sites
is so large and the vapor generation rate so high that the area
between the bubble columns for liquid flow to the surface is
reduced. The relative velocity is then so large that the liquid-
vapor interface becomes unstable, thus essentially starving the
surface of liquid and causing the formation of a vapor blanket
(Kutateladze (1951), Zuber (1958), Chang and Snyder (1960), Moissi
and Berenson (1963)). This has also been visualized as a flow
pattern transition where the liquid filaments break up into drops

which are suspended or "fluidized" by the vapor stream (Wallis (1961)). Numerous semi-empirical correlations have been proposed utilizing these models. Several examples can be cited:

Rohsenow and Griffith (1956)

$$q"_{crit} = 143 \ h_{fg} \ \rho_v \left[\frac{\rho_\ell - \rho_v}{\rho_v}\right]^{0.6} (\frac{a}{g})^{1/4} \ \frac{Btu}{hr \ ft^2} \qquad (7.33)$$

Zuber (1958, 1961)

$$q"_{crit} = 0.131 \ h_{fg} \ \rho_v \left[\frac{\sigma(\rho_\ell - \rho_v)g}{\rho_v^2}\right]^{1/4} \qquad (7.34)$$

The constant in equation (7.34) is generally considered to be low; the value of 0.18 has been recommended (Rohsenow (1973)).

While these correlations describe quite well selected sets of data, they are not accurate for all systems. As shown in Figure 7.15, for example, the predicted critical heat fluxes are widely divergent for boiling of liquid oxygen. In this case, the difficulty is probably due to the fact that the constants and/or exponents in the correlations were established from a rather narrow data base.

Considerable progress has been made in accounting for the effects of heater geometry. Lienhard and coworkers (Sun and Lienhard (1970), Lienhard and Dhir (1973), Lienhard, Dhir and Riherd (1973)) re-examined the Zuber model and found that the vapor-removal configuration varies according to the heater geometry and "size". The general correlation is given as follows:

$$\frac{q"_{crit}}{q"_{crit_F}} = f(L') \qquad (7.35)$$

where $q"_{crit_F}$ is given by equation (7.34) and L' is a characteristic length,

$$L' = L \left[g(\rho_\ell - \rho_v)/\sigma\right]^{1/2} \qquad (7.36)$$

The final semiempirical results, summarized in Figure 7.16, indicate that the correlation curves (infinite horizontal flat plate, horizontal cylinder, vertical plate with both sides heated, vertical plate with one side insulated, and sphere) vary according to geometry for "small" heaters and are also generally different for "large" heaters. These correlation curves were based on over 1000 data points; the typical scatter of the data about the recommended curves is ±20%. It should be pointed out that heaters of a practical size are included in the small category. For saturated water at 1 atm boiling under standard gravity, the value of L' at the inflection, or transition, point corresponds to a ¼-in.-diameter cylinder and a 1-in.-diameter sphere.

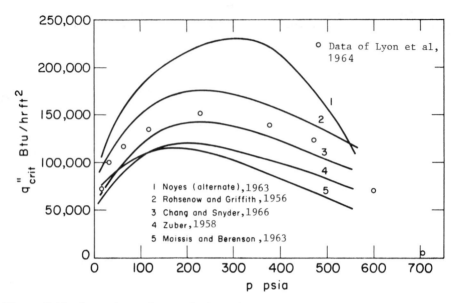

Figure 7.15: Comparison of correlations for critical heat flux of oxygen
(Seader (1965)).

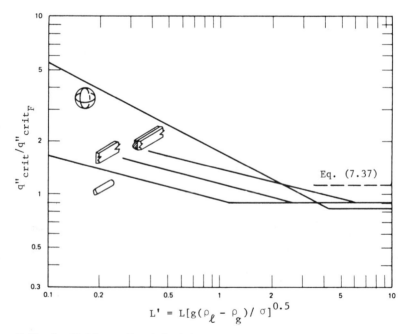

Figure 7.16: Predictions of critical heat flux for various heater configurations
(Bergles (1975)).

Although the experimental verification is difficult, data for large flat plates (facing up) tend to substantiate the semi-empirical prediction for an infinite flat plate:

$$\frac{q"_{crit}}{q"_{crit_F}} = 1.14 \qquad\qquad (7.37)$$

Nucleate boiling and the associated hydrodynamic instabilities vanish for $L' < 0.01$; as heat flux is increased, natural convection proceeds directly into film boiling when the first bubble is nucleated. The region $0.01 < L' < 0.1$ is considered to be a transition region in which the hydrodynamic mechanisms reestablish themselves. However there still seems to be some question as to the behavior of data for $L' \simeq 0.1$. The data of Sun and Lienhard suggest a distinct maximum in $q"_{crit}$ at $L' \simeq 0.1$, whereas the data of many investigators summarized by Tachibana et al. (1967) indicate that $q"_{crit}$ simply increases to the large-cylinder asymptote at $L' \simeq 0.1$.

Additional effects not accounted for in the critical heat flux correlations which are available are usually a *reduction* in $q"_{crit}$ with low thermal capacity heaters

    low thermal capacity heaters

    vertical orientation of long heaters (Bernath (1960))

    non-wetted surfaces

    restriction of the liquid circulation to the heater

    mechanical or metallurgical defects in the heater

and an *increase in* $q"_{crit}$ with

    small amounts of certain additives

    use of d.c. power instead of a.c. power (Scarola (1964))

    rapid increases in power

    certain types of surface fouling

Discussions of those factors not referenced are given by Bergles (1975) or Hsu and Graham (1976).

7.7 **Transition and Film Boiling**

Stable Film Boiling

Film boiling is the most tractable regime analytically since the flow pattern is relatively simple. Exact solutions can be obtained for laminar film boiling in a saturated pool by solving the momentum and energy equation separately, and relating the solutions by means of a heat balance. The result for vertical surfaces can be obtained by what is essentially an inversion of the classifical Nusselt equation for laminar film condensation (Bromley (1950)). The average heat transfer coefficient for a plate of height L is given by

$$h = 0.943 \left[\frac{k_v^3 \, g \, h_{fg}' \, (\rho_\ell - \rho_v)\rho_v}{L \, \mu_v \, (T_w - T_{sat})} \right]^{\frac{1}{4}} \tag{7.38}$$

where

$$h_{fg}' = h_{fg} \left(1 + 0.5 \, \frac{c_v \, (T_w - T_{sat})}{h_{fg}} \right) \tag{7.39}$$

and it is suggested that k_v, ρ_v, and μ_v be evaluated at the mean film temperature $(T_w + T_{sat})/2$.

Under certain conditions, it is possible to apply this expression to horizontal tubes by making the substitution $L = \pi D/2$. A more accurate representation, which accounts for interfacial shear and curvature effects, is given by Breen and Westwater (1962):

$$h = (0.59 + 0.069 \, \frac{\lambda_c}{D}) \left[\frac{k_v^3 \, g \, h_{fg}' \, (\rho_\ell - \rho_v)\rho_v}{\mu_v \, (T_w - T_{sat})\lambda_c} \right]^{\frac{3}{4}} \tag{7.40}$$

where the minimum wavelength for Taylor instability, λ_c, is given by

$$\lambda_c = 2\pi \left[\frac{g_c \, \sigma}{g(\rho_\ell - \rho_v)} \right]^{\frac{1}{2}} \tag{7.41}$$

Film boiling from flat horizontal surfaces was investigated by Berenson (1961), who suggests the following equation:

$$h = 0.67 \left[\frac{k_v^3 \, g \, h_{fg}' \, (\rho_\ell - \rho_v)\rho_v}{\mu_v \, (T_w - T_{sat}) \, \lambda_c} \right]^{\frac{1}{4}} \tag{7.42}$$

A change in film boiling behavior, apparently analogous to the transition to turbulence, has been noted at a critical value of a modified Rayleigh number (Frederking and Clark (1962)):

$$Ra* = \left[\frac{\rho_v \, D^3 g \, (\rho_\ell - \rho_v)}{\mu_v^2} \right] \left[\frac{c\mu}{k} \right]_v \left[\frac{h_{fg}}{c_v \, (T_w - T_{sat})} + 0.5 \, \frac{a}{g} \right] \tag{7.43}$$

The correlation for the turbulent regime, $Ra* > 5 \times 10^7$, is

$$Nu = 0.15 \, (Ra*)^{1/3} \tag{7.44}$$

and is apparently valid for vertical plates, horizontal tubes and spheres.

Since the temperature level is generally quite high in film boiling, it is necessary to account for radiation. A simple superposition of heat-transfer coefficients is not adequate,

however, since the radiant heat transfer results in an increase
in the film thickness. The following equation is recommended for
the resultant heat transfer coefficient:

$$h_T = h\left(\frac{h}{h_T}\right)^{1/3} + h_r \cong h + 0.75\, h_r \tag{7.45}$$

where

$$h_r = \frac{\sigma F (T_w^4 - T_{sat}^4)}{(T_w - T_{sat})} \tag{7.46}$$

and the interchange factor $F \cong 1.0$.

Leidenfrost Point

The lower limit of stable film boiling corresponds to the
breakdown of the continuous insulating film and the onset of
liquid-solid contact. Numerous analyses have been made to predict
this condition, generally based on hydrodynamic stability theory
similar to that employed in determining the critical heat flux.
The result for a flat horizontal surface in terms of the minimum
heat flux is

$$q''_{min} = C\, h_{fg} \rho_v \left[\frac{\sigma g\, (\rho_\ell - \rho_v)}{(\rho_\ell + \rho_v)^2} \right]^{\frac{1}{4}} \tag{7.47}$$

where σ is variously given as 0.177 (Zuber (1959)), 0.13 (Zuber
(1958)) or 0.09 (Berenson (1961)).

For the case of small-diameter tubes, it is necessary to
account for curvature effects, in particular, the effect of
surface tension in the transverse direction upon the Taylor
instability of the interface. Lienhard and Wong (1964) have
suggested the following semi-empirical equation:

$$q''_{min} = 0.057 \frac{\rho_v h_{fg}}{R} \left[\frac{2\, g\, (\rho_\ell - \rho_v)}{(\rho_\ell + \rho_v)} + \frac{\sigma}{(\rho_\ell + \rho_v)^2} \right]^{\frac{1}{2}}$$

$$\times \left[\frac{g\, (\rho_\ell + \rho_v)}{\sigma} + \frac{1}{2R^2} \right]^{-3/4} \tag{7.48}$$

which should probably be restricted to tube sizes considered in
the investigation, i.e., $R < 0.03$ in. For larger tubes it appears
that equation (7.47) may be applied; however, Kovalev (1966) notes
that this equation seriously overpredicts the data for water on
clean surfaces at pressures above atmospheric.

In any case, these equations cannot be relied upon for normal
systems where the liquid contains impurities and the surface
exhibits some degree of contamination. Oxidation (Kovalev (1966)),
increased wettability, (Berenson (1962)) and surface scaling
(Bergles and Thompson (1970)) produce significant increases in

q"min. Test conditions are also extremely important; for example, in an electrically heated test section, axial conduction to the power leads will produce break-up of the film at a higher apparent q"min.

Transition Boiling

Due to the transient character and ill-defined flow pattern of transition boiling, no theory has been formulated for this regime. An accurate representation of the average heat transfer characteristics can be obtained by linearly interpolating between the peak nucleate heat flux point and the Leidenfrost point on the boiling coordinates (log-log).

7.8 Influence of Subcooling and Velocity on the Boiling Curve

Subcooled pool boiling is generally a transient condition observed as the pool is heating up. In certain cases, however, the system heat loss may be high or some cooling means is employed so as to achieve a subcooled bulk condition. Subcooling may be visualized as a perturbation of the saturated pool boiling curve discussed in detail in the preceding sections.

In the natural convection region, the procedure of section 7.3 can be utilized where now

$$q" = h \left[(T_w - T_{sat}) + (T_{sat} - T_b) \right] \tag{7.49}$$

Incipient boiling may be visualized as discussion in an earlier section. Heat-transfer in the isolated bubble region can be treated by the method outlined in the section "*Prediction of the boiling performance of a surface*", if the subcooling is low. Under conditions of high subcooling, however, the bubble mechanics is somewhat different. The bubble may expand beyond the super-heated layer whereupon condensation occurs at the top of the bubble while evaporation is occurring near the base, thus reducing the growth rate. The inertia of the liquid is still sufficient to carry the bubble away from the surface and it condenses in the cold bulk fluid.

Fully established boiling generally deviates to some extent from saturated boiling with a similar system. This behavior is indicated in Figure 7.17 where it is seen that the boiling curve for a tubular heater shifts to the right with increased subcooling. The direction and magnitude of this shift appears to be dependent on pool convective conditions, since the opposite trend was observed with a flat, horizontal heating surface (Duke and Schrock (1961)).

The critical heat flux is strongly dependent upon the degree of subcooling. The data are usually represented as a linear function of the subcooling as exemplified by the simple expression recommended by Ivey (1962):

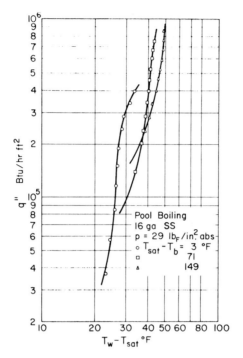

Figure 7.17: Effect of subcooling on nucleate pool boiling (Bergles and Rohsenow (1964)).

$$\frac{q''_{crit,\ sub}}{q''_{crit,\ sat}} = \left[1 + 0.1\ (\frac{\rho_v}{\rho_\ell})^{1/4} \left[\frac{c_\ell \rho_\ell\ (T_{sat} - T_b)}{h_{fg}\ \rho_v}\right]\right]\ (\frac{a}{g})^{0.273}$$

$$(7.50)$$

Ponter and Haigh (1969) found that the following equation should be used for water at subatmospheric pressures:

$$\frac{q''_{crit,\ sub}}{q''_{crit,\ sat}} = 1.06 + 0.015\ (T_{sat} - T_b)/p^{0.474} \qquad (7.51)$$

where $(T_{sat} - T_b)$ is in $^\circ C$, p is in torr and equation (7.34) is used as the reference.

Subcooled film boiling has not been investigated extensively; however, general experimental evidence indicates that heat-transfer coefficients are increased in the stable film boiling region as the subcooling increases. Quantitative predictions can be obtained from the analysis of Sparrow and Cess (1962) for an isothermal vertical plate. The analysis predicts that the heat transfer coefficient for large subcooling will be identical to that obtained for free convection of the liquid alone.

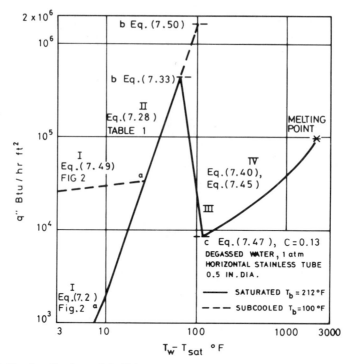

Figure 7.18: Prediction of boiling curve for saturated and subcooled pool boiling.

Subcooling increases q''_{min} and transition boiling heat fluxes; however, no reliable data appear to be available to quantitatively establish this region of the boiling curve.

It has been demonstrated experimentally that pool boiling heat-transfer can be improved by any of the schemes which mechanically agitate the pool bulk liquid, including propeller-type stirrers, pumps, and injected vapor (e.g., Pramuk and Westwater (1956)). As might be expected, heat transfer in the nonboiling convective region is improved, there is little affect on fully-developed nucleate boiling, the peak heat flux is elevated, and coefficients in the transition and film boiling regions are elevated. Quantative prediction of these effects is difficult, however, due to the inability to predict the velocity field in the vicinity of the heater. If the velocity is known, there is an increasing body of literature which can be consulted for correlations applicable to various regions of the boiling curve, e.g. Yilmaz and Westwater (1979).

7.9 Construction of the Complete Boiling Curve

In recapitulation, it is appropriate to develop the complete boiling curve from the preceding equations. The result is shown in Figure 7.18 for a tubular test section which might represent, for example, a heating element in a pool evaporator system. This

is the approach that would be taken to predict the performance
without performing tests on the actual equipment. As emphasized
in the preceding discussion, significant deviations are possible
in all regions of the boiling curve. The generated curve, however,
represents the best estimate based on the current state of the art.

Nomenclature

a	– acceleration
C, C_1, C_{sf}	– constants
c	– constant pressure specific heat
D	– tube diameter
D_b	– bubble departure diameter
f	– frequency of bubble formation
G	– mass velocity
g	– acceleration of gravity
h	– heat transfer coefficient
h_{nc}	– heat transfer coefficient for natural convection
h_{fg}	– latent heat of vaporization
h_T	– resultant film boiling heat-transfer coefficient
k	– thermal conductivity
L	– characteristic length for heat transfer with natural convection
L'	– characteristic length defined by equation (7.36)
m, n	– constants
n	– number of nucleation sites per unit surface area
N	– number of nucleation sites
p	– pressure
p_g	– partial pressure of noncondensible gas in bubble
p_{vr}	– pressure on vapor side of a curved interface
q	– heat transfer rate
q''	– heat flux
q''_{bc}	– heat flux due to bubble induced bulk convection
q''_{crit}	– critical heat flux
q''_{min}	– minimum heat flux for stable film boiling
q''_{nc}	– heat flux due to natural convection
q''_{tr}	– heat flux at transition from isolated bubbles to bubble columns
R	– tube radius
r	– bubble radius
r_c, r'_c	– cavity radii
T	– temperature

T_b — bulk fluid temperature

T_{sat} — saturation temperature

T_w — temperature of heated surface

T_∞ — temperature of superheated pool

t — time

t_d, t_w — characteristic bubble times (noted in Figure 7.9)

v — specific volume

v_{fg} — specific volume change for vaporization

y — normal distance from heating surface

Gr — Grashof number = $L^3 \beta g (T_w - T_b) \rho^2/\mu^2$

Ja — Jakob number = $(T_w - T_{sat}) c_\ell \rho_\ell / h_{fg} \rho_v$

Nu — Nusselt number = $h L/k$

Pr — Prandtl number - $c\mu/k$

β — contact angle; temperature coefficient of expansion

μ — dynamic viscosity

$\emptyset$ — angle of conical cavity

ρ — density

σ — surface tension; Stefan-Boltzmann constant

Subscripts

b — based on bubble characteristics

f — evaluated at film temperature

ℓ — liquid condition

v — vapor condition

References

Bankoff, S G, (1959) "The prediction of surface temperatures at incipient boiling." *Chemical Engineering Progress Symposium Series*, **55**, No. 29, 87-94.

Bankoff, S G, Hajjar, A J and McGlothin, B B, (1958) "On the nature and location of bubble nuclei in boiling from surfaces." *Journal of Applied Physics*, **29**, 1739-1741.

Bankoff, S G, (1978) "Vapor explosions: A critical review." *Heat transfer 1978, Proceedings of the Sixth International Heat Transfer Conference*, **6**, Hemisphere Publishing Corp., Washington, D.C., 355-360.

Berenson, P J, (1961) "Transition boiling heat transfer from a horizontal surface." *Journal of Heat Transfer*, **83**, 351-358.

Berenson, P J, (1962) "Experiments on pool boiling heat transfer." *International Journal of Heat and Mass Transfer*, **5**, 985-999.

Bergles, A E, (1975) "Burnout in boiling heat transfer. Part 1: Pool-boiling systems." *Nuclear Safety*, **16**, 29-42.

Bergles, A E and Rohsenow, W M, (1964) "The determination of forced-convection surface-boiling heat transfer." *Journal of Heat Transfer,* **86**, 356-372.

Bergles, A E and Thompson, W G, Jr. (1970) "The relationship of quench data to steady-state pool boiling data." *International Journal of Heat and Mass Transfer,* **13**, 55-68.

Bernath, L, (1960) "A theory of local-boiling burnout and its application to existing data." *Chemical Engineering Progress Symposium Series,* **56**, No. 30, 95-116.

Breen, B P and Westwater, J W, (1962) "Effect of diameter of horizontal tubes on film boiling heat transfer." *Chemical Engineering Progress,* 58, No. 7, 67-72.

Bromley, C A, (1950) "Heat transfer in stable film boiling." *Chemical Engineering Progress Symposium Series,* **46**, No. 5, 221-227.

Brown, W T, Jr. (1967) "A study of flow surface boiling." Ph.D. Thesis in Mechanical Engineering, Massachussetts Institute of Technology.

Chang, Y P and Snyder, N W, (1960) "Heat transfer in saturated boiling." *Chemical Engineering Progress Symposium Series,* **56**, No. 30, 25-38.

Clark, H B, Strenge, P S and Westwater, J W, (1959) "Active sites for nucleation." *Chemical Engineering Progress Symposium Series,* **55**, No. 29, 103-110.

Cole, R, (1974) "Boiling nucleation." in *Advances in Heat Transfer,* **10**, Academic Press, New York, 85-166.

Cooper, M G and Lloyd, A J P, (1966) "Transient local heat flux in nucleate boiling." *Proceedings of the Third International Heat Transfer Conference,* **3**, 193-203.

Deane, C W IV and Rohsenow, W M, (1970) "Mechanism of nucleate boiling heat transfer to alkali liquid metals." in *Liquid Metal Heat Transfer and Fluid Dynamics,* ASME, New York, 90-99.

Ded, J S and Lienhard, J H, (1972) "The peak pool boiling heat flux from a sphere." *AIChE Journal,* **18**, 337-342.

Duke, E E and Shrock, V E, (1961) "Void volume, site density and bubble size for subcooled nucleate pool boiling." *Proceedings of the 1961 Heat Transfer and Fluid Mechanics Institute,* Stanford University Press, 130-145.

Eddington, R I, Kenning, D B R and Korneichev, A I, (1978) "Comparison of gas and vapor bubble nucleation on a brass surface in water." *International Journal of Heat and Mass Transfer,* **21**, 855-862.

Eddington, R I and Kenning, D B R, (1979) "The effect of contact angle on bubble nucleation." *International Journal of Heat and Mass Transfer,* **22**, 1231-1236.

Frederking, T H K and Clark, J A, (1962) "Natural convection film boiling on a sphere." in *Advances in Cryogenic Engineering,* **8**, Plenum Publishing Corp., New York, 501-506.

Fritz, W, (1935) "Berechnung des maximal volumens von dampfblasen." *Physikalische Zeitschrift,* **36**, 379-384.

Gaertner, R and Westwater, J M, (1960) "Population of active sites in nucleate boiling heat transfer." *Chemical Engineering Progress Symposium Series*, **56**, No. 30, 39-48.

Graham, R W and Hendricks, R C, (1967) "Assessment of convection, conduction and evaporation in nucleate boiling." NASA TN D-3943.

Griffith, P and Wallis, J D, (1960) "The role of surface conditions in nucleate boiling." *Chemical Engineering Progress Symposium Series*, **56**, No. 30, 49-63.

Han, C Y and Griffith, P, (1965a) "The mechanism of heat transfer in nucleate pool boiling - Part I, bubble initiation, growth and departure." *International Journal of Heat and Mass Transfer*, **8**, 887-904.

Han, C Y and Griffith, P, (1965b) "The mechanism of heat transfer in nucleate pool boiling - Part II, the heat flux - temperature difference relation." *International Journal of Heat and Mass Transfer*, **8**, 905-914.

Heidrich, A, (1931) Dissertation Technische Hochschule, Aachen. Cited in M. Jakob, *Heat Transfer*, **1**, Wiley, New York, 1949, 618-620.

Heled, Y and Orell, A, (1967) "Characteristics of active nucleation sites in pool boiling." *International Journal of Heat and Mass Transfer*, **10**, 553-554.

Hsu, Y Y, (1962) "On the size range of active nucleation cavities on a heating surface." *Journal of Heat Transfer*, **84**, 207-216

Hsu, Y Y and Graham, R W, (1976) "Transport processes in boiling and two-phase systems." Hemisphere Publishing Corp., Washington, D.C.

Ivey, H J, (1962) "Acceleration and the critical heat flux in pool boiling heat transfer." *Chartered Mechanical Engineer*, **9**, 413-427.

Jakob, M and Fritz, W, (1931) "Versuche über den verdampfungsvorgang." *Forschung auf dem Gebiete des Ingenieurwesens*, **2**, 435-447.

Keenan, J H, (1941) "Thermodynamics." Wiley, New York.

Kopp, I Z, (1976) "The effect of surface roughness on boiling heat transfer." *Heat Transfer - Soviet Research*, **8**, No. 4, 37-42.

Kovalev, S A, (1966) "An investigation of minimum heat fluxes in pool boiling of water." *International Journal of Heat and Mass Transfer*, **9**, 1219-1226.

Kutateladze, S S, (1951) "A hydrodynamic theory of changes in boiling process under free convection." *Izv. Akademia Nauk Otdelenie Tekh. Nauk*, **4**, 529-536. AEC-tr-1441, 1954.

Lienhard, J H and Dhir, V K, (1973) "Hydrodynamic prediction of peak pool-boiling heat fluxes from finite bodies." *Journal of Heat Transfer*, **95**, 152-158.

Lienhard, J H and Wong, P T Y, (1964) "The dominant instable wavelength and minimum heat flux during film boiling on a horizontal cylinder." *Journal of Heat Transfer*, **86**, 220-226.

Lienhard, J H, Dhir, V K and Riherd, D M, (1973) "Peak pool boiling heat-flux measurements on finite horizontal flat plates." *Journal of Heat Transfer*, **95**, 477-482.

Lorenz, J J, Mikic, B B and Rohsenow, W M, (1974) "The effect of surface conditions on boiling characteristics." *Heat Transfer 1974, Proceedings of the 5th International Heat Transfer Conference,* **IV**, Hemisphere, Washington, DC, 35-39.

Lorenz, J J, Mikic, B B and Rohsenow, W M, (1975) "A gas diffusion technique for determining pool boiling nucleation sites." *Journal of Heat Transfer,* **97**, 317-319.

Lyon, D N, Kosky, P G and Harman, B N, (1964) "Nucleate boiling heat transfer coefficients and peak nucleate boiling fluxes for pure liquid N_2 and O_2 on horizontal platinum surfaces from below 0.5 atmospheres to the critical pressures." in *Advances in Cryogenic Engineering,* **9**, Plenum Publishing Corp., New York, 77-87.

McAdams, W H, (1954) "Heat transmission." 3rd edition, McGraw-Hill, New York.

Merte, H, Jr and Clark, J A, (1964) "Boiling heat transfer with cryogenic fluids at standard, fractional, and near-zero gravity." *Journal of Heat Transfer,* **86**, 351-359.

Mikic, B and Rohsenow, W, (1969) "Bubble growth rates in non-uniform temperature field." *Progress in Heat and Mass Transfer,* **II**, Pergamon Press, Oxford, 283-

Mikic, B and Rohsenow, W M, (1969) "A new correlation of pool boiling data including the effect of heating surface characteristics." *Journal of Heat Transfer,* **91**, 245-250.

Mikic, B B, Rohsenow, W M and Griffith, P, (1970) "On bubble growth rates." *International Journal of Heat and Mass Transfer,* **13**, 657-666.

Mizukami, K, (1979) "Stability of bubble nuclei and nucleation in isothermal liquid." *Review of Kobe University of Mercantile Marine,* Part II, No. 27, 99-108.

Moissis, R and Berenson, P J, (1963) "On the hydrodynamic transitions in nucleate boiling." *Journal of Heat Transfer,* **85**, 221-229.

Moore, F D and Mesler, R B, (1961) "The measurement of rapid temperature fluctuations during nucleate boiling of water." *AIChE Journal,* **7**, 620-624.

Nail, J P, Jr, Vachon, R I and Morehove, J, (1974) "An SEM study of nucleation sites in pool boiling from 304 stainless steel." *Journal of Heat Transfer,* **96**, 132-137.

Noyes, R C, (1963) "An experimental study of sodium pool boiling heat transfer." *Journal of Heat Transfer,* **85**, 125-129.

Nukiyama, S, (1934) "The maximum and minimum value of the heat Q transmitted from metal to boiling water under atmospheric pressure." *Journal Japan Society of Mechanical Engineers,* **37**, 367-374. Translation in *International Journal of Heat and Mass Transfer,* 9, 1966, 1419-1433.

Peterson, W C and Zaalouk, M G, (1971) "Boiling-curve measurements from a controlled heat transfer process." *Journal of Heat Transfer,* **93**, 408-412.

Peyayopanakul, W and Westwater, J W, (1978) "Evaluation of the unsteady-state quenching method for determining boiling curves." *International Journal of Heat and Mass Transfer,* 21, 1437-1445.

Ponter, A B and Haigh, C P, (1969) "The boiling crisis in saturated and subcooled pool boiling at reduced pressures." *International Journal of Heat and Mass Transfer*, **12**, 429-437.

Porteous, W M and Reid, R C, (1976) "Light hydrocarbon vapor explosions." *Chemical Engineering Progress,* **72** (5), 83-89.

Pramuk, F S and Westwater, J W, (1956) "Effect of agitation on the critical temperature difference for a boiling liquid." *Chemical Engineering Progress Symposium Series,* **52**, No. 18, 79-83.

Rohsenow, W M, (1952) "A method of correlating heat transfer for surface boiling of liquids." *Transactions ASME,* **74**, 969-976.

Rohsenow, W M, (1973) "Boiling." in *Handbook of Heat Transfer*, McGraw-Hill, New York, p. 13-28.

Rohsenow, W M and Griffith, P, (1956) "Correlation of maximum heat transfer data for boiling of saturated liquids." *Chemical Engineering Progress Symposium Series,* **52**, No. 18, 47-49.

Sakurai, A and Shiotsu, M, (1974) "Temperature-controlled pool boiling heat transfer." *Heat Transfer 1974, Proceedings of the 5th International Heat Transfer Conference,* **IV,** Hemisphere, Washington, D.C., 81-85.

Scarola, L S, (1964) "Effect of additives on the critical heat flux in nucleate boiling." S.M. Thesis in Mechanical Engineering, Massachussetts Institute of Technology.

Seader, J D, et al, (1965) "Boiling heat transfer for cryogenics." NASA CR-243.

Singh, A, Mikic, B B and Rohsenow, W M, (1976) "Active sites in boiling." *Journal of Heat Transfer,* **98**, 401-406.

Sparrow, E M and Cess, R D, (1962) "The effect of subcooled liquid on laminar film boiling." *Journal of Heat Transfer,* **84**, 149-156.

Stephan, K and Abdelsalam, M, (1980) "Heat-transfer correlations for natural convection boiling." *International Journal of Heat and Mass Transfer,* **23**, 73-87.

Sun, K H and Lienhard, J H, (1970) "The peak pool boiling heat flux on horizontal cylinders." *International Journal of Heat and Mass Transfer,* **13**, 1425-1439.

Tachibana, F, Akiyama, M and Kawamura, H, (1967) "Non-hydrodynamic aspects of pool boiling burnout." *Journal of Nuclear Science and Technology,* **4**, No. 3, 121-130.

Volmer, M, (1939) "Kenetik der Phasenbildung." Steinkopf, Leipzig; Edwards Bros., Ann Arbor, 1945.

Wallis, G B, (1961) "Two-phase flow aspects of pool boiling from a horizontal surface." AEEW-R 103.

Yilmaz, S and Westwater, J W, (1979) "Effect of velocity on heat transfer to boiling Freon-113." *Journal of Heat Transfer,* **102**, 26-31.

Zuber, N, (1958) "On stability of boiling heat transfer." *Transactions* ASME, **80**, 711-720.

Zuber, N, Tribus, M and Westwater, J W, (1961) "The hydrodynamic crisis in pool boiling of saturated and subcooled liquids." *International Developments in Heat Transfer, Part II*, ASME, New York, 230-235.

Chapter 8

Forced Convective Boiling

J. G. COLLIER

8.1   Introduction

Before entering into a detailed discussion of the heat
transfer and fluid dynamic problems encountered in the power and
process chemical industries, it is essential to have a basic
understanding of the various heat transfer regions when a fluid
is boiled under pressure in a confined channel. Not only is it
necessary to understand the processes which occur during normal
steady-state and off-normal operation but, more important, those
transient processes which occur under accident or fault
conditions.

In this chapter it is proposed to describe boiling under
steady-state conditions in a vertical uniformly-heated tube. Each
of the various heat transfer regions present will be described
briefly, together with simple correlations for the estimation of
heat transfer coefficients in each region and for the transition
from one region to another.

8.2   Regions of Heat Transfer in a Vertical Heated Tube

Consider a vertical tube heated uniformly over its length
with a low heat flux and fed with subcooled liquid at its base.
At such a rate the liquid is totally evaporated over the length of
the tube. Figure (8.1) shows in diagrammatic form, the various
flow patterns encountered over the length of the tube, together
with the corresponding heat transfer regions.

Whilst the liquid is being heated up to the saturation
temperature and the wall temperature remains below that necessary
for nucleation, the process of heat transfer is *single-phase
convective heat transfer to the liquid phase* (Region A). At some
point along the tube, the conditions adjacent to the wall are such
that the formation of vapour from nucleation sites can occur.
Initially, vapour formation takes place in the presence of subcooled liquid (Region B) and this heat transfer mechanism is known
as *subcooled nucleate boiling*. In the subcooled boiling region, B,
the wall temperature remains essentially constant a few degrees
above the saturation temperature, whilst the mean bulk fluid
temperature is increasing to the saturation temperature. The
amount by which the wall temperature exceeds the saturation

temperature is known as the *degree of superheat*, ΔT_{SAT}, and the difference between the saturation and the bulk fluid temperature is known as the *degree of subcooling*, ΔT_{SUB}.

The transition between B and C, the *subcooled nucleate boiling region* and the *saturated nucleate boiling region*, is clearly defined from the thermodynamic viewpoint. It is the point at which the liquid reaches the saturation temperature ($x = 0$) found on the basis of simple heat balance calculations. However, subcooled liquid can persist in the liquid core, even in the region defined as *saturated nucleate boiling*. Vapour generated in the subcooled region is present at the transition between regions B and C ($x = 0$); thus, some of the liquid must be subcooled to ensure that the liquid mean (mixing cup) enthalpy equals that of saturated liquid (h_ℓ). This effect occurs as a result of the radial temperature profile in the liquid and the subcooled liquid flowing in the centre of the channel will only reach the saturation temperature at some distance downstream of the point, $x = 0$. In the regions C to G, the variable characterising the heat transfer mechanism is the thermodynamic mass "quality" (x) of the fluid. The "quality" of the liquid-vapour mixture at a distance, z, is given on a thermodynamic basis as:

$$x(z) = \frac{h(z) - h_\ell}{\Delta h_v} \tag{3.1}$$

In the region $0 < x < 1$ and for complete thermodynamic equilibrium, x represents the ratio of the vapour mass flow-rate to the total mass flow-rate. From the thermodynamic definition, equation (8.1), x may have both negative values and values greater than unity. Although these are sometimes used for convenience, negative values or values greater than unity have no practical significance other than to signify that, in the former case, the condition is that of a subcooled liquid and, in the latter case, that of a superheated vapour. The variable, x, is also often referred to as the "vapour mass fraction".

As the quality increases through the *saturated nucleate boiling region* a point may be reached where a fundamental transition in the mechanism of heat transfer takes place. The process of "boiling" is replaced by the process of "evaporation". This transition is preceded by a change in the flow pattern from bubbly or slug flow to annular flow (regions E and F). In the latter regions the thickness of the thin liquid film on the heating surface is often such that the effective thermal conductivity is sufficient to prevent the liquid in contact with the wall being superheated to a temperature which would allow bubble nucleation. Heat is carried away from the wall by forced convection in the film to the liquid film-vapour core interface, where evaporation occurs. Since nucleation is completely suppressed, the heat transfer process can no longer be called "boiling". The region beyond the transition has been referred to as the *two-phase forced convective region* of heat transfer (Regions E and F).

At some critical value of the quality, the complete evaporation of the liquid film occurs. This transition is known as "dryout" and is accompanied by a rise in the wall temperature for channels operating with a controlled surface heat flux. The area between the dryout point and the transition to *saturated vapour* (Region H) has been termed the *liquid deficient region* (corresponding to the drop flow pattern) (Region G). This condition of "dryout" often puts an effective limit on the amount of evaporation which can be allowed to take place in a channel at a particular value of heat flux.

8.2.1 *Variation of Heat Transfer Coefficient with Quality*

It is useful at this stage to describe, at least qualitatively, the progressive variation of the local surface temperature (or local heat transfer coefficient) along the length of the tube as evaporation proceeds. The local heat transfer coefficient can be established by dividing the surface heat flux (constant over the tube length) by the difference between the wall temperature and the bulk fluid temperature. Typical variations of these two temperatures with length along the tube are shown in Figure 8.1. The variation of heat transfer coefficient with length along the tube for the conditions represented in Figure 8.1 is given in Figure 8.2 (curve (i), solid line). In the *single-phase convective heat transfer region*, the wall temperature is displaced above the bulk fluid temperature by a relatively constant amount, (the heat transfer coefficient is approximately constant) and is modified only slightly by the influence of temperature on the liquid

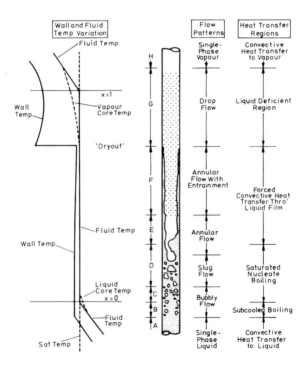

Figure 8.1: Regions of heat transfer in convective boiling.

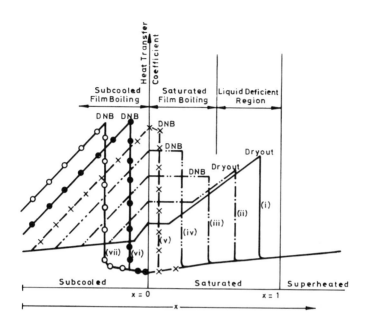

Figure 8.2: Variation of heat transfer coefficient with quality with increasing heat flux as parameter.

physical properties. In the *sub-cooled nucleate boiling region* the temperature difference between the wall and the bulk fluid decreases linearly with length up to the point where x = 0. The heat transfer coefficient, therefore, increases linearly with length in this region. In the *saturated nucleate boiling region* the temperature difference and, therefore, the heat transfer coefficient, remains constant. Because of the reducing thickness of the liquid film in the *two-phase forced convective region* the difference in temperature between the surface and the saturation temperature reduces and the heat transfer coefficient increases with increasing length or mass quality. At the dryout point the heat transfer coefficient is suddenly reduced from a very high value in the forced convective region to a value near to that expected for heat transfer by forced convection to saturated steam. As the quality increases through the *liquid deficient region*, so the vapour velocity increases and the difference in temperature between the surface and the saturation value decreases with the corresponding rise in heat transfer coefficient. Finally, in the single-phase vapour region (x > 1) the wall temperature is once again displaced by a constant amount above the bulk fluid temperature and the heat transfer coefficient levels out to that corresponding to convective heat transfer to a single-phase vapour flow.

The above comments have been restricted to the case where a relatively low heat flux is supplied to the walls of the tube. The effect of progressively increasing the surface heat flux whilst keeping the inlet flow-rate constant, will now be considered with reference to Figures 8.2 and 8.3. Figure 8.2 shows the heat

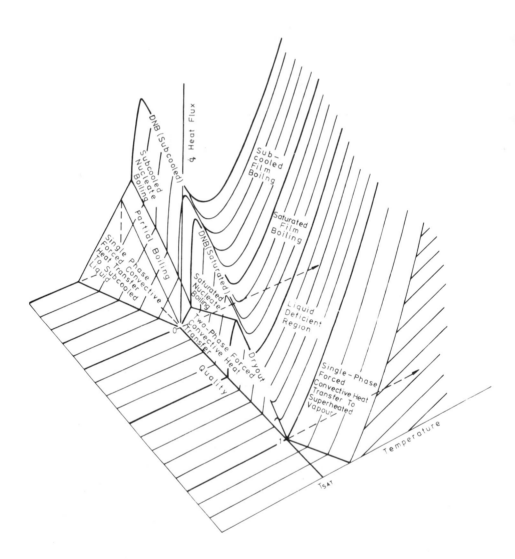

Figure 8.3(a): Forced convection boiling surface (T,x,q̇ co-ordinates)

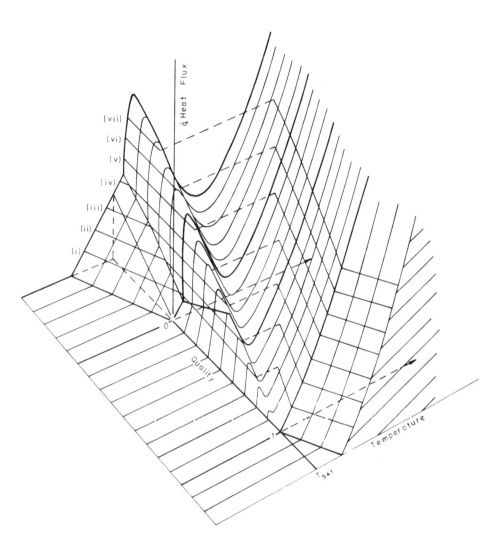

Figure 8.3(b): Forced convection boiling surface (T,x,q̇ co-ordinates)

transfer coefficient plotted against mass quality with increasing
flux as parameter (curves (i)-(vii)). Figure 8.3 shows the various
regions of two-phase heat transfer in forced convective boiling on
a three-dimensional diagram with heat flux, mass quality and
temperature as co-ordinates - "the boiling surface". Curve (i)
relates to the conditions shown in Figure 8.1 for a low heat flux
being supplied to the walls of a tube. The temperature pattern
shown in Figure 8.1 will be recognised as the projection in plan
view (temperature-quality co-ordinates) of curve (i). Curve (ii)
shows the influence of increasing the heat flux. Subcooled
boiling is initiated sooner, the heat transfer coefficient in the
nucleate boiling region is higher but is unaffected in the two-
phase forced convective region. Dryout occurs at a lower mass
quality. Curve (iii) shows the influence of a further increase in
the heat flux. Again, subcooled boiling is initiated earlier and
the heat transfer is again higher in the nucleate boiling region.
As the mass quality increases, before the two-phase forced
convective region is initiated, and while bubble nucleation is
still occurring, an abrupt deterioration in the cooling process
takes place. This transition is essentially similar to the
critical heat flux phenomenon in saturated pool boiling and will
be termed "departure from nucleate boiling" (DNB).

The mechanism of heat transfer under conditions where the
critical heat flux (DNB or dryout) has been exceeded is dependent
upon whether the initial condition was the process of "boiling"
(ie bubble nucleation in the subcooled or low mass quality regions)
or the process of "evaporation" (ie evaporation at the liquid film-
vapour core interface in the higher mass quality areas). In the
latter case, the "liquid deficient region" is initiated; in the
former case the resulting mechanism is one of "film boiling"
(Figure 8.3).

Returning to Figures 8.2 and 8.3, it can be seen that further
increases in heat flux (curves (vi) and (vii)) cause the condition
of "departure from nucleate boiling" (DNB) to occur in the subcooled
region with the whole of the saturated or "quality" region being
occupied by, firstly, "film boiling" and, in the latter stages,
the "liquid deficient region" - both relatively inefficient modes
of heat transfer. In Figure 8.3 the film boiling surface has
been arbitrarily divided into two regions: "subcooled film boiling" and "saturated film boiling". "Film boiling" in forced
convective flow is essentially similar to that observed in pool
boiling. An insulating vapour film covers the heating surface
through which the heat must pass. The heat transfer coefficient
is orders of magnitude lower than in the corresponding region
before the critical heat flux was exceeded, due mainly to the
lower thermal conductivity of the vapour adjacent to the surface.
Figure 8.4 shows the regions of two-phase forced convective heat
transfer as a function of quality with increasing heat flux as
ordinates (an elevation view of Figure 8.2).

In the following sections, criteria will be given whereby
the boundaries delineated in Figures 8.3 and 8.4 - "the boiling
surface" - can be established. In addition, methods for calculating heat transfer rates in each heat transfer region will be given.

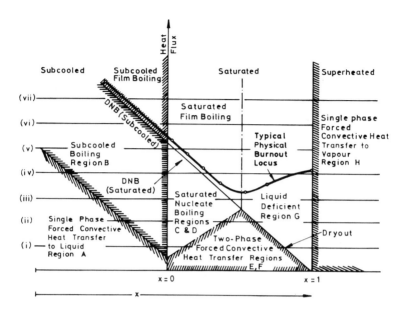

Figure 8.4: Regions of two-phase forced convective heat transfer as a function of quality with increasing heat flux as ordinate.

8.3 Sub-Cooled Boiling

8.3.1 *Single-Phase Liquid Heat Transfer*

Figure 8.5 shows in an idealised form the flow pattern and the variation of the surface and liquid temperatures in the regions designated by A, B and C. The tube surface temperature in region A, *convective heat transfer to single-phase liquid*, is given by:

$$T_W = T_{\ell}(z) + \Delta T_{\ell} \tag{8.2}$$

and

$$\Delta T_{\ell} = \dot{q}/\alpha_{\ell o} \tag{8.3}$$

where ΔT_{ℓ} is the temperature difference between the tube inside surface and the mean bulk liquid temperature at a length z from the tube inlet, α_{fo} is the heat transfer coefficient to single-phase liquid under forced convection. The liquid in the channel may be in laminar or turbulent flow; in either case the laws governing the heat transfer are well established. For turbulent flow the well-known Dittus-Boelter equation has been found satisfactory:

$$\left[\frac{\alpha_{\ell o} D}{\lambda_{\ell}}\right] = 0.023 \left[\frac{\dot{m} D}{\mu_{\ell}}\right]^{0.8} \left[\frac{c_p \mu}{\lambda}\right]^{1/3}_{\ell} \tag{8.4}$$

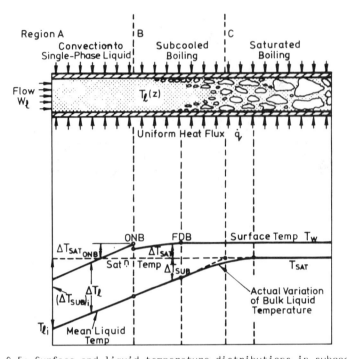

Figure 8.5: Surface and liquid temperature distributions in subcooled boiling.

This relation is valid for heating in vertical upflow in z/D > 50 and $\{\dot{m}D/\mu_\ell\} > 10,000$.

8.3.2 The Onset of Sub-Cooled Nucleate Boiling

Figure 8.6 shows in a qualitative form, the variation of tube inside surface temperature at a point z as the heat flux is steadily increased at a given inlet sub-cooling and mass velocity. Three regions are shown as the *single-phase region AB*, the *partial boiling region BCDE* and the *fully developed subcooled boiling region EF*.

As the heat flux is increased, the surface temperature will follow the line ABD' until the first bubbles nucleate. A higher degree of superheat is necessary to initiate the first bubble nucleation sites at a given heat flux than indicated by the curve ABCDE. When nucleation first occurs the surface temperature drops from D' to D and, for further increases in heat flux, follows the line DEF. The criterion for the onset of boiling can crudely be established as the intersection of line ABD' and the fully-developed boiling curve C"EF.

A more refined treatment of the onset of nucleation can be derived by considering the temperature profile in the region adjacent to the heated wall. Using this approach, Bergles and Rohsenow (1964) obtained an equation for the wall superheat

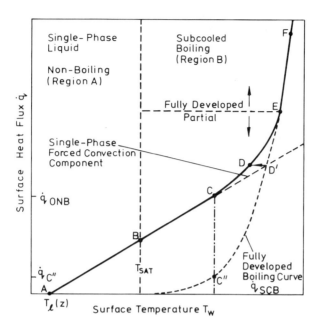

Figure 8.6: Subcooled boiling curve.

required for the onset of subcooled boiling.   Their equation is valid for water only and is:

$$(T_W - T_{SAT})_{ONB} = 0.556 \left[\frac{\dot{q}}{1082\ p^{1.156}}\right]^{0.463p^{0.0234}} \tag{8.5}$$

where $\dot{q}$ is the surface heat flux in W/m^2 and p is the system pressure in bar.   An alternative expression by Davis and Anderson (1966) valid for all fluids is:

$$(T_W - T_{SAT})_{ONB} = \left[\frac{8\sigma\dot{q}T_{SAT}}{\lambda_\ell \Delta h_v \rho_g}\right]^{0.5} \tag{8.6}$$

Equations 8.5 and 8.6 are in good agreement with each other for the case of high pressure water flows.   Frost and Dzakowic (1967) have extended this treatment to cover other liquids.   In their study nucleation was assumed to occur when the liquid temperature, $T_\ell(y)$ was matched to the temperature for bubble equilibrium, T_g, at a distance nr_c where $n = \{c_p\mu/\lambda\}_\ell^2$ rather than r_c.   Thus equation (8.6) becomes:

$$(T_W - T_{SAT})_{ONB} = \left[\frac{8\sigma\dot{q}T_{SAT}}{\lambda_\ell \Delta h_v \rho_g}\right]^{0.5} Pr_\ell \tag{8.7}$$

This equation was compared with experimental data for the onset of boiling for a variety of different fluids. A parameter, X, can be defined as:

$$X = \left[\frac{(T_W - T_{SAT})_{ONB}}{\dot{q}^{0.5} Pr_\ell} \right] = \left[\frac{8\sigma T_{SAT}}{\lambda_\ell \Delta h_v \rho_g} \right]^{0.5} \qquad (8.8)$$

Figure (8.7) shows values of $\left[X_{EXP}/X_{REF} \right]$ plotted against reduced pressure. X_{EXP} was calculated from reported values of $(T_W - T_{SAT})_{ONB}$ and $\dot{q}_{ONB}$; X_{REF} was evaluated from fluid physical properties reduced pressure of 0.05. Values of X_{REF} for various fluids are given in Table (8.1).

The above treatments can only predict the onset of nucleate boiling accurately if there is a sufficiently wide range of "active" cavity sizes available on the heating surface. This is not always the case and for this reason equation (8.8) represents the lower bound of the experimental results shown in Figure (8.7), ie boiling in many instances was not initiated until higher values of surface temperature than estimated from equation (8.8) were reached. The optimum size of "active" cavity r_C may be evaluated from:

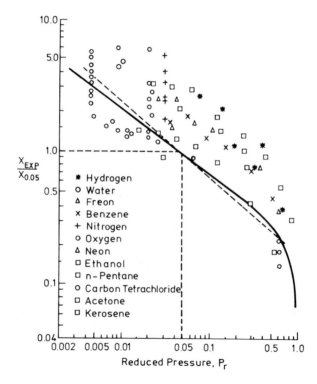

Figure 8.7: Experimental data for the onset of boiling compared with equation (7.8) (Frost and Dzakowic).

TABLE 8.1

Reference Values of X Evaluated at a Reduced Pressure of 0.05

Fluid	X_{REF} $^{\circ}K/(W/m^2)^{0.5}$	Fluid	X_{REF} $^{\circ}K/(W/m^2)^{0.5}$
Ammonia	0.00432	Oxygen	0.00507
Carbon dioxide	0.01186	Benzene	0.01339
Carbon tetrachloride	0.01364	Water	0.00441
Mercury	0.00397	Freon 12	0.01417
Neon	0.00241	Ethanol	0.00891
n-Pentane	0.01214	Acetone	0.01120
Para-hydrogen	0.00338	Kerosene (JP-4)	0.01267
Nitrogen	0.00519	Helium-4	0.00231
Propane	0.00960	Argon	0.00601

$$r_c = \left[\frac{2\sigma T_{SAT} \lambda_\ell}{\dot{q} \Delta h_v \rho_g} \right]^{0.5} \qquad (8.9)$$

The size of "active" nucleation sites actually present on the heated surface may be estimated by methods given earlier. If r_c exceeds these typical values, then evaluate the wall superheat for the onset of boiling from:

$$(T_W - T_{SAT})_{ONB} = \left[\frac{2\sigma T_{SAT}}{r_{MAX} \Delta h_v \rho_g} \right] + \left[\frac{\dot{q} r_{MAX}}{\lambda_\ell} \right] \qquad (8.10)$$

where r_{MAX} is the estimated maximum size of active nucleation site on the surface.

For the subcooled boiling region, Equations (8.6) or (8.7) must be solved simultaneously with the heat transfer equation:

$$\dot{q} = \alpha_{\ell o} \{ T_W - T_\ell(z) \}$$

$$= \alpha_{\ell o} \{ (T_W - T_{SAT})_{ONB} + (T_{SAT} - T_\ell(z)) \} \qquad (8.11)$$

to give the heat flux, $\dot{q}$, and $(T_W - T_{SAT})_{ONB}$ required for the onset of boiling. The boundary between region A and region B shown in Figure (8.4) can be derived from such a simultaneous solution and it is noted that at the condition $(T_{SAT} - T_\ell(z)) = 0$:

$$(\dot{q}_{ONB})_{\Delta T_{SUB}(z) = 0} = \left[\frac{8 \alpha_{\ell o}^2 \sigma T_{SAT}}{\lambda_\ell \Delta h_v \rho_g} \right] \qquad (8.12)$$

For heat fluxes *below* that given by equation (8.12) the bulk liquid must be superheated before vapour generation can take place at the heating surface, ie there is no subcooled boiling region. Such conditions often lead to unstable operation of heat transfer equipment and should be avoided. The above treatment may also be extended to cover the influence of dissolved gases upon nucleation.

8.3.3 Fully-Developed Subcooled Boiling

When boiling is first initiated only a limited number of nucleation sites are operating so that a proportion of the heat will continue to be transferred by normal single-phase processes between patches of bubbles. This transition region is referred to as *partial boiling*. As the surface temperature increases, so the number of bubble sites also increases and the area for single-phase heat transfer decreases. Finally, the whole surface is covered by bubble sites, boiling becomes *fully-developed* and the single-phase component reduces to zero. In the *fully developed* boiling region, velocity and subcooling, both of which have a strong influence on single-phase heat transfer, have little or no effect on the surface temperature. In subcooled boiling the surface temperature is primarily a function of the surface heat flux and the system pressure for a given fluid. The influence of surface condition should be less for forced convective boiling than for pool boiling because the higher heat fluxes and wall superheats shift the range of active nucleation sites to the smaller sizes which should be readily available on most surfaces. However, there is little direct experimental evidence to support this contention.

It would appear that the *form* of the equations found suitable for the correlation of pool boiling data is also suitable for the representation of forced convective fully-developed, subcooled boiling data. The values of the slope and intercept may, however, need to be adjusted from the values found appropriate for pool boiling.

One such correlation is due to Rohsenow (1952):

$$\left[\frac{c_{p\ell}\Delta T_{SAT}}{\Delta h_v}\right] = C_{sf}\left[\frac{\dot{q}}{\mu_\ell \Delta h_v}\left(\frac{\sigma}{g_n(\rho_\ell-\rho_g)}\right)^{\frac{1}{2}}\right]^{0.33}\left[\frac{c_p\mu}{\lambda}\right]_\ell^{1.7} \qquad (7.28)$$

Values of C_{sf} derived from forced (and natural) convection subcooled boiling data are given in Table (8.2). For water and only for water the exponent on the Prandtl number in equation (7.28) should be unity.

Jens and Lottes (1951) summarised experiments on subcooled boiling of water flowing upwards in vertical electrically heated stainless steel or nickel tubes, having inside diameters ranging from 3.63 to 5.74mm. System pressures ranged from 7 bars to 172 bars, water temperatures from 115°C to 340°C, mass velocities from 11 to 1.05×10^4 kg/m^{2}s and heat fluxes up to 12.5 MW/m^2. These data were correlated by a dimensional equation *valid for water only*.

$$\Delta T_{SAT} = 25 \; \dot{q}^{0.25} \; e^{-p/62} \qquad (8.13)$$

where p is the absolute pressure in bar, ΔT_{SAT} (= $T_W - T_{SAT}$) is in $^{\circ}$C and $\dot{q}$ is in MW/m^2. More recently Thom (1965) reported that the values of ΔT_{SAT} estimated from equation (8.13) were consistently low over the range of his experiments. A modified equation of the same form and *valid only for water* was suggested:

$$\Delta T_{SAT} = 22.65 \; \dot{q}^{0.5} \; e^{-p/87} \qquad (8.14)$$

The values of ΔT_{SAT} and α calculated from these equations are not markedly different from those obtained from pool boiling correlations. Therefore, in the absence of other information use may be made of pool boiling data to evaluate fully-developed subcooled boiling curves for fluids other than water.

TABLE 8.2

Values of C_{sf} obtained in the Reduction of the Forced
(and natural) Convection Subcooled Boiling Data of Various Investigators

Reference	Heating Surface	C_{sf}	Fluid-Surface
Rohsenow-Clark (1951)	Vertical tube (4.56mm id)	0.006	Water-nickel
Krieth-Summerfield (1949)	Horizontal tube (14.9mm id)	0.015	Water-stainless steel
Piret-Isbin (1953)	Vertical tube (27.1mm id)	0.013 0.013 0.0022 0.0030 0.00275 0.0054	Water-copper Carbon tetrachloride-copper Isopropyl alcohol-copper n-butyl alcohol-copper 50% K$_2$CO$_3$-copper 35% K$_2$CO$_3$-copper
Bergles-Rohsenow (1964)	Horizontal tube (2.39mm id)	0.020	Water-Stainless Steel

8.3.4 *Partial Boiling*

In the *partial boiling region* nucleation and single-phase convection processes occur simultaneously. Rohsenow (1952) proposed the superposition of a single-phase forced convection component and a subcooled boiling component:

$$\dot{q} = \dot{q}_{\ell o} + \dot{q}_{SCB} \qquad (8.15)$$

The single-phase convection heat flux is evaluated from:

$$\dot{q}_{\ell o} = \alpha_{\ell o} \, (T_W - T_\ell(z)) \qquad (8.16)$$

and the nucleate boiling heat flux, $\dot{q}_{SCB}$, from equation (8.7). An alternative proposal by Bowring (1962) is discussed in detail by Collier (1972). A simple method of constructing the transition between the single-phase convective region and the fully-developed boiling region has been suggested by Bergles and Rohsenow (1964). The procedure can be summarised by reference to Figure (8.6). The fully-developed subcooled boiling curve, $\dot{q}_{SCB}$, FED'C" is established from experimental data at high values of $(T_W - T_{SAT})$ or constructed using empirical equations of the form of equation (8.13). The single-phase forced convection curve ABCD' is constructed from equation (8.16).

The flux, $\dot{q}_{ONB}$, for the onset of nucleation (point C on figure (8.6)) can be calculated from the methods given previously. The fixing of point C enables a point C" to be found on the extrapolated fully-developed boiling curve at the same surface temperature as point C. A simple interpolation formula is given which satisfies the characteristics of the boiling curve between C and E - thus:

$$\dot{q} = \dot{q}_{\ell o} \left[1 + \left\{ \frac{\dot{q}_{SCB}}{\dot{q}_{\ell o}} \left(1 - \frac{\dot{q}_{c"}}{\dot{q}_{SCB}} \right) \right\}^2 \right]^{\frac{1}{2}} \qquad (8.17)$$

Equation (8.17) is then applied by picking off values of $\dot{q}_{\ell o}$ and $\dot{q}_{SCB}$ at various values of T_W to fill in the section CE. This method is recommended for rapid calculation of the complete forced convective subcooled boiling curve.

8.4 Void Fraction and Pressure Drop in Subcooled Boiling

A knowledge of the amount of vapour present in the channel during subcooled boiling is needed to establish the influence on the pressure gradient. In the representation of subcooled boiling shown in Figure (8.8) it will be seen that, for high subcoolings, just after the onset of nucleation, the vapour generated remains as discrete bubbles attached to the surface, whilst growing and collapsing; voidage in this region is essentially a wall effect. At somewhat lower subcoolings, bubbles detach from the surface condensing only slowly as they move through the slightly subcooled

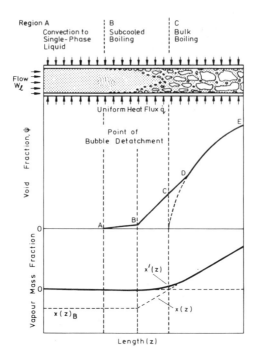

Figure 8.8: Void fraction in subcooled boiling.

liquid; voidage in this region is a bulk fluid effect.

        Over the region AB the void fraction usually remains low and
can be neglected. If it is necessary to calculate a value at point
B, the approximate expression due to Levy (1967) can be used:

$$\psi_B \simeq \frac{\pi}{6} \frac{Y_B}{D_h}$$

(8.18)

where Y_B is the distance from the heated wall to the tip of the
vapour bubble and D_h is the heated equivalent diameter of the
channel. Y_B is given by:

$$Y_B = 0.015 \left[\frac{\sigma D_e}{\tau_W} \right]^{0.5}$$

(8.19)

where τ_W is the wall shear stress and D_e is the hydraulic equivalent
diameter. The wall shear stress can be evaluated from:

$$\tau_W = \left[\frac{f_{\ell o} \dot{m}^2}{2 \rho_\ell} \right]$$

(8.20)

where $f_{\ell o}$ is the single-phase Fanning fraction factor corresponding to a relative roughness of $\varepsilon/D_e = 10^{-4}$ (smooth drawn tubing).

Saha and Zuber (1974) have proposed a simple method to calculate the point of "net vapour generation" which can be assumed to be coincident with point B, the point of bubble detachment, shown in Figure (8.8). At low flowrates the bubble detachment is assumed to be thermally-controlled occurring at a fixed value of the Nusselt number $\{\dot{q} \, D/\lambda_\ell \, (T_{SAT} - T_\ell(z)_B)\}$. At high flowrates bubble departure is hydrodynamically-induced and occurs at a fixed Stanton number $\{\dot{q}/\dot{m} \, c_{p\ell} \, (T_{SAT} - T_\ell(z)_B)\}$. Figure (8.9) shows experimental data from three fluids plotted as Stanton number versus Peclet number $\{\dot{m} \, D \, c_{p\ell}/\lambda_\ell\}$. Two regions can be distinguished. Below a Pe value of 70,000 the data fall on a line of slope minus one. Above this value St is a constant.

If Pe < 70,000, the value of $\Delta T_{SUB}(z)_B$ ($= T_{SAT} - T_\ell(z)_B$) as point B is given by:

$$\Delta T_{SUB}(z)_B = 0.0022 \left[\frac{\dot{q}D}{\lambda_\ell}\right] \tag{8.21}$$

for Pe > 70,000, the value of $\Delta\theta_{SUB}(z)_B$ at B is given by:

$$\Delta T_{SUB}(z)_B = 153.8 \left[\frac{\dot{q}}{\dot{m} \, c_{p\ell}}\right] \tag{8.22}$$

For the region BCD both the void fraction and the pressure drop can be evaluated using the relationships given earlier by Delhaye provided the actual vapour quality can be established. A suitable empirical method has been suggested by Levy (1967). It is assumed that the actual or "true" vapour mass fraction, x'(z), is related to the "thermodynamic" vapour mass fraction, x(z), by the relationship:

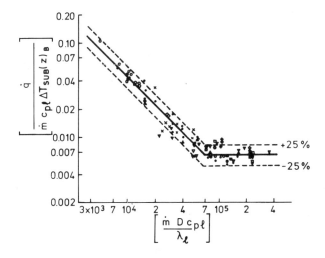

Figure 8.9: Conditions for point of bubble detachment (point B) (Saha and Zuber).

$$x'(z) = x(z) - x(z)_B \exp\{\frac{x(z)}{x(z)_B} - 1\} \qquad (8.23)$$

where $x(z)_B$ is the (negative) "thermodynamic" vapour mass fraction at the point of bubble detachment given by:

$$x(z)_B = -\left|\frac{c_{p\ell}\Delta T_{SUB}(z)_B}{\Delta h_v}\right| \qquad (8.24)$$

Equation (8.23) is a simple relationship which satisfies the following boundary conditions:

(i)   at $x(z) = x(z)_B$ then $x'(z) = 0$

(ii) $\frac{d}{dz}(x'(z))$ at point B must be zero

(iii) $x'(z) \rightarrow x(z)$ for $x(z) >> |x(z)_B|$

Other methods are available but Levy's method is recommended as being simple to use for a wide range of fluids.

8.5   Saturated Boiling

8.5.1   Saturated Nucleate Boiling Region

The mechanism of heat transfer in this region is essentially identical to that in the subcooled region. A thin layer of liquid near to the heated surface is superheated to a sufficient degree to allow nucleation. The heat transfer coefficient and heater surface temperature variation is smooth and continuous through the thermo-dynamic boundary ($x = 0$) marking the onset of saturated boiling. The methods and equations used to correlate experimental data in the subcooled region remain valid for this region with the proviso that $T_\ell(z) = T_{SAT}$. Just as the heat transfer mechanism in the subcooled region is independent of the subcooling and, to a large degree, the mass velocity, so it may be inferred that the heat transfer process in this region is independent of the "mass quality" $x(z)$ and the mass velocity ($\dot{m}$). This, indeed, is found to be the case for experimental studies of "fully-developed nucleate boiling". Because the bulk temperature is constant in this region, the heat transfer coefficient is also constant since ΔT_{SAT} is fixed for a given heat flux and system pressure.

8.5.2   Suppression of Saturated Nucleate Boiling

To maintain nucleate boiling on the heater surface, it is necessary that the wall temperature exceed the critical value for a specified heat flux. If the wall superheat is less than given by equation (8.6) for the imposed surface heat flux then nucleation does not take place; the value of ΔT_{SAT} ($=T_W - T_{SAT}$) for comparison with this equation is calculated from the ratio ($\dot{q}/\alpha_{TP}$), where α_{TP} is the two-phase heat transfer coefficient in the absence of nucleation.

$$\dot{q}_{ONB} = \left|\frac{8\alpha_{TP}^2 \sigma T_{SAT}}{\lambda_\ell \Delta h_v \rho_g}\right| \qquad (8.25)$$

Equation (8.25) represents an equation for the boundary
between the *saturated nucleate boiling* and the *two-phase forced
convective region* (Figure 8.4)). As for the case of subcooled
nucleate boiling, this relationship is valid only on the basis of
a complete range of sizes of "active" cavities on the heating
surface. By analogy with the subcooled region, there is a similar
partial boiling transition between *fully-developed saturated
nucleate boiling* and the *two-phase forced convective* regions. In
this transition region, both the forced convective and nucleate
boiling mechanisms are significant.

Direct observations of nucleation in annular flow have been
reported by Hewitt et al (1963). The geometry used was an
internally heated annulus with a climbing film on the inner heated
surface only. The results obtained are shown in Figure (8.10) and
qualitatively confirm the expected trends. However, the values of
$\dot{q}_{ONB}$ calculated from equation (8.25) are very much lower th'n
measured. This is because of the restricted range of "active"
cavity sizes available on the stainless steel surface employed as
the heater. For such circumstances equation (8.10) should be used
to evaluate $(T_W - T_{SAT})_{ONB}$, together with the recommendations given
earlier regarding the evaluation of r_c. For the experimental
conditions shown in Figure (8.10) a maximum size of "active" cavity
present might be ~ 1 μm radius.

8.5.3 *The Two-Phase Forced Convection Region*

The *two-phase forced convective region* is most likely to be
associated with the annular flow pattern. Heat is transferred by
conduction or convection through the liquid film and vapour is
generated continuously at the liquid film/vapour core interface.
Extremely high heat transfer coefficients are possible in this
region; values can be so high as to make accurate assessment
difficult. Typical figures for water of up to 200 kW/m^2°C have
been reported.

Following the suggestion of Martinelli, many workers have

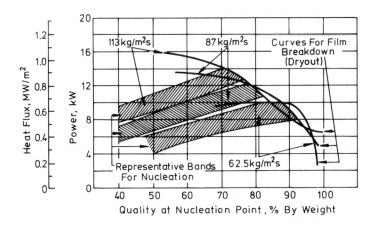

Figure 8.10: Suppression of nucleation in annular flow.

correlated their experimental results for heat transfer rates in
the two-phase forced convective region in the form:

$$\frac{\alpha_{TP}}{\alpha_{\ell o}} \left(\text{or } \frac{\alpha_{TP}}{\alpha_\ell}\right) = fn\left(\frac{1}{X_{tt}}\right) \tag{8.26}$$

where $\alpha_{\ell o}$ (and α_ℓ) is the value of the single-phase liquid heat
transfer coefficient based on the total (or liquid component) flow
and X_{tt} is the Martinelli parameter $[= (\Delta p_\ell/\Delta p_g)^{\frac{1}{2}}]$ for turbulent-
turbulent flow (see Chapter equation). A number of relationships of
the form of equation (8.26) have been proposed and in some cases
these have also been extended to cover the *saturated nucleate boiling region*. Table (8.3) summarises some of the correlations proposed. These correlations do, however, have a high mean error ($\pm$ 30%).

It can be shown that:

$$\frac{\alpha_{TP}}{\alpha_\ell} = \frac{D}{4\delta} = \frac{1}{(1-\psi)} \tag{8.27}$$

where δ is the liquid film thickness on the heating surface and
$(1-\psi)$ is the liquid hold-up. Martinelli successfully correlated
the liquid hold-up $(1-\psi)$ using the parameter X. Thus, it might
be expected that an approach such as that implied by equation
(8.26) might also be successful. However, the use of equation
(8.27), together with actual or predicted void fraction data,
yields estimates of α_{TP} which are about 50% higher than values
measured experimentally.

Chen (1966) has proposed a correlation which has been
generally accepted as one of the best available. The correlation
covers both the *saturated nucleate boiling region* and the *two-phase
forced convective region*. It is assumed that both nucleation and
convective mechanisms occur to some degree over the entire range
of the correlation and that the contributions made by the two
mechanisms are additive:

$$\alpha_{TP} = \alpha_{NcB} + \alpha_c \tag{8.28}$$

where α_{NcB} is the contribution from nucleate boiling and α_c the
contribution from convection through the liquid film.

An earlier, very similar approach which has been used
successfully for design in the process industries over a number of
years was given by Fair (1960).

In the Chen correlation, α_c, the convective contribution is
given by:

$$\alpha_c = 0.023 \left[\frac{\dot{m}(1-x)D}{\mu_\ell}\right]^{0.8} \left[\frac{\mu c_p}{\lambda}\right]_\ell^{0.4} \left[\frac{\lambda_\ell}{D}\right] (F) \tag{8.29}$$

TABLE 8.3

Various Correlations for Heat Transfer in the
Two-Phase Forced Convection and Nucleate Boiling Regions

Author, date reference	Two-phase system fluid	Equation for two-phase forced convective region	Criterion for initiation of nucleation	Modification for nucleate boiling in liquid film
Dengler & Addoms (1956)	2.54cm × 6.1m vertical steam heated tube (water)	$\dfrac{\alpha_{TP}}{\alpha_{\ell o}} = 3.5 \left(\dfrac{1}{X_{tt}}\right)^{0.5}$	$(\Delta_{SAT})_{ONB} = 7.9\, u_\ell^{0.5}$ where u_ℓ is the local liquid velocity (m/s) given by $u_\ell = \dfrac{\dot{m}(1-x)}{\rho_\ell(1-\psi)}$	$\dfrac{\alpha_{TP}}{\alpha_{\ell o}} = (F)\, 3.5 \left(\dfrac{1}{X_{tt}}\right)^{0.5}$ $F = 0.67\left[\dfrac{(\Delta\theta_{SAT} - (\Delta\theta_{SAT})_{ONB})}{\left(\dfrac{dp}{dT_{SAT}}\dfrac{D}{\sigma}\right)_W}\right]^{0.1}$
Guerrieri and Talty (1956)	1.96cm×1.83m vertical tube (various organics)	$\dfrac{\alpha_{TP}}{\alpha_{\ell o}} = 3.4 \left(\dfrac{1}{X_{tt}}\right)^{0.45}$	$\dfrac{r^*}{\delta} > 0.049$ r^* is the radius of an equilibrium bubble corresponding to the wall superheat δ is the thickness of the laminar film given by $\delta = \dfrac{10\mu_\ell}{\rho_\ell}\left[\dfrac{4\rho_\ell}{(dp/dz)D}\right]^{0.5}$	$\dfrac{\alpha_{TP}}{\alpha_\ell} = (E)\, 3.4 \left(\dfrac{1}{X_{tt}}\right)^{0.45}$ $E = 0.187 \left(\dfrac{r^*}{\delta}\right)^{-5/9}$

Table 8.3 cont.

Author, date reference	Two-phase system fluid	Equation for two-phase forced convective region	Criterion for initiation of nucleation	Modification for nucleate boiling in liquid film
Bennett et al (1961)	Internally heated vertical annulus - (water)	$\dfrac{\alpha_{TP}}{\alpha_\ell} = 0.564 \left(\dfrac{1}{X_{tt}}\right)^{0.74} \dot{q}^{0.11}$ $\dot{q}$ in W/m²		Used Rohsenow equation for nucleate boiling
Schrock and Grossman (1959) (1962)	Vertical tubes heated electrically - (water)	These authors proposed one equation for both regions of heat transfer $\dfrac{\alpha_{TP}}{\alpha_\ell} = \gamma\left[\left(\dfrac{\dot{q}}{\dot{m}\Delta h_v}\right) + m\left(\dfrac{1}{X_{tt}}\right)^n\right]$ where $\gamma = 7.39 \times 10^3$, $m = 1.5 \times 10^{-4}$ and $n = 0.66$ However, Collier et al (1962) found better agreement with their data when the modified values given by Wright (1961) ($\gamma = 6.70 \times 10^3$, $m = 3.5 \times 10^{-4}$, $n = 0.66$) were used.		

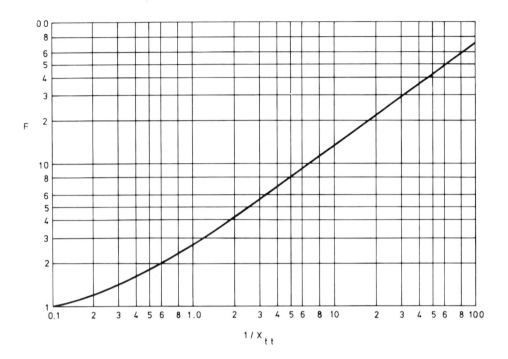

Figure 8.11: Convective boiling factor, F.

The parameter F is a function of the Martinelli parameter, X_{tt} (Figure (8.11)). The equation of Forster and Zuber (1955) was taken as the basis for the evaluation of the "nucleate boiling" component, α_{NcB}. Their pool boiling analysis was modified to account for the thinner boundary layer in forced convective boiling and the lower effective superheat that the growing vapour bubble sees. The modified Forster-Zuber equation becomes:

$$\alpha_{NcB} = 0.00122 \left[\frac{\lambda_\ell^{0.79} c_{p\ell}^{0.45} \rho_\ell^{0.49}}{\sigma^{0.5} \mu_\ell^{0.29} \Delta h_v^{0.24} \rho_g^{0.24}} \right]$$

$$\Delta T_{SAT}^{0.24} \Delta p_{SAT}^{0.75} (S) \qquad (8.29)$$

where S is a suppression factor defined as the ratio of the mean superheat seen by the growing bubble to the wall superheat (ΔT_{SAT}) S is represented as a function of the local two-phase Reynolds number Re_{TP} ($= Re_\ell \times F^{1.25}$) (Figure (8.12)).

Curve fits to the functions shown in Figure (8.11) and (8.12) for F and S respectively are:

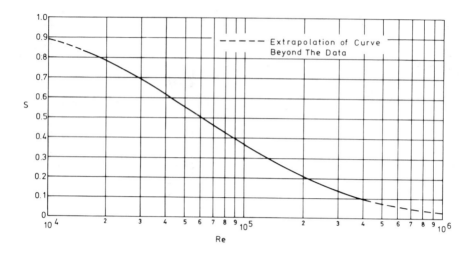

Figure 8.12: Nucleate boiling suppression factor, S.

$$F = 1 \text{ for } 1/X_{tt} \leqslant 0.1 \tag{8.30}$$

$$F = 2.35 \ (1/X_{tt} + 0.213)^{0.736} \text{ for } 1/X_{tt} > 0.1$$

$$S = 1/(1 + 2.53 \times 10^{-6} \ Re_{\ell}^{1.17}) \tag{8.31}$$

This correlation fits the available experimental data remarkably well (with a standard deviation of 11% on heat transfer coefficient).

To calculate the heat transfer coefficient (α_{TP}), at a known heat flux ($\dot{q}$), mass velocity and quality the steps are as follows:

(a)   calculate $1/X_{tt}$ (see Delhaye) or use:

$$X_{tt} \simeq \left(\frac{1-x}{x}\right)^{0.9} \left(\frac{\rho_g}{\rho_\ell}\right)^{0.5} \left(\frac{\mu_\ell}{\mu_g}\right)^{0.1}$$

(b)   evaluate F from Figure (8.11)

(c)   calculate α_c from equation (8.29)

(d)   calculate Re_{TP} from Re_ℓ ($= \dot{m}(1-x)D/\mu_\ell$) and $F(Re_{TP} = F^{1.25}Re_\ell)$

(e)   evaluate S from Figure (8.12) using the calculated value of

Re_{TP}

(f)   calculate α_{NcB} for a range of values of ΔT_{SAT} $(=T_W-T_{SAT})$

(g)   calculate α_{TP} from equation (8.28) for the range of $\Delta\theta_{SAT}$ values.

(h)   plot $\dot{q}$ $(= \alpha_{TP}\,\Delta T_{SAT})$ for the ΔT_{SAT} range against α_{TP} and interpolate (α_{TP}), at $\dot{q}$.

The correlation developed by Chen is the best available for the saturated forced convective boiling region in vertical ducts and is recommended for use with all single component non-metallic fluids. An alternative correlation by Shah, which is also valid for saturated forced convective boiling in horizontal ducts, will be discussed later.

Ananiev et al (1961) proposed a particularly simple relation-ship to correlate their data for condensation of steam in a horizontal tube at elevated pressure. It originates from the analogy between liquid film flow and single-phase flow in a pipe and may be used to get a quick estimate of the heat transfer coefficient in evaporation as well as for condensation:

$$\alpha_{TP} = \alpha_{\ell o}\sqrt{\frac{\rho_\ell}{\bar{\rho}}} \qquad (8.32)$$

where $\alpha_{\ell o}$ is the heat transfer coefficient to liquid flowing at the same total flowrate as the vapour-liquid mixture in the tube and $\bar{\rho}$ is the homogeneous mean density of the vapour-liquid mixture given by:

$$\frac{1}{\bar{\rho}} = \left[\frac{x}{\rho_g} + \frac{(1-x)}{\rho_\ell}\right] \qquad (8.33)$$

The average coefficient $\bar{\alpha}_{TP}$ for complete evaporation of a liquid is given (approximately) by:

$$\bar{\alpha}_{TP} = \frac{\alpha_{\ell o}}{2}\left[1 + \left(\frac{\rho_\ell}{\rho_g}\right)^{\frac{1}{2}}\right] \qquad (8.34)$$

These relationships will only give very approximate values of heat transfer coefficient and should not be used when (ρ_ℓ/ρ_g) is greater than 50.

8.6 Conclusions

In general the rates of heat transfer to water under conditions where the critical heat flux is not exceeded can be well represented by the correlations given in this lecture. These correlations have however not been extensively tested against liquids other than water.

Nomenclature

C_{sf} - fluid-surface constant in Rohsenow correlation   (-)

c_p - specific heat   (J/kg $^\circ$K)

D - tube diameter   (m)

F - Chen factor   (-)

f - friction factor   (-)

g_n - acceleration due to gravity   (m/s^2)

h - mass enthalpy   (J/kg)

Δh_v - latent heat of vaporisation   (J/kg)

$\dot{m}$ - mass velocity (flux)   (kg/m^{2}s)

$\dot{q}$ - heat flux   (W/m^2)

Δp - pressure drop   (N/m^2)

p - pressure   (N/m^2) (bar)

r_c - cavity radius   (m)

S - Chen suppression factor   (-)

T - temperature   ($^\circ$K)

ΔT - temperature difference   ($^\circ$K)

x' - true vapour mass quality   (-)

x - thermodynamic vapour mass quality   (-)

X - parameter defined by equation (8.8)   (-)

    - Martinelli parameter   (-)

Y_B - distance from wall to bubble tip   (m)

z - distance along channel   (m)

Greek

α - heat transfer coefficient   (W/m^2 $^\circ$K)

ε - surface roughness   (m)

λ - thermal conductivity   (W/m $^\circ$K)

δ - film thickness   (m)

μ - viscosity   (Ns/m^2)

σ - surface tension   (N/m)

$\bar{\rho}$ - homogeneous density   (kg/m^3)

ρ - density   (kg/m^3)

ψ - void fraction   (-)

τ - shear stress   (N/m^2)

Subscripts

B - pertaining to the point B

C   - cavity
    - convection
C" - pertaining to the point C"
EXP - experimental
e   - hyraulic equivalent diameter
g   - vapour or gas
h   - heated equivalent diameter
ℓ   - liquid
ℓo - total flow assumed to be liquid
MAX - maximum
NcB - nucleate boiling
ONB - onset of boiling
REF - reference
SAT - saturation
SCB - fully developed subcooled boiling
SUB - subcooling
W   - wall
TP - two-phase
tt - turbulent-turbulent

Dimensionless Groups

Pr - Prandtl Number
Pe - Peclet Number
Re - Reynolds Number
St - Stanton Number

References

Ananiev, E P, Boyko, L D and Kruzhilin, G N, (1961) "Heat transfer in the presence of steam condensation in a horizontal tube", *Int. Heat Transfer Conf., Bouler, Colorado, USA, Int. Developments in Heat Transfer, Pt. II*, Paper 34, 290-295.

Bennett, J A R, Collier, J G, Pratt, H R C and Thornton, J D, (1961) "Heat transfer to two-phase gas-liquid systems. Pt I: Steam-water mixtures in the liquid dispersed region in an annulus". *Trans. Inst. Chem. Engrs.*, **39**, 113, also *AERE-R-3519*.

Bergles, A E and Rohsenow, W M, (1964) "The determination of forced convection surface boiling heat transfer", *Trans. ASME, J. of Heat Transfer*, **86c**, 365.

Bowring, R W (1962) "Physical model based on bubble detachment and calculation of steam voidage in the subcooled region of a heated channel", *OECD Halden Reactor Project Report, HPR-10*.

Chen, J C, (1966) "Correlation for boiling heat transfer to saturated liquids in convective flow". *Int. Eng. Chem. Process Design and Development*, **5**, 322.

Collier, J G, (1972) "Convective boiling and condensation", Published by McGraw Hill Book Co (UK) Ltd.

Collier, J G and Pulling, D J, (1962) "Heat transfer to two-phase gas-liquid systems. Pt. II: Further data on steam/water mixtures in the liquid dispersed region in an annulus". *AERE-R3809*.

Davis, E J and Anderson, G H, (1966) "The incipience of nucleate boiling in forced convection flow". *A.I.Ch.E. Journal*, **12** (4), 774-780.

Dengler, C E and Addoms, J N, (1956) "Heat transfer mechanism for vaporisation of water in a vertical tube". *Chem. Eng. Prog. Symp. Series*, **52** (18), 95-103.

Fair, J R, (1960) "What you need to design thermosyphon reboilers". *Petroleum Refiner*, **39** (2), 105-124.

Forster, H K and Zuber, N, (1955) "Dynamics of vapour bubbles and boiling heat transfer". *A.I.Ch.E. Journal*, **1** (4), 531-535.

Frost, W and Dzakowic, G S, (1967) "An extension of the method of predicting incipient boiling on commercially finished surfaces". *Paper presented at ASME A.I.Ch.E. Heat Transfer Conf.* Paper 67-HT-61, Seattle.

Guerrieri, S A and Talty, R D, (1956) "A study of heat transfer to organic liquids in single tube natural circulation vertical tube boilers". *Chem. Engng. Prog. Symp. Series, Heat Transfer, Louisville.* **52** (18), 69-77.

Hewitt, G F, Kearsey, H A, Lacey, P M C and Pulling, D J, (1963) "Burnout and nucleation in climbing film flow". *AERE-R-4374.*

Jens, W H and Lottes, P A, (1951) "Analysis of heat transfer burnout, pressure drop and density data for high pressure water". *ANL-4627.*

Krieth, F and Summerfield, M, (1949) "Heat transfer to water and high flux densities with and without surface boiling". *Trans. ASME*, **71** (7), 805-815.

Levy, S, (1967) "Forced convection subcooled boiling prediction of vapour volumetric fraction". *Int. J. Heat Mass Transfer*, **10**, 951-965.

Piret, E L and Isbin, H S, (1953) "Two-phase heat transfer in natural circulation evaporators". *A.I.Ch.E. Heat Transfer Symposium, St. Louis. Chem. Engng. Prog. Symp. Series*, **50** (6), 305.

Rohsenow, W M, (1952) "A method of correlating heat transfer data for surface boiling of liquids". *Trans. ASME*, 74, 969.

Rohsenow, W M, (1952) "Heat transfer with evaporation". *Heat Transfer. A symposium held at the University of Michigan during the summer of 1952,* Published by University of Michigan Press, 101-150.

Rohsenow, W M and Clarke, J A, (1951) "Heat transfer and pressure drop data for high heat flux densities to water at high sub-critical pressure". *1951 Heat Transfer and Fluid Mechanics Institute*, Stanford University Press, Stanford, California.

Saha, P and Zuber, N, (1974) "Point of net vapour generation and vapour void fraction in subcooled boiling". *5th Int. Heat Transfer Conf. Tokyo,* Paper B4.7.

Schrock, V E and Grossman, L M, (1959) "Forced convection boiling studies". Forced convection vaporization project - final report 73308-UCZ21821, Nov. 1952 University of California, Berkeley, USA.

Schrock, V E and Grossman, L M, (1962) "Forced convection boiling in tubes". *Nuclear Science and Engng.* 12, 474-480.

Thom, J R S, Walker, W M, Fallon, T A and Reising, G F S, (1965) "Boiling in subcooled water during flow up heated tubes or annuli". *Paper presented at the Symposium on Boiling Heat Transfer in Steam Generating Units and Heat Exchangers,* Manchester 15-16 September 1965 by Inst. of Mech. Engrs. (London). Paper No 6.

Wright, R M, (1961) "Downward forced convection boiling of water in uniformly heated tubes". *UCRL-9744.*

Chapter 9

Burnout

G. F. HEWITT

9.1 Introduction

For the purposes of this chapter, "burnout" is defined as follows:

(1) For a surface with a controlled heat flux (eg with electrical heating, radiant heating or nuclear heating), burnout is defined as that condition under which a small increase in the surface heat flux leads to an *inordinate* increase in wall temperature.

(2) For a surface whose wall temperature is controlled (eg one heated by a condensing vapour), burnout is defined as that condition in which a small increase in wall temperature leads to an *inordinate* decrease in heat flux (or heat transfer coefficient).

The term "burnout" is taken as being synonymous with the terms "critical heat flux (CHF)", "departure from nucleate boiling (DNB)", "boiling crisis" and "dryout". There is no implication that the rise in surface temperature is sufficient to cause melting of the surface, though this may sometimes happen.

In the first part of this chapter, the effects of various system parameters on burnout will be described and this is followed by a discussion of correlations for burnout. The need for better correlations has given greater emphasis to the study of physical mechanisms of burnout and these mechanisms are discussed in the next part of the chapter. Finally, developments in the analytical prediction of burnout in annular flow are discussed.

9.2 Data for Burnout: Effect of System Parameters

No attempt will be made, in the context of this chapter, to give a comprehensive survey of burnout data. Further information is given, for instance, in the book by Collier (1972) and in the review by Hewitt (1978). However, it is useful to set the scene for a discussion of the burnout phenomenon by describing briefly the effects of some system parameters on the onset of burnout.

Parametric effects on onset of burnout in vertical round tubes:

In the normal burnout experiment, with a uniformly heated tube,
burnout occurs first at the end of the tube. Figure 9.1 shows
conceptually the effect of the various system parameters for a
given fluid at a given pressure. The following parametric effects
are illustrated:

(a) For a fixed mass flux (G), tube length (L) and tube inside
     diameter (d_o) the burnout heat flux (ϕ_{BO}) varies approximately
     linearly with the inlet sub-cooling (ie the difference in enthalpy
     between saturated liquid and inlet liquid) Δi_{sub}. This linear
     relationship is obeyed over fairly wide ranges, but it has no
     fundamental significance. If a very wide range of inlet sub-
     cooling is used, then departures from linearity are observed.

(b) For fixed L and Δi_{sub}, ϕ_{BO} rises approximately linearly with
     G at low values of G, but then rises much less rapidly for
     high G values.

(c) For fixed G, d_o and Δi_{sub}, burnout heat flux decreases with
     increasing tube length L as shown in Figure 9.1. However,
     the power input required for burnout, P_{BO}, increases at first
     rapidly, and then less rapidly, also as shown in Figure 9.1.
     For very long tubes, the power to burnout may appear to
     asymptote to a constant value independent of tube length in
     some cases. Again, this only applies over a limited range of
     length.

(d) For fixed G, Δi_{sub} and L, ϕ_{BO} increases with tube diameter
     d_o, the rate of increase decreasing as the diameter increases
     (see Figure 9.1).

The parametric effects illustrated in Figure 9.1 are typical
of those encountered for upwards flow; experimental data for
downwards flow shows surprising little difference to up flow as
illustrated by the results of Bertoni et al. (1976) shown in
Figure 9.2.

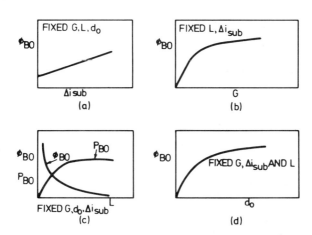

Figure 9.1: Parametric effects on burnout in upwards flow in a round tube.
Sub-cooled liquid fed to the entrance of the tube, uniform heat
flux, fixed pressure, given fluid.

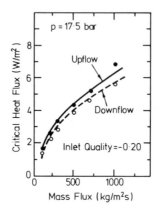

Figure 9.2: Comparison of up flow and down flow critical heat flux (burnout)
            values for the evaporation of Refrigerant-12 in a vertical round
            tube. Bertoni et al. (1976).

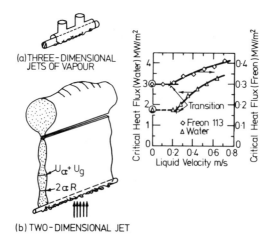

Figure 9.3: Burnout in cross flow over tubes. Lienhard & Eichorn (1976).

Burnout in cross flow over tubes: This case was investigated
by Lienhard & Eichorn (1976) and their results are illustrated in
Figure 9.3. For low cross flow velocities, the vapour release
from the tube followed the characteristic 3-dimensional jet form
also occuring in pool boiling. The critical (burnout) heat flux
was, in this region, close to its value for pool boiling as shown
in Figure 9.3. At high liquid velocities, the pattern of vapour
release changes to a 2-dimensional form as shown and the critical
heat flux begins to increase with increasing liquid velocity.

Burnout in annuli and rod bundles. A wide variety of data has been obtained for burnout in annuli. The two main reasons for the interest in this geometry are:

(1) In the nuclear industry, the main interest is in burnout in rod bundle geometries. The annulus can be regarded as a "single rod bundle".

(2) By making the outer channel wall transparent, it is sometimes possible to view the processes occurring on the inner (heated) surface, throwing light on the mechanism of burnout.

Most of the data obtained have been for the case where the inner surface only is heated. However, more recently, data has begun to appear where both surfaces are heated and the fraction of the power input to, say, the outer surface is varied. Typical of this later data is that of Jensen & Mannov (1974), some of which is illustrated in Figure 9.4. For a fixed inlet sub-cooling, the critical quality (quality at burnout) initially increases as the fraction of power on the outer surface is increased. In this region, burnout occurs first on the *inner* surface. As the fraction of power on the outer surface is further increased, a maximum burnout quality is reached, and beyond this point burnout begins to occur first on the outer surface and the critical quality decreases with increasing fractional power on that surface.

A wide range of data is available for rod bundles of various geometries. A listing of recent data sources is given by for instance, Hewitt (1980). An interesting example of measurements of burnout power in rod bundles is the so-called "odds-and-evens" tests of Macbeth (1974). These tests were carried out with refrigerant-12 with a fixed outer pressure tube diameter but with different numbers of rods and with different spacings. Some of the results are illustrated in Figure 9.5. Of special interest is the fact that the dryout power for a 30-rod uniformly spaced cluster of rods was higher than that for a 36-rod cluster. Macbeth explained this on the basis of oscillations occurring in the outer sub-channels. Here, a sub-channel is defined as the flow area bounded by a line through the centres of two adjacent outer rods, by lines through the centres of these two outer rods normal to the outer wall and by the outer wall itself. If there are an *even* number of these outer sub-channels (as occurs in this case of the

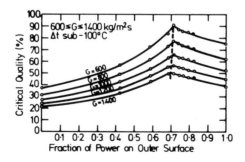

Figure 9.4: Burnout in bi-laterally heated annular channels showing the effect of power distribution between the inner and outer surface on critical quality. Jensen & Mannov (1974).

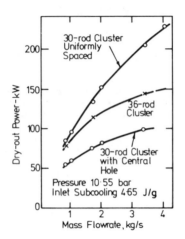

Figure 9.5: Data for burnout (dryout) power for refrigerant-12 flow over multi-
rod bundles. The "odds-and-evens" tests of Macbeth (1974).

30-rod cluster) then transverse oscillations occur leading to
better mixing between adjacent sub-channels and, hence, to a higher
burnout power. With an *odd* number of outer sub-channels, the
oscillations do not take place and, as in the 36-rod cluster, the
burnout power is consequently lower. An alternative explanation
for the difference illustrated in Figure 9.5 is that the effect
of the rod support grids is different in the cases of the 36- and
30-rods clusters respectively. However, this particular set of
data represents a critical test of any modelling methods and we
shall return to discuss it again in that context in the final part
of this chapter.

    Burnout in horizontal and horizontal hair pin tubes: In
horizontal tubes, the burnout heat flux is very much lower than
that observed for the same conditions in vertical flow. This is
illustrated by the data of Merilo (1977) which are reproduced in
Figure 9.6. Note that this figure contains data for both water
and refrigerant, the refrigerant data being "scaled" to be
equivalent to water data using the method of Ahmad (1973) which
is described in section 9.3 below. Studies of dryout in serpentine
tube evaporators are reported by Fisher & Yu (1975). The test
section was in the form of a "hair pin" with the bend separating
the two horizontal tubular sections being in the vertical plane.
Fisher and Yu found that burnout could occur immediately after the
bend, with rewetting of the tube being subsequently re-established
and, in some circumstances, a further dryout (burnout) occuring at
the end of the upper horizontal leg. Fisher and Yu plotted their
results in terms of a map (in the co-ordinates of local quality
and number of tube diameters from the bend) in which the contours
represented the boundaries between regions in which the tube was
completely wet and regions in which the tube was partially dry,
for various levels of heat flux as illustrated in Figure 9.7. The
particular application of this kind of evaporator was to gas-cooled
nuclear reactor boilers. Although the dry-out phenomenon did not
lead to physical melting of the tubes, it could have serious
implications on the occurrence of localised concentrations of

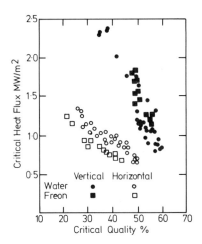

Figure 9.6: Comparison of vertical and horizontal critical heat flux data for a 1.26cm bore tube. Water pressure 6.9 MPa freon data modelled by Ahmad (1973) criteria (Merilo, 1977).

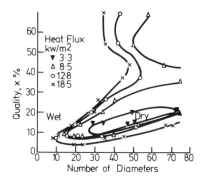

Figure 9.7: Effect of heat flux on development of dry regions on top of horizontal leg of a serpentine tube. Refrigerant-12 data of Fisher & Yu (1975).

dissolved salts and consequent enhanced corrosion rates.

9.3 Correlation of Burnout Data

The technological importance of the burnout phenomenon has led to the development of a very large number of alternative burnout correlation methods. It is beyond the scope of the present chapter to give details of all these alternatives, but an attempt is made to put the methods into perspective by discussing first of all the types of correlation adapted for burnout in tubes, by briefly surveying the alternative methods for sub-channel analysis and, finally, by referring briefly to prediction methods for non-aqueous fluids.

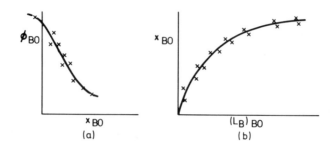

Figure 9.8: Flux/quality and quality/boiling length correlation approaches.

Correlation methods for burnout in round tubes: For a given
mass flux, fluid physical properties (ie pressure for a given
fluid), tube diameter and for uniform heat flux, it is found that
the data for a range of tube length and inlet sub-cooling can be
represented approximately by a single curve of burnout heat flux
ϕ_{BO} against burnout quality x_{BO} as sketched in Figure 9.8(a).
Indeed, Figure 9.6 is an example of such a plot. The superficial
implication of the relationship between burnout flux and quality
at burnout is that the local quality conditions govern the magnitude of the burnout heat flux *at that locality;* this is termed the
"local conditions hypothesis".

The *same data* can also be plotted in terms of burnout
quality x_{BO} and "boiling length" at burnout $(L_B)_{BO}$. The boiling
length is defined as the distance from the point in the channel
at which bulk saturation (ie zero quality) conditions are attained.
This type of plot is illustrated in Figure 9.8(b). This plot can
be regarded as indicating a relationship between the fraction
evaporated at burnout (x_{BO}) as a function of the boiling length
to burnout, indicating the possibility of some "integral" rather
than "local" phenomenon. In fact, it is easy to transform the ϕ/x
relationship into the x/L_B relationship; the boiling length is
given from a heat balance as

$$L_B = d_o \; G \; \lambda \; x/4\phi \qquad\qquad (9.1)$$

where d_o is the tube diameter, G the mass flux, λ the latent heat
of evaporation, x the local quality and ϕ the heat flux. If the
burnout flux is related to the burnout quality by the expression:

$$\phi_{BO} = f_1 \; (x_{BO}) \qquad\qquad (9.2)$$

then it follows that:

$$(L_B)_{BO} = d_o \; G \; \lambda \; x_{BO}/4 \; f_1 \; (x_{BO}) \qquad\qquad (9.3)$$

For the given tube diameter and mass flux, it follows, therefore, that:

$$x_{BO} = f_2 \, ((L_B)_{BO})$$ (9.4)

The vast majority of correlations for burnout fall either into the ϕ/x or the x/L_B categories. An example of the flux/quality type of correlation is that of Bowring (1972) and an example of the quality/boiling length type of correlation is that of Hewitt et al. (1969). Since one form of correlation can be readily transformed into the other, the choice should always be based on the range of data covered and the accuracy with which it is predicted. Thus, the Bowring (1972) correlation for water up flow in round tubes has the advantage of a fairly wide data base and rather low r.m.s. error. However, with all multi-parameter fit correlations of this type, it is important to critically examine the data base wherever possible. Often, very large deviations can occur when correlations are used outside the range of data on which they were based and, sometimes, physically impossible predictions are made. The burnout heat flux should be expected to lie between the limits corresponding to the flux required to bring the wall up to the saturation temperature and the flux required to evaporate all the fluid. Any values calculated outside this range must be wrong!

Although the ϕ/x and x/L_B correlations are equivalent for *uniformly* heated channels, they give quite different results when the heat flux is *non uniform*. The question obviously arises as to which of the two forms is best suited to the prediction of burnout with non-uniform heating, the case which occurs most often in practical application.

Figure 9.9 shows a comparison between data for uniform heating

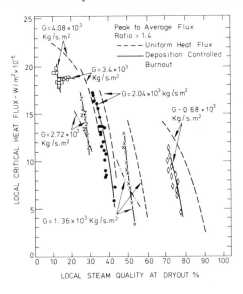

Figure 9.9: Data for critical heat flux for a cosine heat flux distribution compared with data for uniform flux (Keeys et al., 1971).

and non-uniform (cosine distribution) heating where *local* values of burnout flux and burnout quality are plotted. As will be seen, there is a considerable difference in heat flux for burnout at a given quality for the uniform and non-uniform heated tubes respectively. Note that, with the non-uniform heating, burnout could occur first *up-stream* of the end of the channel.

These discrepancies in the ϕ/x representation of burnout, led Tong et al. (1966) to propose the so-called "F-factor" method for the prediction of burnout flux in non-uniformly heated channels. Their method is illustrated schematically in Figure 9.10; the three curves shown in this figure are respectively:

(a)   The flux profile $\phi(z)$.

(b)   The burnout flux $\phi_{BO}(z)_u$ predicted for the local quality condition using a standard correlation for uniform heating.

(c)   The flux to burnout $\phi_{BO}(z)_{nu}$ for the non-uniform heating situation calculated by multiplying $\phi_{BO}(z)_u$ by the F-factor defined as:

$$F = \phi_{BO}(z)_{nu}/\phi_{BO}(z)_u \qquad\qquad (9.5)$$

The calculation of the conditions for the onset of burnout proceeds by progressively adjusting the level of flux, calculating the curve for $\phi_{BO}(z)_u$, calculating F from an appropriate correlation and thus calculating the curve for $\phi_{BO}(z)_{nu}$, and testing to see whether the flux profile touches this latter curve, in which case burnout is predicted to occur. Tong et al. give the following semi-theoretical expression for F:

$$F = \frac{\Omega}{1-\exp(\Omega\, z_{BO})} \int_0^{z_{BO}} \frac{\phi(z)}{\phi(z_{BO})_u} \exp\left[\Omega\,(z_{BO}-z)\right]dz \qquad (9.6)$$

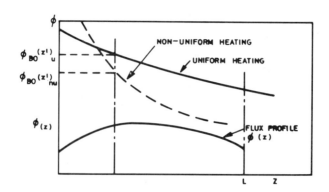

Figure 9.10: Basis of the Tong F-factor method for prediction of burnout with non-uniform heat flux (Tong et al., 1966).

where z_{BO} is the distance from the channel inlet to the point at which the burnout flux is being predicted and Ω is given by the empirical expression:

$$\Omega = 0.44 \frac{\left[1-x\ (z_{BO})\right]^{7.9}}{(G/10^6)^{1.72}} \qquad (9.7)$$

where G is in $lb/h.ft^2$.

The F-factor method is somewhat tedious to apply though it is probably the best approach for low-quality (typically less than 5-10%) conditions. For higher qualities, it transpires that the x/L_B approach fits both non-uniform and uniform heat flux data. This is illustrated for some Harwell data in Figure 9.11. Thus, for qualities above, say, 10%, the burnout conditions for non-uniform heating are probably best calculated by applying the x/L_B form of correlation on a local basis. A further discussion of procedures is given by Hewitt (1980).

Correlation of burnout for rod bundles (sub-channel analysis): It is possible to correlate data for burnout in rod bundles geometries by using overall, mixed flow, types of correlations and these are moderately successful. However, much more precise predictions can be made by using correlations based on "sub-channel analysis". In this form of analysis, the rod bundle is sub-divided into flow zones or sub-channels and conditions are calculated for these individual sub-channels, taking account of cross-mixing between adjacent channels and of cross-flows generated by pressure differences between the sub-channels. Two types of sub-channel have been employed as illustrated in Figure 9.12:

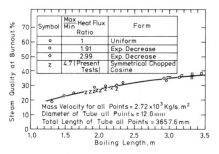

Figure 9.11: Data for various flux shapes represented on a quality/boiling length basis (Keeys et al., 1971).

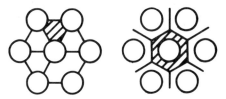

(a) "Classical" Sub-channels (b) "Rod Centred" Sub-channels

Figure 9.12: Types of sub-channel division used in rod-bundle critical heat flux analysis.

(a)   "Classical" sub-channels which are bounded by lines joining
      the centres of individual rods as illustrated. A discussion
      of the application of this form of sub-channel analysis to
      the prediction of burnout in rod bundles is given by, for
      instance, Weisman and Bowring (1975).

(b)   "Rod centred" sub-channels in which the flow zone is bounded
      by lines of symmetry as illustrated in Figure 9.12(b). The
      application of this form of sub-channel analysis is discussed,
      for instance, by Gaspari et al. (1974).

     In the classical sub-channel method, the usual practice is to
devise specific burnout correlations for the sub-channel; in the
application of the rod-centred sub-channel method, Gaspari et al.
were able to employ a standard round-tube burnout correlation to
the sub-channels (on an equivalent diameter basis) with reasonable
success.

     Correlations for non-aqueous fluids: scaling laws: Methods
of prediction of burnout for non-acqueous fluids are of interest in
two main contexts:

(1)   In applications to evaporation of non-aqueous fluids such
      as those encountered in, for instance, thermosyphon reboilers
      in the petroleum and chemical industries.

(2)   In the use of non-aqueous fluids (and in particular refriger-
      ant) as "scaling fluids" for water. It is often convenient
      to use refrigerants to carry out experiments simulating water
      at much higher pressures and heat fluxes, thus reducing the
      cost of the experimental equipment and power. The refrigerant
      burnout data is related to that for the equivalent water
      conditions by "scaling laws", and this subject has met with a
      considerable amount of attention in the nuclear industry.

     Perhaps the most successful set of scaling laws are those
proposed by Ahmad (1973). For fixed vapour-to-liquid density ratio
(ρ_G/ρ_L), length-to-diameter ratio (L/d_o) and fixed inlet sub-cooling
to latent heat ratio $(\Delta i_{sub}/\lambda)$, Ahmad suggests that there should be
a relationship between the "boiling number" $(\phi/G \lambda)$ and the dimensionless group:

$$\psi = (\frac{G d_o}{L}) (\frac{\sqrt{\gamma} \ \mu_L}{\sqrt{\rho_L} \ d_o})^{2/3} (\frac{\mu_L}{\mu_G})^{1/8} \tag{9.8}$$

where μ_L and μ_G are the viscosities of the liquid and gas phases
respectively, d_o the tube diameter, G the mass flux and γ the rate
of change (along the saturation line) of the density ratio with
pressure p:

$$\gamma = [\partial (\rho_L/\rho_G)/\partial p] \tag{9.9}$$

A plot of data by Hauptmann (1973) in terms of the above dimensionless groups is illustrated in Figure 9.13.

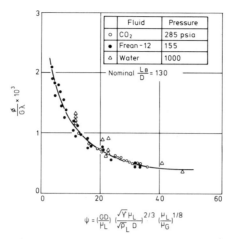

Figure 9.13: Application of Ahmad (1973) scaling method for water, CO_2 and refrigerant data (Hauptmann et al., 1973).

To calculate the burnout heat flux for an organic fluid, values for water for the same L/d_o, ρ_G/ρ_L and $\Delta i_{sub}/\lambda$ are either obtained from available data banks, or are calculated using a correlation for water, and are then plotted in the format shown in Figure 9.13. The organic fluid values may then be calculated by estimating the appropriate values of ψ (equation 9.8) and reading off equivalent values for the boiling number, $\phi/G \lambda$.

9.4 Mechanisms of Burnout

An understanding of the mechanism of burnout is useful firstly in the development of improved correlation and prediction methods, and secondly in devising means for avoiding the occurrence of the phenomenon. A great deal of work has been done in this area and detailed reviews are presented, for instance, by Tong & Hewitt (1972) and Hewitt (1980).

Burnout mechanisms and their regions of operation: A large number of alternative mechanisms for burnout have been proposed but the four which appear to have been reasonably well established experimentally are illustrated in Figure 9.14 and are as follows:

(a)   Formation of hot spot under growing bubble (Figure 9.14(a)). Here, when a bubble grows at the heated wall, a dry patch forms underneath the bubble as the micro-layer of liquid under the bubble evaporates. In this dry zone, the wall temperature rises due to the deterioration in heat transfer; when the bubble departs, the dry patch may be rewetted and the process repeats itself. However, if the temperature of the dry patch becomes too high, then rewetting does not take place and gross local over heating (and hence burnout) occurs. This mechanism is proposed, for instance, by Kirby (1967).

(b)   Near-wall bubble crowding and inhibition of vapour release (Figure 9.14(b)). Here, a "bubble boundary layer" builds up

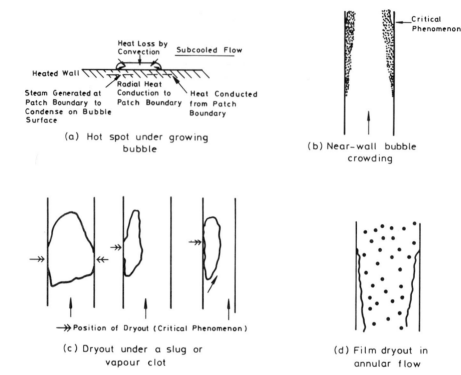

Figure 9.14: Mechanisms of burnout.

     on the surface and vapour generated by boiling at the surface must escape through this boundary layer. When the boundary layer becomes too crowded with bubbles, vapour escape is impossible and the surface becomes dry and overheats giving rise to burnout. This mechanism is discussed, for instance, by Tong et al. (1966).

(c)   Dryout under a slug or vapour clot. In plug flow, the thin film surrounding the large bubble may dry out giving rise to localised overheating and hence burnout. Alternatively, a stationary vapour slug may be formed on the wall with a thin film of liquid separating it from the wall; in this case, localised drying out of this film gives rise to overheating and burnout. This mechanism has been investigated, for instance, by Fiori & Bergles (1968).

(d)   Film dryout in annular flow (Figure 9.14(d)). Here, in annular flow, the liquid film dries out due to evaporation and due to the partial entrainment of the liquid in the form of droplets in the vapour core. This mechanism is discussed in more detail below.

     A tentative map of the regions of operation of the above mechanisms was suggested by Semeria & Hewitt (1974) and is illustrated in Figure 9.15, the regions of operation of the

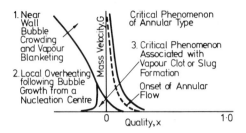

Figure 9.15: Tentative map of regions of operation of various critical heat
            flux mechanisms (Semeria & Hewitt, 1974).

respective mechanisms being shown as a function of mass flux G
and quality x.

      Burnout mechanisms in annular flow: Burnout in annular flow
is the most important mechanism from a practical point of view
since, for channels of reasonable length, the first occurrence of
burnout is likely to occur in this region. Perhaps the most
important tool in investigating annular flow burnout is the
measurement of film flow rate. This is achieved by extracting
the liquid film through a porous wall section; the entrained
droplets are not extracted though, to achieve complete separation,
some of the vapour must also be withdrawn. The technique was
discussed briefly in Chapter 5 and a more detailed discussion is
given by Hewitt (1980).   Film flow rate measurements have been
used to investigate burnout mechanisms in two different ways:

(1)   Measurement of the film flow rate at the end of a heated
      channel as a function of power input to the channel. Results
      obtained using this approach are exemplified in Figure 9.16.
      As will be seen, the film flow rate at the end of the channel
      decreases with increasing power and the onset of burnout
      corresponds quite closely to the point at which the film flow
      rate becomes zero. This confirms the film evaporation (as

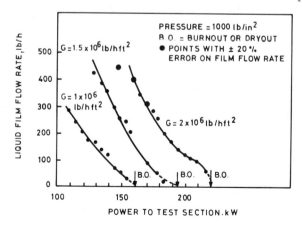

Figure 9.16: Variation of film flow with input power at the end of a uniformly
            heated round tube in which water is being evaporated at 70 bars
            (Hewitt, 1970).

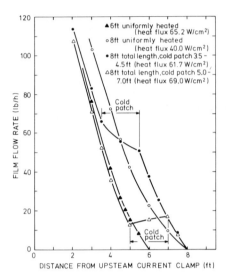

Figure 9.17: Variation of film flow rate with distance with a uniformly and non-
uniformly heated round tube. Water evaporation at low pressure.
Burnout condition occurring at the end of the tube in each case.
Data of Bennett et al. (1966).

distinct from a film boiling, for instance) mechanism for
burnout.

(2)   Although measurements of film flow rate at the end of the
test section are useful in demonstrating the mechanism, they
do not demonstrate the conditions *along the test section* which
have led to the occurrence of the burnout phenomenon.   To
achieve this, measurements of film flow rate may be made along
the test section for a heat flux corresponding to the burnout
heat flux.   This is achieved by first determining the burnout
heat flux and then, for constant inlet conditions, and for
the known burnout heat flux, making measurements of the film
flow rate at the end of channels of various heated lengths,
less than the length of the channel for which the burnout
conditions had been determined.   Results of this form are
illustrated in Figure 9.17; for channels with uniform heating,
the film flow rate decreases along the channel length, going
to zero at the end of the channel where burnout occurs.   Also
shown on Figure 9.17 are the results for variation of film
flow rate with distance for cases where zones ("cold patches")
exist along the channel where the heat flux is zero.   As will
be seen from Figure 9.17, the rate of change of film flow rate
with length is different in these zones and can even change
in sign (when the unheated zone is near the end of the test
section).   An explanation of these effects follows from a
replot of the data shown in Figure 9.17 in the form of
entrained liquid flow rate (ie total liquid flow rate minus
the measured film flow rate) against *local quality*.   This
plot is shown in Figure 9.18.   The dotted line represents
the condition of "hydrodynamic equilibrium" where the rate
of entrainment of droplets are equal and opposite to the rate

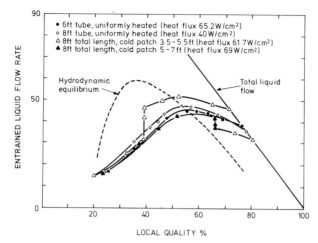

Figure 9.18: Plot of entrained liquid flow as a function of quality for the evaporation of water in a tube at low pressure. (Data of Bennett et al. 1966).

of deposition. This hydrodynamic equilibrium is only achieved in very long tubes at constant quality. As will be seen from Figure 9.18, systems in which the quality is changing (ie with evaporation) do not usually have hydrodynamic equilibrium conditions. In the lower part of the channel, the entrained liquid flow rate is less than that for equilibrium and in the upper part of the channel it is greater. Burnout corresponds to the point at which the entrained liquid flow is equal to the total liquid flow, with the film flow being zero. On Figure 9.18, the "cold patch" regions show an approach towards the equilibrium at constant quality, the entrained flow increasing at the lower quality or decreasing at the higher quality as shown. This leads to the situation in which the quality for burnout (and, thus, the power input) can be actually greater if part of the tube is unheated. This occurs when there is net deposition in the cold patch zone.

Further data of the above types is gradually being accumulated and generally confirms the interpretation.

Effect of heat flux on deposition/entrainment: The results illustrated in Figures 9.17 and 9.18 indicate that burnout occurs when the processes of droplet entrainment, droplet deposition and evaporation lead to a condition in which the film flow rate becomes zero. The rate of evaporation can be calculated from the local heat flux, provided the latent heat is known; for adiabatic flows, the rates of entrainment and deposition can be calculated using the methods described for annular flow in Chapter 5. The important question which arises is whether the rate of entrainment and/or deposition is affected by the presence of a heat flux normal to the surface. The effects which could be caused by a heat flux include:

(1)   The effect of nucleate boiling in the film giving rise to
      additional entrainment due to the bursting of bubbles through
      the film surface and

(2)   The effect of a vapour flux away from the surface inhibiting
      drop deposition onto the surface in the presence of evapora-
      tion.

One type of experiment which might be expected to throw
light on the heat flux effect is that in which a known liquid film
flow rate is injected at the start of a heated channel and the
film flow rate at the end of the channel measured as a function
of the power input. If the film flow rate at the end of the
channel falls more rapidly than can be expected due to pure
evaporation of the film, then it might be deduced that the evaporation is causing enhanced entrainment and vice versa. Results of
this type (obtained for a vertical annulus with a film flow on
the inner surface) were obtained by Hewitt et al. (1963) and are
illustrated in Figure 9.19. Both positive and negative variations
with respect to "pure evaporation" were observed, depending on
the initial quality. A somewhat analogous experiment was undertaken recently by Styrikovich et al. (1978), in this case for a
horizontal film flow in a rectangular channel. Unfortunately,
the results from experiments of this type are somewhat equivocal.
This can be illustrated by considering the experiments in terms
of the "entrainment diagram" (cf. Figure 9.18). Results of type
A and C (Figure 9.19) are represented conceptually in the entrainment diagram form in Figure 9.20. For a given inlet quality x_{in},
the loci of local entrained liquid flow rate can be plotted for
various heat flux levels up to and including that heat flux level
at which burnout occurs, giving an outlet quality x_C. At the low
quality condition (corresponding to curve A in Figure 9.19) the
shift in quality from x_{in} gives rise to a higher value for
entrained liquid flow rate for *hydrodynamic equilibrium* as
illustrated. This gives rise to an increase in actual entrainment
before burnout, this increase being associated with a change in
the system hydrodynamics rather than a *direct* effect of heat flux
(eg bubble nucleation). Similarly, at high quality, the shift in

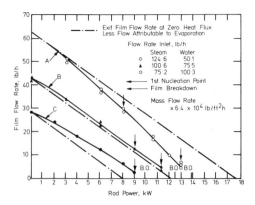

Figure 9.19: Variation of film flow with increasing power for the evaporation
             of a climbing liquid film on the inner surface of an annulus at
             low pressure (data of Hewitt et al. 1963).

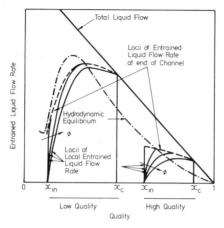

Figure 9.20: Conceptual entrainment diagram for cases where liquid film is injected (initial entrainment zero).

the hydrodynamic equilibrium gives rise to a decrease in entrained liquid flow rate at the end of the channel as illustrated and this would be the kind of effect shown in curve C in Figure 9.19.

A more direct evaluation of the effect of heat flux is obtained by determining the entrained liquid flow rate (ie a difference between the total liquid flow rate and the measured film flow rate) as a function of length in a channel, the first part of which is heated and the second part unheated. Any variation of the net rate of entrainment with heat flux would be detected in a plot of entrained flow rate versus length. Experiments of this type are reported by Bennett et al. (1966) and the results are illustrated in Figure 9.21. Depending on the point at which the unheated zone started, the entrained liquid flow rate could either continue to increase, remain approximately constant or continue to decrease in the unheated zone. In no case was there a marked change in the net rate of deposition or entrainment. Similar results are reported for high pressure steam-water flows by Keeys et al. (1970). Thus, it would seem that the rate of

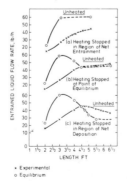

Figure 9.21: Effect of heat flux on rate of change on entrained liquid flow with length. Data of Hewitt (1970).

entrainment or deposition is relatively unaffected by heat flux
for the range of conditions studied in these experiments. This is
not, of course, to say that heat flux effects are *never* important,
but it does give some confidence in applying prediction methods
(such as those described in section 9.5 below) which take no
account of the direct effect of heat flux.

Burnout mechanisms in horizontal channels: Mechanisms in
horizontal channels are discussed by Fisher et al. (1978). Based
on studies over a wide range of pressures and qualities, it was
concluded that there are three principal modes of burnout.

(a)   At very low qualities, a relatively stable stratification of
      the flow occurs with the formation of "ribbons" of vapour at
      the upper part of the channel, leading to over heating in that
      region.

(b)   At low and intermediate qualities, Fisher et al. describe the
      mechanism introduced by Coney (1974) and illustrated in
      Figure 9.22. A "frothy surge" passes along the channel and
      wets the upper surface of the tube. The film thus deposited
      drains away and (in the case of the heated tubes) is also
      evaporated. If the drainage and evaporation are such that
      complete film removal (dryout) occurs before the arrival of
      the next frothy surge then an intermittent dryout and over-
      heating (and subsequent corrosion) can occur.

(c)   At high qualities, annular flow occurs but, for horizontal
      tubes, the liquid tends to concentrate at the bottom of the
      tube, the film at the upper surface being much thinner.
      Thus, dryout occurs at much lower heat fluxes as was illus-
      trated in Figure 9.6. Fisher et al. describe a model for a
      film replenishment and drainage in horizontal annular flow.

9.5   Prediction of Burnout in Annular Flow

As was indicated in Chapter 6, the form of burnout most
widely encountered in practice is that occurring in the annular
flow regime resulting from depletion of the liquid film due to
liquid entrainment and evaporation. Since, over a range of
conditions, the entrainment/deposition processes are relatively
unaffected by the presence of a heat flux (see Section 9.4 above),
the annular flow prediction methods described in Chapter 5 can be
applied to burnout prediction in the annular flow region. The
procedure is to write a mass balance equation for the liquid
film and to integrate this equation along the channel from known
(or assumed) boundary conditions, and determine the point at
which the film flow rate goes to zero.

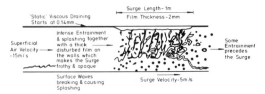

Figure 9.22: Splashing and draining mechanism suggested by Coney (1974).

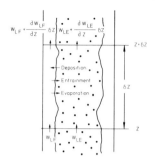

Figure 9.23: Basis of mass balances for liquid film in annular flow.

The principle of the mass balance equation is illustrated in Figure 9.23 where the changes in film flow rate W_{LF} (or entrained liquid flow rate, W_{LE}) over incremental length δz are shown. Defining a liquid film superficial mass flux G_{LF} as W_{LF}/A where A is the channel cross-section, we have the following equation resulting from the mass balance:

$$\frac{A}{S} \frac{dG_{LF}}{dz} = D - E - \frac{\phi}{\delta} \qquad (9.10)$$

where S is the channel periphery, D the deposition rate per unit peripheral area, E the entrainment rate per unit peripheral area, ϕ the heat flux and λ the latent heat of vaporisation. For the specific case of a round tube (illustrated in Figure 9.23, we have:

$$\frac{dG_{LF}}{dz} = \frac{4}{d_0} [D - E - \frac{\phi}{\lambda}] \qquad (9.11)$$

where d_0 is the tube diameter.

The above equations must be integrated from a known boundary condition. This implies that the liquid film flow rate must be known at some point in the channel but, fortunately, for channels of reasonable length, the burnout prediction is relatively insensitive to the boundary condition chosen. It has been found that an assumed value of $G_{LF} = 0.01G$ at $x = 0.01$ gives reasonably consistent results and this value has been used in the calculations described below. This has the implication (which is naturally incorrect) that annular flow exists at a quality of 0.01 but, as was stated, the calculated results for burnout are not very sensitive to these assumptions.

The application of the above methods to the prediction of burnout in various geometries will now be discussed.

Prediction of burnout in round tubes: Clearly, the prediction method for annular flow burnout is most ideally suited to prediction of the boiling length and quality for burnout at a given heat flux. To make this prediction, a simple integration of the equations is carried out, introducing the expressions for D and E which were described in Chapter 5. Predictions of this

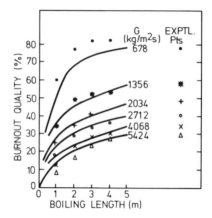

Figure 9.24: Predicted burnout quality as a function of boiling length for
             burnout in the evaporation of water at 70 bar in a 12.7mm bore
             tube (Whalley et al. 1974).

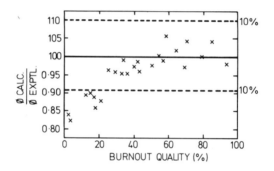

Figure 9.25: Prediction of burnout data for the evaporation of refrigerant-11
             at 23.3 bar in a round tube (Whalley, et al. (1974).

type are reported by Whalley et al. (1974) whose results are
illustrated in Figure 9.24. As will be seen, rather good agreement is obtained with the experimental data, the effect of mass
flux being correctly predicted.

     In most practical prediction problems, the requirement is to
predict the burnout heat flux for a given tube length. Here, the
calculation is done iteratively, successive predictions of the
length-to-burnout being made until the predicted length equals
the tube length. Typical predictions for the evaporation of
refrigerant-11 at 23.3 bar are illustrated in Figure 9.25. As
will be seen, excellent predictions are obtained, particularly
at moderate or high qualities.

     The results and predictions illustrated in Figures 9.24 and
9.25 are for uniform heat flux. In most practical situations,
the heat flux is non-uniform. However, the method can be applied
to axially non-uniform cases simply by introducing the correct

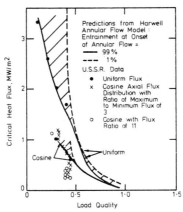

Figure 9.26: Comparison of the data of Doroschuk et al. (1978) with annular
        flow prediction model (Hewitt, 1978).

form of flux/length dependency into the above equations. The
application of the annular flow prediction method to the non-
uniform data of Doroschuk et al. (1978) are reported by Hewitt
(1978) and are illustrated in Figure 9.26. Reasonable prediction
of the cosine/flux data is obtained, though there are some
discrepancies at high flux ratios.

    Burnout prediction in complex geometries: In complex
geometries such as annuli or rod bundles, mass balances have
to be carried out on the films on each respective surface. For
instance, in an annulus, a mass balance is carried out on the outer
surface and on the inner surface respectively. In order to
calculate the rate of entrainment, it is necessary (in the case of
the annulus) to determine the interfacial shear stresses on both
interfaces and, thus, to determine the position of the plane of
zero shear. This can be done by a transformation method as
described by Whalley (1974) whose predictions for annulus burnout
are compared with the data of Jensen & Mannov (1974) in Figure
9.27. Good qualitative and quantitative agreement is observed.

    For rod bundle predictions, the annular flow prediction
method can be used within the framework of the rod-sub-channel

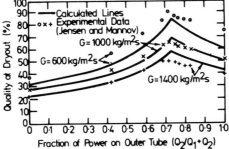

Figure 9.27:   Comparison of predicted and experimental burnout data for an
        annulus (Whalley, 1974).

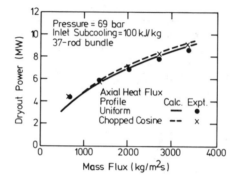

Figure 9.28: Comparison of experimental and predicted dryout power for the evaporation of water in a 37-rod bundle (Whalley, 1978).

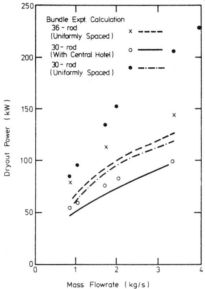

Figure 9.29: Comparison of predicted and experimental dryout power for the evaporation of refrigerant-11 in various rod bundle configuration ("odds-and-evens" experiments of Macbeth).

model mentioned above. Whalley (1978) reports predictions of various rod bundle systems using annular flow modelling and these are exemplified by the results for a 37-rod bundle illustrated in Figure 9.28. The predictions for the "odds-and-evens" tests of Macbeth are shown in Figure 9.29 and are less successful. Clearly, the effects revealed specifically in these tests are worthy of further examination.

Nomenclature

A	- channel cross sectional area	(m^2)
d_o	- tube diameter	(m)
D	- droplet deposition rate (per unit peripheral area)	$(kg/m^2 s)$
E	- droplet entrainment rate (per unit peripheral area)	$(kg/m^2 s)$
F	- Tong F-factor defined by equation 9.5	$(-)$
G	- mass flux	$(kg/m^2 s)$
G_{LF}	- liquid film mass flux (film flow rate per unit total cross sectional area)	$(kg/m^2 s)$
Δi_{sub}	- inlet subcooling (difference between saturated liquid enthalpy and inlet liquid enthalpy)	(J/kg)
L	- channel length	(m)
L_B	- boiling length (length over which saturated conditions prevail)	(m)
$(LB)_{BO}$	- boiling length at burnout	(m)
p	- pressure	(N/m^2)
P_{BO}	- power input for burnout	(W)
S	- channel periphery	(m)
x	- quality	$(-)$
x_{BO}	- quality at burnout	$(-)$
z	- axial distance along channel	(m)
z_{BO}	- axial distance to burnout	(m)
ψ	- Ahmad parameter (equation 9.8)	$(-)$
γ	- parameter defined by equation 9.9	(m^2/N)
λ	- latent heat of evaporation	(J/kg)
μ_G	- vapour viscosity	(kg/ms)
μ_L	- liquid viscosity	(kg/ms)
Ω	- parameter defined by equation 9.7	$(-)$
ϕ	- heat flux	(W/m^2)
$\phi(z)$	- heat flux at axial distance z	(W/m^2)
ϕ_{BO}	- burnout heat flux	(W/m^2)
$\phi_{BO}(z)_u$	- burnout heat flux at axial distance z for uniform heating	(W/m^2)
$\phi_{BO}(z)_{nu}$	- burnout heat flux at axial distance z for non-uniform heating	(W/m^2)
ρ_G	- vapour density	(kg/m^3)
ρ_L	- liquid density	(kg/m^3)

References

Ahmad, S Y, (1973) "Fluid to fluid modelling of critical heat flux: A compensated distortion model". *Int. J. Heat Mass Transfer*, **16**, 641-661.

Bennett, A W, Hewitt, G F, Kersey, H A, Keeys, R K F, and Pulling, D J, (1967) Studies of burnout in boiling heat transfer. *Trans. Inst. Chem. Eng.* **45** T319-333.

Bertoni, R, Cipriani, R, Cumo, M, et al., (1976) "Up-flow and down-flow burnout". *CNEN Report* No. CNEN/RT/ING(76)24.

Bowring, R W, (1972) "A simple but accurate round tube, uniform heat flux, dryout correlation over the pressure range 0.7-17 MN/m^2". *AEEW-R789*.

Collier, J G, (1972) "Convective boiling and condensation." McGraw-Hill Book Company, Maidenhead, England.

Coney, M W E, (1974) "The analysis of a mechanism of liquid replenishment and draining in horizontal two-phase flow". *Int. J. Multiphase Flow*, Pergamon Press, **1**, No. 5, 647-670.

Doroschuk, V E, Levitan, L L, Lantzman, E P, Nigmatulin, R I, and Borevsky, L Ya, (1978) "Investigations into burnout mechanism in steam-generating tubes". Paper presented at the *6th International Heat Transfer Conference*, Tokyo.

Fiori, M P, and Bergles, A E, (1968) "Model of critical heat flux in sub-cooled flow boiling." *MIT Report* - DSR70281-56.

Fisher, S A, Harrison, G S, and Pearce, D L, (1978) "Premature dryout in conventional nuclear power station evaporators". Paper presented at the *6th International Heat Transfer Conference*, Toronto.

Fisher, S A, and Yu, S K W, (1975) "Dryout in serpentine evaporators". *Int. J. Multiphase Flow.* **1**, No. 6, 771-791.

Gaspari, G P, Hassid, A and Lucchini, F, (1974) A rod centred sub-channel analysis with turbulent (enthalpy) mixing for critical heat flux prediction in rod clusters cooled by boiling water. *Proc. 5th Int. Heat Transfer Conf.*, **4**, 295-299.

Hauptmann, E G, Lee, V. and McAdam, D, (1973) "Two-phase fluid modelling of the critical heat flux". *Proc. of Inst. Meeting, Reactor Heat Transfer*, Karlsruhe, (9-11 Oct. 73), 557-576.

Hewitt, G F, (1978) "Critical heat flux in flow boiling." Keynote Lecture, *6th International Heat Transfer Conference*, Toronto.

Hewitt, G F, (1980) "Burnout." Chapter in *Handbook of Multiphase Systems* (Editor G Hetsroni), to be published by Hemisphere Publishing Corporation.

Hewitt, G F, Kersey, H A, and Collier, J G, (1969) "The correlation of critical heat flux for the vertical flow of water in uniformly heated tubes, annuli and rod bundles." *AERE-R5590*.

Hewitt, G F, Kearsey, H A, Lacey, P M C, and Pulling, D J, (1963) "Burnout and nucleation in climbing film flow". *UKAEA Report* No. AERE-R4374.

Jensen, A, and Mannov, G, (1974) "Measurements of burnout, film flow, film thickness, and pressure drop in a concentric annulus 3500 x 26 x 17mm with heated rod and tube". Paper presnted at *European Two-Phase Flow Group Meeting,* Harwell. Paper No. A5.

Keeys, R K F, Ralph J C, and Roberts, D N, (1970)"The effect of heat flux on liquid entrainment in steam-water flow in a vertical tube." AERE-R6294.

Keeys, R K F, Ralph, J C, and Roberts, D N, (1972) "Post burnout heat transfer in high pressure steam-water mixtures in a tube with cosine heat flux distribution". *UKAEA Report* No. AERE-R6411, (1971), Published in *Progress in Heat and Mass Transfer,* **6**, 99-118.

Kirby, J G, Staniforth, R, and Kinneir, L H, (1967)"A visual study of forced convective boiling. Part II: Flow patterns and burnout for a round test section." AEEW-R506.

Lienhard, J H, and Eichhorn, R, (1976) "Peak boiling heat flux on cylinders in a cross flow". *Int. J. Heat Mass Transfer,* **19**, No. 10, 1135-1141.

Macbeth, R V, (1974) "Odds and evens. A formula for enhancing the dryout power in boiling water reactor fuel channels". Paper presented at *European Two-Phase Flow Group Meeting,* Harwell, Paper No. C10.

Merilo, M, (1977) "Critical heat flux experiments in a vertical and horizontal tube with both Freon-12 and water as coolant". *Nucl. Eng. Des.* **44**, No. 1, 1-16.

Semeria, R, and Hewitt, G F, (1974)"Aspects of heat transfer in two-phase one component flows". In *Heat Exchangers: Design and Theory Source Book.* Edited by M Afgan and E U Schlunder, Scripta Book Company, Washington, 319-385.

Tong, L S, Currin, H B, and Larsen, T S, (1966)"Influence of axially non-uniform heat flux on DNB." WCAP-2767. Published in *CEP Symp. Ser.,* **62** (64).

Tong, L S, and Hewitt, G F, (1972)"Overall view point of film boiling CHF mechanisms." *ASME* Paper No. 72-HT-54.

Weisman, J and Bowring, R W, (1975)"Methods for detailed thermal and hydraulic analysis of water-cooled reactors." *Nucl. Sci. Eng.* **57**, (4), 225-276.

Whalley, P B, (1974) "The calculation of dryout in a heated annulus". AERE-M2661.

Whalley, P B, (1978) "The calculation of dryout in a rod bundle - a comparison of experimental and calculated results". AERE-R8799.

Whalley, P B, Hutchinson, P, and Hewitt, G F, (1974) "The calculation of critical heat flux in forced convection boiling." *Heat Transfer,* **4**, 290-294, Scripta Book Company, New York.

Chapter 10

Post Dryout Heat Transfer

J. G. COLLIER

10.1 Summary

This chapter is an up-dated version of a review presented to the NATO Advanced Study Institute on Two-Phase Flow and Heat Transfer in 1976 and subsequently published (Collier (1977)).

Recent published literature on the subject of post-dryout heat transfer is reviewed with the object of indicating the general "state of the art". Significant improvements in understanding have been made over the past 4-5 years. In particular, the development of experimental techniques has allowed the determination of the complete forced convection boiling curve, including the "transition boiling" region for pre-selected values of local vapour quality and mass velocity.

Improved correlations of post dryout heat transfer rates have been developed which take into account the departure from thermo-dynamic equilibrium in this region. In addition, detailed semi-theoretical physical models have been produced which cover the various separate identifiable mechanisms of heat transfer. These include evaporation of droplets striking the heat transfer surface, evaporation of droplets entrained in the vapour phase, and super-heating of the vapour phase.

Rather less is known about the forced convective film boiling process at low vapour qualities and under subcooled conditions. Further experimental work is needed in this region under up-flow, down-flow and reverse flow conditions.

10.2 Introduction

The research work in relation to the safety of water-cooled nuclear reactors continues to increase and as we understand more and more about the physical events following a hypothetical loss-of-coolant, so there exists a continuing need for more accurate information about fuel element temperatures during the course of the accident. At some stage during the accident it is possible for the critical heat flux condition to exist at the heat transfer surface. At this point there is a sharp reduction in the heat transfer coefficient which, in turn, leads to a rise in the fuel element surface temperature. In this paper recent published literature on the subject of heat transfer rates in the post-CHF (or post-dryout) region is reviewed with the object of indicating the general "state-of-the-art".

Significant improvements in understanding have been made over the past few years. In particular, the development of experimental techniques has allowed the determination of the complete forced convection boiling curve including the "transition boiling" region for pre-selected values of local vapour quality and mass velocity.

Although much of the research work is relevant to the water-reactor safety programme this increased understanding will be beneficial to the design of other equipment in the power and cryogenics industries (Bailey (1973)).

Advances have also been made in improving the accuracy of post-dryout heat transfer correlations. The need for simple correlations which take into account the departure from thermo-dynamic equilibrium in this region has been recognised. In addition, detailed semi-theoretical physical models have been produced which cover the various separate identifiable mechanisms of heat transfer. These include evaporation of droplets striking the heat transfer surface, evaporation of droplets entrained in the vapour phase and, finally, direct surface-to-vapour heat transfer which results in superheating of the vapour phase.

Most of the work reviewed covers vapour qualities in excess of 10% by wt and situations where the pre-dryout flow pattern is "annular dispersed". Rather less is known about the forced convective film boiling process at low vapour qualities and under subcooled conditions. Further experimental work is needed in this region under both steady up-flow conditions and under reversing and down-flow conditions.

10.3 A Physical Description of Post-Dryout Heat Transfer

The physical processes occurring in the post-dryout region can be discussed in terms of the experimental techniques used. The simplest experiment which may be undertaken involves establishing a particular liquid flow condition to a long vertical tube heated uniformly (Figure 10.1(a)). Dryout is initiated at the downstream end of the tube either by increasing the heat flux or reducing the flowrate (condition A). Increasing the heat flux will drive the dryout point upstream and the downstream section of the tube will enter the post-dryout region (condition B). If we plot the variation of surface temperature at a given position, z, then we obtain the variation shown in Figure 10.1(b). Bennett et al (1967) carried out this experiment and showed that there was no hysteresis associated with the curve in Figure 10.1(b) when the heat flux was being increased or decreased. This is in contrast to the situation for pool boiling where the return from film boiling conditions to nucleate boiling does not occur until a considerably lower heat flux than the CHF value.

Further information about the physical processes in the post-dryout region can be obtained from a test section in which one section can be operated at a higher or lower heat flux from the rest of the tube. Consider a tubular test section in which the heat flux on the upstream section (ϕ_1) can be adjusted separately from that on a short downstream section (ϕ_2) (Figure 10.2(a)).

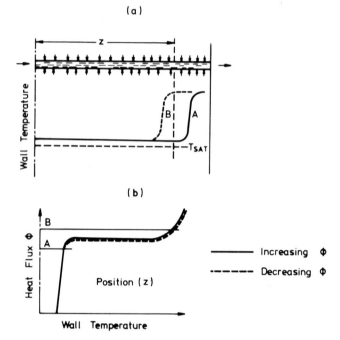

Figure 10.1: Dryout in a uniformly heated tube.

Initially, the two sections are operated at equal values of heat
flux ($\phi_1 = \phi_2$) and dryout is established over the whole of the
downstream section in the identical manner as for the uniformly
heated tube (condition A).   Dryout is now maintained at the downstream end of the upstream section by holding ϕ_1 fixed.   The flux
on the downstream section ϕ_2 is then progressively reduced and it
is found that the wall temperature at position z falls smoothly
with decreasing heat flux in the manner shown in Figure 10.2(b).
At some value of the heat flux (ϕ_2) "re-wetting" occurs and the
wall temperature drops sharply to the nucleate boiling condition.
This experiment was carried out by Bailey (1972) and the picture
below was confirmed.   Similar behaviour can be induced by having,
in the upstream section of the tube, a short length over which the
local heat flux is considerably higher than the remainder of the
tube.   CHF is initiated at this heat flux spike and it is found
that post-dryout behaviour can be studied on the downstream
sections at the lower heat flux.   This technique has been used by
Groeneveld (1974) with Freon and by workers at MIT (Plummer (1974)
and Iloeje (1974)) with nitrogen.

     Figure 10.2(b) depicts a much greater similarity to the pool
boiling situation and indicates that significant hysteresis can
occur for situations where re-wetting by an advancing liquid front
is prevented.   This finding has considerable significance in
relation to the conditions expected on a fuel pin during a loss-of-
coolant accident.   In particular, if parts of the fuel pin remain
"wetted" during the transient we may expect "re-wetting" of those
areas where CHF has previously occurred to take place soon after
the local surface heat flux falls below the CHF value.   The time to

(a)

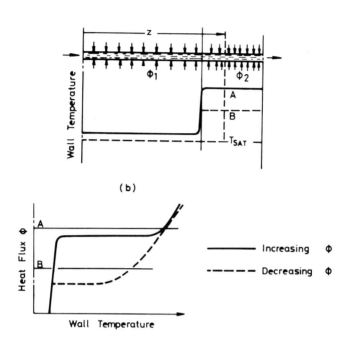

(b)

Figure 10.2: Dryout in a two-section heated tube.

re-wet will be, however, limited by the conduction-controlled
velocity of the "quench" front. On the other hand, if the fuel
pin has "dried-out" completely during the transient, or if an
upstream section persistently indicates dryout, then the pin
surface heat flux must fall significantly below the CHF value (eg
to 10-20% of the CHF value) before "re-wetting" can be initiated.

Having established a considerable similarity between the
forced convective boiling curve and pool boiling curve one is left
with the question "is there a 'transition boiling' region?"
Bailey and Collier (1970) speculated on whether stable operation
in this region was possible using a fluid-heated system and
concluded that experimental information using a constant temperature heat source was required. Recently, however, workers at MIT
have been able to establish the remaining part of the curve using
a novel transient technique. This technique, which has also been
used in the study of cool-down of cryogenic transfer lines (Prasad
(1974) Srinivasan (1974)), again involves the use of a test section
consisting of two parts; a long uniformly heated upstream section
and a short thick-walled high heat capacity downstream section
(Figure 10.3(a)). Dryout is established at the downstream end of
the longer section and the short thick-walled section heated to a
high initial temperature. A cooling or "quench" curve is then
produced (Figure 10.3(b)) as the short downstream section is
cooled to the saturation temperature. Provided the thick wall
section can be treated as a lumped system (no axial or radial
temperature profiles) then the local heat flux can be established

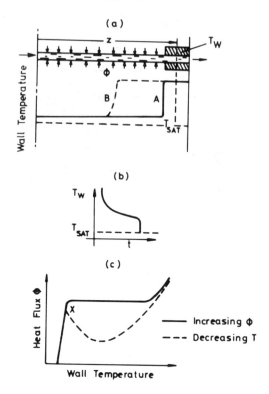

Figure 10.3: Dryout in a transient conduction downstream section.

from the gradient of section temperature with time (Figure 10.3(b)).
The resulting curve of heat flux versus wall temperature is shown
in Figure 10.3(c). Note that the point X does not generally
coincide with the critical heat flux because, in general, both the
flow in the liquid *and* the droplet deposition flux contribute to
the critical heat flux. Measurements by Hewitt (1970) have shown
that X may normally be about 80% of the CHF value rising on some
occasions to as much as 100%, in which case point X and the CHF
value are coincident.

    In addition, similar experiments may be carried out with
dryout occurring at any location in the upstream heated section
(condition B (Figure 10.3(a))).

    The MIT studies showed that the increase in the length of the
dried-out section decreased the heat transfer coefficient at a
given mass velocity and quality due to the thermal non-equilibrium
effects. No influence of heater material was noted between copper,
Inconel 600 and aluminium. Increased surface roughness and the
presence of an oxide film both increased the post-dryout heat
transfer coefficient. The effect of surface roughness is believed
to be due to the increased convective heat transfer to the bulk
vapour whilst the influence of the oxide film is believed to be
associated with the better wettability of the oxide during droplet/
wall contacts.

10.4 Post-Dryout Heat Transfer Correlations

Three types of correlating procedure have been adopted. They are:

(a) Correlations of an empirical nature which make no assumption whatever about the mechanisms involved in post-dryout heat transfer, but solely attempt a functional relationship between the heat transfer coefficient (assuming the coolant is at the saturation temperature) and the independent variables.

(b) Correlations which recognise that departure from a thermodynamic equilibrium condition can occur and attempt to calculate the "true" vapour quality and vapour temperature. A conventional single-phase heat transfer correlation is then used to calculate the heated wall temperature.

(c) Semi-theoretical models where attempts have been made to look at and write down equations for the various individual hydrodynamic and heat transfer processes occurring in the heated channel and relate these to the heated wall temperature.

Groeneveld (1973) has compiled a bank of carefully selected data drawn from a variety of experimental post-dryout studies in tubular, annular and rod bundle geometries for steam/water flows. Table 10.1 taken from Groeneveld's paper lists the individual sources, together with the ranges of the independent variables examined.

Empirical Correlations

A considerable number of empirical equations have been presented by various investigators for the estimation of heat transfer rates in the post-dryout region. Almost all of these equations are modifications of the well-known Dittus-Boelter type relationship for single-phase flow and take no account of the non-equilibrium effects discussed above. Various definitions of the "two-phase velocity" and physical properties are used in these modified forms. Table 10.2 again taken from Groeneveld's paper, lists sixteen such correlations. Each of these correlations is based on only a limited amount of experimental data and Groeneveld therefore proposed a new correlation for each geometry optimised using his bank of selected data.

The Groeneveld correlations for tubes and annuli have the form:

$$Nu_g = a \left[Re_g \left\{ x + \frac{\rho_g}{\rho_\ell} (1 - x) \right\} \right]^b Pr_{g,w}{}^c \, Y^d \qquad (10.1)$$

where $Y = 1 - 0.1 \left(\frac{\rho_1}{\rho_g} - 1 \right)^{0.4} (1 - x)^{0.4}$ \qquad\qquad (10.2)

and the coefficients a, b, c and d are given in Table 10.3, together with the range of independent variables on which the correlations are based.

J G Collier

TABLE 10.1

Post-Dryout Data for Water
(Groeneveld)

Geometry	Reference	Range of Data					Comments
		P bar	ϕ W/cm^2	G g/cm^{2}s	X	De cm	
Tube	Parker (1961)	2	1-6	5-10	0.89-1.00	2.54	50cm heated length rewetting of wall?
	Bennett (1967)	69	30-190	40-530	0.25-1.00	1.27	275cm heated length
	Bailey (1972)	160-180	8-80	67-270	0.20-1.00	1.28	U-tube
	Keeys (1971)	69	80-150	70-410	0.15-0.90	1.27	cosine heat flux distribution
	Bishop (1964)	167-215	30-200	70-325	0.10-0.95	0.254, 0.508	
	Mueller (1967)	69	50-85	70-100	0.62-1.00	1.57	also measured vapour superheat
	Polomik (1967)	69	55-110	70-135	0.80-1.00	1.57	also measured vapour superheat
	Brevi (1969)	50	38-150	47-300	0.40-1.00	0.65, 0.93	
	Bertoletti (1964)	70	10-160	100-400	0.40-0.90	0.50, 0.90	non-steady temperature

TABLE 10.1 (cont'd)

Geometry	Reference	Range of Data					Comments
		P bar	φ W/cm²	G g/cm²s	X	De cm	
	Era (1967)	70	10-130	110-300	0.20-1.00	1.6	
	Herkenrath (1967)	138-250	10-205	70-325	0-1.00	1.0, 2.0	data in graphical form
	Swenson (1961)	207	28-57	95-135	0.20-0.90	1.07	
	Schmidt (1959)	215	32-85	42	0.10-0.90	0.8	data in graphical form vertical and horizontal flow
	Lee (1970)	140-180	30-140	100-400	0.30-0.70	0.9, 1.3	indirectly heated (sodium-heated steam generator)
	Miropol'skiy (1963)	40-220	25-110	100-200	0.20-1.00	0.8	data must be derived from graphs
Annulus	Polomik (1961)	55-97	60-220	100-256	0.15-1.00	0.152, 0.305	
	Polomik (1971)	69	75-230	35-270	0.15-0.65	0.338	results affected by spacers
	Bennett (1964)	35-69	60-180	70-270	0.20-1.00	0.33	
	Era (1967)	70	13-104	80-380	0.30-1.00	0.20, 0.50	effect of spacers was studied

TABLE 10.1 (cont'd)

Geometry	Reference	Range of Data					Comments
		P bar	ϕ W/cm²	G g/cm²s	X	De cm	
	Era (1971)	50	20-60	60-220	0.20-0.90	0.30	uniformly and non-uniformly heated
	Groeneveld (1969)	41-83	50-140	135-410	0.10-0.50	0.405	two heated sections separated by unheated section
Bundles	Hench (1964)	41-97	45-190	39-270	0.20-0.90	1.03	2-rod
	Kunsemiller (1965)	41-97	55-100	39-135	0.30-0.70	1.12	3-rod
	Groeneveld (1970)	63	33-116	110-220	0.30-0.60	0.344	3-rod, effect of crud in-reactor experiment
	Adorni (1966)	50-55	20-150	80-380	0.20-0.90		7-rod, mainly unsteady temperatures
	Matzner (1963)	69	80-235	70-270	0.17-0.60	0.830	19-rod, mainly unsteady temperatures
	McPherson (1971)	109-217	60-145	70-410	0.28-0.53	0.780	28-rod, mainly unsteady temperatures
	Matzner (1968)	34-83	78-115	70-140	0.23-0.38	0.670	19-rod segmented bundle

TABLE 10.2

Post Dryout Correlations
(from Groeneveld)

Equation and References	Range of Applicability of Equation				Eq agrees with data from ref	Comments
	p MPa	ṁ kg/m²s	x	Geometry		
1. $Nu_f = 0.00136[Re_f]^{0.853} Pr_f^{\frac{1}{3}} (\frac{x}{1-X})^{0.147} (\frac{\rho_\ell}{\rho_g})^{\frac{2}{3}}$ Polomik (1961)	55-100	100-245	0.40-0.70	Annulus	Polomik (1961)	Exponents and coefficients for equations 1-3 obtained from least error analysis using data from Polomik (1961)
2. $Nu_f = 0.416 Re_f^{0.509} Pr_f^{\frac{1}{3}} (\frac{\rho_g}{\rho_\ell})^{0.208} (\frac{1-X}{X})^{0.616}$	55-100	100-245	0.40-0.70	-	Polomik (1961)	
3. $Nu_f = 0.023 Re_f^{0.292} Pr_f^{\frac{1}{3}} (\frac{1-X}{X})^{0.01} \phi^{0.417} (\frac{\rho_\ell}{\rho_g})^{0.091}$ ϕ in Btu/h·ft² Polomik (1961)	55-100	100-245	0.40-0.70	-	Polomik (1961)	
4. $Nu_G = 0.00115 Re_G^{0.9} Pr_G^{0.3} (\frac{T_w}{T_{sat}} -1)^{-0.15}$ T in °F Polomik (1967)	40-100	700-270	0.20-1.00	2-rod	Hench (1964)	Has additional temperature parameter to allow for property variations at the heater wall ±20% variation with data from Hench (1964)

TABLE 10.2 (cont'd)

292   J G Collier

Equation and References	Range of Applicability of Equation				Eq agrees with data from ref	Comments
	p MPa	kg/m²s	X	Geometry		
5. $$Nu_G = 0.0039 \left[Re_G \left(x + \frac{\rho_G}{\rho_L}(1-x) \right) \right]^{0.9}$$ Polomik (1967)	40-100	70-270	0.20-1.00	–	Hench (1964)	+10% variation with data from Hench (1964) Better than above equation
6. $$\dot{q}\left[De^{0.2}/(\dot{m}X)^{0.8} \right] = C(T_w - T_{sat})^m$$ $m = 1.284 - 0.00312\,\dot{m}$ $C = \frac{1}{389} e^{0.01665\,\dot{m}}$ Collier (1962)	69	>100	0.15-1.00	Round tubes and annuli	Polomik (1961) Bertoletti (1961)	$T_w - T_{sat}$ must be below 200°C
7. $$\dot{q}\left[De^{0.2}/(\dot{m}X)^{0.8} \right] = 0.018 (T_w - T_{sat})^{0.921}$$ Collier (1962)	69	see comments	0.15-1.00	–	Polomik (1961) Bertoletti (1961)	Correlations may be used if either $\dot{m} < 1000$ kg/m²s or $T_w - T_{sat} > 200°C$
8. $$Nu_f = 0.0193\, Re_f^{0.8}\, Pr_f^{1.23} \left(\frac{\rho_G}{\rho_L} \right)^{0.068} \left[x + \frac{\rho_G}{\rho_L}(1-x) \right]^{0.68}$$ Bishop (1965)	40-215	70-340	0.07-1.00	Round tubes	Polomik (1961) Bishop et al (1965) Swenson et al (1961)	

TABLE 10.2 (cont'd)

Equation and References	Range of Applicability of Equation				Eq agrees with data from ref	Comments
	p MPa	ṁ kg/m²s	X	Geometry		
9. $$Nu_f = 0.033 \, Re_w^{0.8} \, Pr_w^{1.25}\left(\frac{\rho_G}{\rho_L}\right)^{0.197}\left[X + \frac{\rho_G}{\rho_L}(1-X)\right]^{0.738}$$ Bishop (1968)	40-215	70-340	0.07-1.00		Bishop et al (1962) Bishop et al (1964) Miropolskii (1963)	
10. $$Nu_w = 0.098\left[Re_w \cdot \frac{\rho_w}{\rho_G}\left(X + \frac{\rho_G}{\rho_L}(1-X)\right)\right]^{0.8}(Pr_w)^{0.83}\left(\frac{\rho_G}{\rho_L}\right)^{0.50}$$ Bishop (1964)	165-215	135-340	0.10-1.00	-	Swenson (1961) Bishop et al (1964) Miropolskii (1963)	Based on high pressure data
11. $$Nu_w = 0.055\left[(Re_w)\frac{\rho_w}{\rho_G}\left(X + \frac{\rho_G}{\rho_L}(1-X)\right)\right]^{0.82}(Pr_w)^{0.96}\left(\frac{\rho_G}{\rho_L}\right)^{0.34}\left(1 + \frac{26.9}{L/D}\right)$$ Bishop (1964)	165-215	135-340	0.10-1.00	-	Bishop et al (1964)	

TABLE 10.2 (cont'd)

Equation and References	Range of Applicability of Equation				Eq agrees with data from ref	Comments
	p MPa	ṁ kg/m²s	X	Geometry		
12. $T_w - T_{sat} = 1.975\left[\dfrac{\dot{q}}{\dot{m}\left(X + \dfrac{1-X}{4.15}\right)}\right]^2$ Lee (1970)	140-180	100-400	0.30-0.75	-	Lee (1970)	Based on data obtained on 13mm ID indirectly heated tube
13. $Nu_G = 0.023\left[Re_G\left(X + \dfrac{\rho_G}{\rho_L}(1+X)\right)\right]^{0.8} Pr_w^{0.8} Y$ $Y = 1 - 0.1\left(\dfrac{\rho_L}{\rho_G} - 1\right)^{0.4}(1-X)^{0.4}$ Miropolskii (1963)	40-220	70-200	0.06-1.00	-	Swenson (1961) Miropolskii (1963) Schmidt (1960)	Factor Y was determined empirically
14. $Nu_G = 0.0089\left(Re_f \cdot \dfrac{X}{\alpha}\right)^{0.84}(Pr_f)^{\frac{1}{3}}\cdot\left(\dfrac{1-X_{DO}}{X-X_{DO}}\right)^{0.124}$ Brevi (1969)	50	50-300	0.40-1.00	-	Brevi (1969)	
15. $Nu = 0.06\left\{Re_w\left(X + \dfrac{\rho_G}{\rho_L}(1-X)\right)\left[\dfrac{\rho_w}{\rho_G}\right]\left[Pr_w\right]\right\}^{0.8}\left(\dfrac{p}{p_c}\right)^{2.7}\left(\dfrac{\dot{m}}{m_o}\right)^{0.4}$ $m_o = 1000$ kg/m²s Herkenrath (1969, 1970)	140-220	75-410	0.10-1.00	-	Herkenrath (1967) Swenson (1961) Bishop et al (1964)	

TABLE 10.2 (cont'd)

Equation and References	Range of Applicability of Equation				Eq agrees with data from ref	Comments
	p MPa	ṁ kg/m²s	X	Geometry		
16. $Nu_w = 0.005 \left(\dfrac{de \, U_m \, \rho_w}{\mu_m} \right)^{0.8} (Pr_G)^{0.5}$ Tong (1964)	> 138	> 70	< 0.10	–	Bishop (1962, 1964)	Design equation. To be used only at low quality sub-cooled conditions

TABLE 10.3

Empirical Post-Dryout Correlations

Groeneveld

Geometry	a	b	c	d	No. of points	RMS error
Tubes	1.09×10^{-3}	0.989	1.41	- 1.15	438	11.5%
Annuli	5.20×10^{-2}	0.688	1.26	- 1.06	266	6.9%
Tubes and annuli	3.27×10^{-3}	0.901	1.32	- 1.50	704	12.4%

Slaughterback

Geometry	a	b	c	e	f	RMS error
Tubes	1.16×10^{-4}	0.838	1.81	0.278	- 0.508	12.0%

Range of data on which correlations are based

Geometry	Tube	Annulus
Flow direction	vertical and horizontal	vertical
De, cm	0.25 to 2.5	0.15 to 0.63
P, bars	68 to 215	34 to 100
G, $g/cm^2 s$	70 to 530	80 to 410
X, fraction by weight	0.10 to 0.90	0.10 to 0.90
ϕ, W/cm^2	12 to 210	45 to 225
Nu_g	95 to 1770	160 to 640
$[Re_g(X+(1-X)\rho_g/\rho_\ell)]$	6.6×10^4 to 1.3×10^6	1.0×10^5 to 3.9×10^5
Pr_w	0.88 to 2.21	0.91 to 1.22
Y	0.706 to 0.976	0.610 to 0.963

Slaughterback (1973a) (1973b) has examined the trends of Groeneveld's and other post-dryout correlations. His conclusion was that there were significant discrepancies between the various correlations and between the correlations and experimental data. Figure 10.4 shows a comparison of the various correlations with experimental data for one particular condition. This conclusion led Slaughterback (1973b) to undertake a statistical regression of the Groeneveld data bank which resulted in a modified form of the Groeneveld correlation which did not produce artificially high coefficients at low pressures.

$$Nu_g = a \left[Re_g \left\{ x + \frac{\rho_g}{\rho_1} (1-x) \right\} \right]^b Pr_{g,w}^c \phi^e \left(\frac{k_g}{k_{CRIT}} \right)^f \qquad (10.3)$$

Values of the coefficients for this modification are given in Table 10.3. The revision makes a considerable difference at low pressure as seen in Figure 10.4.

Under the auspices of the European Two-Phase Flow Group, four laboratories (Cumo 1974) have carried out post-dryout experiments using identical test sections and instrumentation. The experiments were carried out at a pressure of 70 bar on a tube 10mm ID and 4m heated length. The scatter on measured heat transfer coefficients

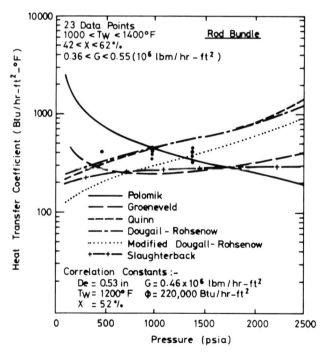

Figure 10.4: Heat transfer coefficients as a function of pressure (Slaughterback (1973a, 1973b).

was between 10 and 25% with a mean of 17.9%. It is stated that
agreement between these data and some published post-dryout
correlations was poor. However, the CNEN correlation:

$$\left(\frac{hD}{k_{g,f}}\right) = 0.0089 \left(\frac{Gx\ D}{\alpha\ \mu_{g,f}}\right)^{0.84} \left(\frac{c_p\mu}{k}\right)^{1/3}_{g,f} \left(\frac{1\ -\ x_{DO}}{x\ -\ x_{DO}}\right)^{1/8} \tag{10.4}$$

gave good agreement.

Mattson et al (1974) have extended the multiple linear
regression analysis used by Slaughterback to cover the annulus and
rod bundle geometries and the "transition" film boiling region as
well as the stable film boiling region. They recommended the
following equations:

(i)   for film boiling in an internally heated annulus

$$h = 34.5\ Re_g^{0.244}\ Pr_{g,w}^{2.54}\ D_{eq}^{-0.304}\ k_g^{0.334} \tag{10.5}$$

(British Engng. Units)   (RMS error 17.5%)

(ii)   for film and transition film boiling in a tube or annular
        geometry

$$h = 9.09 \times 10^4\ \exp(-0.5\ \sqrt{\Delta T}) + 20.8\ Re_g^{0.269}$$

$$Pr_{g,w}^{3.67}\ D_{eq}^{-0.319}\ k_g^{0.306}\ x^{-0.091} \tag{10.6}$$

(British Engng. Units)   (RMS error 28.0%)

(iii) for film and transition film boiling in a rod bundle

$$h = 2.93 \times 10^4\ \exp(-0.5\ \sqrt{\Delta T}) + 1.22\ Re_g^{0.505}$$

$$Pr_{g,w}^{4.56}\ D_{eq}^{-0.160}\ k_g^{0.189}\ x^{-0.113} \tag{10.7}$$

(British Engng. Units)   (RMS error 25.8%)

The use of the above correlations outside the range of variables
for the respective geometries given in Table 10.1 is strongly
discouraged.

10.5 **Transition Boiling**

Various attempts have been made to produce correlations for
the "transition" film boiling region corresponding to the left of
the minimum in the boiling curve (Figure 10.3(c)). These correlations have usually been combined with expressions for the stable
film boiling region as, for example, equations (10.6) and (10.7)
above.

Groeneveld and Fung (1977) have tabulated the various

correlations available for forced convection transition boiling of water. The earliest experimental study was that of McDonough et al (1960) who measured transition film boiling heat transfer coefficients for water boiling over the pressure range 800-2000 psia inside a 0.15" id tube heated by NaK. The correlation they offered was:

$$\left[\frac{\phi_{DO} - \phi(z)}{T_W(z) - T_{DO}} \right] = 730 \exp \left[576/p \right] \tag{10.8}$$

(British Engng. Units)

Tong (1967) suggested the following equation for combined transition and stable film boiling at 2000 psia with wall temperatures less than $800^{O}F$ $\left[(T_W(z) - T_{SAT}) \leqslant 300^{O}F \right]$.

$$h_{TB} = 890 + 16,860 \exp \left[- 0.01 (T_W(z) - T_{SAT}) \right] \tag{10.9}$$

(British Engng. Units)

This, in turn, was revised (Tong (1970)) to cover both transition and film boiling regions

$$h = 7000 \exp \left[- 0.008 \, \Delta T \right] + 0.023 \left(\frac{k_g}{D_{eq}} \right) \exp (- 190/\Delta T)$$
$$\times Re_{g,f}^{0.8} \, Pr_{g,f}^{0.4} \tag{10.10}$$

(British Engng. Units)

This equation was derived from 1442 data covering a pressure of 1000 psia, G from 0.28 x 10^6 to 3.86 x 10^6 lb/hr.ft^2 and ΔT (= $T_W(z) - T_{SAT}$) from 65 to $985^{O}F$. A similar but separate equation was proposed for low pressure reflooding situations.

More recently, Ramu and Weiseman (1974) have attempted to produce a single correlation for post-dryout and reflood situations. They propose that the transition film boiling heat transfer coefficient should have the form:

$$h_{TB} = 0.5 \, S \, h_{DO} \left[\exp \{- 0.0078 \, (\Delta T - \Delta T_{DO}) \} \right.$$
$$\left. + \exp \{- 0.0698 \, (\Delta T - \Delta T_{DO}) \} \right] \tag{10.11}$$

(British Engng. Units)

where h_{DO} and ΔT_{DO} are respectively the heat transfer coefficient and the wall superheat corresponding to the pool boiling CHF condition and S is the Chen nucleation suppression factor.

Tong and Young (1974) have revised the Westinghouse transition film boiling region once again. The revised correlation is in terms of the surface heat flux in the transition boiling region (ϕ_{TB}).

$$\phi_{TB} = \phi_{NB}\ exp\left[- 0.001\ \frac{x^{\frac{2}{3}}}{(dx/dz)}\left(\frac{\Delta T}{100}\right)^{(1+0.0016\Delta T)}\right] \qquad (10.12)$$

(British Engng. Units)

where ϕ_{NB} is the nucleate boiling heat flux (presumably equated with the dryout heat flux). Tong states that the data of Bennett et al (1967) was used to develop this equation. The writer finds this claim surprising in view of the fact that:

(a)   Bennett and his co-workers believed they had little or no droplet-wall impingement;

(b)   for the reasons described earlier, the experiments they carried out could not possibly have included data to the left of the minimum in the boiling curve.

Tong and Young combined equation (10.12) with a stable film boiling heat transfer term and claim an RMS error of 29.2% when comparing the correlation with 507 data points from internally heated annuli and rod bundles. This accuracy is comparable with that established by Mattson for his relation (equation (10.6) and (10.7)).

Despite the shortcomings in its derivation, this correlation can be recommended as the most comprehensively useful of all the empirical correlations.

10.6   Correlations allowing for Departure from Thermodynamic Equilibrium

It has long been known that wall temperatures in the post-dryout region are bounded by two limiting situations, viz:

(i)   *complete departure from equilibrium*. The rate of heat transfer from the vapour phase to the entrained liquid droplets is so slow that their presence is simply ignored and the vapour temperature $T_g(z)$ downstream of the dryout point is calculated on the basis that all the heat added to the fluid goes into superheating the vapour. The wall temperature $T_W(z)$ is calculated using a conventional single-phase heat transfer correlation (Figure 10.5(a)).

(ii)   *complete thermodynamic equilibrium*. The rate of heat transfer from the vapour phase to the entrained liquid droplets is so fast that the vapour temperature $T_g(z)$ remains at the saturation temperature until the energy balance indicates all the droplets have evaporated. The wall temperature $T_W(z)$ is again calculated using a conventional single-phase heat transfer correlation, this time with allowance made for the increasing vapour velocity resulting from droplet evaporation (Figure 10.5(b)).

It is also known that post-dryout heat transfer behaviour tends towards situation (i) at low pressures and low velocity,

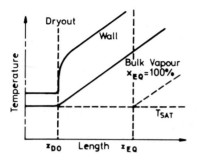

(a) Complete Lack of Thermodynamic Equilibrium

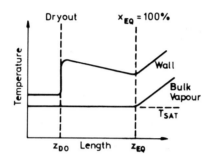

(b) Complete Thermodynamic Equilibrium

Figure 10.5: Limiting conditions for post-dryout heat transfer.

whilst at high pressure (approaching the critical condition) and high flowrates (> 3000 kg/m²s) situation (ii) pertains.

It might have been expected that correlations allowing for various degrees of non-equilibrium behaviour between these two limiting situations would have been published analogous to those available for subcooled void fraction. It is, however, only in the past few years such correlations have appeared.

Consider Figure 10.6, which represents the physical situation in the post-dryout region for a vertical tube diameter D heated uniformly with a heat flux ϕ. Dryout occurs after a length z_{DO} and it is assumed that thermodynamic equilibrium exists at the dryout point. An energy balance indicates that all the liquid is evaporated by z_{EQ} if equilibrium is maintained. However, the actual situation depicted in Figure 10.6 is one in which, following dryout, a fraction (ε) of the surface heat flux is used in evaporating liquid, whilst the remainder is used to superheat the bulk vapour. The liquid is completely evaporated a distance z* from the tube entrance (downstream from z_{EQ}).

Let the surface heat flux $\phi(z)$ be divided into two components $\phi_L(z)$, the heat flux associated with droplet evaporation, and $\phi_g(z)$, the heat flux associated with vapour superheating.

$$\phi(z) = \phi_L(z) + \phi_g(z) \qquad (10.13)$$

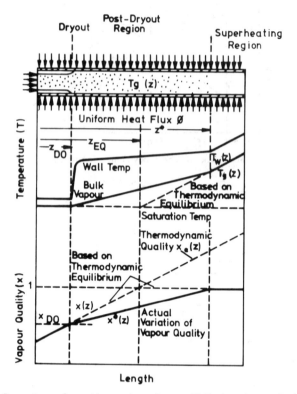

Figure 10.6: Departure from thermodynamic equilibrium in post-dryout region.

Let $\varepsilon = \phi_T(z)/\phi(z)$; ε, in general, can be considered a function of tube length (z) (or alternatively local vapour quality x^*) but to keep the analysis simple assume ε is independent of tube length. This leads to the linear profiles of "actual" bulk vapour temperature and vapour quality with tube length shown in Figure 10.6, rather than the smooth curves obtained in experiment (Figure 10.7).

The "thermodynamic" vapour quality variation with length is given by:

$$(x(z) - x_{DO}) = \frac{4\phi}{DG\,i_{fg}}\,(z - z_{DO}) \qquad (10.14)$$

for $z < z_{EQ}$

and z_{EQ} is given by

$$z_{EQ} = \left[\frac{DG\,i_{fg}}{4\phi}\,(1 - x_{DO})\right] + z_{DO} \qquad (10.15)$$

The "actual" vapour quality variation with length is given by:

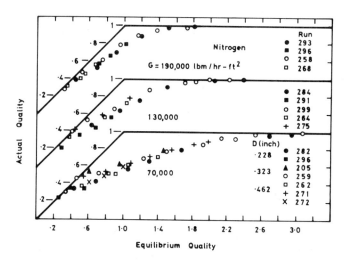

Figure 10.7: Actual vs equilibrium quality values for nitrogen at several mass fluxes (from Forslund (1966)).

$$(x^*(z) - x_{DO}) = \frac{4\phi_\ell}{DG\ i_{fg}}\ (z - z_{DO}) = \frac{4\varepsilon\phi}{DG\ i_{fg}}\ (z - z_{DO}) \qquad (10.16)$$

for $z < z^*$

and z^* is given by:

$$z^* = \left[\frac{DG\ i_{fg}}{4\varepsilon\phi}\ (1 - x_{DO})\right] + z_{DO} \qquad (10.17)$$

Thus, from equations (10.14) and (10.16) ε is also given by:

$$\varepsilon = \left[\frac{x^*(z) - x_{DO}}{x(z) - x_{DO}}\right] = \left[\frac{z_{EQ} - z_{DO}}{z^* - z_{DO}}\right] \qquad (10.18)$$

Also, the "actual" bulk vapour temperature ($T_g(z)$) is given by:

$$T_g(z) = T_{SAT} + \left[\frac{4(1-\varepsilon)\ \phi(z-z_{DO})}{G\ c_{pg}\ D}\right] \qquad (10.19)$$

for $z < z^*$ and

$$T_g(z) = T_{SAT} + \left[\frac{4\phi(z - z_{EQ})}{G\ c_{pg}\ D}\right] \qquad (10.20)$$

for $z > z^*$

The two limiting conditions referred to above are clearly
recognised by setting ε equal to zero and unity respectively.
Plummer et al (1974) have considered the variation of ε with the
major independent variables and have suggested the following
correlations:

For water

$$\varepsilon = 0.402 + 0.0674 \ln \left[G\left(\frac{D}{\rho_g \sigma}\right)^{0.5} (1 - x_{DO})^5 \right] \tag{10.21}$$

For Freon

$$\varepsilon = 0.236 + 0.0811 \ln \left[G\left(\frac{D}{\rho_g \sigma}\right)^{0.5} (1 - x_{DO})^5 \right] \tag{10.22}$$

G is measured in $(lb_m/hr.ft^2)$, D in (ft), ρ_g in $(lb_m/ft^3$ and σ in
(lb_m/hr^2). Data for nitrogen lay between the two equations with
the high mass velocity and low quality data lying close to the
"water" line and the low mass velocity and high quality data
lying nearer to the Freon line (Figure 10.8).

Groeneveld and Delorme (1976) have also used a similar
representation but have expressed their correlation in terms of the
difference between the "actual" vapour enthalpy, $i_g*(z)$ and the
"equilibrium" vapour enthalpy, $i_g(z)$. From equations (10.16) and
(10.19) we have:

$$\left[\frac{i_g*(z) - i_g(z)}{i_{fg}} \right] = x(z) - x*(z) = (1-\varepsilon) \left\{ x(z) - x_{DO} \right\} \tag{10.23}$$

where $x(z)$ is the "equilibrium" quality and $x*(z)$ is the "actual"

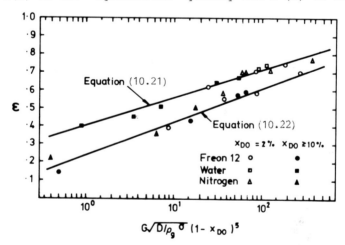

Figure 10.8: Correlation of best fit value of ε vs $G\sqrt{D/\rho_g \sigma}\ (1-x_{DO})^5$ (Plummer
         et al (1974).

vapour quality. In this case a distinction is made between the
"equilibrium" quality, $x(z)$ which has a maximum value of unity and
the "thermodynamic" quality, $x_e(z)$ which can have a value in excess
of unity (Figure 10.6).

$$x_e(z) = 1 + \left(\frac{4\phi(z - z_{EQ})}{D\,G\,i_{fg}} \right) \qquad (10.24)$$

Note that $x(z) = x_e(z)$ for $z < z_{EQ}$ and $i_g(z)$ is equal to the
saturated vapour enthalpy for $x_e(z) < 1$.

The difference $[x(z) - x^*(z)]$ is expressed as a function of
a number of independent variables:

$$x(z) - x^*(z) = \exp(-\tan \psi) \qquad (10.25)$$

$$\text{where } \psi = a_1 \, Pr^{a_2} \, Re_{TP}^{a_3} \left[\frac{\phi D c_p}{k i_{fg}} \right]^{a_4} \sum_{i=0}^{i=2} b_i \, x_e(z)^i \qquad (10.26)$$

if $\psi < 0$, $\quad \psi = 0$

$\psi > \pi/2$, $\quad \psi = \pi/2$

and

$$Re_{TP} = \left[\frac{GD}{\mu_g} \left\{ x(z) + \left(\frac{\rho_g}{\rho_f} \right) (1 - x(z)) \right\} \right]$$

The coefficients to be inserted in equation (10.26), the ranges of
the experimental data points used and the accuracy are given in
Table 4. The Groeneveld/Delorme method is a significant improvement over the empirical correlations discussed earlier but does
appear to suffer from the disadvantage of discontinuities in the
slope of the $x^*(z)$ curve both at the dryout point and at $x(z) = 1$
(Figure 10.9).

A whole range of such correlations are, however, possible
analogous to those developed for subcooled boiling void fraction.
For example, following Kroeger and Zuber (1968) it would be
possible to define:

$$X^* = \left[\frac{x^*(z) - x_{DO}}{1 - x_{DO}} \right] \quad \text{and } z^+ = \left[\frac{z - z_{DO}}{z_{EQ} - z_{DO}} \right] \qquad (10.27)$$

Various relationships between X^* and z^+ might be tested, for
example:

$$X^* = 1 - \alpha \exp\left[- z^+ \right]$$

or $\quad X^* = \beta \tanh z^+$

TABLE 10.4

*Groeneveld and Delorme (1976) Correlation
of Non-Equilibrium Vapour Quality*

Coefficient	Value
a_1	0.13864
a_2	0.2031
a_3	0.20006
a_4	-0.09232
b_0	1.3072
b_1	-1.0833
b_2	0.8455

	Geometry	Tubes
	p(bars)	41 - 205
Range of Data	G(g/cm^{2}s)	27 - 515
for	ϕ (W/cm^2)	15 - 270
Correlation	x_e(wt.fr)	0.10 - 1.50
	T_W(^{o}C)	400 - 845
	No. of points	1402
	RMS error on T_W(%)	6.7

where α and β are parameters to produce a family of curves and are functions of the independent variables $(1 - x_{DO})$, G, pressure, etc.

Gardner (1974) has considered in some detail the conditions under which a water droplet of a given size (δ) under equilibrium conditions will be completely evaporated in a superheated steam flow. If the tube diameter is D and the wall heat flux is ϕ then

$$\delta \ (\frac{\phi}{D})^{\frac{1}{2}} = fn(Z) \qquad (10.28)$$

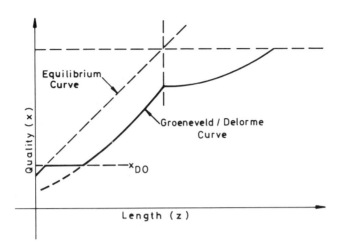

Figure 10.9: The Groeneveld/Delorme Correlation.

where $Z = \Delta T_g \exp (0.209p - 3.76)$ (10.29)

and $fn(z) = 0.06\ Z \left[1 - 0.509\ z^{0.109}\right]$ (10.30)

       (SI units)

     SI units are employed in the above equations, except that
the system pressure p is in MN/m^2. ΔT_g is the vapour superheat
corresponding to the position z*, ie $(T_g(z*) - T_{SAT})$. The value
of $(T_g(z*) - T_{SAT})$ can therefore be conservatively calculated by
estimating the maximum stable droplet size at the dryout condition
and using equation (10.28). This equation is valid for water
only and is accurate for pressure from 4 to 18 MN/m^2 and superheats
up to 50°C.

10.7 Semi-Theoretical Models

     A comprehensive theoretical model of heat transfer in the
post-dryout region must take into account the various paths by
which heat is transferred from the surface to the bulk vapour
phase. Six separate mechanisms can be identified:

(a)   heat transfer from the surface to liquid droplets which impact
      with the wall ("wet" collisions)

(b)   heat transfer from the surface to liquid droplets which enter
      the thermal boundary layer but which do not "wet" the surface
      ("dry" collisions)

(c)   convective heat transfer from the surface to the bulk vapour

(d)   convective heat transfer from the bulk vapour to suspended
      droplets in the vapour core

(e)   radiation heat transfer from the surface to the liquid
       droplets

(f)   radiation heat transfer from the surface to the bulk vapour.

     One of the first semi-theoretical models proposed was that of
Bennett et al (1967) which is a one-dimensional model starting from
known equilibrium conditions at the dryout point.   It was
originally assumed that there was negligible pressure drop along
the channel but Groeneveld (1972) revised the equations to allow
for pressure gradient and flashing effects.   It was also assumed
that droplets could no longer approach the surface.   Therefore,
mechanisms (a) and (b) were not considered.

     More recently Iloeje et al (1974) have proposed a three step
model taking into account mechanisms (a), (b) and (c).   The
physical picture postulated by Iloeje is shown in Figure 10.10.
Liquid droplets of varying sizes are entrained in the vapour core
and have a random motion due to interactions with eddies.   Some
droplets arrive at the edge of the boundary layer with sufficient
momentum to contact the wall even allowing for the fact that, as
the droplet approaches the wall, differential evaporation coupled
with the physical presence of the wall leads to a resultant force
trying to repel the droplet (Gardner 1974).   When the droplet
touches the wall a contact boundary temperature is set up which
depends on the initial droplet and wall temperatures and on
$\sqrt{(k\rho c)}_l/\sqrt{(k\rho c)}_w$.   If this temperature is less than some limiting
superheat for the liquid then heat will be transferred, firstly by
conduction until a temperature boundary layer is built up sufficient

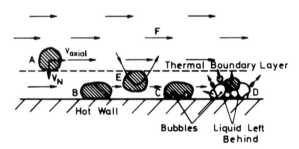

(a) Dispersed Flow Heat Transfer Process

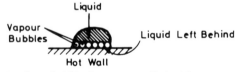

(b) Idealized Bubble Geometry at End of Bubble
Growth – Square Array

Figure 10.10: Dispersed Flow Heat Transfer Model (Iloeje et al (1974)

to satisfy the conditions necessary for bubble nucleation. Bubbles will grow within the droplet Rao (1974) ejecting part of the liquid back into the vapour stream. The remaining liquid is insufficient in thickness to support nucleation and therefore remains until it is totally evaporated. The surface heat flux transferred by this mechanism can be arrived at by estimating the product of the heat transferred to a single drop and the number of droplets per unit time and per unit area which strike the wall. Iloeje et al attempt to quantify the various mechanisms identified above and finally arrive at a somewhat complex expression for the droplet-wall contact heat flux (ϕ_{dc}).

The droplet mass flux to the wall can be estimated from one of a number of turbulent deposition models Hutchinson (1971). It is important to appreciate that at very low values of ($T_W - T_{SAT}$) the heat flux (ϕ_{dc}) predicted from the Iloeje model must coincide with the droplet deposition flux contribution of the Hewitt film flow model of dryout. The trends with vapour quality and mass velocity appear qualitatively correct but a rigorous quantitative comparison with, for example, the prediction of Whalley et al (1973) is needed for a range of working fluids.

Two basic approaches have been taken to estimate the heat flux (ϕ_{dow}) to droplets entering the thermal boundary layer but which do not touch the wall. This heat flux can be estimated as the product of the heat flux that would occur across a vapour film separating the droplet from the heating surface and the fractional area covered by such droplets. This approach has been adopted with slight modifications by Iloeje (1974), Groeneveld (1972) and, Plummer (1974).

Alternatively, Forslund (1966), Hynek (1966) and Course (1974) have assumed the heat transfer coefficient to a single droplet in the spheriodal state condition on a flat heated plate Rao (1974) and then multiplied this by the number of droplets approaching the surface per unit time per unit area.

Finally, the various treatments of mechanisms (c) and (d) offered by various workers follow that proposed by Bennett et al (1967) fairly closely. In some cases slightly differing assumptions are made concerning the shattering of droplets due to the droplet Weber number increasing above a critical value.

Figures 10.11 and 10.12, taken from Iloeje (1974) show, firstly, the variation of each separate heat flux component for a particular value of vapour quality and mass velocity for nitrogen at atmospheric pressure and, secondly, the variation of the total heat flux curve at a given mass velocity but varying vapour qualities. It is important to note that the minimum in the boiling curve occurs at varying values of ($T_W - T_{SAT}$) dependent upon the values of the individual heat flux contributions. This concept of the minimum in the film boiling curve differs somewhat from that proposed by Henry (1974).

10.8 Post-CHF Heat Transfer Coefficients for Low Steam Quality at or Near a Flow Reversal

So far we have considered situations where the flow pattern prior to CHF was annular corresponding therefore to relatively high

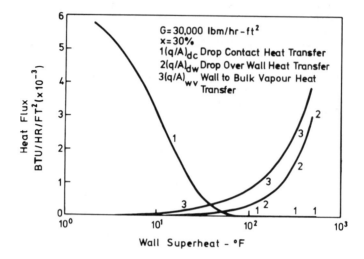

Figure 10.11: Behaviour of components of total heat flux with quality at
30,000 lbm/hr-ft^2 fluid-nitrogen (Iloeje et al (1974))

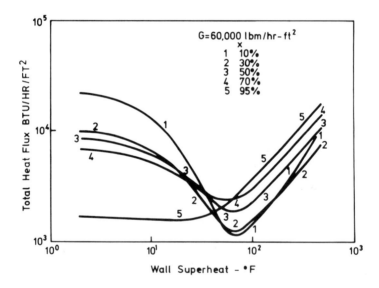

Figure 10.12: Theoretical boiling curves at 60,000 lbm/hr-ft^2 fluid-nitrogen
(Iloeje et al (1974))

vapour qualities at the dryout point (x > 10% by wt). Much less
is known about post-CHF heat transfer coefficients at low vapour
qualitites and under subcooled condition. This is because with
water, at least, it is impossible to study such situations
experimentally except under transient conditions. With water in
the pressure range up to approximately 100 bar, the critical heat
flux values are in excess of 3 MW/m^2 for subcooled boiling, while

typical values of the post-CHF heat transfer coefficients in this
region are 150-500 $W/m^2 {}^{\circ}C$. Any attempt to pass through the
critical heat flux under steady state conditions would produce
heater surface temperatures in excess of $2000 {}^{\circ}C$ and would cause
failure of most common metal surface - physical burnout.

A schematic representation of the conditions under consideration is given in Figure 10.13. The study of forced convective
film boiling inside vertical tubes made by Dougall (1963) included
visual observations of the flow structure. At low qualities and
mass flowrates the flow regime would appear to be an "inverted
annular" one with liquid in the centre and a thin vapour film
adjacent to the heating surface. The vapour-liquid interface is
not smooth, but irregular. These irregularities occur at random
locations, but appear to retain their identity to some degree as
they pass up the tube with velocities of the same order as that of
the liquid core. The vapour in the film adjacent to the heating
surface would appear to travel at a higher velocity. The liquid
core may be in up-flow, stationary, or in down-flow. Vapour
bubbles may be present in the liquid core, but will have little
influence except to modify the core velocity and density which,
in turn, determine the interfacial shear stress.

10.9 Simple Theories of Film Boiling

Because the liquid is displaced from the heating surface by a
vapour film and the uncertainties associated with bubble nucleation
are removed, film boiling is very amenable to analytical solution.
In general, the problem is treated as an analogue of film-wise
condensation and solutions are available for horizontal and
vertical flat surfaces, and also inside tubes under both laminar

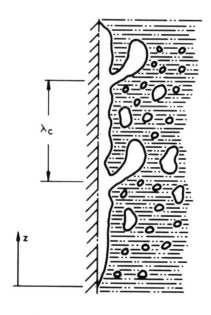

Figure 10.13: Schematic representation of low quality post-CHF condition

and turbulent conditions with and without interfacial shear. It is not, however, proposed to review all these solutions here and the reader is referred to the comprehensive reviews of Clements and Colver (1970), of Hsu (1972) and of Bressler (1972), together with the comments of Wallis and Collier (1968).

The simplest solution is obtained by assuming the vapour film is laminar and that the temperature distribution through the film is linear (Figure 10.14). For a vertical flat surface various boundary conditions may be imposed upon such an analysis, viz

(1)   zero interfacial shear stress (τ_i = 0)

(2)   zero interfacial velocity   (u_i = 0)

(3)   zero wall shear stress   (τ_w = 0)

For the first of these boundary conditions, the local heat transfer coefficient h(z) at a distance z up the surface from the start of film boiling is given by:

$$\left[\frac{h(z)z}{k_g}\right] = \left[\frac{z^3 g \rho_g (\rho_f - \rho_f) i_{fg}'}{4 \ k_g \ \mu_g \ \Delta T}\right]^{\frac{1}{4}}$$

(10.31)

The average coefficient $\bar{h}(z)$ over the region up to a distance z is given by:

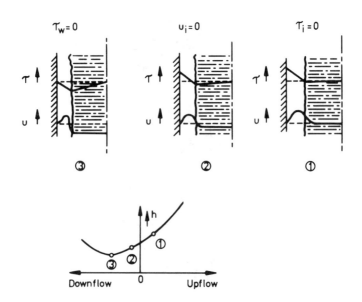

Figure 10.14: Film boiling laminar flow solutions near a flow reversal

$$\left[\frac{\bar{h}(z)z}{k_g}\right] = 0.943 \left[\frac{z^3 g\rho_g(\rho_f - \rho_g)i'_{fg}}{k_g\,\mu_g\,\Delta T}\right]^{\frac{1}{4}} \quad (10.32)$$

If we define

$$Pr* = \left[\frac{\mu_g i'_{fg}}{k_g\,\Delta T}\right] \quad \text{and} \quad Gr* = \left[\frac{z^3 g\rho_g(\rho_f - \rho_g)}{\mu_g^2}\right]$$

$$\bar{N}u_z = \left[\frac{\bar{h}(z)z}{k_g}\right]$$

then $\bar{N}u_z = C\left[Pr*\ Gr*\right]^{\frac{1}{4}}$ (10.33)

Note the analogy with natural convection. The value of C for the first of the above boundary conditions is 0.943 and for the second is 0.667. Typical values of heat transfer coefficient have been calculated from equation (10.31) for water over the pressure range 1-70 bar for a distance z assumed to be 0.5m (same order of magnitude as grid spacing on a water reactor fuel element) and a temperature difference ΔT of $200^{\circ}K$ (Figure 10.15).

A similar analysis (Wallis (1968)) may be carried out for the assumption of turbulent flow in the vapour film. In this case, for a vertical flat surface:

$$\left[\frac{\bar{h}(z)z}{k_g}\right] = 0.056\ Re_g^{0.2}\left[Pr\ Gr*\right]^{\frac{1}{3}} \quad (10.34)$$

Note that the coefficient is approximately constant, independent of distance z in this equation. Typical values of heat transfer coefficient calculated from equation (10.34) for a value of Re_g of 10^4 are also shown in Figure 10.15. Due to the neglect of the resistance between the vapour film and the interface and the effect of interfacial drag the heat transfer coefficients computed from equation (10.34) may be somewhat higher than measured.

In the US loss-of-coolant code RELAP-4 (Moore and Rettig (1973)) the Berenson (1961) correlation for film boiling on a *horizontal* flat plate is used for a stagnation condition. This correlation is very similar in form to equation (10.31).

$$h = 0.425\left[\frac{k_g^3 g\rho_g(\rho_f - \rho_g)i'_{fg}}{\mu_g\,\Delta T\left(\frac{\lambda_c}{2\pi}\right)}\right]^{\frac{1}{4}} \quad (10.35)$$

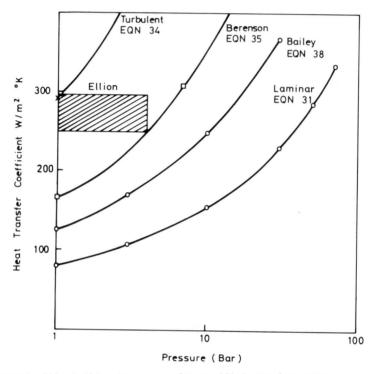

Figure 10.15: Film boiling heat transfer coefficients for water

The thin vapour film over the horizontal surface is unstable and large bubbles form and break away. The characteristic spacing of these bubbles (λ_c) is determined by a balance of surface tension and gravitational forces and is given by:

$$\frac{\lambda_c}{2\pi} = \left[\frac{\sigma}{g(\rho_f - \rho_g)} \right]^{\frac{1}{2}}$$

(10.36)

Values calculated for water from equation (10.35) are given on Figure 10.15.

On *vertical* heated surfaces, the vapour film also appears to break away from the wall in large globular voids and various attempts have been made to produce an equation which is more consistent with this physical behaviour than equation (10.31). Bailey (1971) suggested that the globular voids are formed by a process of "varicose" instability of a hollow gas cylinder within a denser liquid as proposed by Chandrasekhar. The wavelength for maximum growth for a surface tension governed instability of this type is:

$$\frac{\lambda_c}{2\pi} = \frac{r}{0.484}$$

(10.37)

Substitution of λ_c from this equation for z in equation (10.31) leads to:

$$h = \left[\frac{k_g^3 g \rho_g (\rho_f - \rho_g) i_{fg}'}{\mu_g \, \Delta T \, r} \right]^{\frac{1}{4}}$$

(10.38)

where r is the fuel rod radius. Values of λ calculated for water assuming a fuel rod radius of 8mm are given in Figure 10.15. An alternative suggestion which predicts heat transfer coefficients about 33% higher is the use of the original Bromley equation for film boiling on the outside of *horizontal* tubes:

$$h = 0.62 \left[\frac{k_g^3 g \rho_g (\rho_f - \rho_g) i_{fg}}{\mu_g \, \Delta T \, D} \right]^{\frac{1}{4}}$$

(10.39)

Some evidence that this gives the correct order of magnitude for the film boiling coefficient during reflood experiments was given by Amm and Ulrych (1974).

Suryanarayana and Merte (1972) have provided a turbulent boundary layer analysis of film boiling on vertical surfaces which assumes a zero vapour-liquid interface velocity ($u_i = 0$) and empirically takes into account the effect of interfacial oscilla-tions by the use of an "enhancement factor" B which is a measure of the increase in heat transfer over and above the smooth inter-face solution. B is related to the dimensionless amplitude of the interface oscillation b(z) by the relationship:

$$B = \frac{h}{h_{smooth}} = \frac{1}{\sqrt{1 - b(z)^2}}$$

(10.40)

An empirical correlation of B with vapour film Reynolds number (Re_g) was attempted which gave values of B around 2-2.5 for Reynolds numbers of 10^3.

Another recent study of vertical film boiling which tries to take account of the time-varying vapour thickness as the large vapour bubbles are released from the film is that of Greitzer and Abernathy (1972). This study also indicates that the smooth interface laminar flow solution (equation (10.31)) predicts a value of approximately two too low. The effects of subcooling and liquid velocity are also established for methanol over a wide range of values. Typically, increasing the liquid velocity from 0 to 15 ft/sec or, alternatively, the subcooling from 0 to 40°F might double the heat transfer coefficient in each case.

The evidence at the present time is that laminar film boiling with a smooth interface only occurs over relatively short distances ($\sim$ 2") downstream of the dryout or "rewet" front. At longer distances the coefficient becomes independent of distance and takes on a value considerably (approximately a factor of two) higher than the laminar solution would indicate. The exact amount

is a relatively weak function of the vapour film Reynolds number.
If this picture is correct we might expect any effect of flow
direction on the coefficient such as indicated in Figure 10.14 to
be mainly in the short region just downstream of the dryout or
quench front. The experiments of Papell (1971) with hydrogen
indicate that this is, in fact, the case and that the coefficients
in down-flow for this particular region could be a factor of two
or so lower than for up-flow.

10.10   Quenching

The need to guarantee the cooling of water-cooled reactor
fuel elements during a loss-of-coolant accident (LOCA) has prompted
a considerable amount of reserach into the quenching process.
Quenching is a transient process which occurs when an initially
very hot metal surface is exposed to a coolant. In the case of a
tube, an annulus or a rod bundle fuel element, the coolant may be
admitted either from above - *top flooding* - or from the base of
the channel - *bottom flooding*. Sometimes both methods are used
simultaneously.

Figure 10.16 depicts the physical processes which occur in
the case of *top flooding*. An initial cooling at the top of the
channel allows a liquid film to form. This liquid film begins to
run down the surface. However, at some point the film is violently
thrown off the surface. This ejection of the film from the rod
has been termed "sputtering". The "sputtering" point of the
quench front divides the dry wall high temperature region of the
channel from the wet wall low temperature region. It is observed
that the quench front proceeds down the tube at a uniform velocity
which is a function of the initial surface temperature (T_W) amongst
other factors.

The vapour formed by the quenching process escapes upwards
out of the channel. However, if the channel is restricted and
the vapour generation rate large, then the rate of access of liquid

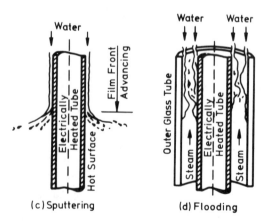

Figure 10.16: Physical processes occurring during top flooding of a hot rod

to the channel may be limited by the "flooding" mechanism, as indicated in Figure 10.16.

Figure 10.17 shows experimental measurements by Bennett et al (1966) of the quench front velocity for a water film advancing down a vertical rod. The inverse quench velocity (1/u) was found to vary with initial rod temperature and with the system pressure. Extrapolation of the inverse quench front velocity lines intercept the horizontal axis at a finite wall temperature. It was suggested that this temperature was related to the so-called "sputtering" temperature just at the quench front. In fact, Bennett et al extrapolated not to the horizontal axis but to a value of 4 s/m where the quench rate became independent of wall temperature.

The physical processes occurring with bottom flooding are different depending on whether the rate of admission of coolant to the channel is high or low (Figure 10.18). At low flooding rates an annular flow region is formed with the quench front being marked by the *dryout* transition between the hot wall liquid deficient region and the climbing liquid film. Other nucleate boiling regimes occur upstream of the quench front. At high flooding rates the liquid level in the channel rises more rapidly than the quench front and a region of film boiling (inverted

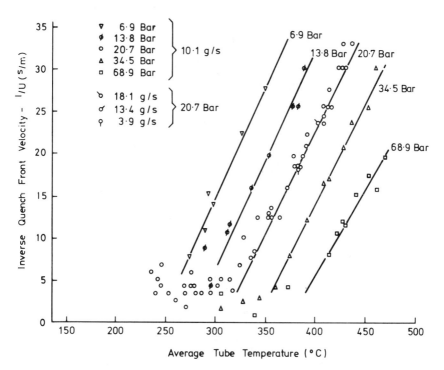

Figure 10.17: Experimental quench front velocity results of Bennett et al (1966) for top flooding

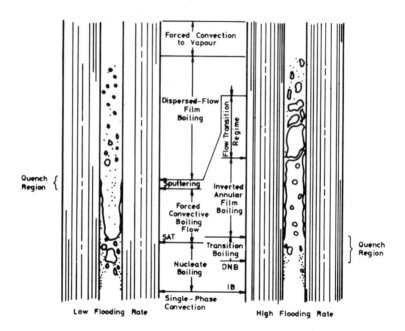

Figure 10.18: Physical processes occurring during bottom flooding of a hot rod

annular region) is formed. The quench front is marked in this
instance by the point of collapse of the thin vapour film. In both
the low and high flooding rate situations a significant amount of
entrainment of the coolant in the vapour stream occurs. This
entrained coolant is ineffective in that it is carried out of the
channel and takes no part in the quenching process.

     At high flooding rates a significant amount of coolant of the
tube occurs in the film boiling region just downstream of the
quench front. Hsu (1975) has correlated experimental heat transfer
data during reflooding experiments and his equation is:

$$h = 1.33 \times 10^5 \ p^{0.558} \ \exp \{- 0.016 \ p^{0.1733} \ \Delta T_{SAT}\} \qquad (10.41)$$

where h is in $kW/m^2 \,^{O}K$, ΔT_{SAT} in K and p in MN/m^2

This equation is based on the assumption that there is no axial
conduction and therefore should only be used with 1-D rewetting
models (see below). The equation can be combined with a fully-
developed film boiling relationship as shown in Figure 10.19.

     Analytical models of the quenching process all involve a
solution of the Fourier heat conduction equation for specified
boundary conditions representative of the heat transfer processes
occurring at the surface of the solid. In order to make the
problem handleable, various simplifications are introduced such as:

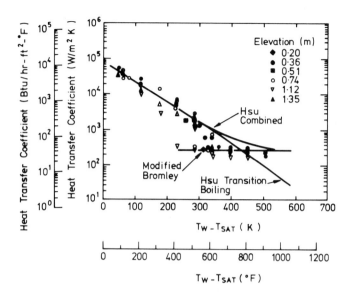

Figure 10.19: Hsu (1975) correlation of transition and film boiling during bottom reflooding

(a)   the solid is assumed to be a homogeneous slab (Figure 10.20) of infinite height, ie in the direction of the advancing quench front (z direction), of uniform thickness δ (in the y direction) and having constant (temperature independent physical properties).

(b)   it is assumed that the dry side of the slab is insulated (y = 0) and the wet side (y = δ) is cooled and that there is no heat generation within the solid.

(c)   the liquid is assumed to wet the surface up to the axial location (z =0) where the surface equals the *sputtering* temperature (T_O) which is assumed to be a constant independent of time or position.

(d)   the quench front velocity is assumed to be constant.

These assumptions lead to the following simplified conduction equation:

$$\frac{\partial^2 T}{\partial y^2} + \frac{\partial^2 T}{\partial z^2} = -\frac{U}{\kappa}\frac{\partial T}{\partial z}$$

(10.42)

where κ is the thermal diffusivity of the solid.

The boundary conditions are:

   T = T_S, the liquid coolant temperature at z = -∞

   T = T_W, the initial wall temperature at z = +∞

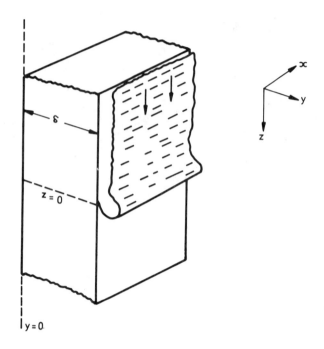

Figure 10.20: Co-ordinate system used in rewetting models

$$\frac{\partial T}{\partial y} = 0 \text{ at } y = 0$$

$$- k\frac{\partial T}{\partial y} = h(z) \ (T - T_S) \text{ at } y = \delta \qquad (10.43)$$

If it is assumed that the boundary conditions are such that the
slab temperature is uniform across its thickness at any value of z,
then a one-dimensional solution may be obtained. In this case:

$$\frac{\partial^2 T}{\partial y^2} \simeq - \frac{h(z)}{k\delta} \ (T - T_S) \qquad (10.44)$$

and equation (10.42) reduces to:

$$\frac{d^2 T}{dz^2} + \frac{U}{\kappa}\frac{dT}{dz} - \frac{h(z)}{k\delta} \ (T - T_S) = 0 \qquad (10.45)$$

This can be non-dimensional and the general solution to equation
(10.45) has been given by Elias and Yadigaroglu (1977). The
various one-dimensional solutions differ only in the assumptions
made about the variation of the heat transfer coefficient h(z) in
front of and behind the quench front. If just two axial regions
are assumed, a "dry" region ahead of the quench front where h(z)
= 0, and a "wet" region behind the quench front where h(z) is

high but constant, then the solution to equation (10.45) is:

$$\frac{1}{U} = \frac{\rho\, c_p}{2} \left(\frac{\delta}{h(z)k}\right)^{\frac{1}{2}} \left\{\left[\frac{2\,(T_W - T_O)}{(T_O - T_S)} + 1\right]^2 - 1\right\}^{\frac{1}{2}} \tag{10.46}$$

an equation given by Yamanouchi (1968). However, to fit experimental data on measured quench front velocities he had to introduce very high values of $h(z)$ in the "wet" region ($\sim 10^6$ W/m^2°K). Such extreme values are not encountered in steady state boiling situations.

Later analyses have introduced more complex variations of $h(z)$ with position relative to the quench front. The most recent model is that of Elias and Yadigaroglu (1977) which generalises this approach and uses as many as twelve axial regions. Figure 10.21 shows some typical results for Zircaloy-4, Inconel-600 and stainless steel and emphasises the importance of the thermal diffusivity of the heating surface. Other models have used a two-dimensional conduction solution.

One outstanding item relates to the *sputtering* temperature itself which must be specified as an input to these models. Various theories relating this temperature to the Leidenfront temperature, the minimum in the classical pool boiling curve and the critical temperature, have been put forward but none is generally accepted. Moreover, it appears that the *sputtering temperature* depends to an extent on the surface condition (ie oxide film, etc) of the heating surface.

10.11 Conclusions

The last few years have seen considerable advances in our

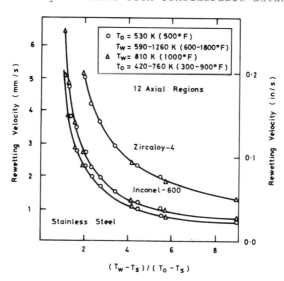

Figure 10.21: Rewetting velocity calculated by the model of Elias and Yadigaroglu

knowledge of heat transfer in the post-CHF region, particularly
at high qualities. A major advance has been the development and
testing of a satisfactory transient technique to allow the whole
of the forced convective boiling curve, including the transition
boiling region, to be covered. Many empirical correlations based
on the statistical regression of large amounts of data are now
available for tubular, annular and rod bundle geometries. Such
correlations do have a use in thermohydraulic safety codes but
only until more soundly based physical models of the heat transfer
processes are built up. These semi-theoretical models are complex
and need further validation. In particular, it is important to
ensure they are physically consistent with our knowledge of the
non-equilibrium hydrodynamic conditions upstream of the dryout
point. We can expect a period of rationalisation whilst these
questions are examined quantitatively. Since it is unlikely that
the large thermohydraulic safety codes can accept detailed models
of this type, an intermediate model which allows for the lack of
thermodynamic equilibrium in a relatively simple manner is being
developed by various workers.

The state of knowledge in the low quality and subcooled film
boiling regions still lags behind that of the post-dryout region.
The new transient experimental techniques should, however, now
allow actual measurements for water over a range of conditions of
interest. It is clear that the classical laminar film boiling
equation is only valid over a very short distance downstream of
the CHF location. Beyond this region vapour is released in large
globular voids which ensures a time varying but thinner vapour
film. It is recommended that equations (10.35), (10.38) or (10.39)
be used for film boiling in water reactor safety codes. A strong
influence of buoyancy which will greatly affect the heat transfer
coefficient at or near a flow stagnation can be expected in the
short region where the laminar film theory applies and there is
experimental evidence to support this view. However, it is strongly
recommended that an experimental study of film boiling at or near
flow stagnation conditions be made using large thermal inertia
heaters to allow entry into the film boiling region during a CHF
transient.

Nomenclature

$a,b,c,d,e,f,$	-	coefficients in power law expressions
B	-	"enhancement factor" (equation (10.40))
$b(z)$	-	dimensionless amplitude of interface oscillation
c_p	-	specific heat
D,De	-	diameter, hydraulic diameter
g	-	acceleration due to gravity
G	-	mass velocity
$h(z)$	-	heat transfer coefficient
i_{fg}	-	latent heat of vaporisation
i'_{fg}	-	modified latent heat of vaporisation
k	-	thermal conductivity
p	-	system pressure

r	−	rod radius
S	−	Chen nucleation suppression factor
T	−	temperature
u	−	velocity
x	−	quality
X*	−	defined in equation (10.27)
Y	−	correction factor given by equation (10.2)
z	−	distance along heated channel
z*	−	distance along heated channel at which last droplets disappear
z^+	−	defined in equation (10.27)
Z	−	defined in equation (10.29)
Nu	−	$(h\,D_e/k)$
$\bar{Nu}_z$	−	$(\bar{h}(z)z/k_g)$
Gr*	−	$(z^3 g\rho_g(\rho_f - \rho_g)/\mu_g^2)$
Re	−	(GD_e/μ)
Pr	−	$(\mu\,c_p/k)$
Pr*	−	$(\mu_g i_{fg}/k_g \Delta T)$

Greek

δ	−	droplet diameter
α	−	void fraction
ρ	−	density
μ	−	viscosity
τ	−	shear stress
φ	−	heat flux
λ_c	−	minimum wavelength for Tayler instability
ΔT	−	wall superheater $(T_W(z) - T_{SAT})$
ε	−	$\phi_L(z)/\phi(z)$
ψ	−	parameter defined by equation (10.26)
σ	−	surface tension

Subscripts

CRIT	−	value at thermodynamic critical point
dc	−	direct contact
DO	−	dryout
dow	−	absence of contact
e	−	thermodynamic
eq	−	equivalent

EQ	- equilibrium
f	- film
g	- gas, vapour, superheating
i	- interface
L	- latent heat, evaporating
l	- liquid
NB	- nucleate boiling
Smooth	- smooth interface
TB	- transition boiling
TP	- two-phase
W	- wall

Superscripts

*	- corresponding to actual vapour quality
-	- average over a distance z

References

Adorni, N, (1966) "Heat transfer crisis and pressure drop with steam-water mixtures: Experimental data with seven rod bundles at 50 and 70 kg/cm². *CISE* R-170.

Amm, H and Ulrych, G, (1974) "Comparison of measured heat transfer coefficients during reflooding a 340-rod bundle and those calculated from current heat transfer correlations". Paper E7, *European Two-Phase Flow Meeting, Harwell.*

Bailey, N A, (1971) "Film boiling on submerged vertical cylinders". AEEW-M1051.

Bailey, N A, (1972) "The interaction of droplet deposition and forced convection in post dryout heat transfer at high subcritical pressures". *European Two-Phase Flow Group Meeting, Rome.*

Bailey, N A and Collier, J G, (1970) "The estimation of tube wall temperatures in the evaporator region of sub-critical once-through sodium heated steam generators". AEEW-M1000.

Bailey, N A, Collier, J G and Ralph, J C, (1973) "Post-dryout heat transfer in nuclear and cryogenic equipment". AERE-R7519.

Bennett, A W, (1964) "Heat transfer to mixtures of high pressure steam and water in an annulus". AERE-R4352.

Bennett, A W, Hewitt, G F, Kearsey, H A and Keys, R K F, (1966) "The wetting of hot surfaces by water in a steam environment at high pressure". AERE-R5146.

Bennett, A W, Hewitt, G F, Kearsey, H A and Keys, R K F, (1967) "Heat transfer to steam water mixtures flowing in uniformly heated tubes in which the critical heat flux has been exceeded". AERE-R5373.

Berenson, P J, (1961) "Film boiling heat transfer from a horizontal surface". *Trans. AS,E, Journal of Heet Transfer,* **83**, Series C, No 3, 351-358.

Bertoletti, S, (1961) "Heat transfer and pressure drop with steam-water spray". CISE R-36.

Bertoletti, S, (1964) "Heat transfer to steam-water mixtures". CISE R-78.

Bishop, A A, Efferding, L E and Tong, L S, (1962) "A review of heat transfer and fluid flow of water in the supercritical region and during 'once-thru' operation". WCAP-2040.

Bishop, A A, Sandberg, R O and Tong, L S, (1964) "High temperature supercritical pressure water loop. Part V: Forced convection heat transfer to water after the critical heat flux at high subcritical pressures". WCAP-2056 (Pt.5).

Bishop, A A, Sandberg, R O and Tong, L S, (1965) "Forced convection heat transfer at high pressure after the critical heat flux". ASME 65-HT-31.

Bressler, R G, (1972) "A review of physical models and heat transfer correlations for free convection film boiling". *Adv. Cryogenic Engng.,* **17**, 382-406.

Brevi, R, Cumo, M Palmieri, A and Pitimada, D, (1969) "Heat transfer coefficient in post-dryout two-phase mixtures". *Paper presented at the European Two-Phase Group Meeting, Karlsruhe.*

Brevi, R, Cumo, M, Palmieri, A and Pitimada, D, (1969) "Post dryout heat transfer with steam/water mixtures". *Trans. Am. Nucl. Soc.,* **12**, 809-811.

Clements, L D and Colver, C P, (1970) "Natural convection film boiling heat transfer". *Industr. Engng. Chem.* **62** (9), 26-46.

Collier, J G, (1962) "Heat transfer and fluid dynamic research as applied to fog cooled power reactors". AECL-1631.

Collier, J G, (1977) "Post-dryout heat transfer - A review of the current position". *Two-Phase Flows and Heat Transfer* II, 769-813, *Proceedings of NATO Advanced Study Institute, Istanbul, Turkey.*

Course, A F and Roberts, H A, (1974) "Progress with heat transfer to a steam film in the presence of water drops - a first evaluation of Winfrith SGHWR cluster loop data". AEEW-M1212. *Paper presented at European Two-Phase Flow Group Meeting, Harwell.*

Cumo, M and Urbani, G C, (1974) "Post-burnout heat transfer (attainable precision limits of the measured coefficient)". CNEN/RT/ING(74)24.

Dougall, R S and Rohsenow, W M, (1963) "Film boiling on the inside of vertical tubes with upward flow of the fluid at low qualities". *Dept. of Mech. Engng., Engng. Project Lab.,* MIT Report No 9079-26.

Durga Prasad, K A, Srinivasan, K and Krishna Murthy, M V, (1974) "Cool-down of foam insulated cryogenic transfer lines". *Cryogenics* **14** (11), 615-617.

Elias, E and Yadigaroglu, G, (1977) "A general one-dimensional model for conduct-controlled rewetting of a surface". *Nuclear Engng. and Design,* **42**, 185-194.

326   J G Collier

Era, A, (1967) "Heat transfer data in the liquid deficient region for steam-water mixtures at 70 kg/cm^2 flowing in tubular and annular conduits". CISE R-184.

Era, A, Gaspari, G P, Protti, M and Zavatarelli, Z, (1971) "Post dryout heat transfer measurements in an annulus with uniform and non-uniform axial heat flux distribution". *European Two-Phase Group Meeting, Risø.*

Forslund, R P and Rohsenow, W M, (1966) "Thermal non-equilibrium in dispersed flow film boiling in a vertical tube". MIT Report 75312-44.

Gardner, G C, (1974) "Evaporation and thermophertic motion of water drops containing salt in a high pressure steam environment". *Paper presented at European Two-Phase Flow Conference, Harwell, England.*

Greitzer, E M and Abernathy, F H, (1972) "Film boiling on vertical surfaces". *Int. J. Heat Mass Transfer*, **15**, 475-491.

Groeneveld, D C, (1969) "An investigation of heat transfer in the liquid deficient regime". AECL-3281.

Groeneveld, D C, (1972) "The thermal behaviour of a heated surface at and beyond dryout". AECL-4309.

Groeneveld, D C, (1973) "Post-dryout heat transfer at reactor operating conditions". AECL-4513. *Paper presented at the Nat. Topical Meeting on Water Reactor Safety, ANS, Salt Lake City, Utah.*

Groeneveld, D C, (1974) "Effect of a heat flux spike on the downstream dryout behaviour". *J. of Heat Tra2sfer.* 121-125.

Groeneveld, D C and Delorme, G G J, (1976) "Prediction of non-equilibrium in the post-dryout regime". *Nucl. Eng. Des.* **36** (1), 17-26.

Groeneveld, D C and Fung, K K, (1977) "Heat transfer experiments in the unstable post-CHF region". *Presented at the Water Reactor Safety Information Meeting, Washington.*

Groeneveld, D C, Thibodeau, M and McPherson, G D, (1970) "Heat transfer measurements on trefoil fuel bundles in the post dryout regime - with data tabulation". AECL-3414.

Hench, J E, (1964) "Multi-rod (two rod) transition and film boiling in forced convection to water at 1000 psia". GEAP-4721.

Hench, J E, (1964) "Transition and film boiling data at 600, 1100 and 1400 psi in forced convection heat transfer to water". GEAP-4492.

Henry, R E, (1974) "A correlation for the minimum film boiling temperature". *AIChE Symp. Series*, **70** (138), 81-90.

Herkenrath, H, Mork-Morkenstein, P, Jung, U and Weckermann, F J, (1967) "Warmeubergang an wasser bei erzwungener stromung in druckbereich von 140 bis 250 bar". EUR-3658d.

Herkenrath, H, et al, (1969) "Die warmeubergangskrise von wasser bei erzwungener stromung unter hohen drucken, teil 2, der warmeubergang im bereich der Krise". *Atomkernenergie*, **14**, 403-407.

Herkenrath, H and Mörk-Mörkenstein, P, (1970) "The heat transfer crisis for water with forced flow at high pressures. Part II, Heat transfer in the crisis region". AERE-Trans. 1129.

Hewitt, G F and Hall-Taylor, N S, (1970) "Annular two-phase flow". *Pergamon Press*, 237.

Hsu, Y Y, (1972) "A review of film boiling". *Adv. in Cryogenic Engng.* **17**, 361-381.

Hsu, Y Y, (1975) "A temperature correlation for the regime of transition boiling and film boiling during reflood". *3rd Water Reactor Safety Research Information Meeting, USNRC, Washington.*

Hutchinson, P, Hewitt, G F and Dukler, A E, (1971) *Chem. Eng. Sci.* **26**, 419.

Hynek, S J, Rohsenow, W M and Bergles, A E, (1966) "Forced convection dispersed vertical flow film boiling". MIT report 70586-63.

Iloeje, O C, Plummer, D N, Rohsenow, W M, Griffith, P, (1974) "A study of wall rewet and heat transfer in dispersed vertical flow". *MIT Dept. of Mech. Engng.* Report 72718-92.

Keeys, R K F, Ralph, J C and Roberts, D N, (1971) "Post-burnout heat transfer in high pressure steam-water mixtures in a tube with cosine heat flux distribution". AERE-R6411.

Kroeger, P G and Zuber, N, (1968) "An analysis of the effects of various parameters on the average void fractions in sub-cooled boiling". *Int. J. Heat Mass Transfer*, **11**, 211-233.

Kunsemiller, D F, (1965) "Multi-rod, forced flow transition and film boiling measurements". GEAP-5073.

Lee, D H, (1970) "Studies of heat transfer and pressure drop relevant to sub-critical once-through evaporators". IAEA-SM-130/56, *IAEA Symp. on Progress in Sodium-Cooled Fast Reactor Engineering, Monaco.*

Mattson, R J, Condie, K G, Bengston, S J and Obenchain, C F, (1974) "Regression analysis of post-CHF flow boiling data". *Paper B3.8, **4**, Proc. of 5th Int. Heat Transfer Conference, Tokyo.*

Matzner, B, (1963) "Basic experimental studies of boiling fluid flow and heat transfer at elevated pressures". *Columbia University Monthly Progress Report* MPR-XIII-2-63.

Matzner, B, Casterline, J E and Kokolis, S, (1968) "Critical heat flux and flow stability in a 9ft 19-rod test section simulating a string of short discrete rod bundles". *Columbia University Topical Report* No 10.

Matzner, B and Neill, J S, (1963) "Forced-flow boiling in rod bundles at high pressures". DP-857.

McDonough, J B, Milich, W and King, E C, (1960) *AIChE Preprint* No 29, *4th US National Heat Transfer Conference.*

McPherson, G D, Matzner, B, Castellana, F, Casterline, J E and Wikhammer, G A,

(1971) "Dryout and post-dryout behaviour of a 28-element fuel bundle". *Paper presented at American Nuclear Society Winter Meeting, Miami Beach.*

Miropol'skiy, Z L, (1963) "Heat transfer in film boiling of a steam water mixture in steam generating tubes". *Teploenergetika.*

Moore, K V and Rettig, W H, (1973) "RELAP-4 - A computer program for transient thermal-hydraulic analysis". ANCR-1127.

Mueller, R E, (1967) "Film boiling heat transfer measurements in a tubular test section". EURAEC-1871/GEAP-5423.

Papell, S S, (1971) "Film boiling of cryogenic hydrogen during upward and downward flow". NASA-TMX-67855. *Paper presented at 13th Int. Congress on Refrigeration, Washington.*

Parker, J D and Grosh, R J, (1962) "Heat transfer to a mist flow". ANL-6291.

Plummer, D N, Iloeje, O C, Rohsenow, W M, Griffith, P and Ganic, E, (1974) "Post critical heat transfer to flowing liquid in a vertical tube". *MIT Dept. of Mech. Engng. Report* 72718-91.

Polomik, E E, (1967) "Transition boiling, heat transfer program, final summary report on program for Feb/63 - Oct/67". GEAP-5563.

Polomik, E E, (1971) "Deficient cooling". *7th Quarterly Report,* GEAP-10221-7.

Polomik, E E, Levy, S and Sawochka, S G, (1961) "Heat transfer coefficients with annular flow during once-through boiling of water to 100% quality at 800, 1000 and 1400 psi". GEAP-3703.

Ramu, K and Weisman, J, (1974) "A method for the correlation of transition boiling heat transfer data". Paper B.4.4, **4,** *Proc. of 5th Int. Heat Transfer Conference, Tokyo.*

Rao, P S V K and Sarma, P K, (1974) "Partial evaporation rates of liquid particles on a hot plate under film boiling conditions". *Con. J. of Chem. Engng.* **52** (3), 415-419.

Schmidt, K R, (1960) "Thermodynamic investigations of highly loaded boiler heating surfaces". AEC-Tr-4033.

Slaughterback, D C, Vesely, W E, Ybarrondo, L J, Condie, K G and Mattson, R J, (1973) "Statistical regression analyses of experimental data for flow film boiling heat transfer". *Papers presented at ASME-AIChE Heat Transfer Conference, Atlanta.*

Slaughterback, D C, Ybarrondo, L J and Obenchain, C F, (1973) "Flow film boiling heat transfer correlations - parametric study with data comparisons". *Paper presented at ASME-AIChE Heat Transfer Conference, Atlanta.*

Srinivasan, K, Seshargin, Rao, V and Krishna Murthy, M V, (1974) "Analytical and experimental investigation on cool-down of short cryogenic transfer lines". *Cryogenics* **14** (9), 489-494.

Suryanarayana, N V and Merte, H, (1972) "Film boiling on vertical surfaces". *J. of Heat Transfer,* **94**, Series C, No 4, 377-384.

Swenson, H S, Carver, J R and Szoeke, G, (1961) "The effects of nucleate boiling versus film boiling on heat transfer in power boiler tubes". ASME 61-WA-201.

Tong, L S, (1964) "Film boiling heat transfer at low quality of subcooled region". *Proceedings 2nd Joint USAEC-EURATOM Two-Phase Flow Meeting, Germantown.* Report CONF-640507, p.63.

Tong, L S, (1967) "Heat transfer in water-cooled nuclear reactors". *Nuc. Engng. and Design,* **6**, 301.

Tong, L S, (1972) "Heat transfer mechanisms in nucleate and film boiling". *Nuc. Engng. and Design,* **21**, 1-25.

Tong, L S and Young, J D, (1974) "A phenomenological transition and film boiling heat transfer correlation". Paper B.3.9, **4**, *Proc. of 5th Int. Heat Transfer Conference, Tokyo.*

Wallis, G B and Collier, J G, (1968) "Two-phase flow and heat transfer". *Notes for a Summer Course, July 15-26, Thayer School of Engineering, Dartmouth College, Hanover, NH, USA.* **III**, 33-46.

Whalley, P B, Hutchinson, P and Hewitt, G F, (1973) "The calculation of critical heat flux in forced convection boiling". AERE-R7520.

Yamanouchi, A, (1968) "Effect of core spray cooling in transient state after loss-of-coolant accident". *J. Nuclear Sci. Technol.* **5** (11), 547-558.

Chapter 11

Heat Transfer in Condensation

J. G. COLLIER

11.1 Preface

This chapter is largely based upon Chapter 8 of the book
"Convective Boiling and Condensation" (Collier (1972)). The
original material has been adapted and in some cases updated to
suit this course. For example, the basic processes of droplet
nucleation and growth and the interfacial resistance at a vapour-
liquid interface have been omitted since these have been covered
in an earlier chapter by Hewitt.

11.2 Introduction

Condensation is defined as the removal of heat from a system
in such a manner that vapour is converted into liquid. This may
happen when vapour is cooled sufficiently below the saturation
temperature to induce the nucleation of droplets. Such nucleation
may occur homogeneously within the vapour or heterogeneously on
entrained particulate matter, for example, within the low-pressure
stages of a large steam turbine. Heterogeneous nucleation may also
occur on the walls of the system, particularly if these are cooled
as in the case of a surface condenser. In this latter case there
are two forms of heterogeneous condensation, drop-wise and film-
wise, corresponding to the analogous cases in evaporation, of
nucleate boiling and film boiling. Film-wise condensation occurs
on a cooled surface which is easily wetted. On non-wetted surfaces
the vapour condenses in drops which grow by further condensation
and coalescence and then roll over the surface. New drops then
form to take their place. This chapter will mainly deal with film-
wise condensation since reliable theories of drop-wise condensation
have not yet been established.

11.3 Drop-Wise Condensation

Drop-wise condensation occurs on non-wettable surfaces. The
necessary reduction in the interfacial tension between the condensate and the surface may be brought about as a result of contamination, accidental or deliberate, ie removal of wettable oxide films
or by electroplating. The condensate appears in the form of droplets. Subsequently, they roll or fall from the surface under the
action of gravity or aerodynamic drag forces. New droplets then
appear on the exposed clear surface.

The mechanism of drop-wise condensation is still a mystery and quite contradictory statements are common in the literature. The amount of published information is extensive since drop-wise condensation offers the attraction of increasing condensing heat transfer coefficients by an order of magnitude over film-wise condensation.

Two fundamentally different models have been proposed. In the first it is postulated that condensation initially occurs in a film-wise manner on a thin unstable liquid film covering all or part of the surface. On reaching a critical thickness the film ruptures and the liquid is drawn into droplets by surface tension forces. This process then repeats itself. This mechanism was first put forward by Jakob (1936) and has been reiterated since in a number of modified forms by Kast (1963) and Silver (1964) amongst others. The model has been supported by the findings of a number of studies, including Baer and McKelvey (1958), Welch and Westwater (1961) and Sugawara and Katsuta (1966), which show condensation occurring entirely between drops in a very thin film. Welch and Westwater examined the process by taking high speed cine film through a microscope. They concluded that droplets large enough to be visible (0.01mm) grew mainly by coalescence leaving a 'lustrous bare area'. This lustre quickly faded and they explained this in terms of the build-up of the thin film which fractured at a thickness of 0.5 to 1 μm.

Silver (1964) has attempted a quantitative treatment of drop-wise condensation based on radially inward drainage of condensate towards a droplet. This model gives the ratio between drop-wise and film-wise condensation rates as:

$$\left[\frac{j_D}{j_F}\right] = \left[\frac{\rho_f^2 D^2 g}{24.2 \ \mu_f j_F}\right]^{1/9} \tag{11.1}$$

where j_D is the drop-wise condensation mass flux and j_F is the film-wise condensation mass flux under identical conditions. For steam at atmospheric pressure for a value of j_F equal to 1.36 x 10^{-2} kg/m²s this ratio is 6.5 and the mean thickness of the thin film is about 2 μm.

In the alternative model of drop-wise condensation it is assumed that droplet formation is fundamentally a heterogeneous nucleation process. This concept was first proposed by Eucken (1937). Studies by Umur and Griffith (1965) and by Erb (1965) support the nucleation mechanism. Using an optical method with polarised light to indicate changes in the thickness of liquid films of molecular dimensions, Umur and Griffith were able to establish that, for low temperature differences at least, the area between drops has no liquid film greater than a monolayer in thickness and that no net condensation takes place in that area.

McCormick and Baer (1963) have suggested that innumerable submicroscopic droplets are randomly nucleated at active sites on the condenser surface. These active sites are wetted pits and grooves in the surface which are continually being exposed by numerous drop coalescences and by large drops falling from the surface. Evidence to support this picture has been accumulated

(McCormick and Baer (1963), Gose et al (1967). Grigull has
produced photographic evidence that preferred nucleation sites
occur on the surface around the circumference of a previously
departed condensate drop.

The growth rate of a drop on an active site may be assumed
to be limited by the heat conduction through the drop. For the
case of a hemispherical drop this leads to:

$$\left(\frac{dr}{dt}\right)^2 = 4.15 \left[\frac{k_f(T_{gi}-T_W)}{i_{fg}\,\rho_f}\right]$$ (11.2)

Gose, Mucciardi and Baer (1967) have developed a model for
drop-wise condensation which accounts for drop nucleation and
growth, coalescence with neighbours, removal and renucleation on
sites exposed by the removal and coalescence mechanisms. The
model was simulated on a computer. In order to account for the
heat transfer coefficients found in steady state drop-wise
condensation, the minimum nucleation site density must be 5 x 10^4
sites/cm^2, a figure which is not unreasonable. It was also shown
that large drops greater than 0.01 cm in diameter grow chiefly by
coalescence. A considerable improvement in drop-wise condensation
was predicted if the nucleation site density could be increased
to 10^8 sites/cm^2 or if droplets could be mechanically removed at
diameters of approximately 10^{-4} cm. One feature not simulated by
this model was the sweeping action of a large drop running down
a surface absorbing all other droplets in its path.

A very considerable amount of work remains to be carried out
to establish satisfactory explanations of all the published
experimental results on drop-wise condensation. Whilst the droplet
nucleation mechanism is certainly the more likely at low condensation rates (ie temperature differences up to 5°C) it is possible
that at higher condensation rates there may be a film disruption
mechanism as an intermediate stage before establishing fully
developed film-wise condensation.

Figure 11.1 shows the results of Takeyama and Shimizu (1974)
which indicate the transition to film-wise condensation at high
temperature differences.

At low temperature differences Le Fevre and Rose (1966)
report drop-wise heat transfer coefficients which increase slightly
with increasing temperature difference (h_D = 0.2 MW/m^2 °C at 1°C
and h_D = 0.27 MW/m^2 °C at 4°C). At high temperature differences
Welch and Westwater and Tameyama and Shimizu indicate a falling
heat transfer coefficient with increasing temperature difference.
This may be due to a change in mechanism or equally well to the
presence of non-condensible gases in the vapour phase.

The influence of pressure on the drop-wise coefficient for
steam at a fixed heat flux is shown in Figure 11.2. For pressures
above 3 bar, O'Bara et al (1967) report that drop-wise condensation
is gradually replaced by a combination of drop and film condensation. This is probably brought about by the reduction in the
surface tension of water at higher temperatures. These same
workers also report on the influence of vapour velocity. For a

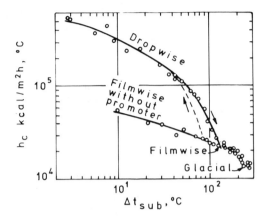

Figure 11.1: Heat transfer coefficients for condensation of steam on a short
vertical copper surface, from Takeyama and Shimizu (1974).
(Palmitic acid used as promoter for drop-wise condensation)

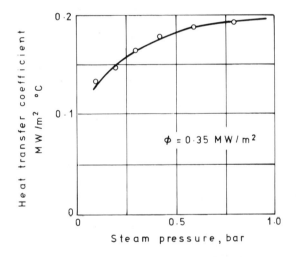

Figure 11.2: Dropwise heat transfer coefficients as a function of steam pressure
(Lefevre and Rose)

given temperature difference, the heat transfer coefficient
increased up to a maximum value with increased vapour velocity up
to 2 m/s and then started to decrease. This was believed to be
caused by the increased coalescence between droplets blanketing
the surface with a film of condensate.

11.4 Film-Wise Condensation

In film-wise condensation the condensate wets the surface
and forms a continuous liquid film. The behaviour of this liquid

film is governed by the normal fluid dynamic laws with gravity, vapour shear and surface tension forces, either singly or together, acting to promote removal of the condensate film.

The various resistances to heat transfer during condensation are shown diagramatically in Figure 11.3. For a pure saturated vapour the resistance at the vapour/liquid interface is small and, to a first approximation, may be neglected. Methods for calculating an effective interfacial heat transfer coefficient have, however, been developed and have been discussed in an earlier chapter by Hewitt.

For a superheated vapour, a vapour containing a non-condensible gas or a multi-component vapour mixture (see later chapter by Collier) there will be a significant resistance to both heat and mass transfer at the vapour/liquid interface. Mass and heat transfer processes occur in parallel within the gas phase and must be considered together to establish the respective driving forces.

Heat must also be transferred through the condensate liquid film to the wall and thence to the coolant. In the case of a pure saturated vapour this temperature drop across the liquid film represents the prime resistance to heat transfer. Techniques which reduce the condensate film thickness or promote a higher "effective" conductivity will therefore increase the condensing side heat transfer coefficient. In the case of a vapour containing incondensibles or with a multi-component vapour mixture, the gas phase resistance is often that dominating the overall condensing side coefficient and increases in the condensate film coefficient may not promote a comparable increase in overall condensation rate.

The Influence of Non-Condensibles on Interfacial Resistance

The presence of even a small quantity of non-condensible gas in the condensing vapour has a profound influence on the resistance to heat transfer in the region of the liquid-vapour interface. The

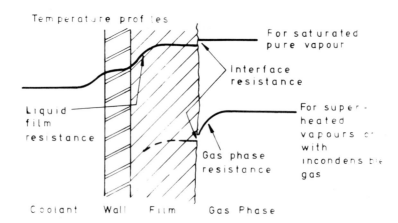

Figure 11.3: Resistances to heat transfer during condensation

non-condensible gas is carried with the vapour towards the interface where it accumulates. The partial pressure of gas at the
interface increases above that in the bulk of the mixture, producing a driving force for gas diffusion away from the surface. This
motion is exactly counterbalanced by the motion of the vapour-gas
mixture towards the surface. Since the total pressure remains
constant the partial pressure of vapour at the interface is lower
than that in the bulk mixture providing the driving force for
vapour diffusion towards the interface. This situation is
illustrated in Figure 11.4 which also shows the variation of
temperature in the region of the interface. It is usual to assume
that the temperature at the interface (T_{gi}) corresponds to the
saturation temperature equivalent to the partial pressure of
vapour (p_{gi}) at the interface. The molar flux of non-condensible
gas ($\dot{n}_A$) passing through a plane parallel to and at a distance y
from the interface is:

$$\dot{n}_A = \dot{n}\tilde{y}_a + D_{AG}\,\tilde{c}\,\frac{d\tilde{y}_A}{dy} = 0 \qquad (11.2)$$

Likewise, the molar flux of vapour ($\dot{n}_g$) is:

$$\dot{n}_g = \dot{n}\tilde{y}_g + D_{AG}\,\tilde{c}\,\frac{d\tilde{y}_g}{dy} \qquad (11.3)$$

where D_{AG} is the binary diffusion coefficient, $\tilde{c}$ is the total molar
concentration, $\dot{n}$ is the mixture molar ("drift") flux towards the

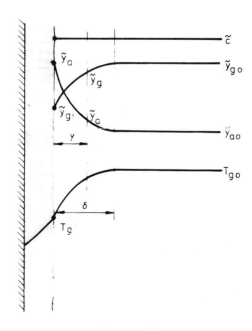

Figure 11.4: The influence of non-condensibles on interfacial resistance

interface, $\tilde{y}_A, \tilde{y}_g$ are the mol fractions of non-condensible gas and condensible vapour respectively ($\tilde{y}_A = p_A/p$ etc).

$$\frac{d\tilde{y}_g}{dy} = - \tilde{y}_A \frac{d\tilde{y}_A}{dy} \; ; \; \tilde{y}_g = 1 - \tilde{y}_A \tag{11.4}$$

Thus:

$$\dot{n}_g = \dot{n}\tilde{y}_g - D_{AG} \tilde{c} \frac{d\tilde{y}_A}{dy} \tag{11.5}$$

and eliminating $\dot{n}$ between equations (9.5) and (9.2)

$$\dot{n}_g = - D_{AG} \tilde{c} \frac{d\tilde{y}_A}{dy} \left[\frac{\tilde{y}_G}{\tilde{y}_A} + 1 \right] \tag{11.6}$$

Integrating the above equation between the interface ($y = 0$) and the edge of the diffusion layer ($y = \delta$):

$$\dot{n}_g = \frac{D_{AG}\tilde{c}}{\delta} \ln \left\{ \frac{1 - \tilde{y}_{gi}}{1 - \tilde{y}_{go}} \right\} \tag{11.7}$$

where $\tilde{y}_{gi}$ is the mol fraction of vapour at the interface and $\tilde{y}_{go}$ is the mol fraction of the vapour within the bulk mixture.

Equation (11.7) can be rewritten in terms of the *mass* condensation flux j_g:

$$j_g = K_g \rho_g \ln \left\{ \frac{1 - \tilde{y}_{gi}}{1 - \tilde{y}_{go}} \right\}$$

or

$$j_g = \frac{K_g \rho_g}{P_{am}} (p_{go} - p_{gi}) \tag{11.8}$$

where p_{go}, p_{gi} are the respective partial vapour pressures in the bulk and at the interface, p_{am} is the log mean of the partial pressure of non-condensible gas at the interface and in the bulk mixture given by:

$$P_{am} = \left[\frac{p_{ai} - p_{ao}}{\ln (p_{ai}/p_{ao})} \right] \tag{11.9}$$

K_g is a mass transfer coefficient, defined by:

$$K_g = \frac{D_{AG}}{\delta} \tag{11.10}$$

The heat transfer rate is also affected by the occurrence of mass transfer. The total heat flux at a distance y from the interface is made up of two components: a sensible heat transfer (conductive) term and a sensible heat transport (convective) term:

$$\phi_g = k_g \frac{dT}{dy} + j_g \, c_{pg} \, (T-T_{gi})$$ (11.11)

Integrating from y = 0 to y = δ and using $h_g = k_g/\delta$

$$\phi_g = \frac{a}{1-e^{-a}} \left[h_g \, (T_{go}-T_{gi}) \right] = h'_g \, (T_{go}-T_{gi})$$ (11.12)

where

$$a = \left[\frac{j_g \, c_{pg}}{h_g} \right]$$

h'_g and h_g are the sensible heat transfer coefficients with and without mass transfer respectively. This correction is often small and may then be ignored.

The transfer of heat from the bulk mixture to the interface is made up of two components; firstly, the sensible heat transferred through the diffusion layer to the interface and, secondly, the latent heat released due to condensation of the vapour reaching the interface. This model of heat transfer in the vapour phase was first expressed by Colburn and Hougen (1934) as:

$$\phi = \phi_v + \phi_g$$

$$= \frac{K_g \, i_{fg} \, \rho_g}{P_{am}} (P_{go}-P_{gi}) + h'_g \, (T_{go}-T_{gi})$$ (11.13)

For the case where the vapour-gas mixture is stagnant only the first term in equation (11.13) need be considered. However, the use of the simple diffusion model for the mass transfer coefficient, K_g, results in an underestimate of the heat flux for a given temperature gradient. This discrepancy has been attributed by Akers et al (1960) and Sparrow and Eckert (1961) to the presence of natural convection effects.

Akers et al (1960) concluded that the ratio of the mass transfer boundary layer thickness, δ, to some characteristic length, L, could be expressed in terms of the product of the Grashof and Schmidt numbers (Gr.Sc) thus:

$$\frac{L}{\delta} = \left\{ \frac{LK_g}{D_{AG}} \right\} = fn \left[\frac{g \, L^3 \, \rho^2}{\mu^2} \left(\frac{\rho_o}{\rho_i} - 1 \right) \frac{\mu}{\rho D} \right]$$ (11.14)

Expressing the function as a power law and using experimental data to produce empirical relationships led to:

$$\left\{\frac{LK_g}{D_{AG}}\right\} = C_1 \left[\frac{g \ L^3 \ \rho}{\mu D}\left(\frac{\rho_o}{\rho_i}\right) - 1\right]^{0.373} \tag{11.15}$$

over the range $10^3 < (Gr.Sc) < 10^7$

where C_1 is 1.02 for vertical surfaces and 0.62 for horizontal tubes; ρ_i and ρ_o are the densities of the fluid adjacent to the interface and in the bulk mixture respectively.

For the case where the vapour-gas mixture flows parallel to the interface the heat transfer coefficient h_g in equation (11.13) is evaluated from the correlation appropriate to the flow regime (ie laminar, transition, or turbulent flow) and flow geometry (ie flow within a tube, over a flat plate, or a tube bank in cross flow). The mass transfer coefficient K_g in the first term of equation (11.13) may be evaluated from the heat transfer coefficient h_g by making use of the analogy between heat and mass transfer. Thus,

$$K_g = \left(\frac{h_g}{\rho_y \ c_{pg}}\right)\left(\frac{Pr}{Sc}\right)^{2/3} \tag{11.16}$$

For the tubulent flow of a condensing vapour-gas mixture within a tube the well-known Dittus-Boelter equation could be used with the physical properties evaluated for the bulk vapour-gas mixture. However, actual heat and mass transfer coefficients will be higher than predicted by this method mainly as a result of the disturbed nature of the film interface. This effect may be allowed for in an approximate manner by multiplying the smooth tube value of the coefficient by the ratio of the interfacial friction factor, f_i, to the smooth tube value, f_g.

The application of equation (11.13) requires some comment since the conditions at the interface are usually unknown. Consider the situation where condensation occurs from a vapour-gas mixture to a vertical wall under conditions where the wall surface temperature, T_W, the bulk temperature of the vapour-gas mixture, T_{go}, and the partial pressure of gas in the bulk vapour, p_{ao}, are known. The various stages to arrive at the overall heat transfer rate are as follows:

(a)   Guess a condensate film interface temperature, T_{gi}, inter-
      mediate between T_{go} and T_W. The heat transfer across the
      liquid condensate film is given by:

$$\phi = h_f(T_{gi} - T_W) \tag{11.17}$$

The heat transfer coefficient, h_f, across the condensate film is evaluated using the methods described below.

(b)   Since the interface is assumed to be a saturation condition
      for the vapour, the choice of T_{gi} allows the partial pressure
      of vapour at the interface (p_{gi}) to be evaluated. The partial
      pressure of vapour in the bulk mixture, $p_{go} = (p - p_{ao})$ is

known. Values of h_g and K_g, the heat and mass transfer coefficients respectively, are evaluated from the appropriate equations. Use is made, where necessary, of the analogy between heat and mass transfer (equation (11.16) and the correction for sensible heat transfer (equation (11.12).

(c) The rate of heat transfer through the diffusion layer may now be calculated from equation (11.13). In general, this heat flux, ϕ_2, will differ from that found from equation (11.17) ϕ_1. The correct solution occurs when a value of interface temperature, T_{gi}, is reached which allows ϕ_1 to equal ϕ_2. This may be achieved by an iterative procedure in which successively better guesses for the interfacial temperature are made. Alternatively, ϕ_1 and ϕ_2 may be evaluated for three or more values of interface temperature and plotted against T_{gi}. The intersection of the two curves gives the correct solution for ϕ and T_{gi}.

Such a calculation is only valid at one point in the condenser and the computation must be repeated for other locations.

More sophisticated boundary layer treatments of the influence of non-condensible gas on forced convection condensation have been published by Minkowycz and Sparrow (1966) and Sparrow et al (1967). These treatments include interfacial resistance, the process of mass transfer as a result of a temperature gradient (thermal diffusion) and energy transport as a result of a concentration gradient. For the case where steam is the condensing vapour and air is the non-condensible gas these processes are unimportant.

Figure 11.5 prepared from the results of Minkowycz and Sparrow (1966) and Sparrow et al (1967), shows the influence of non-condens-

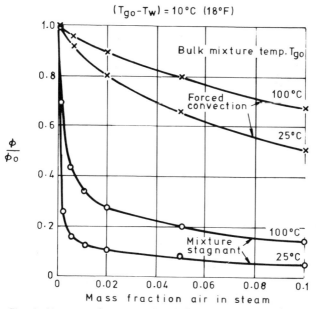

Figure 11.5: The influence of non-condensible air in steam (from Minkowycz and Sparrow (1966) and Sparrow et al (1967).

ible air concentrations on the heat transfer rate in the case when
the steam-air mixture is either stagnant or flowing. The temperature difference between the bulk steam-air mixture and the cooling
surface is held constant at $10^{\circ}C$ ($18^{\circ}F$). The ordinate represents
the ratio of the heat flux with non-condensible air present to
that which would have occurred at the same temperature difference
with pure steam. The presence of air is most marked when the bulk
mixture is stagnant. Reductions in heat transfer rate of more
than 50% are caused by air mass fractions as low as 0.5%. The
reductions become more marked as the total pressure is reduced.
These findings are in qualitative and also reasonably quantitative
agreement with the experimental results of Othmer (1929).

The heat transfer rate in forced convection condensation is
much less sensitive to the presence of non-condensible air although
the reduction in heat transfer becomes more serious as the
pressure is reduced. The reductions in heat transfer rate for
both stagnant and forced convection conditions increase slightly
as the temperature difference $(T_{go}-T_W)$ is increased.

11.5 Film Condensation on a Planar Surface

The Nusselt Equation for a Laminar Film

The first attempt to analyse the film-wise condensation
problem was that of Nusselt (1916) who made the following
assumptions:

(a) the flow of condensate in the film is laminar

(b) the fluid properties are constant

(c) subcooling of the condensate may be neglected

(d) momentum changes through the film are negligible: there is
     essentially a static balance of forces

(e) the vapour is stationary and exerts no drag on the downward
     motion of the condensate

(f) heat transfer is by conduction only.

The first step is to calculate the velocity distribution in a
liquid film flowing down a plane surface of unit width inclined
at an angle θ to the horizontal (Figure 11.6). At a distance z
from the top of the surface the thickness of the film is δ.
Consider a force balance on an element of film lying between y
and δ.

$$(\delta-y) \ dz \ (\rho_f-\rho_g) \ g \sin \theta = \mu_f \left(\frac{du_y}{dy}\right) dz \qquad (11.18)$$

Rearranging and integrating with the boundary condition $u_y = 0$ at
$y = 0$, then:

$$u_y = \frac{(\rho_f-\rho_g) \ g \sin \theta}{\mu_f} \left[y\delta - \frac{y^2}{2}\right] \qquad (11.19)$$

The mass flowrate per unit width Γ is given by:

$$\Gamma = \rho_f \int_0^\delta u_y dy = \left[\frac{\rho_f(\rho_f-\rho_g)g \sin \theta \, \delta^3}{3\mu_f}\right] \qquad (11.20)$$

Consider now the case where condensation occurs building up the film from zero thickness at the top of the plane surface. The rate of increase of film flow-rate with film thickness $(d\Gamma/d\delta)$ is therefore:

$$\frac{d\Gamma}{d\delta} = \left[\frac{\rho_f(\rho_f-\rho_g)g \sin \theta \, \delta^2}{\mu_f}\right] \qquad (11.21)$$

If the film surface temperature is T_{gi} and the wall temperature is T_W the heat transferred by conduction to an element of length dz is:

$$d\phi = \frac{k_f}{\delta} (T_{gi} - T_W)dz \qquad (11.22)$$

The mass rate of condensation on this area $(d\Gamma)$ is therefore:

$$d\Gamma = \frac{k_f}{\delta i_{fg}} (T_{gi} - T_W)dz \qquad (11.23)$$

Substituting equation (11.23) into (11.21) separating variables and integrating (noting $\delta = 0$ at $z = 0$):

$$\mu_f k_f(T_{gi} - T_W)z = \rho_f(\rho_f-\rho_g)g \sin \theta \, i_{fg}\left(\frac{\delta^4}{4}\right) \qquad (11.24)$$

Thus the film thickness δ is given by:

$$\delta = \left[\frac{4\mu_f k_f z (T_{gi}-T_W)}{g \sin \theta \, i_{fg} \, \rho_f(\rho_f-\rho_g)}\right]^{\frac{1}{4}} \qquad (11.25)$$

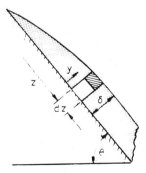

Figure 11.6: Flow of a laminar film over an inclined surface (Nusselt)

The heat transfer coefficient $h_f(z)$ at any point z is given by:

$$h_f(z) = \frac{k_f}{\delta} = \left[\frac{\rho_f(\rho_f - \rho_g)g \sin\theta\, i_{fg} k_f^3}{4\mu_f z (T_{gi} - T_w)}\right]^{\frac{1}{4}} \qquad (11.26)$$

and the Nusselt number is given by:

$$Nu(z) = \frac{h_f(z)z}{k_f} = \left[\frac{\rho_f(\rho_f - \rho_g)g \sin\theta\, i_{fg} z^3}{4\mu_f k_f (T_{gi} - T_w)}\right]^{\frac{1}{4}} \qquad (11.27)$$

Equations (11.26) and (11.27) give the local values of heat transfer coefficient and Nusselt number. The mean value of heat transfer coefficient over the whole surface, $\bar{h}_f$ is given by:

$$\bar{h}_f = \frac{1}{z}\int_0^z h_f(z)\,dz = 0.943 \left[\frac{\rho_f(\rho_f - \rho_g)g \sin\theta\, i_{fg} k_f^3}{\mu_f z (T_{gi} - T_w)}\right]^{\frac{1}{4}} \qquad (11.28)$$

The equation for $\bar{h}_f$ can be expressed in an alternative form:

$$\bar{h}_f = \frac{\Gamma_z i_{fg}}{z (T_{gi} - T_w)} \qquad (11.29)$$

where Γ_z is the condensate flow at a distance z from the top of the plane surface. Combining equation (11.29) with (11.23).

$$\delta = \frac{k_f \Gamma_z dz}{\bar{h}_f z d\Gamma} \qquad (11.30)$$

Combining equation (11.30) with (11.20) yields:

$$k_f \left[\frac{\rho_f(\rho_f - \rho_g)g \sin\theta}{3\mu_f}\right]^{\frac{1}{3}} \frac{dz}{z} = \frac{\bar{h}_f \Gamma^{\frac{1}{3}} d\Gamma}{\Gamma_z} \qquad (11.31)$$

Integrating over z:

$$\bar{h}_f = 0.925 \left[\frac{\rho_f(\rho_f - \rho_g)g \sin\theta\, k_f^3}{\mu_f \Gamma_z}\right]^{\frac{1}{3}} \qquad (11.32)$$

The Reynolds number of a liquid film at a distance z below the top of the plane surface is given by:

$$Re_\Gamma = \frac{4\Gamma_z}{\mu_f} \qquad (11.33)$$

Substituting into equation (11.32) and rearranging for a vertical plane (sin θ = 1):

$$\frac{\bar{h}_f}{k_f}\left[\frac{\mu_f^2}{\rho_f(\rho_f-\rho_g)g}\right]^{\frac{1}{3}} = 1.47 \ \text{Re}_\Gamma^{-\frac{1}{3}} \tag{11.34}$$

Since $h_f(z)$ varies as $z^{-\frac{1}{4}}$ (equation (11.26)) the average coefficient $\bar{h}_f$ is 4/3 times the local coefficient $h_f(z)$.

Improvements to the original Nusselt Theory

A number of papers have made significant improvements to the Nusselt theory. The original anlaysis was extended by Bromley (1952) who considered the effects of subcooling the condensate and by Rohsenow (1956) who also allowed for the non-linear distribution of temperature through the film due to energy convection. Rohsenow (1956) showed that the latent heat of vaporization, i_{fg}, in equation (9.28) should be replaced by:

$$i_{fg}' = i_{fg}\left[1 + 0.68\left(\frac{c_{pf}\Delta T_f}{i_{fg}}\right)\right] \tag{11.35}$$

where $\Delta T_f = (T_{gi} - T_W)$

Sparrow and Gregg (1959), using a boundary layer treatment, removed assumption (d) and considered momentum changes in the film. For common fluids with Prandtl numbers around unity the results obtained show that momentum effects are indeed negligible but for liquid metals with very low Prandtl numbers the heat transfer coefficient falls below the Nusselt prediction with increasing $(c_{pf}\Delta T_f/i_{fg})$.

Chen (1961), Koh, Sparrow and Hartnett (1961) and others have considered the influence of the drag exerted by the vapour on the liquid film. There again the assumption made by Nusselt (assumption (e)) appears justified at Prandtl numbers around unity. For the condensation of liquid metals, however, the inclusion of the interfacial shear effect does cause a further substantial reduction of heat transfer. The ratio of the revised average heat transfer coefficient $(\bar{h}_f)$ to that given by equation (11.28), $(\bar{h}_f)_{Nu}$, can be computed from the following approximate equation given by Chen (1961) which includes both interfacial shear and momentum effects:

$$\frac{\bar{h}_f}{(\bar{h}_f)_{Nu}} = \left[\frac{1+0.68\ A+0.02\ AB}{1+0.85\ B-0.15\ AB}\right]^{\frac{1}{4}} \tag{11.36}$$

where $A = \left[c_{pf}\Delta T_f/i_{fg}\right]$ and $B = \left[k_f\Delta T_f/\mu_f i_{fg}\right]$

Equation (11.36) is valid for $A < 2$, $B < 20$, and for liquids with Prandtl numbers larger than unity or less than 0.05. A number of workers have considered the influence of variations in physical properties across the condensate film (assumption (b)). One method

of introducing this influence is to use the Nusselt's equation
with the property values taken at some effective condensate film
temperature, T_{film}. Thus:

$$T_{film} = T_W + F(T_{gi} - T_W) \qquad (11.37)$$

The value of F given by Drew (1954) is 0.25 and by Minkowycz and
Sparrow (1966) is 0.31.

The Influence of Turbulence

Even at relatively low film Reynolds numbers, the assumption
that the condensate layer is in viscous flow is open to some
question. Experiments aimed at measuring the average thickness
of liquid films flowing down vertical surfaces do confirm the
Nusselt equations but examination of the surface structure of the
flow indicates considerable waviness. This waviness may account
for the observed differences (Drew (1954)) between experimental
and theoretical values. The results of most experimental studies
give heat transfer coefficients up to 40-50% higher than the
theoretical values and it is recommended that coefficients
evaluated from equation (11.38) should be multiplied by a correction
factor 1.2 to allow for this discrepancy in design calculation.

For long vertical surfaces it is possible to obtain condensation rates such that the film Reynolds number exceeds the critical
value at which turbulence begins. Under such circumstances
Kirkbride (1933) found heat transfer coefficients much greater
than given by equation (11.34) By plotting the experimental data
in terms of the dimensionless groupings of equation (11.34) he was
able to obtain an empirical relation for heat transfer in the case
of combined viscous and turbulent flow on the outside surface of
a vertical tube. Colburn (1933) attempted to predict the heat
transfer coefficient for the case of turbulent flow by analogy
with the flow of liquids through pipes under conditions where the
liquid completely filled the pipe but where the pressure drop for
flow was purely due to gravitational forces. The resulting
relation for the local condensing heat transfer coefficient at any
point, z is:

$$\frac{h_f(z)}{k_f} \left[\frac{\mu_f^2}{\rho_f (\rho_f - \rho_g) g} \right]^{\frac{1}{3}} = 0.056 \left(\frac{4\Gamma_z}{\mu_f} \right)^{0.2} \left(\frac{c_{pf} \mu_f}{k_f} \right)^{\frac{1}{3}} \qquad (11.38)$$

Integration of the local coefficient over the tube length using
equation (11.26) up to a Reynolds number of 2000 and equation (11.38)
above 2000 resulted in the relationship shown in Figure 11.7.
Seban (1954) applied the Prandtl-Karman analogy to the problem of
turbulent film condensation with no vapour shear. The results
were presented graphically for a range of Prandtl numbers. In
this work the critical value of the film Reynolds number was taken
as 1600. Figure 11.7 shows the predictions. It will be noted that
there is reasonable agreement with Colburn's analysis at high
Prandtl numbers but that, at low Prandtl numbers, the heat transfer
rate can fall below the Nusselt prediction at high Reynolds number,
presumably due to the increased thickness of the turbulent film.

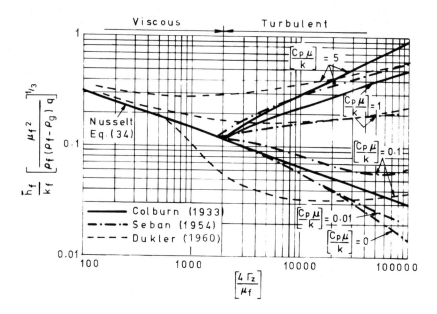

Figure 11.7: Condensation on a vertical surface, no interfacial shear

Condensation on Horizontal Tubes

Laminar film condensation on the outside of a single horizontal tube was first examined by Nusselt, who, using a derivation exactly similar to that outlined in section 11.4 obtained the following relationships:

$$h_f(\alpha) = 0.693 \left[\frac{\rho_f(\rho_f - \rho_g)g \sin \alpha \; k_f^{3}}{\Gamma_\alpha \mu_f} \right]^{\frac{1}{3}}$$ (11.39)

where α is the angle between any radius of the horizontal tube and the vertical, and Γ_α is the local mass flow-rate of condensate per unit length of tube;

$$\bar{h}_f = 0.725 \left[\frac{\rho_f(\rho_f - \rho_g)g \; i_{fg}' \; k_f^{3}}{D \; \mu_f (T_{gi} - T_w)} \right]^{\frac{1}{4}}$$ (11.40)

where D is the outside diameter of the tube, i_{fg}' is the modified latent heat of vaporization given by equation (11.35):

$$\frac{\bar{h}_f}{k_f} \left[\frac{\mu_f^{2}}{\rho_f(\rho_f - \rho_g)g} \right]^{\frac{1}{3}} = 1.51 \; Re_\Gamma^{-\frac{1}{3}}$$ (11.41)

where the Reynolds number is defined by equation (11.33) with Γ_z equal to the mass flow-rate of condensate from the tube per unit length. This equation is valid up to values of Re_Γ of 3200.

In testing Nusselt's equations it is important to see that the experimental conditions comply with the requirements of the theory. The results of most experimental studies for horizontal tubes are within about 15% of equation (11.40). It is interesting to compare the performance of vertical versus horizontal tubes in stagnant vapour. From equations (11.28) and (11.40) it will be seen that for the same temperature difference and fluid conditions the same average heat transfer coefficient will exist if:

$$\frac{0.943}{z^{\frac{1}{4}}} = \frac{0.725}{D^{\frac{1}{4}}} \quad \text{or} \quad z = 2.87 \ D \qquad (11.42)$$

In a bank of horizontal tubes it is possible for the condensate to run off the bottom of the upper tube onto the next tube below (Figure 11.8). Jakob (1949) has considered this problem and concluded that the mean heat transfer coefficient for a vertical column of n horizontal tubes with the same temperature difference is given by equation (11.40) replaced by nD. Thus:

$$\frac{\bar{h}_{fN}}{h_{f1}} = n^{-1/4} \qquad (11.43)$$

and

$$\frac{h_{fN}}{h_{f1}} = n^{3/4} - (n-1)^{3/4} \qquad (11.44)$$

(a) Idealised   (b) Actual

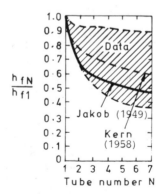

Figure 11.8: Film condensation on horizontal tubes

where $\bar{h}_{fN}$ is the mean coefficient for n tubes and h_{fN} is the coefficient for the nth tube in the column

Experimental results suggest that the coefficients may well be higher than given by equations (11.43) and (11.44) because the condensate does not pass from one tube to another as a continuous sheet (Figure 11.8). Kern (1958) has suggested an alternative to equations (11.43) and (11.44) such that:

$$\frac{\bar{h}_{fN}}{h_{f1}} = n^{-1/6} \quad \text{and} \quad \frac{h_{fN}}{h_{f1}} = n^{5/6} - (n-1)^{5/6} \tag{11.45}$$

Although Berman's data encompasses both Kern and Jakob's prediction, the former (equation (11.45)) is preferred as being the best recommendation.

11.6 The Influence of Interfacial Shear

Laminar Flow over Inclined Surface

It has already been noted that the consideration of interfacial shear can considerably modify the laminar flow solution under certain conditions. Nusselt did derive equations for condensation in the presence of high vapour velocities on the assumption that the entire condensate layer remained in laminar flow. Thus, co-current downward flow over an inclined plane surface results in the following equation instead of equation (11.18).

$$(\delta - y)\,dz\,(\rho_f g \sin \theta - (dp/dz)) + \tau_i\,dz = \mu_f \left(\frac{du_y}{dy}\right)\,dz \tag{11.46}$$

Define a fictitious vapour density ρ_g^* by the relationship:

$$\left(\frac{dp}{dz}\right) = \rho_g^* \, g \, \sin \theta \tag{11.47}$$

If the only pressure gradient is due to the vapour head then $\rho_g^* = \rho_g$. Additional pressure gradient terms will occur when condensation takes place inside a tube. These will be discussed later. Substituting equation (11.47) into equation (11.46) and integrating,

$$u_y = \frac{(\rho_f - \rho_g^*)g \sin \theta}{\mu_f}\left[y\delta - \frac{y^2}{2}\right] + \frac{\tau_i y}{\mu_f} \tag{11.48}$$

$$\Gamma = \left[\frac{\rho_t(\rho_t - \rho_g^*)g \sin \theta \; \delta^3}{3\mu_f}\right] + \frac{\tau_i \rho_f \delta^2}{2\mu_f} \tag{11.49}$$

$$\frac{d\Gamma}{d\delta} = \left[\frac{\rho_f(\rho_f - \rho_g^*)g \sin \theta \; \delta^2}{\mu_f}\right] + \left[\frac{\tau_i \rho_f \delta}{\mu_f}\right] \tag{11.50}$$

Repeating the analysis outlined in section 11.5, the following equation corresponding to 11.25 is derived:

$$\left[\frac{4\mu_f \, k_f \, z(T_{gi}-T_w)}{g \sin \theta \; i'_{fg} \; \rho_f(\rho_f-\rho_g*)} \right] = \delta^4 + \frac{4}{3}\left[\frac{\tau_i \delta^3}{(\rho_t-\rho_g*)g \sin \theta} \right] \quad (11.51)$$

Rohsenow, Webber and Ling (1956) have defined three dimensionless groups, a dimensionless film thickness, $\delta*$,

$$\delta* = \delta \left[\frac{\rho_f(\rho_f-\rho_g*)g \sin \theta}{\mu_f^2} \right]^{\frac{1}{3}} \quad (11.52)$$

a dimensionless distance, $z*$,

$$z* = \left(\frac{4k_f z(T_{gi}-T_w)}{i'_{fg} \, \mu_f \delta} \right) \delta* \quad (11.53)$$

and a dimensionless interfacial shear stress, τ_i*,

$$\tau_i* = \left(\frac{\tau_i}{(\rho_f-\rho_g*)\delta g \sin \theta} \right) \delta* \quad (11.54)$$

Equation

$$z* = \delta*^4 + \frac{4}{3}\delta*^3 \, \tau_i* \quad (11.55)$$

also

$$\bar{h}_f* = \frac{\bar{h}_f}{k_f}\left[\frac{\mu_f^2}{\rho_f(\rho_f-\rho_g*)g \sin \theta} \right]^{\frac{1}{3}} = \frac{4}{3}\frac{(\delta_z*)^3}{z*} + 2\frac{\tau_i*(\delta_z*)^2}{z*} \quad (11.56)$$

and

$$\frac{4\Gamma_z}{\mu_f} = \frac{4}{3}(\delta_z*)^3 + 2\tau_i*(\delta_z*)^2 \quad (11.57)$$

The dimensionless group δ_z* may be eliminated between equations (11.56) and (11.57). The results for downward vapour flow are plotted in Figure 11.9. The dotted line shows the approximate boundary at which turbulence effects become important.

Condensation within a Vertical Tube

In order to apply the theory of Rohsenow, Webber and Ling (1956) to the case of laminar condensation with interfacial shear within a tube it is necessary to evaluate ρ_g* (equation (11.47) in terms of the pressure gradient in the vapour phase. For one-dimensional flow in the vapour phase:

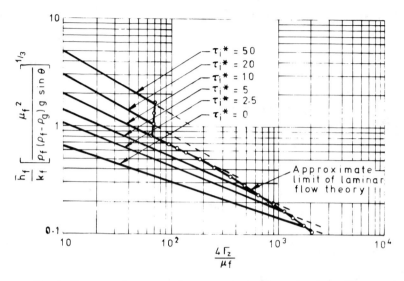

Figure 11.9: The influence of vapour shear on the average laminar flow condensation heat transfer rate (Rohsenow, Webber, and Ling (1956)

$$\frac{dp}{dz} = \rho_g \, g \, \sin\theta + \frac{2\mu_g}{(D-2\delta)} \frac{d\Gamma}{dz} - \frac{2\tau_i}{(D-2\delta)}$$ (11.58)

Therefore

$$\rho_g^* = \rho_g \left[1 + \frac{1}{\rho_g g \sin\theta} \left\{ \frac{2u_g}{(D-2\delta)} \frac{d\Gamma}{dz} - \frac{2\tau_i}{(D-2\delta)} \right\} \right]$$ (11.59)

where

$$\frac{d\Gamma}{dz} = \frac{h_f(z)}{i_{fg}} (T_{gi} - T_W)$$ (11.60)

This method proposed by Rohsenow et al (1956) is in good agreement with a more rigorous treatment presented by Jacobs (1966) provided that the condensation surface is divided axially into a number of short sections over which the interfacial shear stress is assumed constant but of different magnitude to the neighbouring section.

Carpenter and Colburn (1951) also considered the influence of vapour velocity on film condensation inside tubes. In experiments where there was an appreciable co-current downward vapour velocity, heat transfer coefficients lay considerably above the static vapour case. The critical value of the film Reynolds number at which turbulence was initiated was greatly lowered (by a factor of up to ten or so) (see Figure 11.9). Carpenter and Colburn (1951) reasoned, firstly, that the transition from laminar to turbulent flow in the film occurred at much lower Reynolds numbers than in the absence of vapour shear and, secondly, that the major resistance to heat transfer occurred across the laminar sub-layer. They argued that,

as the major force acting on the condensate film was now interfacial shear, the velocity profile in the film might be estimated
from the analogous situation for flow of liquids in pipes.

The equation for the local heat transfer coefficient resulting
from this treatment is:

$$\left[\frac{h_f(z)\mu_f}{k_f\rho_f^{\frac{1}{2}}}\right] = 0.045 \left[\frac{c_p\mu}{k}\right]_f^{\frac{1}{2}} \tau_w^{\frac{1}{2}} \tag{11.61}$$

where τ_w is the shear stress at the outer edge of the laminar sub-
layer (ie the wall shear stress). In calculating a value of τ_w,
account must be taken of the interfacial shear and the gravitational effects in the liquid film. In addition, the change of
momentum of the condensing vapour and the influence of vapour mass
transfer on the interfacial shear stress must be included.

More recently, with the help of further experimental data,
Soliman, Schuster and Berenson (1968) have improved the Carpenter
and Colburn treatment and modified equation (11.61).

$$\left[\frac{h_f(z)\mu_f}{k_f\rho_f^{\frac{1}{2}}}\right] = 0.036 \left[\frac{c_p\mu}{k}\right]_f^{0.65} \tau_W^{\frac{1}{2}} \tag{11.62}$$

The shear stress at the outer edge of the laminar sub-layer τ_W is
made up of the sum of three components:

(a)   the interfacial shear stress, τ_i due to vapour phase friction

$$\tau_i = \frac{D}{4}\left(-\frac{dp}{dz} F\right) \tag{11.63}$$

(b)   the gravitational force τ_z in the liquid film

$$\tau_z = \frac{D}{4}(1-\alpha)(\rho_f-\rho_g)g\sin\theta \tag{11.64}$$

(c)   the momentum change τ_a in the vapour phase

$$\tau_a = \frac{D}{4}\frac{G^2}{\rho_g}\left(\frac{dx}{dz}\right)\left[a_1\left(\frac{\rho_g}{\rho_f}\right)^{\frac{1}{3}} + a_2\left(\frac{\rho_g}{\rho_f}\right)^{\frac{2}{3}} + \ldots a_5\left(\frac{\rho_g}{\rho_f}\right)^{\frac{5}{3}}\right] \tag{11.65}$$

where the coefficients, a_1 to a_5, are given by:

$$\left.\begin{array}{l}
a_1 = 2x-1-\beta x \\[4pt]
a_2 = 2(1-x) \\[4pt]
a_3 = 2(1-x-\beta+\beta x) \\[4pt]
a_4 = (1/x)-3+2x \\[4pt]
a_5 = \beta(2-(1/x)-x)
\end{array}\right\} \tag{11.66}$$

x is the vapour mass quality and β is the ratio of the mean liquid film velocity to the interfacial velocity (β = 2 for laminar flow and 1.25 for turbulent flow).

The void fraction correlation due to Zivi (1964) is used in the derivation of equation (11.65) and strictly should also be used in equation (11.64). In the case of upward flow of a condensing vapour, the gravitational term must be subtracted rather than added. Equation (11.62) has been verified over the Prandtl number range 2 to 10.

For the average heat transfer coefficient, Carpenter and Colburn (1951) suggested a simplified relationship arrived at by neglecting the gravitational and momentum terms and assuming τ_W equal to the interfacial shear stress τ_i.

$$\left[\frac{\bar{h}_f \mu_f}{k_f \rho_f^{\frac{1}{2}}} \right] = 0.065 \left[\frac{c_p \mu}{k} \right]_f^{\frac{1}{2}} \tau_i^{\frac{1}{2}} \qquad (11.67)$$

where

$$\tau_i = f_i^{1} \left[\frac{\bar{G}_g^{2}}{2\rho_g} \right] \qquad (11.68)$$

and f_i^{1} is an "apparent" interfacial friction factor evaluated as for single-phase pipe flow at a mean vapour velocity $\bar{G}_g$ given by:

$$\bar{G}_g = \left[\frac{G_1^{2} + G_1 G_2 + G_2^{2}}{3} \right]^{\frac{1}{2}} \qquad (11.69)$$

G_1 and G_2 are the vapour mass velocities at the beginning and exit of the section considered. If all the vapour is condensed $G_2 = 0$ and $\bar{G}_g = 0.58 G_1$.

It is of interest to note that equation (11.67) (and also (11.62) and (11.61)) can be rearranged in terms of the dimensionless groups h_f^* and τ_i^* thus:

$$\bar{h}_f^{*} = 0.065 \, Pr_f^{\frac{1}{2}} (\tau_i^{*})^{\frac{1}{2}} \qquad (11.70)$$

Equation (11.70) is valid for Prandtl numbers between 1 and 5 and τ_i^* from 5 to 150.

Rohsenow et al (1956) carried out a similar analysis to that of Carpenter and Colburn by extending Seban's treatment to the case of a positive interfacial shear stress. The prandtl-Nikuradse "universal velocity profile" was used and momentum effects within the liquid film were neglected. In addition, it was necessary to make a similar assumption for the critical transition value of the film Reynolds number. The results of this analysis were presented graphically in dimensionless form as h_f^* against film Reynolds number (4 Γ_z/μ_f) with the interfacial shear stress τ_i^* as parameter. In the area where equation (11.67) was applicable there was good agreement with the Carpenter and Colburn analysis.

Dukler (1960) more recently has presented a rather more refined method of predicting the hydrodynamics and heat transfer in vertical film-wise condensation. Working from the definition of eddy viscosity and using the Deissler equation for its variation near a solid boundary, Dukler obtained velocity distributions in the liquid film as a function of the interfacial shear and film thickness. The differential equations were solved digitally by means of a computer. While a single universal velocity distribution is usual in full pipe flow, Dukler found that the velocity distribution curve in the liquid film depends both on the interfacial shear and on the film thickness. Assuming that the ratio of the eddy thermal diffusivity to the eddy viscosity was unity and that the physical properties of the fluid do not change in the direction of heat transfer, equivalent temperature profiles were constructed.

From the integration of the velocity and temperature profiles, liquid film thickness and point heat transfer coefficients were computed. The evaluated results of these two properties were displayed as a function of the Reynolds and Prandtl numbers with the interfacial shear as parameter. Average condensing heat transfer coefficients were evaluated from the point values and are shown in Figure 11.7 for the case of zero interfacial shear. The results agree with the classical Nusselt values at very low Reynolds numbers and with the empirical equations of Colburn (1933) in the turbulent region. There is also good agreement with Seban's analysis at high Reynolds and Prandtl numbers. Figures 11.10 and 11.11 show the local heat transfer coefficients obtained from Dukler's analysis for the case of positive interfacial shear for Prandtl numbers of 1 and 5 respectively.

A number of objections can be raised to Dukler's analysis:

(a)   the velocity distribution used was obtained from experiments in pipes with a different shear stress distribution;

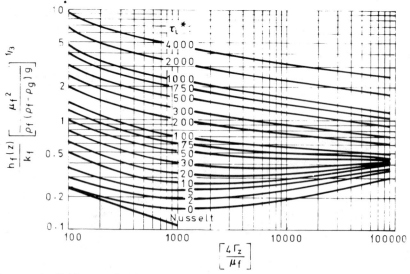

Figure 11.10: Dukler analysis, Prandtl number = 1.0

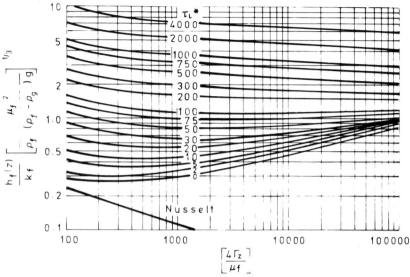

Figure 11.11: Dukler analysis, Prandtl number = 5.0

(b)   in defining the temperature profile at values of y^+ greater
       than 20, the molecular conductivity term is neglected with
       respect to the eddy conductivity.

       Whilst this latter assumption is acceptable for fluids with
values of Prandtl numbers around unity and higher, at low values
of Prandtl number, the fall-off in heat transfer in the turbulent
region is greatly overestimated (Figure 11.7). Lee (1964) has
repeated Dukler's analysis for the case with no interfacial shear
using an improved velocity profile and removing the above
objection concerning the temperature profile. When this is done
Lee obtains a solution which gives results very close to Dukler's
(1960) solution at high Prandtl number and to Seban's (1954) at
low Prandtl number.

       There remains the question of correcting Dukler's analysis for
positive values of interfacial shear stress. An analysis which
does this has been given by Kunz and Yerazunis (1967). In
addition to considering the points (a) and (b) raised above, these
authors have also included interfacial resistance effects and the
variation of the ratio of the eddy diffusivity of heat to the
eddy diffusivity of momentum ($\varepsilon_H/\varepsilon$) with Prandtl and Reynolds
number. Table 11.1 gives values of the local dimensionless heat
transfer coefficient $h_f^*(z)$ given by equation (9.56) as a
function of the local Reynolds number ($4\,\Gamma_z/\mu_f$) and Prandtl number
for the case of zero interfacial shear. The usefulness of these
theoretical predictions is confirmed by the remarkably good
agreement between values for local heat transfer coefficient along
the length of the tube calculated from Dukler's analysis and
measured by Carpenter (1951).

Condensation within a Horizontal Tube

       The problem of condensation within a horizontal tube is best
examined by reference to the flow patterns experienced in

Pr \ $4\,\Gamma_z/\mu_f$	10^2	10^3	10^4	10^5
0.01	0.2661	0.1081	0.0157	0.0212
0.1	0.2661	0.1152	0.0852	0.1062
1.0	0.2700	0.2110	0.2993	0.495
5.0	0.3085	0.3778	0.6210	1.139

Table 11.1: Values of $h_f^*(z)$ as a function of $(4\,\Gamma_z/\mu_f)$ and Prandtl number for the case of zero interfacial shear (Kunz and Yerazunis)

horizontal gas-liquid flow. Bell, Taborek and Fenoglio (1969) have reviewed the available correlations using such a framework. The Baker flow pattern map forms a suitable starting point. A useful suggestion made by Bell et al (1969) concerns the preparation of an overlay for use with the Baker map. This overlay, constructed to the same scale, represents the path followed on the map during the course of progressive condensation or evaporation at a constant mass velocity. If the point marked x = 0.5 on the overlay is centred over the point on the Baker map corresponding to the particular mass velocity G and a vapour mass quality x = 0.5 then the line of the overlay can be used to determine the range of quality over which each flow pattern may occur.

Stratified Flow. The stratified flow pattern corresponds to a region of small vapour velocity and low interfacial shear forces. In these circumstances the Nusselt equation for condensation on the outside of a horizontal tube (equation (11.40)) can be applied in a modified form. It is assumed that a laminar film of condensate runs down the inside upper surface of the tube and collects as a stratified layer of liquid in the lower part of the tube (Figure 11.12).

$$\bar{h}_f = F \left[\frac{\rho_f(\rho_f-\rho_g)g\,i_{fg}\,k_f^3}{D\,\mu_f(T_{gi}-T_w)} \right]^{\frac{1}{4}} \tag{11.71}$$

The factor F allows for the fact that the rate of condensation on the stratified layer of liquid running along the bottom of the tube is negligible. The value of F depends on the angle 2ϕ subtended from the tube centre to the chord forming the liquid level.

Figure 11.12: Laminar condensation within a horizontal tube (after Nusselt)

Thus

$$F = \left(1 - \frac{\phi}{\pi}\right) F^1 \qquad (11.72)$$

where values of F^1 and F are given as a function of ϕ in Table 11.2. In the limit as 2ϕ approaches zero F equals 0.725; cf equation (11.40).

Chato (1962) concluded that the heat transfer coefficient was not particularly sensitive to the angle 2ϕ and suggested a mean value of 120°. In the study by Chato and the earlier analysis of Chaddock (1957) it was assumed that the vapour is stagnant and that the condensate flows under a hydraulic gradient (Figure 11.13a). In this case, the angle 2ϕ decreases with length along the tube. The alternative and perhaps more common situation (Rufer (1966)) is when there is a pressure gradient and the condensate at the tube outlet fills the tube cross-section (Figure 11.13b). In this latter case, the angle 2ϕ increases with length along the tube and may be related to the void fraction α by the following relationship if the condensate film is of negligible thickness:

$$(1-\alpha) = \frac{\phi - \frac{1}{2} \sin 2\phi}{\pi} \qquad (11.73)$$

ϕ^0	F^1	F
0	0.725	0.725
10	0.754	0.712
20	0.775	0.689
30	0.793	0.661
40	0.808	0.629
50	0.822	0.594
60	0.835	0.557
70	0.846	0.517
80	0.857	0.476
90	0.866	0.433
100	0.874	0.389
110	0.881	0.343
120	0.887	0.296
130	0.892	0.248
140	0.896	0.199
150	0.899	0.150
160	0.902	0.100
170	0.903	0.050
180	–	0

Table 11.2. Values of the Factors F and F^1 for Laminar Flow Stratified Condensation in a Horizontal Tube

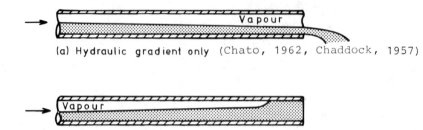

(a) Hydraulic gradient only (Chato, 1962, Chaddock, 1957)

(b) Pressure gradient (Rufer & Kezios, 1966)

Figure 11.13: Exit conditions in stratified condensation

Slug, Plug and Wavy Flow. These flow patterns, although still basically stratified, correspond to situations where the vapour shear forces are significant. Using the nomenclature of Lockhart and Martinelli the flow regime is viscous-turbulent (vt) ($G_g D/\mu_g$ > 2000, $G_f D/\mu_f$ < 2000). Point measurements of the condensing film coefficients for methaol and acetone have been reported by Rosson and Myers (1965) for condensation within a horizontal tube under such conditions. Coefficients were found to vary greatly from the top to the bottom of the tube. Near the top of the tube the film coefficient was independent of the amount of liquid in the tube and was correlated by including the effect of vapour shear on F in equation (11.71). Thus:

$$F = 0.31 \left[\frac{G_g D}{\mu_g}\right]^{0.12} \tag{11.74}$$

Near the bottom of the tube the analogy between heat and momentum transfer was used to predict the coefficient. Thus:

$$\left[\frac{h_f(z)D}{k_f}\right] = \frac{\phi_{fvt}\sqrt{[8(G_f D/\mu_f)]}}{5[1+(1/Pr_f)\ln(1+5Pr_f)]} \tag{11.75}$$

where ϕ_{fvt} is the square root of the two-phase pressure drop multiplier for the case of viscous-turbulent flow.

An example of the variation of heat transfer coefficient around the tube circumference for methanol is given in Figure 11.14. At some angle, θ_m, the local heat transfer coefficient is the arithmetic average of the upper and lower tube coefficients. Rosson and Myers found that the angle θ_m was a function of the liquid and vapour superficial Reynolds numbers and a grouping designated the Galileo number, Ga:

$$Ga = \left[\frac{D^3 \rho_f (\rho_f - \rho_g) g}{\mu_f^2}\right] \tag{11.76}$$

This relation is shown in Figure 11.15 where (θ_m/π) is plotted against the superficial vapour Reynolds number with $\{Ga/(G_f D/\mu_f)^{0.5}\}$ as parameter. Marked on Figure 11.15 is the value of (θ_m/π)

corresponding to the conditions of the example given in Figure
11.14. The slope $(d(h_f(z)/d\theta)$ at θ_m was found to be a function of
the product of the superficial vapour Reynolds number and the mean
temperature drop through the condensate film, $(T_{gi}-T_W)$. Thus

$$\left[\frac{d(h_f(z))}{d\theta}\right]_{at\ \theta_m} = 0.835 \left[\frac{G_g D}{\mu_g}(T_{gi}-T_W)\right]^{-0.6} \qquad (11.77)$$

where $h_f(z)$ is measured in kW/m² deg, θ in radians and $(T_{gi}-T_W)$ in
°C, or:

$$\left[\frac{d(h_f(z))}{d\theta}\right]_{at\ \theta_m} = 12,000 \left[\frac{G_g D}{\mu_g}(T_{gi}-T_W)\right]^{-0.6} \qquad (11.78)$$

where $h_f(z)$ is measured in Btu/h.ft² °F, θ in degrees, and $(T_{gi}-T_W)$
in °F.

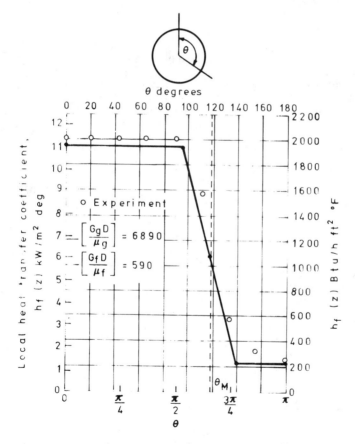

Figure 11.14: Circumferential variation of local condensing heat transfer
coefficient (Rosson and Myers (1965))

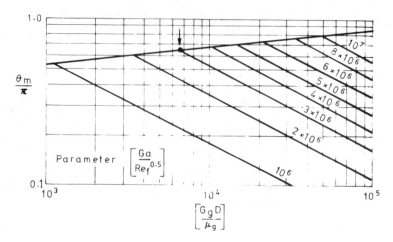

Figure 11.15:   Location of midpoint of heat transfer coefficient versus angle
                (Rosson and Myers (1965))

Annular Flow.   In this flow pattern, well away from the boundaries
with wavy and slug flow, the correlations of Carpenter and Colburn
(1951) (equation (11.61)) of Soliman, Schuster and Berenson (1968)
(equation (11.62)) of Dukler (1960) and of Kung and Yerazunis
(1967) may be used.   In the evaluation of the wall shear stress,
τ_w, the gravitational force, τ_z (of equation (11.64)), is not now
included.   Ananiev et al (1961) propose a particularly simple
relationship to correlate their data for the condensation of steam
in a horizontal tube at elevated pressures.   The analogy between
liquid film flow and single-phase flow in a pipe is used to
correlate their results.

$$h_f(z) = h_{fo}\sqrt{\frac{\rho_f}{\bar{\rho}}} \qquad\qquad (11.79)$$

where h_{fo} is the heat transfer coefficient to liquid flowing at the
same total flowrate in the tube and $\bar{\rho}$ is the homogeneous mean
density of the vapour-liquid mixture.

    The average coefficient $\bar{h}_f$ for complete condensation of a
vapour is approximately given by:

$$\bar{h}_f = \frac{h_{fo}}{2}\left[1 + \left(\frac{\rho_f}{\rho_g}\right)^{\frac{1}{2}}\right] \qquad\qquad (11.80)$$

These relationships are recommended for values of (ρ_f/ρ_g) less than
50.

Condensation outside a Horizontal Tube

    The influence of vapour shear on the heat transfer coefficient
for flow across a horizontal tube has been considered analytically

by Shekriladze and Gomelauri (1966). It was assumed that the interfacial shear occurs solely as a result of the momentum lost by the condensing vapour. Assuming no separation of the laminar vapour boundary layer and laminar flow within the condensate film, the following equation is given for the case where the cross flow is downward over the tube:

$$\left[\frac{\bar{h}_f D}{k_f}\right] = 0.9 \left[\frac{\rho_f u_{g\infty} D}{u_f}\right]^{0.5} \tag{11.81}$$

where $u_{g\infty}$ is the vapour velocity approaching the tube. Equation (11.81) is in reasonable agreement with the results of Berman and Tumanov (1962). At high values of Reynolds number (> 10^6) separation of the boundary layer occurs and the experimental results fall below equation (11.81) and approach a parallel curve corresponding separation at an angle of $\theta = 82^{\circ}$. The equation for this lower curve is the same as equation (11.81) but with the 0.9 replaced by a lower factor (0.59). Condensation, however, tends to inhibit boundary layer separation. For lower velocities where both gravitational and shear forces are important:

$$h_f = \left[\frac{1}{2} h^2_{shear} + (\frac{1}{4} h^4_{shear} + h^4_{grav})^{\frac{1}{2}}\right]^{1/2} \tag{11.82}$$

where h_{shear} is given by equation (11.81) and h_{grav} by equation (11.40).

Honda and Fujii (1974) extended this analysis to include vapour flows at various angles of incidence. Their results for horizontal flows are almost identical to the downflow result. However, at high vapour velocities and low temperature differences Hawes (1976) has observed a marked reduction in the condensing coefficient by as much as a factor two. Butterworth (1977) has suggested that this is due to a dynamic effect whereby local reductions in static pressure occur in the regions of high velocity along the sides of the tube. This would effectively cause a reduction in the local saturation temperature and hence the temperature driving force.

Equations (11.81) and (11.82) can be applied to condensation within tube bundles provided a satisfactory method of estimating $u_{g\infty}$ is available. Butterworth (1977) proposes that this should be evaluated as (u_{gs}/ψ) where u_{gs} is the velocity calculated in the absence of any tubes and ψ is the void fraction for the bundle (ie free volume divided by total volume). This approach was tested using the data of Nobbs and Mayhew (1976) for downflow of atmospheric pressure steam in a tube bank (Figure 11.16). The data tend to lie between the prediction with and without flow separation with the data at the higher temperature differences following the no flow separation line and at low temperature differences the theory allowing for flow separation. Further analytical and experimental work is needed to study vapour shear effects within tube bundles and the interaction with condensate drainage.

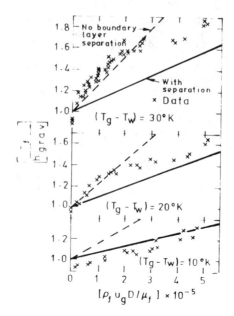

Figure 11.16: Comparison of modified Shekriladze-Gomelauri model with data of
              Nobbs and Mayhew. (Butterworth (1977))

Nomenclature

a_1-a_5	- coefficients in equation (11.66)	(-)
a	- Ackermann correction (given below (equation (11.12)	(-)
A,B	- terms given by equation (11.36)	(-)
c_p	- specific heat	(J/kg^{O}K)
c	- molar concentration	(moles/m^3)
D	- tube diameter	(m)
D_{AG}	- binary diffusion coefficeint	(m^2/s)
F	- factor in equation (11.71)	
F	- weighting factor in equation (11.37)	
f	- friction factor	(-)
g	- acceleration due to gravity	(m/s^2)
h,h(z)	- heat transfer coefficient	(W/m^2 ^{O}K)
i_{fg}	- latent heat of vaporization	(J/kg)
j_D	- dropwise condensation mass flux	(kg/m^{2}s)
j_F	- filmwise condensation mass flux	(kg/m^{2}s)
j	- mass flux	(kg/m^{2}s)
L	- characteristic length	(m)
k	- thermal conductivity	(W/m^2 ^{O}K/m)

K	- mass transfer coefficient	(m/s)
$\dot{n}$	- molar flux of material	(moles/m^{2}s)
n	- number of tubes	(-)
p	- partial pressure	(N/m^2)
r	- droplet radius	(m)
T	- temperature	($^\circ$K)
T_W	- wall temperature	($^\circ$K)
T_{gi}	- interfacial gas temperature	($^\circ$K)
t	- time	(s)
ΔT_{SUB}	- subcooling (T_{SAT} - T)	($^\circ$K)
u	- velocity	(m/s)
$\tilde{y}$	- mol fraction	(-)
x	- vapour quality	(-)
y	- distance from surface	(m)
z	- distance along surface or channel	(m)

Greek

δ	- thickness of boundary layer	(m)
δ	- thickness of liquid film	(m)
α	- angle between any radius of horizontal tube and the vertical	($^\circ$)
ε	- eddy diffusivity	(m^{2}s)
ρ	- density	(kg/m^3)
μ	- viscosity	(kg/ms)
ϕ^2	- two-phase friction multiplier	(-)
ϕ	- heat flux	(W/m^2)
ϕ	- angle substended by interface of stratified liquid (Figure 11.12)	
τ	- shear stress	(N/m^2)
ψ	- void fraction of tube bundle	(-)
θ	- angle of inclination	($^\circ$)
Γ	- flowrate per unit perimeter	(kg/ms)

Dimensionless Numbers

Ga	- Galileo number	(-)
Gr	- Grashof number	(-)
Pr	- Prandtl number	(-)
Sc	- Schmidt number	(-)
Nu	- Nusselt number	(-)
Re_Γ	- Reynolds number based on liquid film flowrate	(-)

δ*   - dimensionless film thickness (equation
          (11.52))   (-)
z*   - dimensionless distance (equation (11.53))
τ_i*   - dimensionless interfacial shear stress
          (equation (11.54))
h*   - dimensionless heat transfer coefficient
          (equation (11.56))

Subscripts

film   - pertaining to liquid film
f   - liquid
a   - accelerational
g   - gas or vapour
gs   - in absence of any tubes
g∞   - velocity approaching tube
D   - dropwise
F   - filmwise
A,a   - non-condensible gas
i   - interface
o   - refers to bulk properties
am   - log mean (given by equation (11.9))
W   - wall
v   - vapour
1,2   - conditions 1,2
y   - value at position y
z   - value at position z
Nu   - as obtained by Nusselt analysis
n   - pertaining to nth tube
shear   - dominated by shear forces
grav   - dominated by gravitational forces

Superscripts

'   - modified value
-       - average value

References

Akers, W W, Davis, S H and Crawford, J E, (1960) "Condensation of a vapour in the presence of a non-condensing gas". *Chem. Ena. Prog Symp.* Series **56** (30), 139-144.

Ananiev, E P, Boyko, L D and Kruzhilin, G N, (1961) "Heat transfer in the presence of steam condensation in horizontal tube". *Int. Develop. in Heat Transfer, Part II*, Paper 34. 290-295. *Int. Heat Transfer Conf. Boulder, Colorado, USA.*

Baer, E and McKelvey, J M, (1958) "Heat transfer in dropwise condensation". *Proc. of Symp. on Heat Transfer in Dropwise Condensation, Newark, University of Delaware, USA.*

Bell, K J, Taborek, J and Fenoglio, F, (1969) "Interpretation of horizontal in-tube condensation heat transfer correlations using a two-phase flow regime map". *AIChE preprint No 18, Presented at 11th Nat. US Heat Transfer Conf. Minneapolis, Minn. USA.*

Berman, L D and Tumanov, Yu A, (1962) "Condensation heat transfer in a vapour flow over a horizontal tube". *Teploenergetika,* **10**, 77-84.

Bromley. L A, (1952) "Effect of heat capacity of condensate". *Ind. Eng. Chem.* **44**, p2966.

Butterworth, D, (1977) "Developments in the design of shell and tube condensers". Paper 77-WA/HT-24 *Presented at Winter Annual Meeting of ASME, Atlanta, Georgia.*

Carpenter, E F and Colburn. A P, (1951) "The effect of vapour velocity on condensation inside tubes". *Proc. of General Discussion on Heat Transfer,* 20-26, *Inst. Mech. Engrs/ASME.*

Chaddock, J B, (1957) "Film condensation of vapours in horizontal tubes". *Refrig. Engrg.* **65**, 36-41 and 90-95.

Chato, J G, (1962) "Laminar condensation inside horizontal and inclined tubes". *ASHRAE Journal.* **4**, 52-60.

Chen, M M, (1961) "An analytical study of laminar film condensation, Part 1 - flat plates". *J. of Heat Transfer*, Series C, **83**, 48-55.

Colburn, A P, (1933) "The calculation of condensation where a portion of the condensate layer is in turbulent motion". *Trans. AIChE,* **30** 187.

Colburn, A P and Hougen, O A, (1934) "Design of cooler condensers for mixtures of vapours with non-condensing gases". *Ind. Eng. Chem.* **26** (11), 1178-1182.

Collier, J G, (1972) "Convective boiling and condensation". *McGraw Hill Book Co., Maidenhead, UK.*

Drew, T B, (1954) See McAdams, W H, *Heat Transmission*, 3rd ed. *McGraw Hill, New York.*

Dukler, A E, (1960) "Fluid mechanics and heat transfer in vertical falling film systems". *Chem. Eng. Prog. Symp.* Series No 30, **56**, 1. Together with appendix supplied by author.

Erb, R A and Thelen, E, (1965) "Dropwise condensation". *First Int. Symp. on Water Desalination, Washington DC.* See also Erb, R A, Ph.D. Diss., Temple Univ.

Euken, A, (1937) *Naturwissenschaften*, **25**, 209.

Gose, E, Macciardi, A N and Baer, E, (1967) "Model for dropwise condensation on randomly distributed sites". *Int. J. Heat Mass Transfer*, **10**, 15-22.

Hawes, R I, (1976) "Effect of vapour cross-flow velocity on condensation". *NEL Report 619*, 55-58. *Nat. Engng. Lab., Glasgow.*

Honda, H and Fujii, T, (1974) "Effect of the direction of an oncoming vapour on laminar filmwise condensation on a horizontal cylinder". *Heat Transfer* 1974, **3**, 299-303. *5th Int. Heat Transfer Conf., Japan.*

Jacobs, H R, (1966) "An integral treatment of combined body force and forced convection in laminar film condensation". *Int. J. Heat Mass Transfer*, **9**, 637-648.

Jakob, M, (1936) *Mech. Engrg.* **58**, 729.

Jakob, M, (1949) "Heat Transfer". John Wiley.

Kast, W, (1963) *Chemie-Ingr. Tech.* **35**, 163.

Kern, D Q, (1958) "Mathematical development of the tube loading in horizontal condensers". *Am. Inst. Chem. Engrs.* J. **4** (2), 157-160

Kirkbride, C G, (1933) "Heat transfer by condensing vapour on vertical tubes". *Trans. AIChE*, **30**, 170.

Koh, J C Y, Sparrow, E M and Hartnett, J P, (1961) "The two-phase boundary layer in laminar film condensation". *Int. J. Heat Mass Transfer*, **2**, 69-82.

Kunz, H R and Yerazunis, S, (1967) "An analysis of film condensation, film evaporation and single phase heat transfer". *Paper presented at 9th ASME-AIChE Heat Transfer Meeting, Seattle*, 7-HT-1.

Le Fevre, E J and Rose, J W, (1966) "A theory of heat transfer by dropwise condensation". *3rd Int. Heat Transfer Conf. Chicago*, ASME-AIChE, **II**, Paper 80, 362-375.

Lee, Jon, (1964) "Turbulent film condensation". *AIChE J.* **10**, 540-544.

McCormick, J L and Baer, E, (1963) "On the mechanism of heat transfer in dropwise condensation". *J. Colloid Sci.* **18**, 208.

Minkowycz, W J and Sparrow, E M, (1966) "Condensation heat transfer in the presence of non-condensibles, interfacial resistance, superheating, variable properties and diffusion". *Int. J. Heat Mass Transfer*, **9**, 1125-1144.

Nobbs, D W and Mayhew, Y R, (1976) "Effect of downward vapour flow and inundation on condensation rates on horizontal tube banks". *NEL-619 op cit.* 39-54.

Nusselt, W, (1916), "Die oberflächenkondensation des wasserdampfes". *Zeitschr. ver. deutsch. Ing.* **60**, 541 and 569.

O'Bara, J T, Killion, E S and Roblee, L H S, (1967) "Dropwise condensation of steam at atmospheric and above atmospheric pressure". *Chem. Eng. Sc.* **22**, 1305-1314.

Othmer, D F, (1929) "The condensation of steam". *Ind. Chem. Eng.* **21**, 576-583.

Rohsenow, W M, (1956) "Heat transfer and temperature distribution in laminar film condensation". *Trans. ASME,* **78**, 1645-1648.

Rohsenow, W M, Webber, J H and Ling, A T, (1956) "Effect of vapour velocity on laminar and turbulent film condensation". *Trans. ASME,* **78**, 1637-1643.

Rosson, H F and Myers, J A, (1965) "Point values of condensing film coefficients inside a horizontal tube". *Chem. Eng. Symp.* Series, **61** (59), "Heat Transfer - Cleveland", 190-199.

Rufer, C E and Kezios, S P, (1966) "Analysis of two-phase one component stratified flow with condensation". *J. of Heat Transfer,* 265-275.

Seban, R A, (1954) "Remarks on film condensation with turbulent flow". *Trans. ASME,* **76**, 299.

Shekriladze, I G and Gomelauri, V I, (1966) "Theoretical study of laminar film condensation of flowing vapour". *Int. J. Heat Mass Transfer,* **9**, 581-591.

Silver, R S, (1964) "An approach to a general theory of surface condensers". *Proc. Inst. Mech. Engrs.* **178**, Part 1, No 14, 339-376.

Soliman, M, Schuster, J R and Berenson, P J, (1968) "A general heat transfer correlation for annular flow condensation". *J. of Heat Transfer,* 267-276.

Sparrow, E M and Eckert, E R G, (1961) "Effects of superheated vapour and non-condensible gases on laminar film condensation". *AIChE J.* **7** (3), 473-477.

Sparrow, E M and Gregg, J L, (1959) "A boundary layer treatment of laminar film condensation". *J. of Heat Transfer,* Series C, **81**, 13.

Sparrow, E M, Minkowycz, W J and Saddy, M, (1967) "Forced convection condensation in the presence of non-condensibles and interfacial resistance". *Int. J. Heat Mass Transfer,* **10**, 1829-1845.

Sugawara, S and Katsuta, K, (1966) "Fundamental study on dropwise condensation". *3rd Int. Heat Transfer Conference, ASME-AIChE.* **II**, 354-361.

Takeyama, I and Shimizu, S, (1974) "On the transition of dropwise-film condensation". *Heat Transfer 1974,* **III**, p274 (paper CS2.5, *5th Int. Heat Transfer Conf. Tokyo*).

Umur, A and Griffith, P, (1965) ASME Paper 64-WA/HT-3, *J. of Heat Transfer,* 275-282.

Welch, J F and Westwater, J W, (1961) "Microscopic study of dropwise condensation". *Int. Developments in Heat Transfer,* ASME, Part II.

Zivi, S M, (1964) "Estimation of steady state steam void fraction by means of the principle of minimum entropy production". *J. of Heat Transfer,* 247-257.

Chapter 12

Augmentation of Two-Phase Heat Transfer

A. E. BERGLES

12.1 Introduction

Energy and materials saving considerations, as well as economic incentives, have led to recent expansion of efforts to produce more efficient heat-exchange equipment. Common thermal-hydraulic goals are to reduce the size of a heat exchanger required for a specified heat duty, to upgrade the capacity of an existing heat exchanger, to reduce the approach temperature difference for the process streams, or to reduce the pumping power.

The study of improved heat transfer performance is referred to as heat-transfer augmentation, enhancement, or intensification. In general, this means an increase in heat-transfer coefficient. Attempts to increase "normal" heat-transfer coefficients have been recorded for over a century and there is a large store of information. A recent report (Bergles et al, 1979) cites 1967 technical publications, excluding patents and manufacturers' literature. The recent growth of activity in this area is clearly evident from the yearly distribution of the publications presented in Figure 12.1. Broad reviews of developments in augmented heat transfer are available (Bergles, 1969, 1978). Other specific reviews apply to phase-change heat transfer (Collier, 1972, Bergles, 1976).

Augmentation techniques can be classified as passive methods, which require no direct application of external power and as active schemes, which require external power. The effectiveness of both types of techniques is strongly dependent on the mode of heat transfer, which could range from single-phase free convection to dispersed flow film boiling. Brief descriptions of passive techniques, as they apply to phase-change heat transfer, follow.

Treated surfaces involve fine-scale alternation of the surface finish or coating (continuous or discontinuous).

Rough surfaces are produced in many configurations ranging from random sand-grain-type roughness to discrete protuberances or cavities. The apparent heat transfer surface area is not appreciably increased.

Extended surfaces are routinely employed in many heat exchangers. Current work of interest in this review is directed

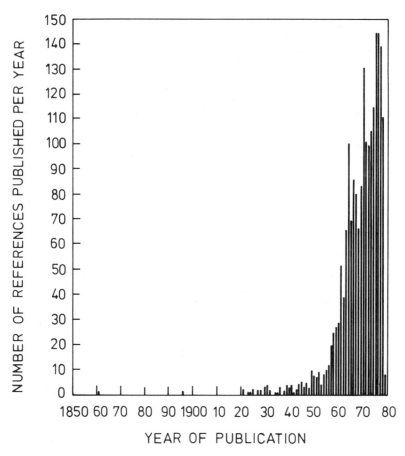

Figure 12.1: References on heat transfer augmentation versus year of publication, to early 1979 (Bergles et al, 1979).

toward development of new types of extended surfaces, such as integral inner-fin tubing.

Displaced enhancement devices are inserted into the flow channel so as to indirectly improve energy transport at the heated surface. They are used with forced flow.

Swirl flow devices include a number of geometrical arrangements or tube inserts for forced flow that create rotating and/or secondary flow: coiled tubes, inlet vortex generators, twisted-tape inserts, and axial core inserts with a screw-type winding.

Surface tension devices consist of wicking or grooved surfaces to direct the flow of liquid.

Additives primarily involve liquid trace additives for boiling systems.

Mechanical aids involve stirring the fluid by mechanical means or by rotating the surface.

Surface vibration at relatively low frequency has been applied to both boiling and condensing.

Fluid vibration is the more practical type of vibration enhancement due to the mass of most heat exchangers. The vibrations range from pulsations of about 1 Hz to ultrasound.

Electrostatic fields (d.c. or a.c.) are applied in many different ways to dielectric fluids. Generally speaking, electro-static fields can be directed to cause greater bulk mixing of fluid in the vicinity of the heat transfer surface.

Suction involves vapor removal in nucleate or film boiling through a porous heated surface.

Two or more of the above techniques may be utilized simul-taneously to produce an enhancement separately. This is termed *compound augmentation*.

It should be emphasized that one of the motivations for studying enhanced heat transfer is to assess the effect of an inherent condition on heat transfer. Some practical examples include roughness produced by standard manufacturing, surface vibration resulting from rotating machinery or flow oscillations, fluid vibration resulting from pumping pulsation, and electrical fields present in electrical equipment.

This chapter will outline the above techniques in more detail. The emphasis is on effective and cost-competitive (proven or potential) techniques which have made the transition from the laboratory to commercial heat exchangers. Industrial applications of augmentation are outlined in Chapter 21.

12.2 Pool Boiling

Selected passive and active augmentation techniques have been shown to be effective for pool boiling and flow boiling. Most techniques apply to nucleate boiling; however, some techniques are applicable to transition and film boiling.

The strong influence of surface treatment on nucleate pool boiling, including the effect of aging, has been discussed in Chapter 7. Numerous types of surface treatment and structuring have been utilized to intentionally reduce the wall-minus-satura-tion temperature difference, ΔT_s. While it is possible to improve nucleate boiling by minor changes in surface finishing procedure (Berenson, 1962), special procedures are most effective. For water, the placement of small non-wetting spots (teflon or epoxy), either splattered on the surface or placed in pits, reduces ΔT_s at constant $\dot{q}$ by a factor of 3-4 (Young and Hummel, 1964). This comparison, as well as others in this Chapter, is based on heat flux evaluated for the surface area of the plain tube.

With wetting fluids (refrigerants, cryogens, organics, alkalai liquid metals), doubly re-entrant cavities are required to insure vapour trapping. The probability of having such active nucleation sites present is increased by machining and/or forming the surface. Furthermore, large cavities can be created which result in steady-state boiling at low ΔT_S (Ragi, 1972, Webb, 1972, Nakayama et al, 1975, Wieland-Werke, 1978). It is possible to minimize ΔT_S by selecting the optimum pore or groove size for the fluid conditions of interest. An alternative technique involves coating the surface with a porous metallic layer by poor welding (Marto and Rohsenow, 1966), sintering or brazing (O'Neill et al, 1972, Danilova et al, 1976), electrolytic deposition (Oktay and Schmeckenbecher, 1974), flame spraying (Dahl and Erb, 1976, Danilova et al, 1976), bonding of particles by plating (Fujii et al, 1979), or metallic coating of a foam substrate (Janowski et al, 1978).

Superheat reductions of up to a factor of 10 have been reported with these machined or coated surfaces. It should be noted that the mechanism of vaporization is different for these surfaces than for normal cavity boiling. Here, the liquid flows to the interior where thin film evaporation occurs over a large surface area; the vapor is then ejected by "bubbling" (Czikk and O'Neill, 1979, Nakayama et al, 1979). Data are given in Figure 12.2 for Thermoexcel tubes having a variety of surface structure dimensions, a typical sintered surface, and a plain tube. The heat flux is based on the bare tube area which is essentially the envelope area of the special tubes. It should be emphasized that

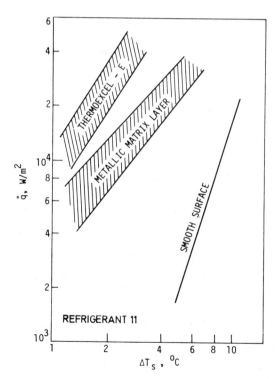

Figure 12.2: Nucleate pool boiling for smooth and structured surfaces on electrically heated copper tubes (Nakayama et al, 1975).

the performance of these special surfaces is very sensitive to surface geometry and fluid condition. Additionally, very low temperature differences are involved; hence, it is necessary to be especially careful, or at least consistent, in measuring wall temperatures and saturation temperatures (pressures). Limited evidence indicates that the critical or burnout heat flux, $\dot{q}_C$, with structured surfaces is usually as high or higher than that with plain surfaces (O'Neill et al (1972).

Low and medium fin tubes with external, circumferential fins are produced by many manufacturers for boiling of refrigerants and organics. Gorenflo (1966) reported single-tube boiling of R-11 which indicated that the heat transfer coefficients based on total area were higher for all six tubes tested than for the reference plain tube. Hesse (1973) found that for pool boiling of R-114, α was higher for finned tubes than for a plain tube; however, $\dot{q}_C$ was lower. Both comparisons were on a total area basis. When referenced to the projected area at the maximum outside diameter, the critical heat fluxes were about equal to those of the plain tube. As discussed by Westwater (1973), large fins, properly shaped or insulated, promote large heat dissipation rates if the base temperature is in the film boiling range.

Additional passive augmentation techniques are discussed in other papers (Bergles, 1969, 1976, 1978). While the use of displaced "egg crate" structures, wicking, or trace additives can be very effective in reducing ΔT_s or elevating $\dot{q}_C$, the practical applications of these techniques are quite limited.

Active augmentation techniques, also discussed in the review papers noted above, include heated surface rotation, surface wiping, surface vibration, fluid vibration, electrostatic fields, and suction at the heated surface. These techniques can be effective in reducing ΔT_s and/or increasing $\dot{q}_C$. Few fundamental studies have been conducted in recent years due to lack of practical applications.

Compound augmentation, which involves two or more techniques applied simultaneously, has also been studied. For instance, the addition of surface roughness to the evaporator side of a rotating evaporator-condenser increased the overall coefficient by 10% (Bromley et al, 1966).

12.3 Boiling Within Tubes

In accordance with the general observation that surface finish has less effect on flow boiling than on pool boiling, the various surface treatments mentioned in Section 12.2 have relatively little influence on boiling within tubes. A prime consideration, of course, is that it is more difficult to modify internal surfaces. Porous boiling surfaces do not improve high-flux, subcooled flow boiling; however, the annoying boiling curve hysteresis with refrigerants is eliminated (Murphy and Bergles, 1972). Various types of rough surfaces increase subcooled $\dot{q}_C$ by only 10% (Burck et al, 1969).

A variety of surface modifications have been proposed to improve forced convection evaporation of water and other power

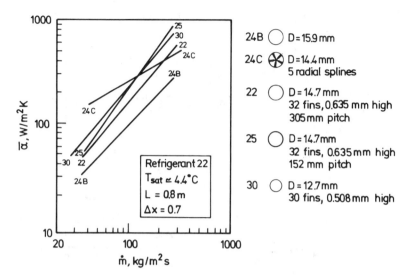

Figure 12.3: Heat transfer coefficients for evaporation in internally finned tubes (Kubanek and Miletti, 1979).

cycle fluids: dual-diameter tubes, slotted helical inserts, helical ribs, machined protuberances, and coiled wire inserts (Bergles, 1969). Most of these configurations are effective in increasing vaporization α, $\dot{q}_c$, and post dryout α.

Tubes with integral or inserted internal fins increase heat transfer rates for refrigerant evaporation by several hundred percent over the smooth tube values (Schlünder and Chwala, 1969, and Kubanek and Miletti, 1979). Data from the latter study are shown in Figure 12.3; the heat transfer coefficients are based on the surface area of the smooth tube of the same diameter. "Enhanced" heat transfer tubes are available for vertical and horizontal evaporators. While the distorted tubes (eg doubly fluted and spirally corrugated) have been developed primarily for augmentation of condensation on the outside wall, it has been demonstrated that heat transfer coefficients for the evaporating fluid on the inside of the tube are also increased (Johnson et al, 1971). In falling film evaporation, Thomas and Young (1970) found that α can be increased by a factor of over ten by use of internally finned tubes.

A variety of devices have been proposed to augment flow boiling by imparting a swirling of secondary motion to the flow. Inlet vortex generators of the spiral-ramp or tangential-slot variety have been used to accommodate very large heat fluxes for subcooled flow boiling of water. One of the highest fluxes on record, $\dot{q}_c = 1.73 \times 10^8$ W/m^2, has been obtained with this technique by Gambill and Greene (1958). Inlet swirl is effective in increasing $\dot{q}_c$ for subcooled boiling of water in a tube (Mayinger et al, 1966) or in an annulus with the inner tube heated (Ornatskiy et al, 1973).

Twisted tapes are quite popular due to their simplicity and

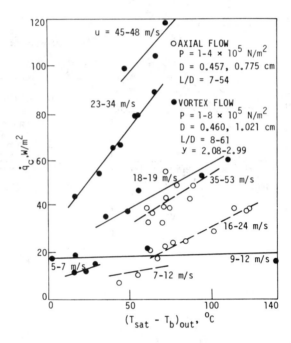

Figure 12.4: Influence of twisted-tape inserts on critical heat flux for
subcooled boiling of water (Gambill et al, 1961)

adaptability to existing heat exchange equipment. They are ideal
for hotspot applications since a short tape can cure the thermal
problem while having little effect on the overall pressure drop.
Boiling curves for subcooled boiling with twisted tapes are
similar to those with empty tubes (Lopina and Bergles, 1973);
however, $\dot{q}_c$ can be increased by up to 100% (Gambill et al, 1961),
as shown in Figure 12.4. Loose-fitting tape inserts have been
used by Sephton (1971) with tubes intended for vertical tube
evaporators for sea-water desalination. These inserts are also
effective for once-through vaporization of cryogenic fluids
(Bergles et al, 1971) or steam (Cumo et al, 1974, Hunsbedt and
Roberts, 1974) since all two-phase regimes are beneficially
affected.

Coiled tube vapor generators have advantages in terms of
packing and generally higher heat transfer performance. The
augmentation of boiling is very sensitive to geometrical and flow
conditions (Hopwood, 1972, Gorlov and Rzaev, 1972). Modest
improvements in α (circumferential average) for forced convection
vaporization are obtained, with the improvement increasing as
coil diameter is decreased. In the subcooled region, $\dot{q}_c$ is lower
than for a comparable straight tube; however, $\dot{q}_c$ or x_c is usually
substantially higher than the straight tube value at outlet qualities of 0.2 and higher. This was recently demonstrated by Jensen
(1980) as shown in Figure 12.5. The post dryout α is also
increased with helical coils.

Trace additives are not particularly effective in subcooled

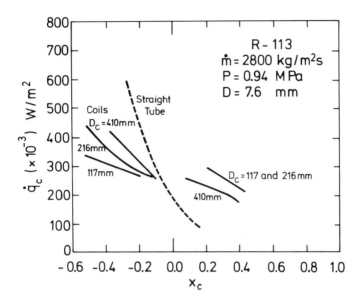

Figure 12.5: Critical heat flux for helical coils of various diameter compared with straight tube (Jensen, 1980).

nucleate boiling; ΔT_s is reduced with some additives, but increased for others (Bergles, 1969). On the other hand, the overall α was doubled when a surfactant was added to seawater evaporating in a vertical (upflow) tube (Sephton, 1975).

Active techniques are difficult to apply to boiling within tubes. The effects of tube vibration on $\dot{q}_c$ are all very small. Fluid vibrations, either sonic or ultrasonic, do not appreciably affect flow boiling. Electric fields do improve certain regimes of boiling; for example, increases in $\dot{q}_c$ of over 100% have been reported. Further details of studies on active techniques can be found in the surveys by Bergles (1969, 1976, 1978).

12.4 Vapor Space Condensation

Selected passive and active augmentation techniques are effective for vapor space condensation and forced convection condensation (Williams (1968), Bergles (1969, 1976, 1978), Collier (1972)). Some of the arrangements demonstrated success-fully in the laboratory have made the transition to commercial condensers. Much of the current interest relates to condensation of organic fluids which have thermophysical properties resulting in comparatively low condensation heat transfer coefficients.

The discussion in this section is directed primarily toward condensation on the outside of horizontal tubes where the vapor is supplied at very low velocity. Dropwise condensation may be considered augmentation of film condensation by surface treatment. The fundamental understanding and advances required for practical application are very well covered by Tanasawa (1978). It should be noted that the only real application is for steam condensers, as nonwetting substances are not available for most other working fluids. For example, no dropwise condensation promoters ("Freon-phobic") have been found for refrigerants (Iltscheff, 1971). The augmentation of dropwise condensation, beyond inducing the process by selection of an effective, durable promoter, is fruitless since the heat transfer coefficients are already so high.

Glicksman et al (1973) showed that average coefficients for film condensation of steam on horizontal tubes can be improved up to 20% by strategically placing strips of Teflon or other non-wetting material around the tube circumference.

Rough surfaces augment condensation primarily by introducing turbulence in the film. Nicol and Medwell (1966) found that the heat transfer coefficient for steam condensing outside a vertical tube could be doubled by knurling the surface.

Surface extensions can be divided into the following types: loosely attached wires, contoured surfaces, and integral fins. Surface-tension-driven flows and condensate drainage are important considerations for all these types. Thomas (1968) demonstrated that vertical surfaces equipped with vertical wires improve coefficients by as much as 800%. The condensate is drawn to the wires so that effective thin-film condensation prevails over much of the surface. The augmentation is limited by flooding.

Condenser tubes may be shaped, as shown in Figure 12.6, to exploit the Gregorig effect, whereby condensation occurs primarily at the tops of convex ridges. Surface tension forces then pull the condensate into the concave grooves where it runs off. The resulting average heat transfer coefficient is substantially greater than that for a uniform film thickness. Several analyses have been proposed; a recent analysis for the optimum Gregorig surface is presented by Webb (1978).

Horizontal tubes with eliptical shape are superior to circular tubes, as shown by the analysis of Moalem and Sideman (1976).

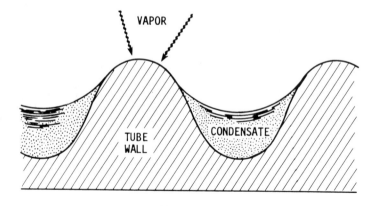

Figure 12.6: Profile of the condensing surface developed by Gregorig (1954).

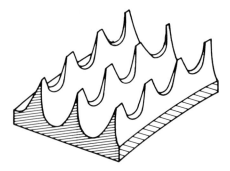

Figure 12.7: Formed and machined condensing surface. (Nakayama et al, 1975).

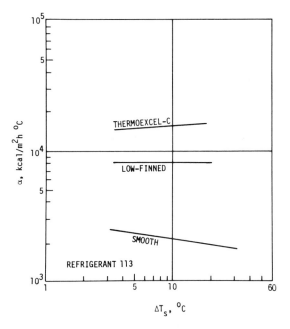

Figure 12.8: Relative performance of condenser tubes according to Nakayama et al
(1975).

Horizontal tubes may also be contoured, and "roped" or "corrugated"
tubes are available for desalination or steam condenser applications. More commonly, low integral circular fin tubes, made by
many manufacturers, are used in shell-and-tube heat exchangers.
The paper by Katz et al (1954) is frequently referred to for
design. Average condensing coefficients (based on total area) for
finned tubes are higher than for plain tubes of similar diameter.

Three-dimensional extended surfaces have been proposed for
horizontal tube condensers, eg pin-finned tubes (Chandran and
Watson, 1976). The surface described by Nakayama et al (1975) is
shown in Figure 12.7. The performance is compared to plain and

finned horizontal tubes in Figure 12.8. Webb and Gee (1979)
presented an analytical basis for the superior performance of
condensing surfaces with three-dimensional fins.

Although there is a substantial literature on augmentation
of vapor space condensation by active techniques (Williams (1968),
Bergles (1969, 1976, 1978), Collier (1972)), the prospects for
employment of these techniques in actual systems are small.
Systems studied include rotating cylinders, disks, and square
tubes; vibrating horizontal tubes; electrical fields applied to a
horizontal tube; removal of condensate film on a vertical surface
by suction; removal of non-condensibles at a condensing interface
by suction; electrical fields applied to vertical plates, inclined
plates, or horizontal tubes; electrical and magnetic fields
applied to vertical plates; rotating pin-finned tubes; a rotating
rough disk; and a rotating disk with suction. The primary use
of the available information would appear to be in situations
where the external influence is naturally present, eg rotating
or vibrating equipment.

12.5 Forced Convection Condensation

This section considers condensation inside tubes with forced
flow. Tubes having rough interior surfaces have been installed
in horizontal tube evaporators where the brine evaporates on the
outside of the condensing-steam-heated tubes. Internally grooved
and knurled tubes were tested by Cox et al (1970). Substantial
increases in condensing performance were observed for several of
the configurations. Optimization of the roughness appears to be
necessary. Luu and Bergles (1980) have found that tubes with
repeated-rib roughness increase condensing coefficients for R-113
by over 100%.

Vrable et al (1974) studied horizontal in-tube condensation
of R-12 in internally-finned tubes. The heat transfer coefficient
based on nominal surface area was increased by about 200%.
Reisbig (1974) also condensed R-12 (with some oil present) in
internally finned tubes having an increased area of up to 175%.
The nominal heat transfer coefficient was increased by up to 300%.
Royal and Bergles (1978a, 1978b) presented heat transfer and
pressure drop with straight or spiralled fins. Up to 150%
increases in average coefficients were observed for complete
condensation (see Figure 12.9: Tubes D, E, F, and G). Similar
improvements were obtained by Luu and Bergles (1979, 1980) with
the same tubes for R-113.

Azer et al (1976) reported data for condensation in tubes
having standard static mixer inserts. These inserts qualify as
a displaced enhancement device due to bulk mixing of the flow
by the elements which have right-and left-hand helices. Substantial improvements in heat transfer coefficients were reported;
however, the increases in pressure drop were very large.

Condensation in coiled tubes was studied by Miropolskii and
Kurbanmukhamedov (1975), and condensation in tube bends was studied
by Traviss and Rohsenow (1973). In both cases, modest increases
in condensation rates were observed relative to straight tubes.

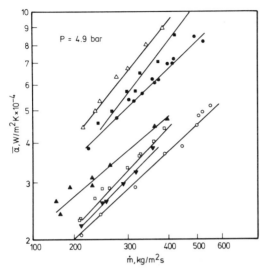

Figure 12.9: Enhancement of in-tube condensation by use of internally finned tubes and tubes with twisted-tape inserts (Royal and Bergles, 1978a).

Royal and Bergles (1978a, 1978b) found that twisted-tape inserts improved in-tube condensation of water by 30% (Figure 12.9: Tubes B and C); however, the pressure drop was also high. Luu and Bergles (1979, 1980) report similar results for R-113.

Shklover and Gerasimov (1963) used an interesting baffling technique to create a spiralling motion in the vapor condensing outside a tube bundle. With vapor velocities approaching sonic velocities, the condensing coefficients were high; however, no base data were included for comparison.

Some data are available on active techniques as discussed in the review papers cited earlier: rotating tubes with acceleration normal to the condenser tube axis, acoustic vibrations directed at downflowing vapor in a tube, electric fields applied to condensate film in a vertical annulus, and rotating tubes with an internal porous coating. Practical applications of these techniques are unlikely due to the additional equipment required and the subsequent reduction in system reliability.

Nomenclature

D — tube inside diameter

D_c — coil diameter

$\dot{m}$ — mass velocity

P — pressure

$\dot{q}$ — heat flux

$\dot{q}_c$ — critical heat flux

ΔT_s - wall-minus-saturation temperature difference
x_c - quality at critical heat flux location
α - heat transfer coefficient
$\bar{\alpha}$ - average heat transfer coefficient

References

Azer, N Z, Fan, L T, and Lin, S T, (1976) "Augmentation of condensation heat transfer with in-line static mixers". *Proceedings of the 1976 Heat Transfer and Fluid Mechanics Institute*, Stanford University Press, 512-526.

Berenson, P J, (1962) "Experiments on pool-boiling heat transfer". *Int. J. Heat Mass Transfer*, **5**, 985-999.

Bergles, A E, (1969) "Survey and evaluation of techniques to augment convective heat and mass transfer". in *Progress in Heat and Mass Transfer*, U. Grigull and E. Hahne (Editors), **1**, 331-424, Pergamon, Oxford.

Bergles, A E, Fuller, W D, and Hynek, S J, (1971) "Dispersed film boiling of nitrogen with swirl flow". *Int. J. Heat Mass Transfer*, **14**, 1343-1354.

Bergles, A E, (1976) "Survey of augmentation of two-phase heat transfer". *ASHRAE Trans.*, **82**, 1, 881-890.

Bergles, A E, (1978) "Enhancement of heat transfer". *Heat Transfer 1978*, **6**, Hemisphere, Washington, 89-108.

Bergles, A E, Webb, R L, Junkhan, G H, and Jensen, M K, (1979) "Bibliography on augmentation of convective heat and mass transfer". HTL-19, ISU-ERI-Ames-79206, Iowa State University.

Bromley, C A, Humphreys, R F, and Murray, W, (1966) "Condensation on and evaporation from radially grooved rotating disks". *J. Heat Transfer*, **88**, 80-93.

Burck, E, Hufschmidt, W, and De Clereq, E, (1969) "Einfluss künstlicher rauigkeit auf die kritische wärmestromdichte bei erzwungener konvection". *Atomkernenergie*, **14**, 305-308.

Chandran, R, and Watson, F A, (1976) "Condensation on static and rotating pinned tubes". *Transactions of the Institution of Chemical Engineers*, **54**, 65-72.

Collier, J G, (1972) "Convective heat and mass transfer". McGraw-Hill, Maidenhead, 348-368.

Cox, R B, Matta, G A, Pascale, A S, and Stromberg, K G, (1970) "Second report on horizontal-tubes multiple-effect process pilot plant tests and design". Office of Saline Water Research and Development Progress Report 592.

Cumo, M, Farello, G, Ferrari, G, and Palazzi, G, (1974) "The influence of twisted tapes in subcritical, once-through vapor generator in counter flow". *J. Heat Transfer*, **96**, 365-370.

Czikk, A M, and O'Neill, P S, (1979) "Correlation of nucleate boiling from porous metal films". in *Advances in Enhanced Heat Transfer*, ASME, 53-60.

Dahl, P J, and Erb, L D, (1976) "Liquid heat exchanger interface method". U.S. Patent 3,990,862, November 9.

Danilova, G N, Guygo, E I, Borishanskaya, A V, Bukin, V G, Dyundin, V A, Kozyrev, A A, Malyugin, G I, (1976) "Enhancement of heat transfer during boiling of liquid refrigerants at low heat fluxes". *Heat Transfer-Sov. Res.*, 8, No. 4, 1-8.

Fujii, M, Nishiyama, E and Yamanaka, G, (1979) "Nucleate pool boiling heat transfer from micro-porous heating surface". in *Advances in Enhanced Heat Transfer*, ASME, 45-51.

Gambill, W R, and Greene, N D, (1958) "A preliminary study of boiling burnout heat fluxes for water in vortex flow". *Chem. Eng. Prog.*, 54, No. 10, 68-76.

Gambill, W R, Bundy, R D, and Wansbrough, R W, (1961) "Heat transfer, burnout, and pressure drop for water in swirl flow tubes with internal twisted tapes". *Chem. Eng. Prog. Symp. Ser.*, 57, No. 32, 127-137.

Glicksman, L R, Mikic, B B, and Snow, D F, (1973) "Augmentation of film condensation on the outside of horizontal tubes". *American Institute of Chemical Engineers Journal*, 19, 636-637.

Gorenflo, D, (1966) "Zum wärmeübergang bei blasenverdampfung an rippenrohren". Dissertation, Technishe Hochschule, Karlsruhe.

Gorlov, I G, and Rzaev, A E, (1972) "Features of hydrodynamic and heat exchanges during the flow of liquids in helically-coiled tubes". *Heat Transfer-Sov. Res.*, 5, No. 4, 80-88.

Gregorig, R, (1954) "Hautkondensation an feingewellten oberflächen bei berücksichtigung der oberflächenspannungen". *Zeitschrift für Angewandte Mathematik und Physik*, 5, 36-49.

Hesse, G, (1973) "Heat transfer in nucleate boiling, maximum heat flux and transition boiling". *Int. J. Heat Mass Transfer*, 16, 1611-1627.

Hopwood, P F, (1972) "Pressure drop, heat transfer and flow phenomena for forced convection boiling in helical coils- a literature review". AEEW R-157.

Hunsbedt, A, and Roberts, J M, (1974) "Thermal-hydraulic performance of a 2MWT sodium heated, forced recirculation steam generator model". *J. Eng. Power*, 96, 66-76.

Iltscheff, S, (1971) "Über einige versuche zur erzielung von tropfkondensation mit fluorierten kältemitteln." *Kältetechnik-Klimatisierung*, 23, 237-241.

Janowski, K R, Shum, M S, and Bradley, S A, (1978) "Heat transfer surface". U.S. Patent 4,129,181, December 12.

Jensen, M K, (1980) "Boiling heat transfer and critical heat flux in helical coils". Ph.D. Dissertation, Iowa State University.

Johnson, B M, Jansen, G, and Otwzarski, P C, (1971) "Enhanced evaporating film heat transfer from corrugated surfaces". ASME Paper No. 71-HT-33.

Katz, D L, Young, E H, and Balekjian, G, (1954) "Condensing vapors on finned tubes". *Petroleum Refiner*, 33, No. 11, 175-178.

Kubanek, G R, and Miletti, D L, (1979) "Evaporative heat transfer and pressure drop performance of internally-finned tubes with refrigerant 22". *J. Heat Transfer*, **101**, 447-452.

Lopina, R F, and Bergles, A E, (1973) "Subcooled boiling of water in tape-generated swirl flow". *J. Heat Transfer*, **95**, 281-283.

Luu, M, and Bergles, A E, (1980) "Augmentation of in-tube condensation of R-113". Heat Transfer Laboratory Report, Iowa State University.

Luu, M, and Bergles, A E, (1979) "Experimental study of the augmentation of in-tube condensation of R-113". *ASHRAE Trans*, **85**, Pt. 2, 132-145.

Marto, P J, and Rohsenow, W M, (1966) "Effects of surface conditions on nucleate pool boiling of sodium". *J. Heat Transfer*, **88**, 196-204.

Mayinger, F, Schad, O, and Weiss, E, (1966) "Untersuchungen der kritischen heizflächenbelastung bei siedendem wasser". MAN Report No. 09.03.01.

Miropolskii, Z L, and Kurbanmukhamedov, A, (1975) "Heat transfer with condensation of steam within coils". *Thermal Engineering*, No. 5, 111-114.

Moalem, D, Sideman, S, (1976) "Theoretical analysis of a horizontal condenser-evaporator elliptical tube". *Journal of Heat Transfer*, **19**, 259-270.

Murphy, R W, and Bergles, A E, (1972) "Subcooled flow boiling of fluorocarbons-hysteresis and dissolved gas effects on heat transfer". *Proc. 1972 Heat Transfer and Fluid Mechanis Institute*, Stanford University Press, Stanford, 400-416.

Nakayama, W, Daikoku, T, Kuwahara, H, and Kakizaki, K, (1975) "High-flux heat transfer surface 'Thermoexcel'". *Hitachi Review*, **24**, No. 8, 329-333.

Nakayama, W, Daikoku, T, Kuwahara, H, and Nakajima, T, (1979) "Dynamic model of enhanced boiling heat transfer on porous surfaces". in *Advances in Enhanced Heat Transfer*, ASME, 31-43.

Nicol, A A, and Medwell, J O, (1966) "The effect of surface roughness on condensing steam". *Canadian Journal of Chemical Engineering*, 170-173.

Oktay, S, and Schmeckenbecher, A F, (1974) "Preparation and performance of dendritic heat sinks". *J. Electrochemical Soc.*, **21**, 912-918.

O'Neill, P S, Gottzmann, C F and Terbot, C F, (1972) "Novel heat exchanger increases cascade cycle efficiency for natural gas liquifaction". *Advances in Cryogenic Engineering*, **17**, 421-437.

Ornatskiy, A P, Chernobay, V A, Vail'yev, A F, and Perkov, S V, (1973) "A study of the heat transfer crisis with swirled flows entering an annular passage". *Heat Transfer-Sov. Res.*, **5**, No. 4, 7-10.

Ragi, E, (1972) "Composite structure for boiling liquids and its formation". U.S. Patent 3,684,007, August 15.

Reisbig, R L, (1974) "Condensing heat transfer augmentation inside splined tubes". ASME Paper No. 74-HT-7 presented at AIAA/ASME Thermophysics and Heat Transfer Conference, Boston, July.

Royal, J H, and Bergles, A E, (1978a) "Augmentation of horizontal in-tube condensation by means of twisted-tape inserts and internally finned tubes". *Journal of Heat Transfer*, 100, 17-24.

Royal, J H, and Bergles A E, (1978b) "Pressure drop and performance evaluation of augmented in-tube condensation". *Heat Transfer 1978*, 2, Hemisphere, Washington, 459-464.

Schlünder, E U, and Chwala, M, (1969) "Örtlicher wärmeübergang und druckabfall bei der strömung verdampfender kältemittel in innenberippten, waggerechten rohren". *Kältetechnik-Klimatisierung*, 21, No. 5, 136-139.

Sephton, H H, (1971) "Interface enhancement for vertical tube evaporator: a novel way of substantially augmenting heat and mass transfer". ASME Paper No. 71-HT-38.

Sephton, H H, (1975) "Upflow vertical tube evaporation of sea water with interface enhancement; process development by pilot plant testing". *Desalination*, 16, 1-13.

Shklover, G G, and Gerasimov, A V, (1963) "Heat transfer of moving steam in coil-type heat exchangers". *Teploenergetika*, 10, No. 5, 62-63.

Tanasawa, I, (1978) "Dropwise condensation - the way to practical applications". *Heat Transfer 1978*, Proceedings of the Sixth International Heat Transfer Conference, 6, Hemisphere Publishing Corp., 393-405.

Thomas, D G, (1968) "Enhancement of film condensation rate on vertical tubes by longitudinal fins". *American Institute of Chemical Engineers Journal*, 14, 644-649.

Thomas, D G, and Young, G, (1970) "Thin film evaporation enhancement by finned surfaces". *Ind. Eng. Chem. Proc. Des. Dev.*, 9, 317-323.

Traviss, D P, and Rohsenow, W M, (1973) "The influence of return bends on the downstream pressure drop and condensation heat transfer in tubes". *ASHRAE Transactions*, 79, Part 1, 129-137.

Vrable, D L, Yang, W J, and Clark, J A, (1974) "Condensation of refrigerant-12 inside horizontal tubes with internal axial fins". *Heat Transfer 1974*, III, The Japan Society of Mechanical Engineers, 250-254.

Webb, R L, (1972) "Heat transfer surface having a high boiling heat transfer coefficient". U.S. Patent 3,696,861, October 10.

Webb, R L, (1978) "A generalized procedure for the design and optimization of fluted Gregoric condensing surfaces". *Proceedings of the Fifth Ocean Thermal Energy Conversion Conference*, 3, VI-123 - VI145.

Webb, R L, and Gee, D L, (1979) "Analytical predictions for a new concept spine-fin surface geometry". *ASHRAE Trans.* 85, Pt. 2.

Westwater, J W, (1973) "Development of extended surfaces for use in boiling liquids". *AIChE Symp. Ser.*, 69, No. 131, 1-9.

Wieland-Werke AG, P.O. Box 4420, Ulm, West Germany, (1978) "High-performance tubes for flooded evaporators". Wieland-Productinformation SAW-15e-06.78.

Williams, A G, Nadapurkar, S S, and Holland, A F, (1968) "A review of methods for enhancing heat transfer rates in surface condensers". *The Chemical Engineer*, No. 223, 367-373.

Young, R K, and Hummel, R L, (1964) "Improved nucleate boiling heat transfer". *Chem. Engng. Prog.*, **60**, No. 7, 53-58.

Chapter 13

Instabilities in Two-Phase Systems

A. E. BERGLES

Abstract

   Many practical heat exchangers involving boiling are suscept-
ible to thermal-hydrodynamic instabilities, which may cause flow
excursions or flow oscillations of constant amplitude or diverging
amplitude. These excursions or oscillations could induce boiling
crisis, disturb control systems, or cause mechanical damage. This
chapter discusses the various types of instabilities according to
a mechanism-based classification. The parametric effects on flow
instability, observed experimentally and in industrial practice,
are systematically presented. Various analytical techniques for
predicting the instability thresholds are noted.

13.1 Introduction

   Flow instabilities are undesirable in boiling, condensing,
and other two-phase flow processes for several reasons. Sustained
flow oscillations may cause forced mechanical vibration of components or system control problems. Flow oscillations can affect the
local heat transfer characteristics, possibly resulting in oscillatory wall temperatures, or may induce boiling crisis (referred to
also as critical heat flux (CHF), departure from nucleate boiling
(DNB), burnout or dryout). Flow stability becomes of particular
importance in water-cooled and water-moderated nuclear reactors and
steam generators. The designer's job is to predict the threshold
of flow instability so that he can design around it or compensate
for it (Firman (1975), Gilli et al (1975)).

   There are areas of confusion regarding the different types of
flow instabilities. This is perhaps understandable since an
instability can start in a heated channel or in a two-phase flow
system, can be a primary or secondary phenomenon, and can be static
or dynamic in nature. Since the behavior and method of prediction
of one type is different from another, the investigator must
recognize the type of flow instability on hand before he can perform
or apply an analysis.

   This chapter is based on a series of previous reviews of
instabilities in two-phase systems (Bouré et al (1971), Bergles
(1972), Bouré et al (1973) and Bergles (1977)). Recent work on
the subject is included.

13.1.1 Classification of Flow Instabilities

Flow instability phenomena can be classified on the basis of the following general definitions:

A *steady flow*, rigorously speaking, is one in which the system parameters are functions of the space variables only. Practically, however, they undergo small fluctuations (due to turbulence, nucleation, or slug flow). These fluctuations play a role in triggering several instability phenomena.

A flow is *stable* if, when momentarily disturbed, its new operating conditions tend asymptotically towards the initial ones.

A flow is subject to a *static instability* if, when the flow conditions change by a small step from the original steady-state ones, another steady state is not possible in the vicinity of the original state. The cause of the phenomenon lies in the steady-state laws; hence, the threshold of the instability can be predicted only by using steady-state laws. A static instability can lead either to a different steady-state condition or to a periodic behavior.

A flow is subject to a *dynamic instability* when the inertia and other feedback effects have an essential part in the process. The system behaves like a servomechanism and the knowledge of the steady-state laws is not sufficient, even for the threshold prediction. The steady-state may be a solution of the equations of the system, but it is not the only solution. The above-mentioned fluctuations in a steady flow may be sufficient to start the instability.

A *secondary* phenomenon is a phenomenon which occurs after the *primary* one. In this paper, the term secondary phenomenon will be used only in the very important particular case when the occurrence of the primary phenomenon is a necessary condition for the occurrence of the secondary phenomenon.

An instability is *compound* when several elementary mechanisms interact in the process and cannot be studied separately. It is said to be *fundamental* (or pure) in the opposite sense.

The various flow instabilities are classified in Table 13.1.

13.2 Physical Mechanisms and Observed Parametric Effects

This section describes the physical mechanisms and summarizes the experimentally observed phenomenon of flow instabilities classified in the previous section. The observed parametric effects are also systematically presented. The parameters considered include:

Geometry - Channel length, size, inlet and exit restrictions, single or multiple channels;

Operation conditions - pressure, inlet subcooling, mass velocity, power input, forced or natural convection;

Boundary conditions - axial heat flux distribution, pressure drop across channels.

TABLE 13.1

Classification of Two-Phase Flow Instabilities

Class	Type	Mechanism	Characteristics
1. Static instabilities 1.1 Fundamental (or pure) static instabilities	1. Flow excursion or Ledinegg instability	Equation (13.1)	Flow undergoes sudden, large amplitude excursion to a new, stable operating condition
	2. Boiling crisis	Ineffective removal of heat from heated surface	Wall temperature excursion and flow oscillation
1.2 Fundamental relaxation instability	1. Flow pattern transition instability	Bubbly flow has less void but higher ΔP than that of annular flow. Condensation rate depends on flow regime.	Cyclic flow pattern transitions and flow rate variations
1.3 Compound relaxation instability	1. Bumping, geysering, or chugging	Periodic adjustment of metastable condition, usually due to lack of nucleation sites	Periodic process of super-heat and violent evaporation with possible expulsion and refilling
	2. Condensation chugging	Bubble growth and condensation, followed by surge of liquid	Periodic interruption of vent steam flow due to condensation and surge of water up the downcomer
2. Dynamic instabilities 2.1 Fundamental (or pure) dynamic instabilities	1. Acoustic oscillations	Resonance of pressure waves	High frequencies (10-100 Hz) related to time required for pressure wave propagation in system

Table 13.1 cont'd

Class	Type	Mechanism	Characteristics
	2. Density wave oscillations	Delay and feedback effects in relationship between flow rate, density, and pressure drop	Low frequencies (~1 Hz) related to transit time of a continuity wave
2.2 Compound dynamic instabilities	1. Thermal oscillations	Interaction of variable heat transfer coefficient with flow dynamics	Occurs in film boiling
	2. Boiling Water Reactor (BWR) instability	Interaction of void reactivity coupling with flow dynamics and heat transfer	Strong only for a small fuel time constant and under low pressures
	3. Parallel channel instability	Interaction among small number of parallel channels	Various modes of flow redistribution or U-tube manometer oscillations
	4. Condensation oscillations	Interaction of condensing interface with pool convection	Occurs with steam injection into vapor suppression pool
2.3 Compound dynamic instability as secondary phenomena	1. Pressure drop oscillations	Flow excursion initiates dynamic interaction between channel and compressible volume	Very low frequency periodic process (~0.1 Hz)

The effects of geometry and boundary conditions are usually interrelated, such as in flow redistribution among parallel channels. With common headers connected to the parallel channels, the flow distribution among channels is determined by the dynamic pressure variations in individual channels, caused by flow instability in each channel. Thus, the boundary conditions of a heated channel *separate* the channel instability from the system instability.

13.2.1 *Fundamental Static Instabilities*

Flow excursion. The flow excursion or Ledinegg instability (Ledinegg (1938)) involves a sudden change in the flow rate to a lower value. It occurs when the slope of the channel demand pressure-drop - flow-rate curve (*internal* characteristic of the channel) becomes algebraically smaller than the loop supply pressure-drop - flow-rate curve (*external* characteristic of the channel). The criterion for this first-order instability, as given in Figure 13.1, is

$$\partial \Delta P / \partial G \big|_{int} < \partial \Delta P / \partial G \big|_{ext} \,. \qquad (13.1)$$

This behavior requires that the channel characteristics exhibit a region where the pressure drop decreases with increasing flow. Physical situations exist where the sum of the component terms - friction, momentum, and gravity - increase with decreasing flow. Consider, for example, pressure drop for subcooled boiling of water at high heat flux (Daleas and Bergles (1965)) as represented in Figure 13.2. As the flow is reduced with other conditions held the same, the pressure drop increases once subcooled boiling is initiated. The increase is particularly rapid at low subcooling due to the friction and the acceleration caused by non-equilibrium voids. It is evident that if the pump characteristic as seen by the heated section, has less of a negative slope than the test section curve, the system is unstable. With pump characteristic A, for example, operation at point b is impossible since a slight decrease in flow will cause a spontaneous shift to point a. The new equilibrium point corresponds to such a low flow rate that burnout occurs. It is apparent that it would not be possible to operate to the left of the minimum in the heated section characteristic with a flat pump characteristic. By installing a throttle valve upstream, the effective pump characteristic can be

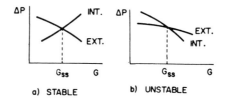

Figure 13.1: Pressure drop characteristics.

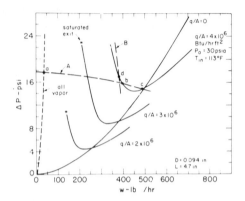

Figure 13.2: Heated-section pressure-drop characteristics. "All-vapor" curve
       is for q/A = 4 x 10⁶ (Daleas and Bergles (1965)).

steepened enough (curve B) to avoid flow excursion and allow stable
operation in the negative sloping region (point d).

     The same low pressure subcooled boiling system was also tested
for excursive instability by paralleling the heated channel with a
large bypass (Maulbetsch and Griffith (1965)). With a constant
pressure drop boundary condition, excursions leading to CHF were
always observed near the minima in the pressure-drop - flow-rate
curves. It is evident from Figure 13.3 that the premature critical heat flux is well below the actual CHF limit for the channel.
It is to be noted that for these subcooled boiling conditions, the
Ledinegg instability represents the limiting condition for a large
bank of parallel tubes between common headers, since any individual
tube sees an essentially constant pressure drop. Stable operation
beyond the minimum and up to the critical heat flux can be achieved
by throttling individual channels at their inlet; however, the
required increase in supply pressure may be considerable. Recent
discussions of the parallel channel problem are given by Ledinegg
and Wiesenberger (1975) and Worrlein (1975).

     Eselgroth and Griffith (1968) have considered the problem of pump
failure in a parallel channels array which approximately models a
nuclear reactor core. As the flow reverts to natural circulation,

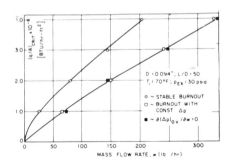

Figure 13.3 Critical heat flux versus mass flow rate for constant pressure drop
       (Maulbetsch and Griffith (1965)).

there is a hydraulic excursion to reverse flow in a number of channels. Dryout would be expected during the flow reversal.

It is appropriate to mention the studies of direct contact condensation reported by Westendorf and Brown (1966). Saturated steam was injected into the core while subcooled water was introduced through an annular injector to form a film at the tube wall. Under certain conditions of flow rates and coolant inlet temperature, an interface excursion was observed, whereby the interface between annular flow and all liquid flow moved downstream out of the test section (Figure 13.15). In effect, complete condensation was not possible within the tube. It is probable that this effect is due to a minimum in the Δp - w curve produced by competing momentum and friction pressure drop terms. Dynamic oscillations observed in this system are discussed in subsequent sections.

Hands (1974) presented an analytical solution to the Ledinegg instability problem. The various terms accounting for the total pressure drop across a heated tube were expressed as simple equations, the pressure drop was differentiated with respect to velocity, and the maximum and minimum points on the ΔP - G curve were presented as a function of mass velocity. The conditions for unconditional stability, i.e., no maximum or minimum points, are readily visualized from plots of the analytical results.

Turning now to instabilities in full-scale industrial equipment Margetts (1972) reports flow instability failures in several feed-water economizers. Boiler feedwater is preheated in a large number of parallel channels by recovering heat from ammonia plant primary flow gases. Calculations suggested that the severely over-heated tube had the highest head load and was operating at the minimum in the pressure-drop - flow-rate curve, which was essentially flat and set by the remaining parallel unheated tubes in the bank. The problem was solved by eliminating one-third of the unheated tubes to increase the driving pressure drop of the heated tubes.

In another case reported by Margetts (1972), pinhole failures were observed in an economizer tube after two years of operation with considerably lower gas temperatures. It was speculated that the channel had experienced some flow excursion, and ultimately failed due to accelerated corrosion in the dry region. The solution to this problem was to install throttling restrictions at the inlet to each tube.

Parallel channel systems in downward flow are subject to a somewhat different type of excursion: that of flow reversal in some tubes (Bonilla (1957)). In heated downcomers, minima in Δp - w curves frequently occur due to interaction of momentum and gravitational pressure drop terms. The flow reversal reported by Giphshman and Levinzon (1966) in a pendent superheater can be explained in terms of the hydraulic characteristic.

Failure of one tube in a downward flow serpentine superheater of the Breed plant has been reported by the American Electric Power Service Corporation (1968). Special instrumentation indicated reverse flow in that tube. Even though the flow was super-critical, large density changes at the transition temperature made the flow behave like two-phase flow. This problem was solved by

changing the unit to upflow so that the gravitational component of
the pressure drop was stabilizing.

Recently, Liquid Metal Fast Breeder Reactor (LMFBR) steam
generators have been examined for excursive instability
(Yadigaroglu (1978) and Chan and Yadigaroglu (1979)). In some
cases they do not exhibit flow excursions in spite of having
unstable hydrodynamic characteristics, apparently due to the
stabilizing effect of the thermal coupling between primary and
secondary fluids. Excursions have been reported in some studies,
however; the flow excursion transient may be long, of the order of
minutes or even hours, in long tubes such as those used in LMFBRs.

Boiling crisis. Boiling crisis is caused by a change of heat
transfer mechanism and is characterized by a sudden rise of wall
temperature. The hydrodynamic and heat transfer relationships
near the wall in a subcooled or low quality boiling have been
postulated by several investigators, e.g., Tong (1968), as boundary
layer separation during the boiling crisis. This separation could
cause a flow oscillation; however, conclusive experimental
evidence is still lacking.

Mathisen (1967) observed that boiling crisis occurred simultaneously with flow oscillations in a boiling water channel at
pressures higher than 870 psia. Dean et al (1971) also observed
that boiling crisis occurred simultaneously with flow oscillations
in a boiling Freon-113 flow with electrical heating on a stainless
steel porous wall and vapor injection through the porous wall.

13.2.2 *Fundamental Relaxation Instability*

Flow pattern transition instabilities have been postulated as
occurring when the flow conditions are close to the point of
transition between bubbly flow and annular flow. A temporary
increase in bubble population in bubbly or slug flow (arising from
a temporary reduction in flow rate) may change the flow pattern to
annular flow with its characteristically lower pressure drop.
Thus the excess driving pressure drop will speed up the flow rate
momentarily. As the flow rate increases, however, the vapor
generated may become insufficient to maintain the annular flow,
and the flow pattern then reverts back to bubbly or slug flow.
The cycle can be repeated. This oscillatory behavior is partly
due to the delay incurred in acceleration and deceleration of the
flow. In essence, each of the hydrodynamically compatible sets of
conditions induces the transition towards the other one. This is
typically a relaxation mechanism, and it results in a periodic
behavior. In general, relaxation processes are characterized by
finite amplitudes at the threshold.

Cyclic flow pattern transitions have been observed in
connection with oscillatory behavior (Fabrega (1964) and Jeglic and
Grace (1965)), but it is not clear whether the flow pattern
transition was the cause (through the above mechanism) or the
consequence of a density wave or pressure drop oscillation.

Bergles et al (1967b) have suggested that low pressure CHF
data may be strongly influenced by the presence of an unstable
flow pattern within the heated section. The complex effects of

length, inlet temperature, mass velocity, and pressure, on the CHF appear to be related to the large scale fluctuations characteristic of the slug flow regime. While this is strictly not a relaxation instability, the problem is of such industrial importance that some discussion is warrented here. Two-phase flow noise generated by unstable flow patterns, particularly slug flow, can introduce mechanical vibration of mechanical components, which, in turn, could amplify the flow disturbances or lead to dryout. For example, it is two-phase noise that is responsible for the vibrations in simulated reactor fuel bundles observed by Gorman (1970) and Cali et al (1970). Grant (1971) has reported that shell-side slug flow was responsible for large-scale pressure fluctuation and exchanger vibration in a model of a segmentally baffled shell-and-tube heat exchanger.

An instability which appears to fall into this classification is found in nuclear reactor safety. During the later stages of blowdown, refill of the lower plenum, and reflood of the core, the subcooled emergency core coolant (EEC) may interact with steam flowing in the cold legs to produce flow and pressure oscillations or rapid transients. Oscillations have been observed in scale model tests (1/30 - 1/3 Pressurized Water Reactor (PWR) scale) of the ECC injection region. It appears that this is primarily a flow pattern transition instability due to the change from dispersed liquid to a liquid slug near the injection point. The heat transfer rate to droplets is high with the result that the pressure is reduced. A plug of injected liquid then moves upstream and covers the injection port, the condensation rate is reduced, and the pressure increases so that the steam from the steam generator flows into the injection area. Rothe et al (1977) report data taken in a transparent 1/20-scale cold leg model. A one-dimensional analysis was presented which identifies the governing dimensionless parameters, predicts flow regime maps for various types of water plug behavior, and predicts the amplitude, frequency, and history of the unsteady flow and pressure. The predictions are compared with these data and with data from other scale models. Flow regimes, including "Oscillating Liquid Plug", are predicted closely, and frequency and pressure amplitude are generally predicted to within 25% of the measured values.

13.2.3 *Compound Relaxation Instability*

Bumping, geysering and chugging. These static phenomena are coupled so as to produce a repetitive behavior which is not necessarily periodic. When the cycles are irregular, each flow excursion can be considered as hydraulically independent of others. Bumping is exhibited in boiling of the alkali metals at low pressure. As indicated in Figure 13.4, an erratic boiling region exists where the surface temperature moves between boiling and natural convection in rather irregular cycles. It has been postulated that this is due to the presence of gas in certain cavities. The effect disappears at higher heat fluxes and higher pressures.

Geysering has been observed in a variety of closed end, vertical columns of liquid which are heated at the base. When the heat flux is sufficiently high, boiling is initiated at the base. In low pressure systems this results in a suddenly increased vapor generation due to the reduction in hydrostatic head, and usually causes an explosion of vapor from the channel. The liquid then returns, the subcooled non-boiling condition is restored, and the

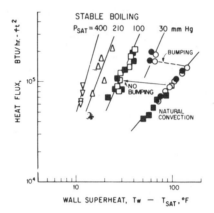

Figure 13.4: Effect of pressure on nucleate boiling of sodium from natural sites in a nickel surface (Deane and Rohsenow (1969)).

cycle starts over again.   Geysering is observed not only in nature but also in rocket propellent storage and feed systems.   In laboratory tests Griffith (1962) observed geysering periods in the approximate range 10-1000 sec.

An alternative mechanism for expulsion is the vapor burst. The vapor burst instability is characterized by the sudden appearance and rapid growth of the vapor phase in a liquid where high values of superheat have been achieved.   It occurs most frequently with the alkali liquid metals and with fluorocarbons, both classes of fluids having near-zero contact angles on engineering surfaces. With liquids having such good wettability, all of the larger cavities may be flooded out, with the result that a high superheat is required for nucleation.   As the heat flux is raised, nucleation and the associated temperature excursion are observed.   The well-known boiling curve hysteresis is obtained when the power is reduced; the wall superheat decreases smoothly.   Hysteresis in flow boiling of Freon-113 is given by Murphy and Bergles (1971).

Research in liquid metal boiling has been directed toward accumulating data and developing models to describe vapor bubble growth in confined geometries, e.g. (Singer and Holtz (1970) and Grolmes and Fauske (1970)).   The objective is to describe coolant behavior following a hypothetical complete blockage of the channel inlet.   Because of the tendency to superheat rather uniformly to high values, the initial vapor bubble that forms will grow rapidly, with probable expulsion of the liquid from the channel.   Experimental observations of the slug expulsion of Freon-113, used to model alkali liquid metals, were reported by Ford et al (1971a) (1971b).   The available experimental evidence suggests that the expulsion may be either a one-shot affair or periodic, whereby the liquid slug decelerates and re-enters the heated section.

The term chugging is usually reserved for the cyclical phenomenon characterized by the periodic expulsion of coolant from a flow channel.   The effects range from simple transitory variations of the inlet and outlet flow rates to a violent ejection of large amounts of coolant, usually through both ends of the channel.   As with the phenomena mentioned previously, the cycle consists of incubation, nucleation, expulsion, and re-entry of the fluid. In the double-ended-expulsion case, when the two liquid slugs entering the channel collide, the vapor phase probably collapses. Although the phrase chugging was used by Fleck (1960) to describe gross instabilities in the SPERT-1 reactor system in 1959, little evidence of similar phenomena has been observed since that time.

Thermosyphon reboilers, which are widely used in the chemical process industry, are subject to a variety of static expulsion instabilities when conditions are not adequate to establish steady circulation. Chexal and Bergles reported observations of instabilities in a small-scale natural circulation loop resembling a thermosyphon reboiler (Chexal and Bergles (1973)). A typical flow regime map is shown in Figure 13.5. Regime II, periodic exit large bubble formation, which resembles chugging, was observed. Regime IV, period exit small bubble formation, and regime V, periodic extensive small bubble formation, exhibit some characteristics of geysering. These instabilities would be expected during start-up of a thermosyphon reboiler. In his survey, Hawtin (1970) has pointed out that similar unstable behavior has been observed in conventional steam boilers being started up at low pressures.

Nakajima et al (1975) reported an instability caused by the periodic discharge of steam bubbles coalesced in a horizontal

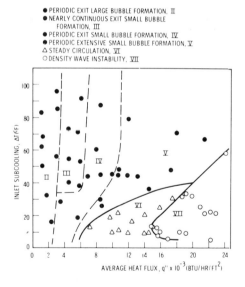

Figure 13.5: Typical flow regime boundaries for a small-scale thermosyphon reboiler (Chexal and Bergles (1973).

leg of a natural circulation riser. The oscillations are of the
type noted in this section.

Aritomi et al (1978) described a "slug excursion instability"
in a heated U-tube. Under certain conditions, a large bubble is
formed near the downcomer exit; this bubble rises up the downcomer
and the flow is reduced until the bubble condenses. This is
followed by a surge of the subcooled feedwater. This phenomenon
can be considered geysering; however, condensation plays an
important role in the process.

Condensation Chugging. An important nuclear safety application of
direct-contact condensation involves venting of air and steam into
a water pool so as to limit the pressure rise in the containment
vessel. At low steam flow rates (or at higher flow rates with low
pool temperatures), water is forced out of the downcomer (injection
pipe) and a steam bubble grows at the end of the downcomer. The
bubble then condenses and water surges up the downcomer in water-
hammer fashion. The pressure pulses from this repetitive process
induce significant loads on some boiling water reactor containments.

Sargis et al (1978, 1979) report deterministic and statistical
analyses. The calculated steam pressure and wall pressure histories
are in qualitative agreement with the irregular, repetitive behavior
observed in a small-scale experiment. Marks and Andeen (1979)
report additional data, including a map of the various condensation
regimes for their particular system.

Chugging is not yet completely understood, due in part to its
extreme sensitivity to many geometrical and fluid variables. It
does appear, however, that the phenomenon is a relaxation instability, somewhat the inverse of the vaporization chugging noted
above. In practice, the safety problems are solved by replacing
large diameter downcomers by multiple smaller pipes.

13.2.4 *Fundamental Dynamic Instabilities*

The mechanism involves the propagation of disturbances, which,
in two phase flow, is itself a very complicated phenomenon.
Roughly it can be said that disturbances are transported by two
kinds of waves: pressure (acoustic) waves and density (void)
waves. In any real system, both kinds of waves are present and
interact; but their velocities differ in general by one or two
orders of magnitude, thus allowing the distinction between these
two kinds of pure, primary, dynamic instabilities.

Acoustic Instability. Acoustic (or pressure wave) oscillations
are characterized by a high frequency, the period being of the same
order of magnitude as the time required for a pressure wave to
travel through the system. Acoustic oscillations have been observed
in subcooled boiling, bulk boiling, and film boiling. Several
investigators have reported pressure oscillations of relatively high
frequency (10 - 100 Hz) when operating under rather highly subcooled
conditions. Although these oscillations can be frequently regarded
as a curiosity, it was demonstrated by Bergles et al (1967) that

pressure drop amplitudes can be very large compared to the steady-state values, and inlet pressure fluctuations can be a significant fraction of the pressure level (Figure 13.6). The oscillations were characterized by a threshold in the negative-sloping region of the pressure-drop - flow curve and a frequency greater than 35 Hz. For constant flow conditions, as the power was increased, the frequency was high at subcooled exit conditions, reached a minimum at low quality, and increased at higher quality. The oscillations were considered to be acoustic in character and induced by the subcooled boiling. The observed frequency behavior was consistent with the propagation of pressure disturbances along the heated channel. It was observed that the amplitude of the oscillations diminished considerably in the bulk boiling region. They were persistent enough, however, to be noticed as fluctuations super-imposed upon low frequency oscillations (density wave) encouraged when the inlet throttle valve was opened. A similar superposition of high frequency acoustic oscillations on low frequency oscilla-tions were reported earlier by Fleck (1961) and Gouse and Andrysiak (1963). It has not been established whether pressure wave and density wave phenomena interact to any significant degree.

Audible frequency oscillations of 1000-10,000 Hz were detected by a microphone in water flow at supercritical pressures (3200 - 3500 psia) by Bishop et al (1965). The minimum heat flux at which the whistle-like sound occurred was 0.84 x 10^6 Btu/hr ft^2. In general, whenever a whistle was heard, the temperature of the fluid at the test section outlet was in the temperature range of 690 - 750°F, and the wall temperature was above the pseudo-critical temperature. The intensity of the whistle appeared to increase as the inlet temperature was reduced and as the flow rate was increased. Similar sounds were also detected at sub-critical pressures during subcooled boiling. A colorful survey of "boiling songs" was given by Firstenberg et al (1960). These acoustic oscillations are usually accompanied by undesirable vibrations of

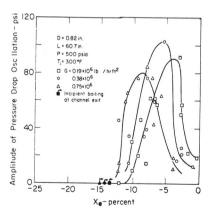

Figure 13.6: Acoustic oscillations for boiling water (Bergles et al (1967a)).

the flow channel.

Flow oscillations at acoustic frequencies are frequently
observed when cryogenic fluids are heated to film boiling conditions
in forced flow under subcritical or supercritical conditions.
Thurston et al (1967) and Edeskuty and Thurston (1967) reviewed the
available evidence and suggested that the oscillations are driven
by the thermal response of the vapor film to flow disturbances.
A pressure wave passing the heated surface compresses the vapor
boundary layer, thereby increasing the thermal conductance of the
film. This enhances the vapor generation rate. Similarly, a
rarefaction wave passes the surface, the boundary layer expands, and
the thermal conductivity momentarily decreases. The process repeats
itself. In this manner the acoustic wave is reinforced upon passing a given point in the channel. The observed frequencies
(generally greater than 10 Hz) were in approximate agreement with
frequencies expected for the first harmonic of a pipe open at both
ends, or for Helmholtz response, depending on the geometry of the
system. The data for first appearance of oscillations of any
type, acoustic or density wave, were correlated by

$$q''/GH_{fg} = 0.005v_f/v_{fg} \, .$$
(13.2)

Acoustic oscillations in the range of 5 - 30 Hz were detected
by Cornelius (1965) with natural circulation of Freon-114 at bulk
temperatures below the critical temperature but at critical
pressure. The amplitude of the pressure oscillations reached
nearly 20% of the pressure level. Experiments suggested that the
wavelength of the oscillations was equal to the length of the loop.
The excitation mechanism suggested was similar to that adopted by
Thurston et al (1967), although the near-critical conditions were
suggested as playing a role.

A somewhat related instability has been postulated as occurring
because of choked flow in the heated section (Gouse et al (1967)).
The acoustic velocity may exhibit a rather sharp transition from a
low value to a high value at the transition from bubbly or slug
flow to annular flow. A periodic transition of flow patterns may
cause periodic flow choking and unchoking. While this is a plaus-
ible flow oscillation mechanism, no observations of this behavior
appear to have been reported.

It is of interest to note that these oscillations have also
been observed during blowdown experiments. Höppner (1971), for
example, found that subcooled decompression of a pressurized
column of hot water was characterized by dampened cyclical pressure
changes resulting from multiple wave reflections.

As shown in Figure 13.15, high frequency fluctuations in the
50 - 200 Hz range have been observed with direct contact condensation (Westendorf (1966)). These irregular oscillations were
attributed to vapor bubble collapse; however, it is probable that
the small scale bubble behavior merely initiated the larger scale
acoustic oscillations which were observed.

Density wave instability. A temporary reduction of inlet flow in a heated channel increases the rate of enthalpy rise, thereby reducing the average density. This disturbance affects the pressure drop as well as the heat transfer behavior. For certain combinations of geometrical arrangement, operating conditions, and boundary conditions, the perturbations can acquire a 180° out-of-phase pressure fluctuation at the exit. The perturbations are then transmitted to the inlet flow rate and become self-sustained (Stenning and Veziroglu (1965)). For boiling systems, the oscillations are due to multiple regenerative feedbacks between the flow rate, vapor generation rate, and pressure drop; hence, the name "flow-void feedback instabilities" has been used (Neal et al (1967)). Since transportation delays are of paramount importance for the stability of the system (Bouré (1966) and Bouré and Mihaila (1967)), the alternative phrase "time-delay oscillations" has also been used.

More practically, these are low frequency oscillations in which the period is approximately one to two times the time required for a fluid particle to travel through the channel. Accordingly, "density wave oscillations" and "density effect mechanism" are also used to describe the phenomena. The term "density wave" appears to be widely used and has been adopted here.

Density wave oscillations are common in a variety of equipment, and have been extensively studied during the past twenty years. Since density wave instability is the most common type of two-phase flow instability, its parametric effects have been often observed. The major effects are described below.

Effect of channel length. The effect of channel length with a constant power density was studied in a Freon loop by Crowley et al (1967). By cutting out a section of heated length at the inlet and restoring the original flow rate, they found that the reduction of the heated length increases the flow stability in forced circulation. A similar effect was found in a natural circulation loop (Mathisen (1967)) as shown in Figure 13.7.

Effect of restrictions. An inlet restriction increases single phase flow friction (which is in phase with the change of inlet flow) and thus it provides a damping effect on the increasing flow. Hence, an inlet restriction increases flow stability. A restriction

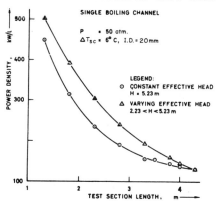

Figure 13.7: Power density at hydrodynamic instability with varying test section lengths (Mathisen (1967)).

at the exit of a boiling channel increases two-phase friction, which is out of phase with the change of inlet flow. A low inlet flow (single phase) increases the void generation and the exit pressure drop. It further slows down the flow. Thus, an exit restriction reduces the flow stability, as observed by numerous investigators (Wallis and Heasley (1961), Anderson et al (1962) and Jain (1965)).

Steam generator instabilities are currently the subject of investigation in a number of commercial and government laboratories. McDonald and Johnson (1970) have indicated that the laboratory version of the B & W Once-Through Steam Generator was subject to oscillations under certain combinations of operating conditions. The observed periods of 4 - 5 sec suggest density wave instability. It was found that flow resistance in the feedwater heating chamber stabilized the unit. This is equivalent to an inlet resistance.

Parallel channels. Collins and Gacesa (1969) have studied the effect of the by-pass ratio on density wave instability in parallel channel flow. They tested the power at the threshold of flow oscillation in a 19 rod bundle with a length of 195 in. by changing the by-pass ratio from 2 to 18. The results shown in Figure 13.5 indicate that a high by-pass ratio destabilizes the flow in parallel channels. These results are predicted by the analysis of Carver (1970).

Density wave oscillations may occur in parallel channels such that the flows oscillate in phase. More commonly, there are interactions between the channels. Aritomi et al (1977a, 1977b, 1979) present analytical and experimental results for density wave oscillations with interactions between two, three, or four parallel channels. It was concluded that the channels either oscillate in phase or are shifted by 180°. System stability was enhanced by varying geometry or heat flux of the individual channels; in this case variable phase shifts were observed. Earlier observations of what appear to be density wave oscillations in small numbers of non-interconnected parallel channels were reported by Dolgov and Sudnitsyn (1965), Crowley et al (1967) and Koshelov et al (1970).

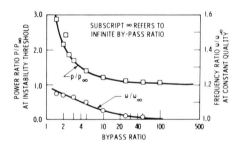

Figure 13.8: Effect of by-pass ratio on threshold of stability and frequency for a typical set of flow conditions (Carver (1969)).

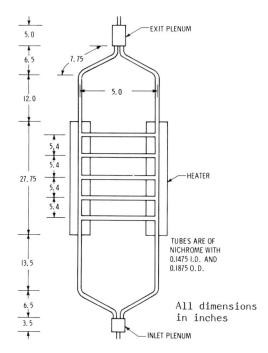

Figure 13.9: Cross-connected parallel channels (Veziroglu and Lee (1971)).

Veziroglu and Lee (1971) have studied density wave instabilities
in a cross-connected, parallel-channel system. It was found that
this system was more stable than either a single channel or
parallel channels without cross-connections (Figure 13.9). This
work confirms the common speculation that rod bundles are more
stable than simple parallel channel test sections.

The final series of tests with the FRIGG loop, which provided
an electrically heated fuel element simulation for the Marviken
boiling heavy water reactor, was reported by Nylund et al (1970).
In general, burnout or dryout preceeded the threshold of instability, presumed to be density wave oscillations. The parametric
trends of the extrapolated stability boundaries are in agreement
with the observations noted in this section. The effects of
radial and axial heat flux variation on stability boundaries were
found to be rather inconclusive.

Collins et al (1971) have described unstable behavior in a
full-scale simulated nuclear reactor core section. As bundle
power was increased to the point where low quality steam was
generated, a variety of oscillations in flow rate were observed.
While it appears that both density wave and pressure drop oscillations were obtained, the major fluctuations were definitely of the
density wave variety. Extensive heater thermocouples indicated
dryout occurring on all heater surfaces during the low flow
portion of the cycle. Detailed measurements of the threshold of
periodic dryout were recorded as a function of system geometry and
flow conditions; however, no attempt was made to analyze or report

data for the oscillation threshold.

   Effect of pressure. The increase of system pressure at a
given power input reduces the void fraction and thus the two-phase
flow friction and momentum pressure drops. These effects are
similar to those of a decrease of power input or an increase of
flow rate, and thus stabilize the system. The increase of pressure
decreases the amplitude of the void response to disturbances.
However, it does not affect the frequency of oscillation significantly.

   The effect of pressure on flow stability during natural
circulation, given by Mathisen (1967) is shown in Figures 13.10
and 13.11. It should be noted in Figure 13.11 that at pressures
higher than 870 psia the boiling crisis occurs simultaneously with
flow oscillation. It should also be noticed in these figures that
increase of power input reduces the flow stability. A possible
remedy is an increase of system pressure to stabilize the flow at
a high power input.

   Effect of inlet subcooling. An increase in inlet subcooling
decreases the void fraction and increases the non-boiling length
and, hence, the transit time. Thus, the increase of inlet sub-

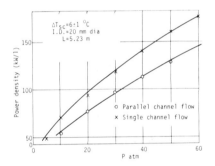

Figure 13.10: Effect of pressure on the power density at hydrodynamic instability
          (Mathisen (1967)).

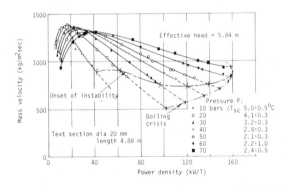

Figure 13.11: Effect of pressure on the flow in a single boiling channel
          (Mathisen (1967)).

cooling stabilizes the two-phase boiling flow at medium or high
subcoolings. At small subcoolings, an incremental change of
transit time is significant in the response delay of void generation from the inlet flow and the increase of inlet subcooling
destabilizes the flow. These stabilizing and destabilizing effects
are competing. Thus, the effect of inlet subcooling on flow
instability exhibits a minimum, as shown in Figure 13.12 for a
natural circulation system. Crowley et al (1967) observed similar
effects of inlet subcooling in a forced circulation loop. Moreover,
they noticed that starting a system with high subcooling can lead
to a large amplitude oscillation.

Effects of mass velocity and power. The mass velocity effect
on the threshold power of flow instability in a single channel, as
reported by Mathisen (1967), is shown in Figure 13.11.

Collins and Gacesa (1969) tested the effects of mass velocity
and power on flow oscillation frequency in a 19 rod bundle with
steam-water flow at 800 psia. They found that the oscillation
frequency increases with power input to the channel, but that the

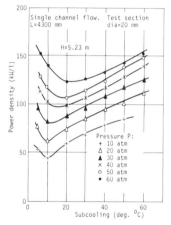

Figure 13.12: Effect of inlet subcooling and pressure on flow instability
(Mathisen (1967)).

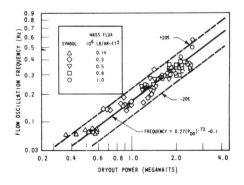

Figure 13.13: Variation of flow oscillation frequency with dryout power (Collins
and Gacesa (1969).

mass velocity has a relatively small effect on the frequency
(Figure 13.13). Dolezal (1975) discusses the minimum rate of
circulation which is required to insure dynamic stability in the
evaporator of forced-circulation boilers.

Effect of cosine heat flux. The effect of cosine heat flux
distribution was tested by Dijkman (1971). He found that the
cosine heat flux distribution stabilizes the flow. The reason
may be due to the decrease in local pressure gradient at the exit
where the heat flux is lower than average.

On the other hand, the data of Biancone et al (1965) suggest
that the cosine distribution is destablizing. Yadigaroglu and
Bergles (1972) also reported a destabilizingeffect of a nonuniform
heat flux distribution. They suggest reasons for the apparently
contradictory experimental observations.

Effect of Tube Wall. It should be noted that most of the
flow instability tests were conducted with electric heaters of very
small thermal capacitances so that wall effects are negligible in
evaluating the density wave oscillations. However, in cases where
the time constant of channel wall is comparable with the period of
oscillation, thermal capacitance effects should be taken into
consideration, as suggested by Neal et al (1967) and Yadigaroglu
and Bergles (1972).

Higher-order oscillations. Yadigaroglu and Bergles (1972)
studied the instability of a Freon-113 flow. Their stability map,
which defines "higher-order" oscillations, is shown in Figure 13.14.
Higher-order is the integral multiplier of the frequency of
oscillation. They observed a high frequency oscillation occurring
prior to a low-frequency oscillation. The horizontal lines represent the experimental conditions as the mass velocity is reduced
at constant power. The circle o is the threshold of flow instabilit
and the open triangle Δ is the transition point from one order to
another of oscillation. The presence of the higher modes is
explained in terms of the dynamic behavior of the boiling boundary.

Steam generator applications. There has been considerable
concern about the stability of once-through steam generators for
LMFBR, and numerous studies of density wave oscillations in this
equipment have been reported. The uniqueness of steam generators
derives from the fact that the heat input is a dependent variable.
Interpretation of experimental observations and analytical formulations must consider coupling between primary and secondary sides of
the heat exchanger.

Wazink and Efferding (1974) reported on the hydrodynamic
stability of a 1-MWt sodium-heated steam generator. The nine 130-
ft tubes were arranged in a serpentine pattern. Traces of the
inception of density wave instabilities were included. Studies of
an alternate design LMFBR steam generator by Hunsbedt and Roberts
(1974) indicated instabilities which appear to be of the density
wave variety.

Cho et al (1976) discussed the stability of the steam generator for a molten salt breeder reactor. The DYNAM code indicated

highly stable flow of the supercritical water/steam tubeside flow. Due to the behavior of the density ratio above the critical point, the stability is enhanced by increasing exit orificing and decreasing pressure.

Cho and co-workers (1978) also examined the thermal-hydraulic behavior of the conceptional design of dry-saturated steam evaporators for a large LMFBR plant. Both static stability and dynamic (density wave) stability were predicted.

Cryogenic system applications. Cryogenic systems frequently exhibit density wave instabilities. In connection with development of superconducting transmission lines, there is concern that density wave instability could result in temperature oscillations large enough to increase the conductor temperature to the transition range, thereby greatly increasing the power requirement. Jones and Peterson (1975) reported a numerical analysis of supercritical helium flow stability. Subsequently, Daney et al (1979) determined the density wave stability boundary for supercritical helium in a very long tube having a choked exit condition.

Maps of density wave instabilities. Analytical considerations suggest that the main effects of fluid properties, gravity, pressure, inlet velocity, inlet enthalpy, and heat flux are taken into account by three dimensionless groups (Fabrega (1964), Bouré (1966), Bouré and Mihaila (1967), and Ishii and Zuber (1970)):

Dimensionless velocity representing flow and heat input effects (subcooled void effects may be included (Saha et al (1976)).

Dimensionless enthalpy representing density and subcooling effects.

Dimensionless gravity representing buoyancy and velocity effects.

Loss coefficients for restrictions, fittings, etc. are also required. The groups vary somewhat among various investigators, and occasionally other dimensionless groups are introduced to account for less important effects. The general concept of using dimensionless groups has been found useful in presenting parametric trends. In effect, presentations such as Figures 13.5 and 13.14 are made more general.

As an example of mapping stability boundaries in nondimensional coordinates, consider Figure 13.15. The coordinates represent dimensionless enthalpy and dimensionless velocity, and the effect of increasing inlet restriction is represented by increasing the inlet orifice coefficient. The data indicate rather clearly that a more stable system results from increased inlet throttling. This is due to having a larger fraction of the test section pressure drop in phase with the inlet velocity.

Using several of the above-mentioned groups, Ünal (1980) recently developed correlations based on experimental observations of the onset of density wave instabilities in both forced circulation (106 points) and natural circulation (110 points) steam generators. Comparison of the correlations with other data will be required to establish the general validity of the equations.

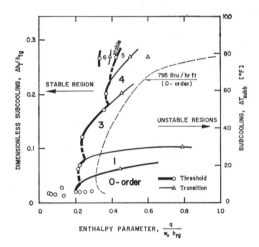

Figure 13.14: Experimental stability map for a Freon-113 system at 265 Btu/hrft
(uniform heat flux distribution). Smoothed threshold line for
zero-order oscillations also shown, as a dotted line (Yadigaroglu
and Bergles (1972)).

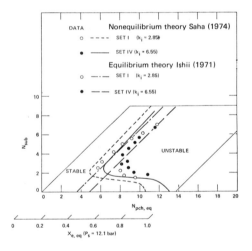

Figure 13.15: Flow stability map, with experimental data and analytical predictions, showing effect of inlet restrictions (Saha et al (1976)).

Condensing instability. Experimental observations of density
wave instabilities have been mostly made on boiling systems; however
they can also occur during condensation. The general behavior of
condensation instability seems to indicate density wave oscillations
It is appropriate to mention the low-frequency pressure and interfac
oscillations observed in connection with direct contact condensation
by Westendorf and Brown (1966). The oscillations are characterized
by the annular condensing length alternately growing to some maximur
then diminishing or collapsing (Figure 13.16).

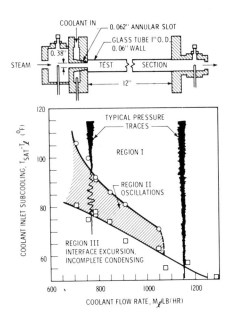

Figure 13.16: Instability regions for direct contact condensation (Westendorf and Brown (1966)).

The instability of a downflow steam condensation inside a 0.292 in. diameter, 8 ft long copper tube was reported by Goodykoontz and Dorsch (1967). The pressure oscillations occurred only when the condensing lengths were between 1.7-3.7 ft. No unstable conditions were encountered at very short or very long condensing lengths.

13.2.5 *Compound Dynamic Instability*.

Thermal oscillations. As identified by Stenning and Veziroglu (1965) these oscillations appear to be associated with the thermal response of the heated wall after dryout. It was suggested that the flow could oscillate between film boiling and transition boiling at a given point, thus producing large amplitude temperature oscillations in a channel wall subject to constant heat flux. An interaction with density wave oscillations is apparently required, with the higher frequency density wave oscillations acting as a disturbance to destabilize the film boiling. With their Freon-11 system, using a nichrome heater (0.08 in. wall), the thermal oscillations had a period of approximately 80 sec.

Thermal oscillations are considered a regular feature of dryout of steam-water mixtures at high pressures. Gaudiosi (1965) and Quinn (1966) recorded wall temperature oscillations of several hundred degrees downstream of the dryout point, with periods ranging from 2 to 20 sec. This was attributed to movement of the dryout point due to instabilities or by small variations in pressure imposed by the loop control system.

An interesting paper presented by Cho et al (1971) describes puzzling changes in performance of a once-through steam generator

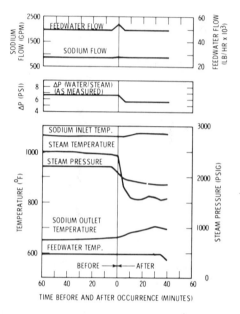

Figure 13.17: Record of flow conditions illustrating a sudden performance
            decrease (flip) in sodium heated steam generator (Cho et al
            (1971)).

heated by liquid sodium (LMBFR simulation). A record of parameters
during a sudden performance decrease ("flip") is given in Figure
13.17; equally sudden performance increases were also observed.
From the authors' description it appears that the performance
changes are due to shifts in the dryout location, generally in the
manner of a thermal oscillation.

     Steam generators represent a truly conjugate problem since the
tube wall characteristics and secondary fluid behavior are
important. There has been concern that for the Clinch River
Breeder Reactor Plant (CRBRP), dryout oscillations could lead to
appreciable fatigue and corrosion damage within the design plant
life of 30 years (Leonard (1976)). Recent work in this area
includes the following studies.

     Efferding et al (1976) reported water-side corrosion tests in
a model of the alternate concept evaporator for the CRBRP.
Particular attention was given to water-side corrosion. A "drift"
behavior was observed where operating conditions shifted from a
high quality dryout to a low quality departure from nucleate
boiling.

     Chu et al (1976) took typical operating conditions and
transient movement of the dryout point and calculated the resulting thermal stresses in clean steam generator tubes. As expected,
the alternating stress is strongly dependent on the dryout
characteristics.

     A major experimental investigation of sodium heated steam

generators has been under way at Argonne National Laboratory. In
one of the most recent reports on this Project, France et al.
(1979) discuss thermal fluctuations in the tube wall under transition boiling conditions. Conceptually, the dryout point is
assumed to be steady and the fluctuating heat transfer coefficient
is due strictly to the alternate nucleate and film boiling
characteristic of transition boiling. Frequencies in the range of
0.3-1.0 Hz were observed, with amplitudes increasing as the sodium
temperature increased and water pressure decreased.

Boiling water reactor instability. In a boiling water nuclear
reactor, the thermo-hydrodynamic instability is complicated due to
the feedback through a void reactivity-power link. The feedback
effect can be dominant when the time constant of a hydraulic
oscillation is close to the same magnitude of the time constant of
the fuel element. Strong nuclear-coupled thermohydrodynamic
instabilities occurred in the early SPERT reactor cores where a
metallic fuel (small time constant) was operated in a low-pressure
boiling water flow. Since modern BWRs are operated at 1000 psia
with uranium oxide fuel having a high time constant of 10 sec, the
problem of flow instability is alleviated (Lahey and Moody (1977)).

Parallel channel instability. Parallel channel interactions
were reported by Gouse and Andrysiak (1963) for a two-channel
Freon-113 system. The flow oscillations, as observed through the
electrically heated glass tubing, were generally 180° out of phase.
Well within the stable region, for density wave oscillations, the
test sections began to oscillate in phase with very large amplitude
flow oscillations. The observed frequencies were in the vicinity
of the manometer natural periods for the system of two or three
tubes. While these oscillations are frequently labeled as mano-
meter-type oscillations, Yadigaroglu (1979) has suggested that they
are thermo-hydrodynamic in nature. The probable cause is the feedbacks between the flooding rate, the vapor generation rate, and the
back pressure created by the mixture leaving from the top of the
heated section. While these characteristics resemble those of
density wave oscillations, a different and complicating factor is
the propagation of the quench front which strongly affects the
steam generation dynamics.

 • During a loss of coolant accident of a pressurized water
reactor, the reflooding rate of the core is controlled by the
balance between the driving head of the injected emergency coolant
and the back pressure in the upper plenum. Simulation tests have
indicated that oscillations of the flow rate can occur which are
especially violent during the initial phases of the transient
(Waring et al (1974) and White and Duffy (1974)).

Condensation oscillations in pressure suppression pools.
Under conditions of high steam flow rates and low pool subcooling
condensation oscillations are observed at the outlet of the downcomer (injection pipe) in pressure suppression pools. The interface remains at the outlet of the downcomer and oscillates, thereby producing pressure oscillations in the pool. According to
Okazaki (1979), the cycle can be described in terms of a) an
increase in steam pressure due to a low condensation rate, b)
subsequent increases in "bubble" volume, c) movement of cooler

water to the interface, d) rapid condensation, with decrease in steam pressure and bubble volume, and e) heating up of the stagnant water around the interface which results in a low condensation rate. The well-timed movement of the pool water can perhaps be explained by the density gradient instability postulated by Marks and Andeen (1979). In any event, this appears to be a compound dynamic instability involving the steam-water interface and pool convection.

13.2.6 *Compound dynamic instability as secondary phenomenon.*

For this class of instabilities, the secondary phenomenon, a dynamic instability is triggered by a static instability phenomenon. The only example of this instability is pressure-drop oscillations which occur in systems having a compressible volume upstream of, or within, the heated section.

Pressure-drop oscillations have been studied in considerable detail by Maulbetsch and Griffith (1965, 1967) for subcooled boiling of water, and by Stenning et al (1965, 1967) for bulk boiling of Freon-11. Maulbetsch and Griffith found that the instability was associated with operation on the negative sloping portion of the pressure drop - flow curve. Due to the high heat fluxes, CHF always occurred during the excursion which initiated the first cycle of oscillation. Daleas and Bergles (1965) demonstrated that the amount of upstream compressibility required for unstable behavior is surprisingly small, and that relatively small volumes of cold water can produce large reduction in the CHF for small diameter, thin walled tubes. The reduction in critical flux is less as subcooling, velocity, and tube size are increased. With tubes having high thermal capacity, oscillations can be sustained without boiling crisis (Aladiev et al (1961)); however, the amplitudes are usually large enough so that preventive measures are required. It was found that when the compressible volume was "large" (size at which there is no further reduction in CHF), the CHF always occurred at the minimum in the pressure-drop - flow rate curve. This suggested that it was necessary to provide a throttle valve between test section and compressible volume so as to effectively stabilize the test section characteristics. Progressive throttling increased the CHF until an asymptotic stable value was reached. The cure is the same as that required for the Ledinegg instability; in this case also, very large throttle valve pressure drops may be required to realize the stable CHF.

Stenning et al (1967) encountered sustained oscillations when operating their Freon-11 system with an upstream gas-loaded surge tank. Large amplitude cycles commenced when the flow rate was reduced below the value corresponding to the minimum in the test section characteristic. A typical pressure-drop oscillation cycle with the indicated 40 sec period is shown superimposed on the steady-state pressure-drop - flow-rate curve in Figure 13.18. As expected, the oscillations could be eliminated by throttling between the surge tank and heated section. Figure 13.18 also illustrates that density wave oscillations occur at flow rates below those at which pressure-drop oscillations were encountered. The frequencies of the oscillations were quite different and it was easy to distinguish the mechanism by casual observation. However, the frequency of the pressure-drop oscillations is

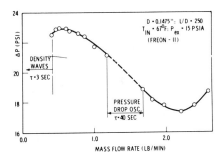

Figure 13.18: Density wave and pressure drop oscillation (Stenning et al (1967)).

determined to a large extent by the compressibility of the surge tank. With a relatively "stiff" system it might be possible to raise the frequency to the point where it is comparable to the density wave frequency, thus making it harder to distinguish the governing mechanism.

Pressure-drop oscillations are associated with operation on the negative-sloping portion of the Δp - w curve. As noted previously, heated downcomers in steam generators have competing gravitational and frictional pressure drop terms which frequently result in multi-valued Δp - w curves. Maulbetsch (1967) has suggested that the instabilities observed in the boiler described by Krasykova and Glusher (1965) were of the pressure drop variety. This three-pass, once-through unit at the Russian Central Boiler-Turbine Institute exhibited large amplitude pulsations when evaporation occurred at the end of the first (upflow) pass. The negative-sloping portion of the curve would then be expected due to the decreased pressure rise in the downflow pass. Instabilities were also observed under supercritical conditions when the transition temperature was reached at the end of the first pass; the associated large changes in specific volume cause the usual two-phase flow behavior.

Collins et al (1971) reported flow fluctuation traces which strongly suggest that pressure drop oscillations with a period of approximately 50 sec are superimposed on density wave oscillations having a frequency of about 3 Hz (Figure 13.19b). As bundle power was increased, only large amplitude density wave oscillations remained (Figure 13.19c).

Veziroglu and Lee (1971) obtained a rather extensive record of pressure drop oscillations for their parallel channel system. Boundaries are mapped in Figure 13.20. It appeared that the density wave oscillations occurred during the low flow rate portion of the pressure drop oscillation cycle. A recent summary of this work is given by Lee et al (1977).

Maulbetsch and Griffith (1965, 1967) have suggested that very

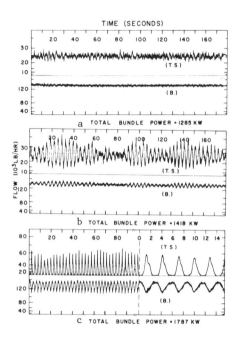

Figure 13.19: Flow fluctuations in a full-scale reactor channel as a function
of bundle power (Collins et al (1971)).

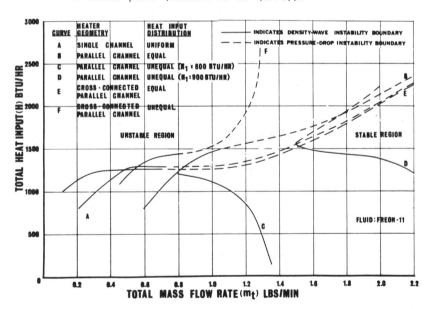

Figure 13.20: Stability boundaries for density wave and pressure drop
oscillations (Veziroglu and Lee (1971)).

long test sections (L/D > 150) may have sufficient internal
compressibility to initiate pressure-drop oscillations. In this
case, no amount of inlet throttling will improve the situation.

13.3 Analysis

13.3.1 *Analysis of Static Instability*

Most static instabilities are induced by primary phenomena such
as the Ledinegg pressure-drop condition, boiling crisis, maximum
superheat for vapor burst, and flow pattern transitions. All of
these primary phenomena can be predicted by using steady state
criteria or correlations. Therefore, the threshold of static
instability can be predicted by using steady-state evaluations.

Analysis of flow excursion. The threshold of flow excursion
can be predicted by evaluating the Ledinegg instability criterion in
a flow system or a loop according to equation (13.1).

Analysis of boiling crisis instability. The threshold of a
boiling crisis instability can be predicted by the occurrence of a boiling crisis. Such predictions are well documented by Tong (1965) and
Collier (1972).

Analysis of flow-pattern transition instability. The boundary
of flow pattern transitions is not sharply defined. It is usually
an operational band, and it changes with channel geometry and
boundary conditions as well as fluid properties. At present, there
is no definite analytical method for predicting the stability of
flow patterns. Further experimental and analytical studies are
required in this field.

Analysis of bumping, geysering, or chugging. The primary phenomenon of this flow instability is rapid vapor formation at the maximum
degree of superheat in the near-wall fluid. The value of maximum
superheat varies with the heated wall surface conditions as well
as the fluid conditions. No reliable methods exist to predict
active nucleation site distributions for engineering surfaces;
hence, these nucleation instabilities cannot be predicted

Considerable effort has been devoted to prediction of
incipient boiling and subsequent voiding for LMFBR hypothetical
loss-of-flow accidents. A number of codes, based on a pure slug
flow model, have been developed to calculate the thermal-hydraulic
instability in a fast reactor core: BLOW (Schlechtendahl (1970)),
NEMI (Pezilli et al (1970)), and SLUG (Cronenberg et al (1971)).
Fauske and Grolmes (1971) present a comprehensive review of the
published models.

13.3.2 *Analysis of Dynamic Instability*

The mathematical model describes the dynamic system, which
consists of the flow and its boundary conditions. The description
of the flow is based on the one hand on the conservation equations
and on the other hand on the so-called constitutive laws. The
constitutive laws define the properties of the system with a
certain degree of idealization, simplification, or empiricism,
e.g., equation of state, steam tables, friction correlation, and

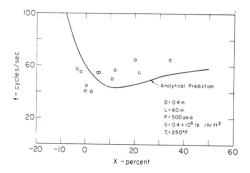

Figure 13.21: Stability boundaries for density wave and pressure drop oscillations
        (Bergles et al (1967)).

heat-transfer correlation. The conservation equations (mass,
momentum, energy) are three, four, five, or six in number, depending on the degree of sophistication of the model.

Analysis of acoustic instability. An analysis of acoustic
oscillations was reported by Bergles et al (1967).   The high
frequencies observed led naturally to the consideration of pressure
wave propagation.   The test section was idealized as an inlet
restriction, a line containing a homogeneous fluid, and a second
restriction representing the test section pressure drop lumped at
the exit.   The pressure drop across the test section was presumed
constant.   The basic equations were solved by standard perturbation
techniques and the method of characteristics.   The condition for
marginal dynamic stability gives the threshold of acoustic oscillations.   This threshold criterion is based on the slope of the
heated section pressure drop characteristic and a frequency, which
is simply the transit time of an acoustic wave in the tube.   A
procedure was developed to obtain a reasonable void profile along
the channel, which, in turn, could be used to calculate the average
speed of sound.   As shown in Figure 13.21 the analytical predictions
are in reasonable agreement with the observed frequency level and
the variation of frequency with exit quality.

Analytical work relating to acoustic instability has been
performed primarily in conjunction with analysis of pressure wave
propagation during hypothetical loss-of-coolant accidents in
pressurized water reactors.   A comprehensive review of existing
theory is presented by Höppner (1971).

Analysis of density wave instability. The analysis of
density wave instability can be divided into two groups: (a)
computer codes, and (b) simplified models.

(a)   *Computer codes.*   Since the computer codes can handle the
      complicated mathematics, most of the compounded and feedback
      effects are built into computer codes for analyzing the dynamic
      instabilities.   Among the well-known codes are STABLE (Jones
      and Dight (1964)), FABLE (Jones and Yarbrough (1964)), DYNAM
      (Efferding (1968)), HYDNA (Currin et al (1961)), RAMONA
      (Solberg and Bakstad (1967)), SABRE-3 (Nahavandi and von
      Hollen (1964)), FLASH (Margolis and Redfield (1966)), LOOP

(Davies and Potter (1967)), NUFREQ (Lahey and Yadigaroglu (1974)), and the recent steam generator codes DEW, CRESCENDO, and SG-EIGEN (Yamakawa and Tsuchiya (1979)).

Analytical efforts have been made to classify density wave instabilities according to the major factors influencing the threshold and frequency, e.g., acceleration in heated section, friction in unheated riser, etc. Recent studies in this area were reported by Fukuda and Kobori (1979) and Hands (1979). The results should be regarded with caution, however, since large numbers of factors interacting in complex fashion makes it difficult to generalize as to which factors are stabilizing or destablizing.

(b) *Simplified models.* Many flow stability analyses have been made by using simplified models instead of comprehensive computer codes. Some of these models are described by Bouré (1966), Bouré and Mihaila (1967), Ishii (1971), Ishii and Zuber (1970), Quandt (1961), Saha (1974), Wallis and Heasley (1961), Yadigaroglu and Bergles (1969). They have been used primarily to gain some physical understanding of the phenomenon. Some models give results which are in good agreement with experimental data. As an example, consider the comparison of experiment and theory given in Figure 13.15. Both the non-equilibrium and approximate equilibrium theories are in rather good agreement with the experimental data. The latter equation is presented below as it provides a rapid estimate of the density wave stability boundary.

$$N_{pch,eq} - N_{sub} = \frac{2\left[k_i + \dfrac{f_s C_m L}{2D} + k_e\right]}{1 + \dfrac{1}{2}\left[\dfrac{f_s C_m L}{2D} + 2k_e\right]} \tag{13.3}$$

Analysis of thermal oscillations. Wedekind and coworkers have carried out a series of investigations of dryout point oscillations in horizontal tube refrigerant evaporators. Wedekind and Beck (1974) theoretically demonstrate the influence of the heat flux and inlet quality on the statistical characteristics of the mixture-vapor transition point oscillations.

Analysis of pressure drop oscillations. Maulbetsch and Griffith (1965, 1967) performed a stability analysis of an idealized model of the system in which pressure-drop instabilities were observed. The test section and upstream compressible volume were treated as lumped parameters. The governing equations for flows in the heated section and into the compressible volume were treated by a perturbation technique and standard stability criteria. At marginal stability, a critical slope of the heated section pressure drop - flow curve was obtained.

A comprehensive discussion of the analytical aspects of instabilities was recently presented by Yadigaroglu (1978).

13.4 Summary

This chapter has introduced the various types of instabilities encountered in two-phase systems. The discussion is based on a classification which takes into account the mechanisms and essential characteristics of the phenomena.

The world literature provides ample evidence of observations of these instabilities in both laboratory and industrial equipment. The experimental evidence is useful in diagnosing the type of instability, estimating the threshold of oscillations of excursions, and providing a cure for the non-steady behavior. On the other hand, simple analytical solutions and complex computer codes are now available for predicting flow instabilities.

Through analyses and experiments, the understanding of two-phase instabilities has progressed to the point where the designer can avoid or minimize this undesirable behavior in most cases.

Nomenclature

C_m = two-phase friction factor

D = diameter of tube

f = frequency of oscillation

f_s = single-phase friction factor

G = mass velocity

H = head for natural circulation boiler

H_{fg} = enthalpy of vaporization

ΔH = subcooled enthalpy

k_e = exit orifice coefficient

k_i = inlet orifice coefficient

L = heated length of channel

M_ℓ = mass flow rate of liquid

m_t = total mass flow rate

$N_{pch,eq} = \dfrac{q''4\ (\rho_\ell\ \rho_v)\ L}{D\ H_{fg}\ \rho_\ell\rho_v\ v_i}$

$N_{sub} = \dfrac{(\rho_\ell - \rho_v)\ \Delta H_i}{H_{fg}\ \rho_v}$

P_{DO} = bundle power at dryout condition

p,P = pressure

$\Delta p,\Delta P$ = pressure drop

q = heat transfer rate

$q/A,q''$ = heat flux

T = temperature

ΔT = inlet subcooling

v = specific volume, velocity

w = flow rate

X	= vapor quality
ρ	= mass density
τ	= period of oscillation
ω	= frequency of oscillation

Subscripts

crit	= critical heat flux condition
e,ex	= exit conditions
eq	= equilibrium conditions
ext	= external to heated channel
in	= inlet conditions
f	= saturated liquid
fg	= transfer from liquid to vapor
i	= inlet
int	= internal to heated channel
ℓ	= liquid phase
v	= vapor phase
o	= initial, steady state, or upstream conditions
∞	= conditions with infinite bypass ratio
w	= properties at wall surface
s	= properties at saturated conditions
sat	= properties at saturated conditions
sc	= subcooling
ss	= steady state

References

Aladiev, I T, Miropolsky, Z L, Doroschuk, V E and Styrikovich, M A, (1961) "Boiling crisis in tubes". *Intern. Developm. in Heat Transfer, ASME, New York.*

American Electric Power Service Corporation, (1961) Symposium on the Commercial Operation of the Breed Plant, *Proc. of the Amer. Power Conf.*

Anderson, R P, Bryant, L T, Carter J C and Marchaterre, J F, (1962) "Transient analysis of two-phase natural circulation systems, *USAEC Report ANL-6653.*

Aritomi, M, Aoki, S, and Inoue, A, (1977a) "Instabilities in parallel channel of forced-convection boiling upflow systems, (I)." Mathematical model, *Journal of Nuclear Science and Technology,* **14**, 22-30.

Aritomi, M, Aoki, S, and Inoue, A, (1979b) "Instabilities in parallel channel of forced-convection boiling upflow systems, (II). Experimental resources, *Journal of Nuclear Science and Technology,* **14**, 88-96.

Aritomi, M, Aoki, S, and Inoue, A, (1979) "System with different flow conditions between two channels, (III)." Journal of Nuclear Science and Technology, **16**, 343-355.

Aritomi, M, Aoki, S, Inoue, A, and Narabayashi, T, (1978) "Instabilities in parallel boiling channels; Part 6. Effects of boiling in downcomer on flow instability." Paper B206, Proceedings of the 15th Annual Symposium, Heat Transfer Society of Japan, 175-177.

Bergles, A E, (1973) "Thermal-hydraulic instability: Most recent assessments." Presented at the Seminar on Two Phase Thermohydraulics, Rome, Italy, 1972. Termoidraulica dei Fluidi Bifase, Comitato Nazionale per L'Energia Nucleare, 203-232.

Bergles, A E, (1977) "Instabilities in two-phase systems." in Two-Phase Flows and Heat Transfer, **I**, Hemisphere Publishing Corp., Washington, 383-422.

Bergles, A E, Goldberg, P, and Maulbetsch, J S, (1967a) "Acoustic oscillations in a high pressure single channel boiling system." EURATOM Report, Proc. Symp. on Two-Phase Flow Dynamics, Eindhoven, EUR 4288e, 535-550.

Bergles, A E, Lopina, R F, and Fiori, M P, (1967b) "Critical-heat flux and flow pattern observations for low pressure water flowing in tubes." J. Heat Transfer, **89**, 69-74.

Biancone, F, et al., (1965) "Forced convection burnout and hydrodynamic instability experiments for water at high pressures, Part I, Presentation of data for round tubes with uniform and non-uniform heat flux." EUR-2490e.

Bishop, A A, Sandberg, R O, and Tong, L S, (1965) "Forced convection heat transfer to water at near-critical temperature and super-critical pressures." USAEC Report WCAP-2056-Part 111B, 1964, and also Paper 2-7, AIChE - I. Chem. E. Joint Meeting, London.

Bonilla, C F, (1957) Nuclear Engineering, Chapter 8, McGraw-Hill Book Co., New York.

Bouré, J A, (1966) "The oscillatory behavior of heated channels, Part I and II." CEA-R 3049.

Bouré, J A, Bergles, A E, and Tong, L S, (1971), "Review of two-phase flow instability." ASME Paper No. 71-HT-42, presented at 12th National Heat Transfer Conference, Tulsa, Oklahoma.

Bouré, J A, Bergles, A E, and Tong, L S, (1973) "Review of two-phase flow instability." Nuclear Engineering and Design, **25**, 165-192.

Bouré, J A and Mihaila, A, (1967) "The oscillatory behavior of heated channels." Proc. Symp. on Two-Phase Flow Dynamics, Eindhoven, EUR 4288e, 695-720.

Cali, G P, Grillo, P, and Testa, G, (1970) "Fuel rod and bundle vibrations induced by coolant." Trans. ANS, **13**(1), 333-335.

Carver, M B, (1969) "Effect of by-pass characteristics on parallel channel flow instabilities." Proc. Instn. Mech. Engrs., **184**, Pt. 3C.

Chan, K C and Yadigaroglu, (1979) "Two-phase flow stability of steam generators." Proceedings of Japan-US Seminar on Two-Phase Flow Dynamics, Kobe, Japan, 43-54.

Chexal, V K, and Bergles A E, (1973) "Two-phase instabilities in a low pressure natural circulation loop." *AIChE Symposium Series,* **69** (131), 37-45.

Cho, S M, Ange, L J, Fenton, R E, and Gardner, K A, (1971) "Performance changes of a sodium-heated steam generator." *ASME* Paper 71-HT-15.

Cho, S M, Chou, H L, and Cox, J F, (1976) "Thermal-hydraulic design of a super-critical, once-through steam generator for the molten-salt breeder reactor plant." *ASME* Paper No. 76-JPGC-NE-9.

Cho, S M, Kao, T T, and Hassett, S E, (1978) "Thermal/hydraulic design of dry-saturated steam evaporators for a large fast breeder reactor plant." *ASME* Paper No. 78-JPGC-NE-7.

Chu, C L, Wolf, S, and Dalcher, A W, (1976) "Oscillatory dryout related thermal stresses in clean steam generator tubes." *ASME* Paper No. 76-JPGC-NE-2.

Collier, J G, (1972) *Convective Boiling and Condensation,* McGraw-Hill, Maidenhead.

Collins, D B and Gacesa, M, (1969) "Hydrodynamic instability in a full-scale simulated reactor channel." *Proc. Instn. Mech. Engrs.,* **184**, Pt. 3C, 115-126.

Collins, D B, Gacesa, M, and Parsons, C B, (1971) "Study of the onset of premature heat transfer crisis during hydrodynamic instability in a full-scale reactor channel." *ASME* Paper 71-HT-11.

Cornelius, A J, (1965) "An investigation of instabilities encountered during heat transfer to a supercritical fluid." *ANL-7032.*

Cronenberg, A W, Fauske, H K, Bankoff, S G, and Eggen, D T, (1971) "A single-bubble model for sodium expulsion from a heated channel." *Nuclear Eng. and Design,* **16**, 285-293.

Crowley, J D, Deane, C, and Gouse Jr, S W, (1967) "Two phase flow oscillations in vertical, parallel heated channels." *EURATOM* Report, *Proc. Symp. on Two-Phase Flow Dynamics,* Eindhoven, EUR 4288e, 1131-1172.

Currin, H B, Hunin, C M, Rivlin, L and Tong, L S, (1961) "HYDNA-digital computer program for hydrodynamic transients in a pressure tube reactor or a closed channel core." *USAEC* Report CVNA-77.

Daleas, R S and Bergles, A E, (1965) "Effect of upstream compressibility on subcooled critical heat flux." *ASME* Paper No. 65-HT-67.

Daney, D E, Ludtke, P R, and Jones, M C, (1979) "An experimental study of thermally-induced flow oscillations in supercritical helium." *J. Heat Transfer,* **101**, 9-14.

Davies, A L and Potter, R, (1967) "Hydraulic stability: An analysis of the causes of unstable flow in parallel channels". *Proceedings of the Symposium on Two-Phase Flow Dynamics,* Eindhoven, EUR 4288e, 1225-1266.

Deane IV, C W and Rohsenow, W M, (1969) "Mechanism and behavior of nucleate boiling heat transfer to the alkalai liquid metals." *MIT Engineering Projects Lab.* Report 76303-65.

Dean, R A, Dougall, R S, and Tong, L S, (1971) "Effect of vapor injection on critical heat flux in a subcooled R-113 (Freon) flow." Paper presented at *Intern. Symp. on Two-Phase Flow Systems,* Haifa, Israel.

Dijkman, F J M, (1971) "Some hydrodynamic aspects of a boiling water channel." Thesis, Technische Hogeschool te Eindhoven, 1969. *Nucl. Eng. and Design,* **16**, 237-248.

Dolezal, R, (1975) "Die umlaufzahl als kriterium der dynamischen stabilität bei verdampfern von zwangsumlaufkesseln." *Brennst. - Wärme-Kraft.* **27**, 287-290.

Dolgov, V V, and Sudnitsyn, O A, (1965) "On hydrodynamic instability in boiling water reactors." *Thermal Engineering,* **12** (3), 65-70.

Edeskuty, F J and Thurston, R S, (1967) "Similarity of flow oscillations induced by heat transfer in cryogenic systems." *Proc. Symp. on Two-Phase Flow Dynamics,* Eindhoven, EUR 4288e, 551-567.

Efferding, L E, (1968) "DYNAM-A digital computer program for the dynamic stability of once-through boiling flow and steam superheat." *USAEC* Report GAMD-8656.

Efferding, L E, Hwang, J Y, and Waszink, R P, (1976) "Interim evaluation of exposure conditions for the waterside corrosion test of a sodium heated steam generator evaporator model employing a duplex tube." *ASME* Paper No. 76-JPGC-NE-5.

Eselgroth, P W and Griffith, P, (1968) "The prediction of multiple heated channel flow patterns from single channel pressure drop data." *MIT Engineering Projects Lab.* Report 70318-57.

Fabrega, S, (1964) "Etude expérimentale des instabilités hydrodynamiques survenant dans les réacteurs nucléaires à ébulitition." *CEA-R-2884.* {9, 21}

Fauske, H K and Grolmes, M A, (1971) "Analysis of transient forced convection liquid-metal boiling in narrow tubes." *Chem. Eng. Prog. Symp. Series,* **67**, No. 119, 64-71.

Firman, E C, (1975) "Waterside flow problems in boilers." *Combustion.* July, 21-28.

Firstenberg, H, Goldmann, K and Hudson, J H, (1960) "Boiling songs and associated mechanical vibrations." NDA 2131-12.

Fleck, Jr, J A, (1960) "The dynamic behavior of boiling water reactors." *J. Nucl. Energy,* Part A, **11**, 114-130.

Fleck Jr, J A, (1961) "The influence of pressure on boiling water reactor dynamic behavior at atmospheric pressure." *Nucl. Sci. and Eng.,* **9**, 271-280.

Ford, W D, Bankoff, S G and Fauske, H K, (1971a) "Slug ejection of Freon-113 from a vertical channel with nonuniform initial temperature profiles." Paper No. 7-11. Presented at *International Symp. on Two-Phase Systems,* Haifa, Israel.

Ford, W D, Fauske, H K and Bankoff, S G, (1971b) "The slug expulsion of Freon-113 by rapid depressurization of a vertical tube." *Int. J. Heat Mass Transfer,* **14**, 133-140.

France, D M, Carlson, R D, Chiang, T and Priemer, R, (1979) "Characteristics of transition boiling in sodium-heated steam generator tubes." *J. Heat Transfer,* **101**, 270-275.

Fukuda, K and Kobori, T, (1979) "Classification of two-phase flow instability hy density wave oscillation model." *Journal of Nuclear Science and Technology*, **16**, 95-108.

Gaudiosi, G, (1965) "Experimental results on the dependence of transition boiling heat transfer on loop flow disturbances." GEAP-4725.

Gilli, P V, Edler, A, Halozan, H and Schaup, P, (1975) "Probleme des wärmeüberganges, druckverlustes und der strömungsstabilität in thermisch hochbeanspruchten dampferzeugerrohren." *VGB Kraftwerkstechnik*, **55**, 589-600.

Giphshman, I N and Levinson, V M, (1966) "Disturbances in flow stability in the pendant superheater of the TPP-110 boiler." *Thermal Engineering*, **13** (5), 57-62.

Goodykoontz, J H and Dorsch, R G, (1967) "Local heat-transfer coefficients and static pressures for condensation of high-velocity within a tube." *NASA Lewis Research Center*, E3498.

Gorman, D J, (1970) "An experimental and analytical investigation of fuel element vibration in two-phase parallel flow." *Trans. ANS*, **13** (1), 333.

Gouse Jr, S W and Andrysiak, C D, (1963) "Fluid oscillations in a closed loop with transparent, parallel, vertical, heated channels." *MIT Engineering Projects Lab*. Report 8973-2.

Gouse Jr, S W and Evans, R G, (1967) "Acoustic velocity in two-phase flow." *Proc. Symp. on Two-Phase Flow Dynamics*, Eindhoven, EUR 4288e, 585-623.

Grant, I D R, (1971) "Shell-and-tube heat exchangers for the process industries." *Chemical Processing*. December.

Griffith, P, (1962) "Geysering in liquid-filled lines." ASME Paper No. 62-HT-39.

Grolmes, M A and Fauske, H K, (1970) "Modeling of sodium expulsion with Freon-11." *ASME* Paper No. 70-HT-24.

Hands, B A. (1974) "A re-examination of the Ledinegg instability criterion and its application to two-phase helium systems." in *Multi-Phase Flow Systems*, Institution of Chemical Engineers, London.

Hands, B A, (1979) "Density wave oscillations: The identification of various modes." AIChE Symposium Series, **75**, No. 189, 165-176.

Hawtin, P, (1970) "Chugging flow." *AERE*-R 6661.

Höppner, G, (1971) "Experimental study of phenomena affecting the loss of coolant accident." Ph.D thesis in Engineering, Univ. of California, Berkeley.

Hunsbedt, A and Roberts, J M, (1974) "Thermal-hydraulic performance of a 2 MWt sodium-heated, forced recirculation steam generator model." *J. Eng. Power*, **96**, 66-76.

Ishii, M, (1971) "Thermally induced flow instabilities in two-phase mixtures in thermal equilibrium." Ph.D. thesis in Mechanical Engineering, Georgia Institute of Technology.

Ishii, M and Zuber, N, (1970) "Thermally induced flow instabilities in two phase mixtures." *4th Int. Heat Transfer Conf.*, Paris, **V**, Paper # B5.11.

Jain, K C, (1965) "Self-sustained hydrodynamics oscillations in a natural circulation two-phase flow boiling loop." Ph.D thesis, Northwestern Univ.

Jeglic, F A and Grace, T M, (1965) "Onset of flow oscillations in forced flow subcooled boiling." *NASA-TN-D 2821.*

Jones, A B and Dight, A G, (1964) "Hydrodynamic stability of a boiling channel." *USAEC* Reports KAPL-2170, 1961, KAPL-2208, 1962, KAPL-2290, 1963, and KAPL-3070, 1964.

Jones, A B and Yarbrough, W M, (1965) "Reactivity stability of a boiling reactor, Part 1: KAPL-3072, 1964; Part 2: KAPL-3093, 1965."

Jones, M C and Peterson, R G, (1975) "A study of flow stability in helium cooling systems." *J. Heat Transfer*, **97**, 521-527.

Koshelev, I I, Surnov, A V and Nikitina, L V, (1970) "Inception of pulsations using a model of vertical water-wall tubes." *Heat Transfer-Soviet Research*, **2**, No. 3.

Krasyakova, L Y and Glusker, B N, (1965) "Hydraulic study of three-pass panels with bottom inlet headers for once-through boilers." *Thermal Engineering*, **12** (8), 19-26.

Lahey Jr, R T and Yadigaroglu, G, (1974) "A Lagrangian analysis of two-phase hydrodynamic and nuclear-coupled density-wave oscillations." *Heat Transfer 1974, Proceedings of the Fifth International Heat Transfer Conference*, **IV**, Tokyo, Japan, 225-229.

Lahey Jr, R T and Moody, F J, (1977) "The thermal-hydraulics of a boiling water nuclear reactor." *The American Nuclear Society*, Hinsdale, Illinois.

Ledinegg, M, (1938) "Instability of flow during natural and forced circulation." *Die Wärme*, 61, 8, 891-898. AEC-tr-1861, 1954.

Ledinegg, M and Wiesenberger, J, (1975) "Strömung in beheizten rohrregistern." *VGB Kraftwerkstechnik*, **55**, 581-589.

Lee, S S, Veziroglu, T N and Kakac, S, (1977) "Sustained and transient boiling flow instabilities in two parallel channel systems." in *Two-Phase Flows and Heat Transfer*, **1**, S. Kakac and F. Mayinger (Editors), Hemisphere Publishing Corp., Washington, 467-510.

Leonard, J R, (1976) "Effect of system variables on departure from nucleate boiling in the Clinch River breeder reactor plant steam generators." *ASME* Paper No. 76-JPGC-NE-4.

Margetts, R J, (1972) "Exursive instability in feedwater coils". AIChE Paper presented at *13th National Heat Transfer Conference*, Denver, Colorado.

Margolis, S G and Redfield, J A, (1966) "FLASH: A program for digital simulation of the loss of coolant accident." *WAPD-TM-534.*

Marks, J S and Andeen, G B, (1979) "Chugging and condensation oscillation." ASME Paper presented at *18th National Heat Transfer Conference*, San Diego, CA.

Mathisen, R P, (1967) "Out of pile channel instability in the loop Skälvan." *Proc. Symp. on Two Phase Flow Dynamics*, Eindhoven, EUR 4288e, 19-64.

Maulbetsch, J S, (1967) "Pressure drop - flow rate instabilities." Notes for M.I.T. Special Summer Program, *Two-Phase Gas-Liquid Flow and Heat Transfer*.

Maulbetsch, J S and Griffith, P, (1965) "A study of system-induced instabilities in forced-convection flows with subcooled boiling." *MIT Engineering Projects Lab.* Report 5382-35.

Maulbetsch, J S and Griffith, P, (1967) "Prediction of the onset of system-induced instabilities in subcooled boiling." *Proc. Symp. on Two-Phase Flow Dynamics*, Eindhoven, EUR 4288e, 799-825.

McDonald, B N and Johnson, L E, (1970) "Nuclear once-through steam generator. Report BR-923, TPO-67, presented to *Amer. Nucl. Soc. Top. Meet.*, Williamsburg, Virginia.

Murphy, R W and Bergles, A E, (1971) "Subcooled flow boiling of fluorocarbons." *MIT Engineering Projects Lab.* Report 71903-72.

Nahavandi, A N and von Hollen, R F, (1964) "A space-dependent dynamic analysis of boiling water reactor systems." *Nucl. Sci. and Eng.* **20**, 392-413.

Nakajima, I, Kawada, A, Fukuda, K and Kobori, T, (1975) "An experimental study on the instability induced by voiding from a horizontal pipe line." *ASME* Paper No. 75-WA/HT-20.

Neal, L G, Zivi, S M and Wright, R W, (1967) "The mechanisms of hydrodynamic instabilities in boiling systems." *Proc. Symp. on Two-Phase Flow Dynamics*, Eindhoven, EUR 4288e, 957-980.

Nylund, O, et al, (1970) "Hydrodynamic and heat transfer measurements on a full-scale simulated 36-rod BHWR fuel element with non-uniform axial and radial heat flux distribution". *AB Atomenergi* Report FRIGG-4.

Okazaki, M, (1979) "Analysis for pressure oscillation phenomena induced by steam condensation in containments with pressure suppression system, (I), Model and analysis for experimental conditions of Marviken reactor." *Journal of Nuclear Science and Technology,* **16**, 30-42.

Pezilli, M, Sacco, A, Scarano, G, Tomassetti, G and Pinchera, G C, (1970) "The NEMI model for sodium boiling and its experimental basis." in *Liquid-Metal Heat Transfer and Fluid Dynamics*, ASME, New York, 153-161.

Quandt, E, (1961) "Analysis and measurement of flow oscillations." *Chem. Eng. Progress Symp. Ser.* 32, No. 57, 111-126.

Quinn, E P, (1966) "Forced-flow heat transfer to high-pressure water beyond the critical heat flux." *ASME* Paper 66-WA/HT-36.

Rothe, P M, Wallis, G B and Block, J A, (1977) "Cold leg ECC flow oscillations." in *Symposium on the Thermal and Hydraulic Aspects of Nuclear Reactor Safety,* **1**: *Light Water Reactors,* O C Jones, Jr. and S G Bankoff, Editors, ASME, New York, 133-150.

Saha, P, (1974) "Thermally induced two-phase flow instabilities including the effect of thermal equilibrium between the phases." Ph.D. Thesis in Mechanical Engineering, Georgia Institute of Technology.

Saha, P, Ishii, M and Zuber, N, (1976) "An experimental investigation of the thermally induced flow oscillations in two-phase systems." J. Heat Transfer, 98, 616-622.

Sargis, D A, Masiello, P J and Stuhmiller, J H, (1979) "A probabilistic model for predicting steam chugging phenomena." ASME Paper presented at 18th National Heat Transfer Conference, San Diego, CA.

Sargis, D A, Stuhmiller, J H and Wang, S S, (1978) "A fundamental thermal-hydraulic model to predict steam chugging phenomena." in Topics in Two-Phase Heat Transfer and Flow, S G Bankoff, Editor, ASME, New York, 123-134.

Schlechtendahl, E G, (1970) "Theoretical investigations on sodium boiling in fast reactors." Nucl. Sci. Eng., 41, 99-114.

Singer, R M and Holtz, (1970) "Comparison of the expulsion dynamics of sodium and nonmetallic fluids." ASME Paper No. 70-HT-23.

Solberg, K O and Bakstad, P, (1967) "A model for the dynamics of nuclear reactors with boiling coolant with a new approach to the vapor generation process." Proc. of Symp. on Two-Phase Flow Dyanmics, Eindhoven, EUR 4288e, 871-934.

Stenning, A H and Veziroglu, T N, (1965) "Flow oscillations modes in forced convection boiling." Proc. 1965 Heat Transfer and Fluid Mech. Inst., Stanford Univ. Press, 301-316.

Stenning, A H, Veziroglu, T N and Callahan, G M, (1967) "Pressure-drop oscillations in forced convection flow with boiling." EURATOM Report, Proc. Symp. on Two-Phase Flow Dynamics, EUR 4288e, 405-427.

Thurston, R S, Rogers, J D and Skoglund, V J, (1967) "Pressure oscillations induced by forced convection heated in dense hydrogen." Adv. in Cryogenic Eng., 12, 438-451.

Tong, L S, (1965) Boiling Heat Transfer and Two-Phase Flow, John Wiley and Sons, Inc., New York.

Tong, L S, (1968) "Boundary layer analysis of the flow boiling crisis." Int. J. Heat Mass Transfer, 11, 1208-1211.

Ünal, H C, (1980) "Correlations for the determination of the inception conditions of density-wave oscillations for forced and natural circulation steam generator tubes." J. Heat Transfer, 102, 14-19.

Veziroglu, T N and Lee, S S, (1971) "Boiling-flow instabilities in a cross-connected parallel-channel upflow system." ASME Paper No. 71-HT-12.

Wallis, G B and Heasley, J H, (1961) "Oscillations in two phase flow system." J. Heat Transfer, 83, 363-369.

Waring, J P, et al, (1974) "PWR FLECHT-SET Phase B1 Data Report." WCAP-8431.

Waszink, R P and Efferding, L E, (1974) "Hydrodynamic stability and thermal performance tests of a 1-MWt sodium-heated once-through steam generator model." *J. Engineering for Power*, **96**, 189-200.

Wedekind, G L and Beck, B T, (1974) "Theoretical model of the mixture-vapor transition point oscillations associated with two-phase evaporating flow instabilities." *J. Heat Transfer*, **96**, 138-144.

Westendorf, W H and Brown, W F, (1966) "Stability of intermixing of high-velocity vapor with its subcooled liquid in cocurrent streams." NASA TND-3553.

White, E P and Duffey, R B, (1974) "A study of the unsteady flow and heat transfer in the reflooding of water reactor cores." *Central Electricity Generating Board* Report RD/B/N3134.

Wörrlein, K, (1975) "Instabilitäten bei der durchflussverteilung in beheizten rohrstängen von dampferzeugern." *VGB Kraftwerkstechnik*, **55**, 513-518.

Yadigaroglu, G, (1978) "Two-phase flow instabilities and propagation phenomena." *von Karman Institute for Fluid Dynamics*, Lecture Series 1978-5.

Yadigaroglu, G and Bergles, A E, (1969) "An experimental and theoretical study of density-wave oscillations in two-phase flow." *MIT* Report No. DSR 74629-3. {31}

Yadigaroglu, G and Bergles, A E, (1972) "Fundamental and higher-mode density-wave oscillations in two-phase flow: The importance of the single-phase region." ASME Paper No. 71-HT-13, 1971, *J. Heat Transfer*, **94**, 189-195.

Yamakawa, M and Tsuchiya, T, (1979) "Prediction of instability in sodium heated steam generator." *Proceedings of the Japan-U.S. Seminar on Two-Phase Flow Dynamics*, Kobe, Japan, 583-603.

Chapter 14

Scaling and Modelling Laws in Two-Phase Flow and Boiling Heat Transfer

F. MAYINGER

14.1 Summary

In two-phase flow scaling is much more limited to very narrowly defined physical phenomena than in single phase fluids. For complex and combined phenomena it can be achieved not by using dimensionless numbers alone but in addition a detailed mathematical description of the physical problem - usually in the form of a computer program - must be available. An important role is played by the scaling of the thermodynamic data of the modelling fluid. From a literature survey and from our scaling experiments the conclusion can be drawn that Freon is a quite suitable modelling fluid for scaling steam - water mixtures. However, without a theoretical description of the phenomena nondimensional numbers for scaling two-phase flow must be handled very carefully.

14.2 Introduction

The transfer of experimental results obtained under scaled conditions to the reality in an industrial plant is an old problem of engineering. There must be a sufficient similarity of the main parameters influencing the precedents in the test equipment and in the original set-up. This similarity can be of geometrical, mechanical, static, dynamic, thermal, thermodynamic, electrical and mechanical nature.

In single phase hydrodynamics and heat transfer we are accustomed to use scaling and modelling laws for many years. The occurrences are analogous in a diabatic flow if

the velocity field,

the temperature field and

the pressure field (supersonic flow)

are similar. In the area of subsonic flow the velocity and temperature fields are described by the well-known conservation laws for mass, energy and momentum. From these laws dimensionless numbers like

$$\frac{x}{1} , \frac{y}{1} , \frac{z}{1}$$

$$Re = \frac{w \cdot l}{\nu} \; , \; Fr = \frac{w^2}{gl} \; , \; Eu = \frac{\Delta p}{\rho w^2}$$

$$Nu = \frac{\alpha \cdot l}{\lambda} \; , \; Gr = \frac{l^3 g \beta \theta}{\nu^2} \; , \; Pr = \frac{\eta \cdot c}{\lambda}$$

are derived and similarity is assumed, if these dimensionless numbers are identical in the test and in the original conditions. Usually it is easy to guarantee the geometrical similarity by choosing the same ratios of length x/l, y/l, z/l, i.e. to make all dimensions of the test object proportional to a characteristic length l of the original set-up. The other dimensionless numbers represent ratios of forces or of energies (e.g. the Nusselt-number) which cannot be made identical in the model and in the original, as can easily be seen from the different exponents of the parameters. The velocity is in the Reynolds-number to the first power and in the Froude-number to the second power. The Reynolds-number is the ratio of inertia and viscous forces and the Froude-number contains the buoyancy force. This means, that in mixed convection either the forced flow or the buoyant induced one can be scaled only. In addition the velocity and the temperature field is influenced by the thermodynamic properties - in single phase flow mainly the viscosity, density, specific heat and thermal conductivity. By comparing the behaviour of different fluids not all these properties can be exactly scaled in general.

The velocity and the temperature field can be described by functions of the form

$$\frac{w}{w_o} = f_w \; (\frac{x}{l}, \; \frac{y}{l}, \; \frac{z}{l}, \; Re, \; Pr, \; Gr) \tag{14.1}$$

$$\frac{\theta}{\theta_o} = f_\theta \; (\frac{x}{l}, \; \frac{y}{l}, \; \frac{z}{l}, \; Re, \; Pr, \; Gr) \tag{14.2}$$

$$Nu = f_\alpha \; (Re, \; Pr, \; Gr) \tag{14.3}$$

and in practice we rule out one of the dimensionless numbers, i.e. the Reynolds-number or the Grashof-number depending whether forced or free convection prevails. Therefore already in single phase flow scaling or similarity laws have a restricted and only approximate validity. This we have to keep in mind when we now discuss similarity in two-phase flow and in boiling heat transfer.

14.3 Dimensionless Numbers for Two-Phase Flow and Boiling Heat Transfer

From a theoretical point of view it is the best and most exact procedure to derive the dimensionless numbers from differential equations completely describing the interesting physical phenomena by arranging the terms in a certain way. Another systematic procedure is to form a matrix from the exponents of

F Mayinger

$$F = w^a \; ; \; \Delta p^b \; ; \; l^c \; ; \; \rho^d \; ; \; \eta^e$$

$$\left(\frac{m}{s}\right)^a \; ; \; \left(\frac{kg}{ms^2}\right)^b \; ; \; (m)^c \; ; \; \left(\frac{kg}{m^3}\right)^d \; ; \; \left(\frac{kg}{ms}\right)^e$$

mass (kg)	0	b	0	d	e	$\cdots$
length (m)	a	-b	c	-3d	-e	$\cdots$
time (s)	-a	-2b	0	0	-e	$\cdots$
$\vdots$	$\vdots$	$\vdots$	$\vdots$	$\vdots$	$\vdots$	

or

	a	b	c	d	e	$\cdots$
kg	0	1	0	1	1	$\cdots$
m	1	-1	1	-3	-1	$\cdots$
s	-1	-2	0	0	-1	$\cdots$
$\vdots$	$\vdots$	$\vdots$	$\vdots$	$\vdots$	$\vdots$	

Figure 14.1: Matrix for forming dimensionless groups.

the influencing parameters and their dimensions as indicated in Figure 14.1. The number of the influencing parameters minus the rank of the matrix gives the number of the dimensionless groups and these again can be elaborated by evaluating the matrix. Finally a third possibility to obtain dimensionless numbers is the method of trial and error by multiplying and dividing selected parameters.

By these methods or sometimes just by intuitive experience, several dimensionless numbers were found for two-phase flow and for thermohydraulic processes with phase change a number of years ago. Figure 14.2 gives some examples of them. They are usually restricted to a single phenomenon, for example, the Weber-number to droplet formation or the Jakob-number to boiling. (Ja represents the heat flux ratio in the liquid and in the vapour from the wall.) A special role is for the Martinelli parameter x, which was found by Martinelli studying two-phase pressure drop, but which seems to have a much more general validity.

Weber number :

$$We = \frac{\rho \cdot d_d \; w^2}{\sigma}$$

Laplace constant :

$$La = \frac{\sigma}{g(\rho_1 - \rho_g)} \quad (m^2)$$

Jakob number :

$$Ja = \frac{c_1 \cdot \rho_1 (\Delta v_w - \Delta v_{bulk})}{hlg \cdot \rho_g}$$

Bubble Reynolds number :

$$Re_b = \frac{\rho_g \cdot wb \cdot db}{\eta_g}$$

Boiling number :

$$N_{Bo} = \frac{\dot{q}}{hlg \cdot \rho_g \cdot w_g}$$

Buoyancy number :

$$N_{Bu} = \frac{\rho_1 - \rho_g}{\rho_1}$$

Subcooling number :

$$N_{sub} = \frac{\Delta h_{sub} \, (\rho_f - \rho_g)}{hlg \cdot \rho_g}$$

Phase change number :

$$N_{ph} = \frac{\dot{m}_g \cdot (\rho_l - \rho_g) \cdot l}{\rho_g \cdot \rho_l \cdot w_l}$$

Slip ratio

$$s = \frac{W_g}{W_1}$$

Drift number

$$N_D = \frac{V_{D,M}}{W_{1\,in}}$$

Martinelli parameter :

$$X = \left(\frac{\rho_g}{\rho_1}\right)^{0,5} \cdot \left(\frac{\eta_1}{\eta_g}\right)^{0,1} \cdot \left(\frac{1-\dot{x}}{\dot{x}}\right)^{0,9}$$

Figure 14.2: Examples for dimensionless numbers in two-phase systems.

In single phase and single component systems the velocity and the temperature field have to be regarded only but in two-phase flow in addition we have to take into account the density distribution - i.e. the void fraction or the quality - in the system. Finally in a multi-component system the concentration field would have to be added. In two-phase flow with and without boiling we must have information on

the velocity field,

the phase distribution field and

the temperature field.

The last one usually can be neglected if we have a single component two-phase system in thermodynamic equilibrium.

The second phase brings with it a lot of complications in our similarity or scaling deliberations. The hydrodynamic and thermodynamic properties in both phases have to be similar and also the phase change behaviour must be comparable. Before discussing whether these conditions are compatible we have briefly to think about, why and what do we want to scale in two-phase flow: In single phase flow usually we have the problem of scaling the size

of an apparatus or of a plant, that is we do the experiments with
smaller, scaled down objects. In two-phase flow there is the
question of the proportion too, but a much more severe problem
rises from the very large heat input, which is needed to do two-
phase and boiling experiments. One would need for example several
MW to make heat transfer experiments with a representative section
of a nuclear reactor fuel element. A geometrically scaling down
of the fuel rod array is almost not possible because the bubbles
keep their size and the ratio between bubble diameter and rod
space would change and thus seriously influence the boiling phenomena. There is another possibility to reduce the power input by
using instead of water a modelling liquid with low latent heat
of evaporization. And so in two-phase flow not so much a scaling
of the geometry but a scaling of the fluid is of interest and
importance.

Due to the fact that thermodynamic properties of the gases
and the liquid phase of the modelling fluid play an important
role, it is essential that

(i)   these properties are well known and

(ii) that they show a similar behaviour as the original fluid.

Besides water the refrigerants (R11, R12, R13) are pretty well
researched and have well known thermodynamic properties. In
addition they offer convenient experimental conditions with respect
to pressure and temperature.

14.4   Thermodynamic Scaling of the Fluids

From a very first point of view the claim to the similarity
of the thermodynamic conditions seems simply to be solved by
selecting the thermodynamic state of the modelling fluid in such
a way that the important influencing properties are comparable.
This can precisely be done for one property, but for the others
only approximately. So we find very often in the literature
that in the scaling tests the thermodynamic state of the modelling
fluid is so selected that the density ratio ρ_f/ρ_g between liquid
and vapour is the same in the model and in the original. This
may be a very serious restriction because there can be other
properties as important as the density.

If there would be a common equation of state for both liquids
which we could write in a dimensionless reduced form with the
help of the thermodynamic consistency we could make use of the
thermodynamic theorem of corresponding states. One of the simplest
equations of state is certainly the Van der Waals-equation, which
we can write as follows:

$$\left(\frac{p}{p_c} + \frac{3}{v/v_c}\right)\left(3\,\frac{v}{v_c} - 1\right) = 8\,\frac{T}{T_c} \tag{14.4}$$

In this form - which is reduced with the critical data - the Van
der Waals-equation contains only universal constants not restricted
to a certain liquid. Certainly the Van der Waals-equation allows

only a very rough and unprecise prediction of the thermodynamic properties of Freon. But it gives the hint that the critical data p_c, T_c and v_c may be useful to get a corresponding state for comparable thermal, caloric and even transport properties, since, as Grigull (1968) showed, there is according to thermodynamic consistency, a strong link between transport data and certain caloric properties.

In a single component system the two phases usually only exist along the saturation line which means that one thermodynamic property fixes the whole state. So using the critical properties we can either scale with the critical pressure ratio

$$\left(\frac{p}{p_c}\right)_{Original} = \left(\frac{p}{p_c}\right)_{Model} \tag{14.5}$$

or with any other critical ratio like

$$\left(\frac{T}{T_c}\right)_0 = \left(\frac{T}{T_c}\right)_M ; \left(\frac{\rho}{\rho_c}\right)_0 = \left(\frac{\rho}{\rho_c}\right)_M ; \left(\frac{\eta}{\eta_c}\right)_0 = \left(\frac{\eta}{\eta_c}\right)_M \tag{14.6}$$

Comparisons have proved that the pressure ratio is the most useful parameter for many applications especially for scaling between water and Freon. As shown in Figure 14.3 there is only a deviation of a few percent in the density ratio ρ_1/ρ_g of liquid and vapour for water and Freon if we scale it with the critical

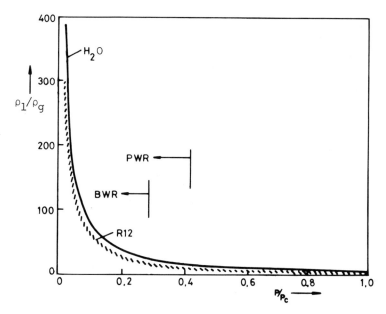

Figure 14.3: Density ratio ρ_l/ρ_g of water and Freon 12 as a function of reduced pressure.

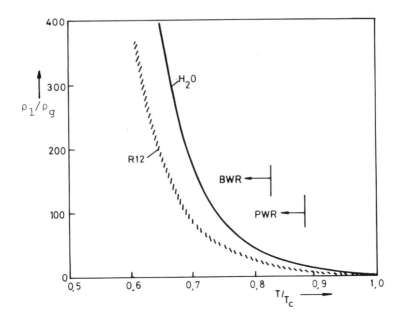

Figure 14.4: Density ratio ρ_l/ρ_g of water and Freon 12 as a function of reduced
        temperature.

pressure ratio. This again means scaling with p/p_c and with ρ_l/ρ_g
give almost the same results. This is not the case if we use
instead of the pressure ratio the critical temperature ratio, as
can be seen from Figure 14.4.

    In many hydrodynamic and heat transfer problems the density
ratio may play the important role but there are other problems
like bubble formation, flashing, entrainment and so on, where
other properties as surface tension, viscosity, thermal conductivity
are governing the phenomenon. A comparison of some properties
scaled with the critical pressure ratio is given in Figure 14.5,
where data for viscosity, thermal conductivity, surface tension,
Prandtl-number and latent heat of evaporization are plotted. We
see that the absolute values of these data for water and for
Freon differ considerably. But we can transform the Freon data
to the water data by simply multiplying with a given factor
as demonstrated with the dotted lines in Figure 14.5. This
factor then has to be taken in account also in the fluid dynamic
scaling. If we deduce the mass flow rate for the modelling
conditions with the help of the Reynolds-number for example we
get the condition

$$Re = \frac{\dot{m}d}{\eta} \; ; \; Re_0 = Re_M; \; \dot{m}_M = \dot{m}_0 \frac{\eta_M}{\eta_0} = k \cdot \dot{m}_0 \qquad (14.7)$$

where the factor k according to Figure 14.5 would be $k_l = 2.15$ and
$k_g = 0.8$. In two-phase flow, even for this simple problem, we

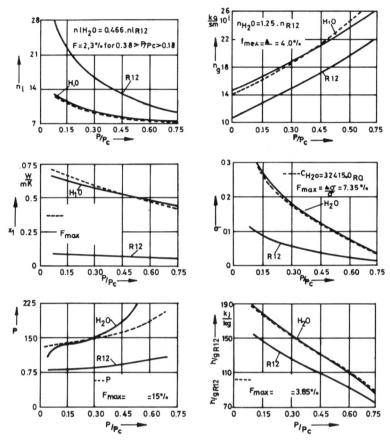

Figure 14.5: Comparison of thermodynamic and transport properties of water and Freon 12.

therefore have two correction factors, one for the liquid and another one for the gaseous phase. One therefore has to decide which phase is the more important one for the special fluid dynamic process we are interested in. For scaling mass flow rate in two-phase flow we can primarily only use the viscosity either of the liquid or of the gas. For the ratio of liquid and vapour mass flow rate an additional condition exists namely the quality $\dot{x}$ for the void fraction ε. That means we have to neglect the influence of the viscosity of one of the phases. Already from this simple example we see that for two-phase scaling we have to subdivide in primary and secondary parameters to a much greater extent than we are used to doing in single phase flow.

In transient conditions, for example with flashing in a sea-water desalination plant or during a loss of coolant accident in a water-cooled nuclear reactor, we need the first derivatives with respect to pressure of some thermodynamic properties in addition. As shown in Figure 14.6 also, these derivatives can be made

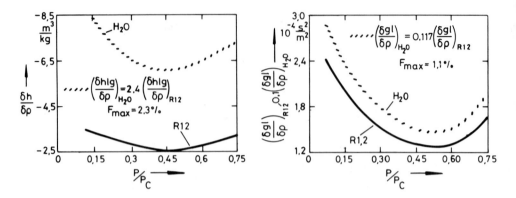

Figure 14.6: Comparison of derived thermodynamic properties (water and Freon 12).

coincident by a simple multiplying factor.

A similar comparison of thermodynamic properties as discussed here but restricted to the surface tension and the viscosity can be found in a paper by Baker (1965) who introduced empirical reference values for surface tension and viscosity. A more detailed comparison is given in Belda's thesis (1975).

14.5 Scaling of Simple Fluid Dynamic Occurrences

In scaling we have to distinguish two steps, namely

(i)   to select the conditions in the model and

(ii) to transfer the results obtained in the test model to original conditions.

Under simple circumstances both steps can be made with the help of dimensionless numbers as is well known for pressure drop and heat transfer problems in single phase flow. For more complicated cases the dimensionless numbers give us the approximate hydrodynamic and thermodynamic conditions for running the tests and the transfer has to be made by the help of a detailed mathematical description of the physical phenomena, for example, in the form of differential equations integrated in a numerical way. Let us first discuss examples of more simple fluid dynamic occurrences.

A first step in studying two-phase flow is the investigation of the flow pattern, i.e. the distribution of the two phases in any cross sections of the channel. Quandt (1963) deduced predictions for flow pattern from the deliberation that the forces acting in the flow must be responsible for the phase distribution. If the forces induced by the pressure drop prevail, annular flow should be present and if the surface tension becomes greater than the forces resulting from the pressure drop and from the buoyancy there should be bubble flow. From these simple ratios of the forces we can derive dimensionless numbers as for example

$$N_{\text{annular flow}} = \frac{\rho \, w^2 \, d}{\sigma} \tag{14.8a}$$

$$N_{\text{bubble flow}} = \frac{\sigma}{\rho \, g \, d^2} \tag{14.8b}$$

where equation (14.8a) is the ratio of the inertia and surface tension forces and equation (14.8b) is the ratio of the surface tension and the buoyancy forces. Equation (14.8a) is nothing else than the well-known Weber-number and equation (14.8b) represents the Laplace-constant divided by the square of a length, for example, the bubble diameter. Unfortunately the method of Quandt was not very successful in predicting the boundaries of the different flow patterns. For horizontal flow another procedure for the prediction of flow patterns was proposed by Baker. He defined the two property parameters

$$\psi_{FB} = \left(\frac{\sigma_{H_2O}}{\sigma}\right)\left[\frac{\eta_1}{\eta_{H_2O}}\left(\frac{\rho_{H_2O}}{\rho_1}\right)^2\right] \tag{14.9a}$$

$$\lambda_{FB} = \left[\left(\frac{\rho_g}{\rho_{air}}\right)\left(\frac{\rho_1}{\rho_{H_2O}}\right)\right]^{0.5} \tag{14.9b}$$

and plotted a flow regime map with $1/\dot{x} = \dot{M}_1/\dot{M}_g$ and the mass flow rate of the gas as additional parameters, as shown in Figure 14.7. The terms in equation (14.9a) and (14.9b) are not dimensionless but we can regard them as a scaling aid. For vertical flow often the flow regime map by Bennett (1965) is used. Zetzmann (1976) demonstrated that in some cases simply by modifying the abscissa

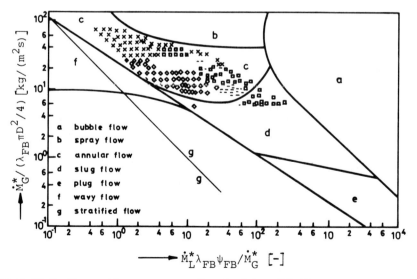

Figure 14.7: Flow pattern map according to Baker (1965).

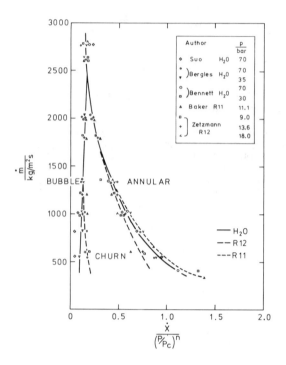

Figure 14.8: Bennett's flow pattern map, modified according to Zetzmann (1976).

of this diagram with the critical pressure ratio the flow regime
map can be generalized for different fluids and pressures as
shown in Figure 14.8.

A little more difficult to survey is the picture in the case
of the void fraction ε and the slip ratio S which are closely
connected if the quality $\dot{x}$ is given:

$$S = \frac{1-\varepsilon}{\varepsilon} \; \frac{\dot{x}}{1-\dot{x}} \; \frac{\rho_1}{\rho_g} \qquad\qquad (14.10)$$

The slip ratio and through it the void fraction are influenced by
the buoyancy forces and by the pressure drop in the channel. Thus
empirical correlations for predicting void fraction, e.g. that
by Kowalczewski (1964) or by Kütükcüglu (1967) contain the Froude-
number

$$Fr = \frac{\dot{m}^2}{g \cdot d_h \cdot \rho_1^2} \qquad\qquad (14.11)$$

as scaling parameter. Hewitt used the Martinelli-parameter

$$X = \sqrt{\frac{(\Delta p/\Delta l)_1}{(\Delta p/\Delta l)_g}} = \left(\frac{\rho_g}{\rho_1}\right)^{0.5} \cdot \left(\frac{\eta_1}{\eta_g}\right)^{0.1} \cdot \left(\frac{1-\dot{x}}{\dot{x}}\right)^{0.9} \qquad (14.12)$$

for his void correlation. That means for doing modelling tests we would have to adapt either the Froude-number or the Martinelli-parameter. But both correlations do not claim general validity for a variety of fluids.

A first but very promising step for generalizing the prediction of void fraction was done by Nabizadeh (1976). Starting from the equation of Zuber and Findlay (1965),

$$\varepsilon = \frac{\dot{x}}{\rho_g}\left[C_0\left(\frac{\dot{x}}{\rho_g} + \frac{1-\dot{x}}{\rho_1}\right) + \frac{1.18}{\dot{m}}\left(\frac{\sigma \cdot g(\rho_1-\rho_g)}{\rho_1^2}\right)^{0.25}\right]^{-1} \qquad (14.13)$$

and supported by a large number of reliable measurements he developed an empirical correlation for the C_0-factor in Zuber-Findlay's equation of the form

$$C_0 = \left(1 + \frac{1-\dot{x}}{\dot{x}} \cdot \frac{\rho_g}{\rho_1}\right)^{-1}\left[1 + \frac{1}{n} \text{ Fr}^{-0.1} \cdot \left(\frac{\rho_g}{\rho_1}\right)^n \cdot \left(\frac{1-\dot{x}}{\dot{x}}\right)^{\frac{11n}{9}}\right] \qquad (14.14)$$

$$n = \sqrt{0.6 \frac{\rho_1-\rho_g}{\rho_1}} \qquad (14.15)$$

Applying this extended and improved Zuber-Findlay equation he gets quite good agreement between correlated and measured data for water, Freon 12 and Freon 113 as can be seen from Figure 14.9. However, we cannot immediately deduce the parameters of a model test for given conditions of an original test from equation (14.13) and (14.14) before looking to the magnitude of influence of the different terms. Nabizadeh points out that it is sufficient to adjust the same values of the density ratio of the quality and of the total mass flow density $\dot{m}$ in the model and in the original plant.

For pressure drop scaling, very early a first step was done by Martinelli (1949). He gave a simple correlation between the two-phase multiplier ϕ_1^2:

$$\left(\frac{\Delta p}{\Delta l}\right)_{2ph} = \phi_1^2 \left(\frac{\Delta p}{\Delta l}\right)_{1 \text{ only}}$$

$$\left(\frac{\Delta p}{\Delta l}\right)_{2ph} = \phi_g^2 \left(\frac{\Delta p}{\Delta l}\right)_{g \text{ only}}$$

$$(14.16)$$

and the parameter X defined by himself as already mentioned in equation (14.12) and Figure 14.2.

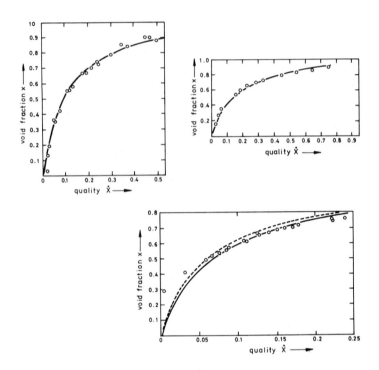

Figure 14.9: Void fraction, comparison of water and Freon data, scaled with
Zuber-Findlay (1965) equation modified by Nabizadeh (1976).

In a detailed experimental and theoretical analysis Friedel
(1975) elaborated scaling rules for two-phase pressure drop. His
measurements prove that the two-phase multiplier, (equation
(14.16)) is the same in water and in Freon if the following
dimensionless groups are equal under test and under layout
conditions:

$$\left.\frac{\left(\eta_{1/\eta_g}\right)^{0.2}}{\left(\rho_{1/\rho_g}\right)^{0.9}}\right|_0 = \left.\frac{\left(\eta_{1/\eta_g}\right)^{0.2}}{\left(\rho_{1/\rho_g}\right)^{0.9}}\right|_M \qquad (14.17)$$

$$\left.\frac{Fr_1}{Re_1^{0.25}}\right|_0 = k\left.\frac{Fr_1}{Re_1^{0.25}}\right|_M \qquad \dot{x}\big|_0 = \dot{x}\big|_M \qquad (14.18)$$

The second group with the Froude- and the Reynold-number contains
a scaling factor k which is a slight function of the pressure and
varies between 1.68 and 1.75. With good approximation, the
value 1.7 can be taken. Friedel adapts the thermodynamic state
by choosing the same viscosity/density-ratio in the model and in

the original. Then the mass flow rate in the test is fixed by
the scaling factor k via the Reynold-number of the liquid. Proceeding thusly, measured data in water and in Freon fall along the
same line as demonstrated in Figure 14.10, where the two-phase
multiplier is plotted versus the quality $\dot{x}$. See Friedel (1975) for
conditions.

Regarding heat transfer in two-phase flow we have to
distinguish several regions

- a first one, in which bubbles are formed at the heated
  wall with low quality in the fluid,

- a second one with a thin liquid film at the wall and
  vaporization at the phase boundary between the liquid
  film and the gaseous bulk flow,

- a third one with a vapour film at the wall, which the
  liquid cannot wet, and low quality in the fluid and

- a fourth one, where the liquid cannot wet the wall either,
  but where the fluid is mainly vapour containing a very small
  quantity of liquid in the form of droplets.

Heat transfer with bubble formation at the wall is a
function of the heat flux density, and as a generalizing parameter
sometimes the Laplace-constant (see Figure 14.2) is used. The
whole bubble boiling process is up to now not well enough understood to derive reliable scaling laws. A promising step to
describe heat transfer with nucleate pool boiling by using
dimensionless numbers

$$Nu = \frac{\alpha \cdot d_{bub}}{\lambda_1} = 0.013 K_1^{0.8} \cdot K_2^{0.4} \cdot K_3^{0.133}$$

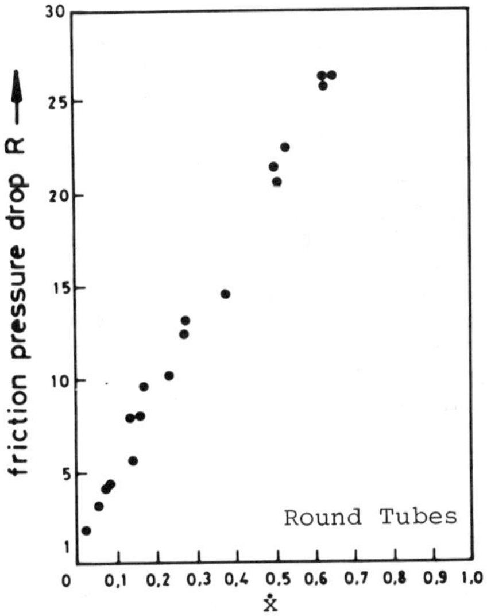

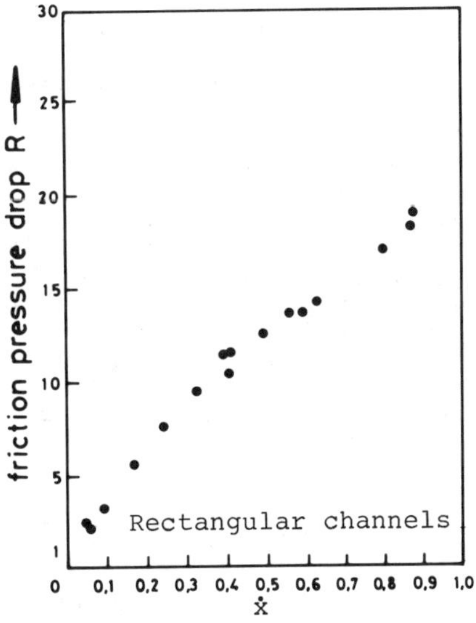

Figure 14.10: Two-phase flow pressure drop multiplier scaled according to
Friedel (1975). R = $(\Delta p (\Delta l) 2ph / (\Delta p / \Delta l) 1ph$

$$K_1 = \frac{\dot{q} \cdot d_{bub}}{\lambda_1 \cdot T_s} \; ; \; K_2 = \frac{d_{bub} \cdot T_s \lambda}{\sigma \cdot \nu_c} \tag{14.19}$$

$$K_3 = \frac{h_{1g} \cdot \rho_g \cdot K}{(f \cdot d_{bub})^2 \rho_1 \cdot d_{bub}}$$

$$f = \text{bubble frequency} \quad K = \text{rel. roughness of surface}$$

was made by Stephan (1964), which also could be used for defining scaling criteria. In the high quality region the situation is a little simpler with the thin liquid film at the wall through which the heat is transported by conduction and convection. Based on a proposal of Schrock and Grossmann (1962) the ratio of the heat transfer coefficients in two-phase flow and in single phase flow is given as a function of the Martinelli-parameter and of the boiling number:

$$\frac{\alpha_{2ph}}{\alpha_{1ph(liquid)}} = C_1 \left[N_{BO} \cdot 10^4 + C_2 \left(\frac{1}{x} \right)^n \right]^m \tag{14.20}$$

The heat transfer of the single phase-liquid flow can be calculated from the well known correlations in the literature usually containing the Reynolds-number and the Prandtl-number. So scaling for this phenomenon should work if we select the Martinelli-parameter, the boiling-number and the quality correctly. But comparing heat transfer coefficients measured in Freon, organic fluids and water the empirical constants in equation (14.20) always have to be adjusted separately, as Calus (1973) showed. Future experiments have to clarify whether there is an additional scaling parameter to be taken into account.

Heat transfer with spray- or fog-cooling, i.e. in the fourth above-mentioned region, sometimes plays an important role in engineering practice as for example in the superheating part of a boiler or during the emergency core cooling of a nuclear reactor. Scaling these heat transfer conditions is a not too serious problem, because the heat transfer coefficient is mainly governed by the vapour flow and the liquid droplets are only cooling the boundary layer near the wall and so improving the heat transport. Dimensionless groups for modelling therefore are a modified Reynolds-number and the Prandtl-number related to the vapour phase. A well-known correlation is the equation of Dougall-Rohsenow.

$$Nu = 0.023 \, Re^{0.8} \, Pr_g^{0.4}$$

$$Re = \frac{\dot{m} d_h}{\eta_g} \left[\dot{x} + \frac{\rho_g}{\rho_1}(1-\dot{x}) \right] \tag{14.21}$$

Scaling criteria are the Reynolds-number as defined in equation
(14.21), the Prandtl-number of the vapour, and the quality of
the fluid. Comparisons with experimental data measured in water
and Freon gave quite good results and good agreement with the
Dougall-Rohsenow correlation. From this example we learn that
always the vapour, which governs the fluiddynamic or heat
transfer process has to be adjusted in its thermodynamic
properties.

In some cases also for problems with interfacial momentum
and heat transfer at the phase boundaries unique correlations and
some scaling criteria can be found. Viecenz (1980) studied the
phase separation, the drift flux and the mean void fraction $<\varepsilon>$
in liquid pools with gas or vapour bubbling through. He
correlated the volume averaged void fraction $<\varepsilon>$ by

$$<\varepsilon> = C \left[\frac{u_o^{"2}}{g\sqrt{\frac{\sigma}{g(\rho'-\rho'')}}} \right]^n \left[\frac{\sqrt{\frac{\sigma}{g(\rho'-\rho'')}}}{d_{cont}} \right]^{0.176}$$

$$\left[\frac{\rho'}{\rho'-\rho''} \right]^{0.585} \cdot \left[\frac{\upsilon'}{\upsilon''} \right]^{0.255}$$

(14.22)

where C and n depend on the superficial velocity $u_o^{"}$ of the gas
in the vessel (assuming no liquid in the vessel) and the Laplace
constant as shown in Figure 14.11. This figure also proves that
there is good agreement between the results predicted by equation
(14.22) and measurements in the literature. Equation (14.22)
represents a simple and reliable scaling law in the form of
dimensionless numbers for phase separation in gasified liquids.

14.6 Scaling Combined Phenomena

The fluiddynamic behaviour in an engineering apparatus like
a boiler, a nuclear reactor or destillation and extraction
columns is a very strongly combined effect of several simple
phenomena, and usually fluid flow and heat transfer are linked
together. The heat transferred to the fluid determines the void
fraction; the produced vapour influences the pressure drop,
which governs the mass flow rate, and this again can affect the
heat transfer. Up to now we regarded only each single simple
phenomenon separately assuming no influence from other processes.

Out of the numerous problems of technical two-phase flow,
for discussion here we select the burnout behaviour and the blowdown conditions during a loss of coolant accident in a nuclear
reactor.

The exact knowledge of the critical heat flux with boiling

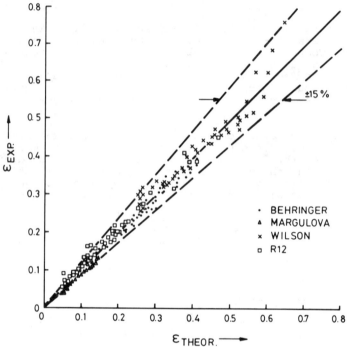

Figure 14.11: Comparison of experimental data with equation (14.22), Viecenz (1980).

(burnout or dryout) is an urgent problem in the layout of the fuel elements of water cooled nuclear reactors. Due to the high heat input and the extensive instrumentation the experiments are very expensive. Therefore it was obvious to look for more convenient and cheaper test conditions. Here Freon as a modelling fluid seemed to offer a possibility. There are several proposals in the literature for scaling critical heat flux from Freon to water. All start from the thermodynamic condition that the density ratio of liquid and vapour has to be the same in the modelling and in the original fluid.

Stevens et al. (1965/66) (1970) preferred a more empirical procedure and made a direct comparison of the burnout data in Freon and water with simple scaling factors. He postulated that in addition to the same density ratio and the identical geometry, the quality (or subcooling) at the inlet and the outlet of the heated length have to be the same in the model and in the original. Therefore only the mass flow rate remains as a free variable to be adjusted in the scaled down conditions. This scaled mass flow rate is got via the scaling factor K by plotting the quality $\dot{x}$ (at the outlet) versus the parameter

$$\left[\dot{m} \, d_h^{0.25} \left(\frac{d_h}{l} \right)^{0.59} \right] \; ;$$

as this is done in Figure 14.12 with the help of measurements in a rod cluster which were carried out by Hein and Kastner (1972).

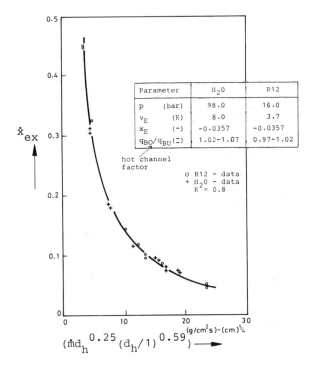

Figure 14.12: Stevens and Staniforth (1965/66) burnout scaling factor for mass
flow checked with data of Hein and Kastner (1972).

Barnett (1963) used dimensional analysis for deriving
his scaling laws. He listed all parameters, which are according
to the status of knowledge mainly governing the physical
behaviour of the phenomenon and others which have a minor
influence. Doing so, he gets a relation of the form

$$q"_{CHF} = f(d,l,\dot{m},p,\Delta h,h_{1g},\rho_1,\rho_g,c_1,\lambda_1,\nu_1,\sigma)\,\frac{d(\rho_1/\rho_g)}{dp} \qquad (14.23)$$

generally describing the critical heat flux. Adapting the
dimensional analysis equation (14.23) can be arranged in the
form of dimensionless groups and he gets

$$\frac{q"_{CHF}}{h_{1g}^{1.5}\rho_g} = f\left(\frac{1}{d}\,,\,\frac{D\cdot\rho_g\cdot h_{1g}^{0.5}c_{p_1}}{\lambda_1}\,,\,\frac{\dot{m}}{\rho_g h_{1g}^{0.5}}\,,\,\frac{\Delta h}{h_{1g}}\,,\,\frac{\rho_1}{\rho_g}\,,\right.$$

$$\left.\frac{\nu\cdot\rho_g\cdot c_{p_1}}{\lambda_1}\,,\,\frac{\sigma\cdot c_{p_1}}{h_{1g}^{0.5}\cdot\lambda_1}\,,\,h_{1g}\cdot\rho_g\cdot\frac{d(\rho_1/\rho_g)}{dP_s}\right) \qquad (14.24)$$

$$q'' = f\left(1, d, \dot{m}, p, \Delta h, hl_g, \rho_1, \rho_g, cp_1, \lambda_1, \nu_1, \sigma \; \frac{d(\rho_1/\rho_g)}{dp_s}\right)$$

$$\gamma = \frac{d(\rho_1/\rho_g)}{dp_s} \qquad \beta = \frac{d\vartheta_s}{dp_s}$$

No	properties	equation $\dot{q} = f(l,d,\dot{m},p,\Delta h)$	F_q	$F_{d,l}$	$F_{\dot{m}}$	$F_{\Delta h}$
4	$h_{lg}, \rho_1, \rho_g, cp_1, \lambda_1, \gamma$	$\dfrac{q'' \cdot \gamma^{0.5}}{\lambda \sqrt{\rho_1}} = f\left(\dfrac{l}{d}, \dfrac{d \cdot cp_1 \cdot \sqrt{\rho_1}}{\lambda_1 \cdot \gamma^{0.5}}, \dfrac{\dot{m} \cdot \gamma^{0.5}}{\sqrt{\rho_1}}, \dfrac{\rho_1}{\rho_g}, \dfrac{\Delta h}{h_{lg}}\right)$	22.43	0.717	1.914	11.72
5	$h_{lg}, \rho_1, \rho_g, \beta, \lambda_1, \gamma$	$\dfrac{q'' \cdot \gamma^{0.5}}{h_{lg}\sqrt{\rho_1}} = f\left(\dfrac{l}{d}, \dfrac{d \cdot h_{lg} \gamma^{0.5}\sqrt{\rho_1}}{\beta \cdot \lambda_1}, \dfrac{\dot{m}\gamma^{0.5}}{\sqrt{\rho_1}}, \dfrac{\rho_1}{\rho_g}, \dfrac{\Delta h}{h_{lg}}\right)$	22.43	0.482	1.914	11.72
10	$h_{lg}, \rho_1, \rho_g, \mu_g, \gamma$	$\dfrac{q'' \cdot \gamma^{0.5}}{h_{lg}\sqrt{\rho_1}} = f\left(\dfrac{l}{d}, \dfrac{d \cdot \sqrt{\rho_1}}{\mu_g \cdot \gamma^{0.5}}, \dfrac{\dot{m} \cdot \gamma^{0.5}}{\sqrt{\rho_1}}, \dfrac{\rho_1}{\rho_g}, \dfrac{\Delta h}{h_{lg}}\right)$	22.43	0.814	1.914	11.72
12	$h_{lg}, \rho_1, \rho_g, cp_g, \lambda_g, \beta$	$\dfrac{q''\sqrt{\beta \cdot cp_g}}{h_{lg}^{1.5} \cdot \rho_1} = f\left(\dfrac{l}{d}, \dfrac{d\sqrt{cp_g \cdot h_{lg}}\rho_1}{\lambda_g \sqrt{\beta}}, \dfrac{\dot{m}\sqrt{\beta \cdot cp_g}}{\sqrt{h_{lg}}\rho_1}, \dfrac{\rho_1}{\rho_g}, \dfrac{\Delta h}{h_{lg}}\right)$	2308	0.317	1.968	11.72

Figure 14.13: Scaling factors for critical heat flux according to Barnett (1963).

From these groups in equation (14.24) Barnett developed several scaling factors N_n depending which thermodynamic properties he was taking in account in addition to the values of latent heat of evaporization h_{lg} and density of liquid and vapour ρ_1, ρ_g.

Examples of sets for scaling factors elaborated by Barnett with this method are shown in Figure 14.13. Barnett offers several possibilities for scaling critical heat flux depending which combination of thermodynamic properties is assumed to be the most appropriate one.

Bouré (1970) was the first, who clearly stated that a comprehensive scaling of the critical heat flux is not possible and that only a restricted similarity of the boiling crisis phenomenon in water and in Freon may exist. Realizing this he emphasized in his study the development of strict rules for a limited similarity.

He distinguishes - similar to Barnett - two groups of parameters: primary parameters and secondary ones. He used the latter ones to determine correction factors for the primary parameters. Doing so he obtained the following 4 dimensionless groups:

$$N_1 = K_1 \frac{\rho_1}{\rho_g} \; ; \quad N_2 = K_2 \frac{\Delta h}{h_{lg}}$$

$$N_3 = K_3 \frac{\dot{q} \cdot l}{h_{lg} \cdot \dot{m} \cdot d} \qquad N_4 = K_4 \frac{\dot{m}}{\rho_1 \sqrt{g_1}}$$

(14.25)

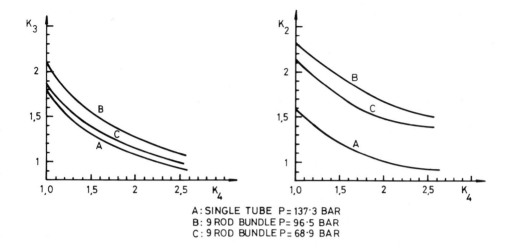

A: SINGLE TUBE P= 137·3 BAR
B: 9 ROD BUNDLE P= 96·5 BAR
C: 9 ROD BUNDLE P= 68·9 BAR

Figure 14.14: Scaling factors for critical heat flux, according to Bouré (1970).

The correction- or scaling factors K_1-K_4 are dependent on the geometry of the channel as well as on the pressure. The values of two of them can be chosen arbitrarily, and by using the same density ratio of liquid and vapour in water and in Freon the factor K_1 becomes unity. The scaling factor K_4 may vary in the range from 1 to 2.5 and by fixing this factor the scaling factors K_2 and K_3 can be taken from Figure 14.14.

Finally we have to mention the method proposed by Ahmad (1971) who also subdivided the parameters in groups, namely system describing, primary and secondary parameters. He gets 13 dimensionless numbers from which he selects 3 for restricted scaling, arranging them as a function of the modelling parameter

$$\psi = f\left(\frac{\dot{m} \cdot d}{\eta_1} , \frac{\eta_1}{\sigma \cdot d \cdot \rho_1} , \frac{\rho_1}{\rho_g}\right) \tag{14.26}$$

In addition he gives a function for the modified boiling number (see Figure 14.2)

$$N_{BO}^* = \frac{\dot{q}}{\dot{m} \cdot h_{lg}} = f\left(\psi, \frac{\Delta h_{in}}{h_{lg}} , \frac{\rho_1}{\rho_g} , \frac{1}{d}\right) \tag{14.27}$$

and from experiments he empirically finds the following relationship between the scaling numbers:

$$\psi = \left(\frac{\dot{m} \cdot d}{\eta_1}\right)\left(\frac{\eta_1}{\sigma \cdot d \cdot \rho_1}\right)^{n_1}\left(\frac{\rho_1}{\rho_g}\right)^{n_2} \tag{14.28}$$

Rearranging the terms in equation (14.28) we realize that the modelling parameter ψ is a function of well-known dimensionless numbers

$$\psi = \underbrace{\left(\frac{\dot{m}^2 \cdot d}{\rho_1 \cdot \sigma}\right)^{0.67}}_{We} \cdot \underbrace{\left(\frac{\dot{m} \cdot d}{\eta_1}\right)^{-0.133}}_{Re_1^*} \cdot \underbrace{\left(\frac{\dot{m} \cdot d}{\eta_g}\right)^{-0.2}}_{Re_g^*} \qquad (14.29)$$

So in addition to Barnett's and Bouré's energy-oriented
dimensionless groups, Ahmad introduced hydrodynamic scaling
numbers.

  Before we check the validity of these scaling laws for
critical heat flux we have to recall that the uncertainty of
burnout measurements in in the order of $\pm$ 10%. The best agreement between scaling and original data can be found for simple
geometries of course. With inside cooling of uniformly heated tubes
agreement within $\pm$ 7% was realized. More interesting for
practical use, but also more complicated due to additional
influencing hydrodynamic parameters, is the scaling of critical
heat flux for rod bundles. Hein and Kastner (1972) compared their
measurements in water and Freon and scaled them according to
Bouré as shown in Figure 14.15. They found that there is an
influence of the mass flow rate on the factor K_3 (see equation
(14.25)), which scales the critical heat flux from Freon to water.
The tests in Figure 14.15 were done with a 9-rod cluster. The
influence of the mass flow rate is relatively small if all 9 rods

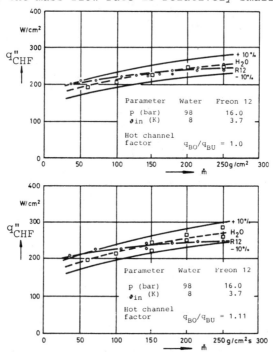

Figure 14.15: Influence of mass flow rate and radial power distribution on
          critical heat flux scaling (Bouré).

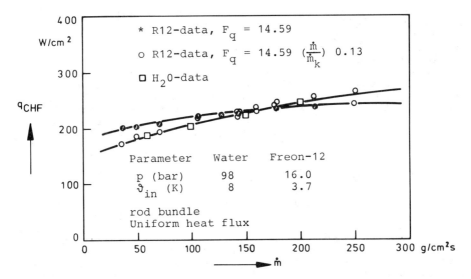

Figure 14.16: Mass flow correction for critical heat flux scaling (Bouré (1970),
          Hein and Kastner (1972).

have the same heat flux, i.e. if there is uniform heat flux
distribution. With increasing hot channel factor the deviation
of the scaling values become greater, which is probably due to a
mixing effect between the subchannels not being taken into
account in the scaling deliberations. This is a clear hint for
the restrictive similarity.

     For uniform heat flux distribution water and Freon data can
be easily brought in agreement by a simple mass flow dependent
correcting function proposed by Hein and Kastner (1972) as
demonstrated in Figure 14.16. It has to be mentioned that in
this comparison of correction taking into account the pressure
dependency of the scaling factor, which was proposed later by
Courtaud and De Bousquet (1973) is not yet considered. For more
detailed information on the agreement between scaling and measured
data of critical heat flux the papers by Hein (1972, 1976) may be
consulted.

     From the comparison in Figure 14.15 we learned that hydro-
dynamic phenomena like mixing may have a strong influence on
critical heat flux scaling. The first reaction on this problem
may be to simulate all hydrodynamic conditions as well as possible,
i.e. to use exactly the same hot channel factor in the water and
in the Freon tests. This, however, presupposes, that the
similarity conditions for mixing and for critical heat flux are
the same or else we can only scale one of both correctly. From this
we can draw the conclusion, that - generally speaking - scaling
with dimensionless numbers only is not sufficient in two-phase
flow. We need a theoretical description of the whole thermodynamic and fluiddynamic process, predicting the behaviour in
the model as well as in the original set-up.

     The circumstances become even more complicated if we consider

transient conditions for example the thermohydraulic behaviour
of the two-phase flow during a blowdown due to a loss of coolant
accident in a reactor. Doing experimental research work for
investigating the fluiddynamic and heat transfer aspects of this
problem we are forced to scale even using water in our experimental set-up, because the rod bundle in the test can only be a
very small section of a whole reactor core, i.e. we have a
geometrical scale down. The number of rods tested is usually
limited by the total power input available to heat them. If we
use Freon instead of water we can multiply the number of rods by
a factor of approximately 15. But then we have to ask how to
scale from Freon to water in addition.

Blowdown tests with Freon were made in our laboratory and
compared with blowdown measurements performed with water by a
German manufacturer of nuclear reactors (KWU, 1974). The
conditions in the Freon loop were very carefully adapted and
scaled to the water test rig. There was a two fold purpose of
these measurements, namely to study the burnout delay time during
blowdown and to determine the heat transfer coefficient under post-
dryout conditions. Depending which of these two phenomena are
intended to be scaled one gets different similarity terms. Here
we just want to discuss in a little more detail the dryout delay
scaling.

The blowdown occurrence is too complex to overcome the scaling
problems simply by using dimensionless numbers. One has to
develop a physical conception of the thermohydraulic events and
formulate it in mathematical expressions as differential
equations. Due to the rapidly reduced pressure,flashing and then
a very high acceleration directed to the break occurs in the fuel
elements. In case of a large break in the primary system of a
pressurized water reactor the pressure would be reduced within
about 20 s from 160 to approximately 5 bar. Emphasis has to be
given to imitate in the Freon loop these very severe flashing
conditions and the resulting acceleration of the fluid. The
acceleration can be guaranteed by adapting the same local and
temporal pressure ratio, which can be done by lowering the outside
pressure of the Freon loop to about 0.1 bar. Then we still have
the problem of critical (sonic) mass flow rate and different heat
storage conditions in the Freon and in the water loop. All these
problems can only be overcome by a detailed theoretical analysis
used as basis for scaling. This analysis has to predict the
results of Freon tests and the water tests as well. The scaling
is not done from one experimental condition to the other but both
test results are compared with theoretical predictions.

The tests were done for boiling water reactor conditions,
where vapour is present already in steady state operation. The
liquid phase is flowing along the heated wall and the vapour is
concentrated in the bulk between the rods. The dryout, i.e. the
disappearance of the liquid layer at the wall then is caused by
the following effects:

(1) Vaporization (heat flux from the rods)

(2) Flashing due to pressure decrease

(3) Entrainment due to high vapour velocity

$$\frac{d}{dz}\left[(\rho_l w_l (1-\epsilon)(1-e)\,\rho_l w_E (1-\epsilon)\cdot e)\right] + \frac{d}{dz}(\rho_g w_g \epsilon) + \frac{d}{dt}\left[(\rho_l (1-\epsilon)(1-e) + \rho_l (1-\epsilon)\cdot e + \rho_g \epsilon)\right] = 0$$

$$\frac{h_l}{h_g}\frac{d}{dz}\left[(\rho_l w_l (1-\epsilon)(1-e) + \rho_l w_E (1-\epsilon)\cdot e)\right] + \frac{d}{dz}(\rho_g w_g \epsilon) + \frac{1}{h_g}\frac{d}{dt}\left[(\rho_l (1-\epsilon)(1-e) + \rho_l (1-\epsilon)\cdot e)h_l + (\rho_g \epsilon\, h_g)\right] = \dot{q}(z)\cdot\frac{U}{A_T}\cdot\frac{1}{h_g}$$

$$\dot{M}_l = \rho_l\, A_T\,(1-\epsilon)(1-e)\cdot w_l$$

$$\dot{M}_E = \rho_l\, A_T\,(1-\epsilon)\cdot e\cdot w_E$$

Figure 14.17: Mass and energy equation for dryout delay during blowdown.

Velocities of phases

$$w_{ph} = \sqrt{\left(\frac{dp}{dz}\right)_{ph}\cdot\frac{2\,dh}{(\xi\cdot\rho)_{ph}}}$$

$$\left(\frac{dp}{dz}\right)_{ph} = \frac{1}{\phi^2_{ph}}\cdot\left(\frac{dp}{dz}\right)_{2-ph}$$

$$\frac{1}{\phi_l^2} = (1-\epsilon)^2 \quad,\quad \frac{1}{\phi_g^2} = \frac{\epsilon^{2.5}}{1+75\cdot(1-\epsilon)}$$

$$w_l = \sqrt{\frac{A_1}{A_T}\cdot\left(\frac{dp}{dz}\right)_{2-ph}\cdot\frac{2\,dh}{\xi\,r\,\rho_l}}$$

$$w_g = \sqrt{\frac{\sqrt{A_T\cdot(A_T-A_1)}}{A_T - 75 A_1}\cdot\left(\frac{dp}{dz}\right)_{2-ph}\cdot\frac{2\,dh}{\xi\cdot\rho_g}}$$

Figure 14.18: Dryout delay during blowdown, equations describing phase velocities.

We can describe the whole procedure with the well-known conservation equations for mass, energy and momentum. In addition one needs information on the slip ratio because the difference in the phase velocity produces entrainment.

Belda (1975) developed an analysis on this basis starting from the equations given in Figure 14.17 and using for the phase velocities the equations derived from pressure drop considerations as pointed out in Figure 14.18. The entrainment behaviour during the transient process was estimated with the aid of investigations carried out by Hewitt (1970). With this mathematical treatment he gets a differential-integral equation describing the film thickness as a function of time and place:

$$\frac{1}{\rho_l\cdot l}\left(\dot{M}_l(l,t)-\dot{M}_l(0,t)+\dot{M}_{in}(l,t)\right) + \frac{\partial A_1}{\partial t} + A_1\left(\frac{1}{\rho_g}\cdot\frac{\partial p_1}{\partial t}\right) +$$

$$\frac{1}{h l_g}\left(\frac{\rho_g}{\rho_1}\cdot\frac{\partial h_g}{\partial t}\cdot\frac{\partial h_1}{\partial t}\right) + A_T\left(\frac{\int\dot{q}(z)\cdot dz\cdot U}{\rho_1\cdot h_{1g}\cdot l\cdot A_T} - \frac{\rho_g}{\rho_1}\cdot\frac{1}{h_{1g}}\cdot\frac{\partial h_g}{\partial t}\right) = 0 \qquad (14.30)$$

In equation (14.30) the local and temporal values of mass flow rate and quality in the fuel element or in the test section have to be known from blowdown calculations, which can be performed for example with the well-known computer codes like RELAP IV or BRUCH. We then can determine from equation (14.30) at what time after the break the liquid layer at a certain position of a rod

in the bundle would become zero, which means that the dryout
occurs. For adapting the Freon tests conditions one can derive
from equation (14.30) and from the equations in Figure 14.17
and 14.18 dimensionless scaling numbers, too. They are listed in
Figure 14.19. From this figure we can see that the system pressure
is four times over determined. We therefore have to decide,
which of these four expressions is the dominant one. It can be
easily seen that the dependence of the liquid enthalpy from
pressure is much smaller than the change of latent heat of
evaporization with pressure. So two of the conditions can be
ruled out and the remaining two fortunately give quite similar
values in the order of 0.19. But there is a 5th condition because
we decided to scale with the critical pressure ratio and from this
we get the value of 0.188. So finally 0.19 could be chosen as a
good compromise.

By scaling the hydrodynamic and thermodynamic conditions in
the Freon loop before and during the blowdown with the dimensionless numbers given in Figure 14.19 we got a quite good agreement
between water and Freon behaviour as can be seen from Figure
14.20, where the temporal course of system pressure, mass flow
rate and void fraction and the pressure drop in the fuel element
are plotted for water and for Freon versus the blowdown time.
This conformity, however, does not mean that the dryout delay
times are the same in Freon and in water because we have only made
the hydrodynamic side of the process analogous and we could not
completely scale thermodynamic details, for example, stored heat
in the construction materials and the complete coincidence between
thermodynamic and hydrodynamic properties. This means we have to
do an indirect scaling by comparing the measured data with the
theoretically predicted ones. As Figure 14.21 demonstrates there
is good agreement between measured and predicted dryout delay
time. The comparison was done for different locations of the
break - in the hot and in the cold leg of the primary system -
for a variety of break areas and for different heat flux densities
in the moment before the accident occurs.

Starting from fundamental deliberations the theoretical
approach should predict the behaviour in water as well as in
Freon. Unfortunately there are available only very few water
tests where the dryout delay time during blowdown was measured
and where in addition the temporal course of the hydrodynamic
conditions in the fuel elements is reliably known. Comparison
with water experimental data at hand show good agreement with the
theory as can be seen from Figure 14.22. This encourages drawing
the conclusion that scaling on the basis of a theoretical
analysis is a promising procedure for combined and complicated
two-phase flow phenomena.

To support this comparison between measured Nusselt-numbers
in water and Freon during blowdown may be shown in Figure 14.23
without discussing the scaling laws for this special problem. In
this figure measurements done with water (KWU, 1974) and with
Freon (Viert & Mayinger, 1975) are compared with correlations.
Keeping in mind that there are many uncertainties arising from
the measuring technique, the agreement is satisfactory. It has to
be added, that the comparison can be made only in the period
between 5 and 12 s after the break appears, because before we

heat flux

$$K_1 = \frac{\dot{q}(z)\cdot|\cdot U_B}{m_T \cdot A_T \cdot h_{1g}}$$

pressure

$$K_{21} = \frac{P_s}{\rho_1}\cdot\frac{\partial\rho_1}{\partial p} \qquad K_{23} = \frac{P_s}{h_{1g}}\cdot\frac{\rho_g}{\rho_1}\cdot\frac{\partial h_1}{\partial p}$$

$$K_{22} = \frac{P_s}{h_{1g}}\cdot\frac{\partial h_1}{\partial p} \qquad K_{24} = \frac{P_s}{h_{1g}}\cdot\frac{\partial h_{1g}}{\partial p}$$

acceleration

$$K_3 = \frac{\Delta p \cdot \rho_1}{\dot{m}_T^2} = Eu_1$$

$$K_4 = \frac{\Delta p \cdot \rho_g}{\dot{m}_T^2}\cdot\frac{\rho_g}{\rho_1} = Eu_g\cdot\frac{\rho_g}{\rho_1}$$

mass flow rate

$$K_5 = \xi_1 \simeq Re = \frac{\dot{m}_T \cdot d}{n_1}$$

geometry

$$K_6 = \frac{\sqrt{A_T}}{T} \simeq \frac{d_h}{T}$$

Figure 14.19: Dimensionless groups for dryout delay scaling during blowdown (Belda, 1975).

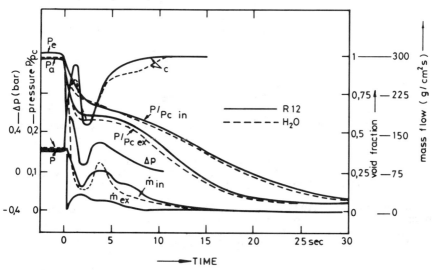

Figure 14.20: Blowdown during loss of coolant accident, comparison between water tests and scaled Freon 12 data.

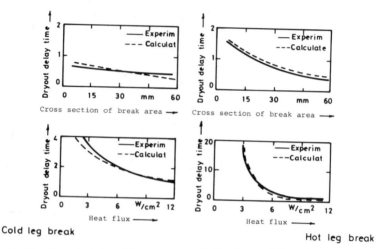

Cold leg break   Hot leg break

Figure 14.21: Dryout delay time, measured (in Freon 12) and calculated data.

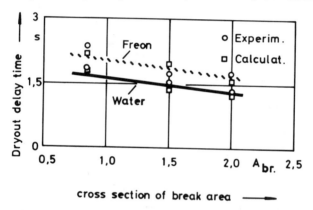

cross section of break area ⟶

Figure 14.22: Dryout delay time, comparison of experimental and theoretical
data for water and for Freon 12.

have transition boiling for which the used scaling model is not
valid and afterwards the hydrodynamic conditions in the water and
in the Freon loop differ considerably.

There are certainly a lot of other complicated two-phase flow
problems which are of great technical interest. These include
the phenomena of subcooled boiling, of flow oscillations and
instabilities and the interchannel mixing.

14.7 Conclusions

Even in single phase flow scaling is restricted under certain
circumstances as we have seen in the example of combined forced
and free convection. This limitation becomes much stronger in
two-phase flow. Whilst in a single phase fluid we have usually

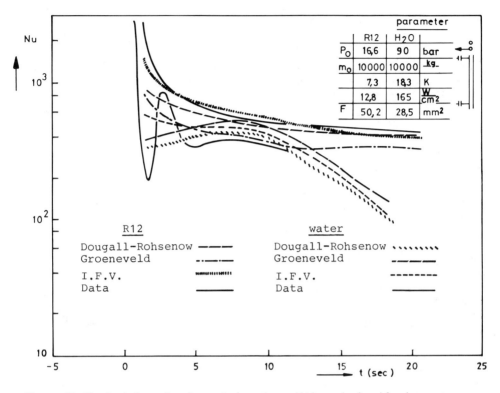

Figure 14.23: Heat transfer in post dryout conditions during blowdown, comparison of water data and scaled Freon data.

similarity for hydrodynamic and heat transfer processes simultane-ously, in two-phase flow each modelling law or scaling number is only valid for a single special phenomenon.

Scaling only with the help of dimensionless numbers is limited in two-phase flow to simple and isolated problems where the physical phenomenon is a unique function of a few parameters. If there is a reaction between two or more physical occurrences, dimensionless scaling numbers mainly serve for selecting the hydrodynamic and thermodynamic conditions of the modelling tests and we have to separate the influencing parameters in primary ones determining the system and secondary ones which are of minor influence.

In not too complicated cases scaling to the original circumstances can be done by empirical correlations considering the important physical laws of the process. For many technical applications in the future we shall be forced to scale via a computer code which analyses and describes the thermohydraulic phenomena as well as possible.

This makes it necessary to get a better understanding of the thermohydraulic behaviour in two-phase flow systems. A better

theoretical approach would also help us to generalize experimental
data and such to make full use of the numerous experimental results
in two-phase flow, which up to now stand to a great extent apart
and uncomparable. So scaling is not only an expedient to save
expense for experimental research work but it also helps to compare
and generalize measured data. To elaborate scaling laws for two-
phase flow, we will have to put more effort to the theoretical
analysis of this process.

Nomenclature

A — cross sectional area
c, c_p — specific heat at constant pressure
d, D — diameter
g — acceleration of gravity
H — height
h — enthalpy
Δh — enthalpy difference
h_{lg} — latent heat of evaporation
l — length
$\dot{m}$ — mass velocity
M — mass flow rate
p, ρ — pressure
Δp — pressure drop
$\dot{q}, q''$ — heat flux
s — slip
T — temperature
t — time
U — perimeter
u_o'' — superficial velocity of vapour
$U_{D,m}$ — vapor drift velocity
v — volume
w — velocity
$\dot{x}$ — quality
x, y, z — length coordinates

Greek symbols

ψ_{FB} — flow pattern number
α — heat transfer coefficient
ε — void fraction
λ_{FB} — flow pattern number
λ — thermal conductivity

η — dynamic viscosity
ϕ — two-phase parameter
ρ — density
σ — surface tension
ν — kinematic viscosity
θ — temperature
$\Delta\theta$ — temperature difference

Dimensionless groups (see also p. 427)

Fr — Froude-Number
Gr — Grashof-Number
N — bubble-Number
N_{BO} — boiling-Number

Nu   - Nusselt-Number
Pr   - Prandtl-Number
Re   - Reynolds-Number

Subscripts

b,bub - bubble
c   - critical
CHF   - critical heat flux
cont - container
d   - drop
ent   - entrainment
ex   - outlet
g   - vapour
h   - hydraulic
H_2O - water
in   - inlet
l   - liquid
lg   - evaporation
m   - model
o   - original
s   - saturation
sub   - subcooling
T   - total

superscript * denotes superficial; ' denotes liquid; " denotes vapour

References

Ahmad, S, (1971) "Fluid to fluid modelling of critical heat flux: a compensated distortion model". AECL-3663.

Baker, J L, (1965) "Flow regime transitions in vertical two-phase flow". ANL-7093.

Barnett, P, (1963) "The scaling of forced convection boiling heat transfer." UKAEA-AEEW-R R134.

Belda, W, (1975) "Dryoutverzug bei kühlmittelverlust in kernreaktoren." Dissertation, T.U. Hannover.

Belda, W, Mayinger, F and Viert, K, (1974) "Blowdownuntersuchungen mit dem kaeltemittel R12." Proceedings der Reaktortagung des DAF, Berlin, S. 7.

Bennett, A, et al, (1965) "Flow visualization studies of boiling at high pressure." AERE R4874.

Bouré, J, (1970) "A method to develop similarity laws for two-phase flows." ASME paper 70-HT-25.

Calus, W F and Denning, R K, (1973) "Heat transfer in a natural circulation single tube reboiler, Part I: Single component liquids." *Chem. Eng. J.*, 233-250.

Courtaud, M and Du Bousquet, J L, (1973) "Freon water modelling of critical heat flux in rod bundles". European Two Phase Flow Group Meeting, Brussels, June 1973, Paper B1.

Friedel, L, (1975) "Modellgesetze fuer den reibungsdruckverlust in der Intiphasenstroemung". VDI Forschungsheft 572.

Crigull, U, u.a. (1968) "Die Eigenschaften von wasser und wasserdampf nach der 1968 IFC-Formulation". *Waerme- und Stoßßuebertragung*, **1**, 202-213.

Hein, D, (1976) "Modellgesetze fuer stationare burnoutvorgaenge." Kolloquium ueber Aehnlichkeitsgesetze in Zweiphasenstroemungen, Hannover, 25 Feb. 1976, (available from Institut fuer Verfahrenstechnik, U Hannover).

Hein, D and Kastner, W, (1972) "Versuche zur heizflaechenbelastung unter stationaeren und instationaeren bedingungen in Freon 12 und wasser." M.A.N.-Bericht 45.03.02.

Hewitt, G F and Hall-Taylor, N S, (1970) "Annular two-phase flow". Oxford.

Kowalczweski, J J, (1964) "Two-phase flow in an unheated and heated tube." Dissertation ETH Zuerich.

Kütükcüoglu, A and Njo, D H, (1967) "Dampfvolumenanteil beim Verdampfen." Technische Mitteilung, TM-JN-307, EIR-Würenlingen.

KWU-R5-2885, (1974) Abschlußbericht ueber die abblaseversuche mit 4-stabbuendein."

Martinelli, R and Lockhart, R, (1949) "Proposed correlation of data for isothermal, two-phase component in pipes." *Chem. Eng. Progr.* **45**, 39.

Nabizadeh, H, (1976) "Uebertragungsgesetze fuer den dampfvolumenanteil zwischen Freon und wasser, Kolloquium ueber aehnlichkeitsgesetze in Zweiphasenstroemungen." Institut fuer Verfahrenstechnik der T.U. Hannover.

Quandt, E, (1963) "Analysis of gas-liquid flow patterns." AIChE, Preprint 47, *6th National Heat Transßer Conßerence*, Boston.

Schrock, V E and Großmann, (1962) "Forced convection boiling in tubes." *Nucl. Sci. and Eng.* **12**, 474-481.

Stephan, K, (1964) "Beitrag fuer Thermodynamik des Waermeueberganges beim Sieden." Abh. des Kaeltetechnischen Ver. Nr. 18, Verlag C.F. Mueller, Karlsruhe.

Stevens, G and Staniforth, R, (1965/66) "Experimental studies of burnout using Freon 12 at low pressures with reference to burnout in water at high pressure." *Proc. oß the Inst. oß Mech. Eng.*, **180**, Part 3, Paper 10.

Stevens, G and Macbeth R, (1970) "The use of Freon 12 to model forced convection burnout in water." ASME paper 70-HT-20.

Viecenz, H J, (1980) "Blasenaufstieg und phasenseparation in behaeltern bei dampfeinleitung und druckentlastung." Dissertation, Universitaet Hannover.

Viert, K and Mayinger, F, (1975) "Vergleich verschiedener waermeuebergangsbeziehungen im filmsiedebereich bei blowdownuntersuchungen." Paper 117, Reaktortagung, Nuernberg.

Zetzmann, K and Mayinger F, (1976) "Stroemungsformen - aehnlichkeitsbeziehungen zwischen wasser und R-12." Proc. des Kolloquiums ueber Aehnlichkeitsgesetze in Zweiphasenstroemungen von Wasser und R 12, Institut fuer Verfahrenstechnik der T.U. Hannover.

Zuber, N and Findlay, J, (1965) "Average volumetric concentration in two-phase flow systems." *Journal oß Heat Transßer*, **87**, 453-468.

Chapter 15

Instrumentation

J. M. DELHAYE

15.1 Introduction

Due to the need for measurements of practical and fundamental parameters, a great deal of work has been expended in the study of measuring techniques to be used in two-phase gas-liquid flows. A decade ago, an in-depth review of two-phase flow instrumentation was edited by Letourneau and Bergles (1969) including already 368 references. More recently, several detailed summaries of measurement techniques were given for particular two-phase flow applications (Table 15.1).

TABLE 15.1

A List of Recent Reviews on Two-Phase Flow Instrumentation

Authors	Date	Topics
Brockett and Johnson	1976	Water reactor safety studies
Jones and Delhaye	1976	Transient and statistical aspects of two-phase flows
Leonchik et al	1976	Dispersed flows
Moore	1977	Wet steam flows
Azzopardi	1977	Drop size
Hewitt	1978	Two-phase flow experimental studies
Solesio	1978	Film thickness
Delhaye	1978	Optical techniques
Hewitt and Whalley	1979	Optical techniques
Veteau	1979	Interfacial area

Apart from the problems encountered in any measuring method used in single-phase flow and which are amply described in the works of Katys (1964), Doebelin (1966), Benedict (1969), Bradshaw (1971), Considine (1971), Dowdell (1974) or Richards (1977), further difficulties appear when attempting to measure quantities in two-phase flow. The majority of these difficulties are related to the presence of the two phases but others are due to the conditions of flow confinement. These difficulties are exemplified in the introduction to the paper by Jones and Delhaye (1976). Difficulties due to the presence of two phases may include the hydrodynamic response time due to the interaction between probes and interfaces, the entrapment of gas bubbles in the manometric lines, the occurrence of significant fluctuations in the quantities to be measured, cavitation effects, ... In addition to these, there are also difficulties due to flow geometries such as opaque metallic walls imposed by temperature and pressure conditions, or small hydraulic diameters such as in rod bundles.

Despite these difficulties, several techniques can be used to measure practical or fundamental parameters with a fair degree of accuracy. The purpose of the following is to present the principles of a few methods which can be considered as well-established. For reviews covering a broader spectrum the reader is referred to the list of Table 15.1.

15.2 Photon Attenuation Techniques

Radiation attenuation methods furnish the line and area void fractions R_{G1} and R_{G2} which were defined in the chapter entitled "Two-Phase Flow Patterns". They are the most frequently used void fraction measurements and can provide detailed information on the flow pattern (Jones 1973; Jones and Zuber, 1975).

15.2.1 Absorption Law

A beam of monochromatic, collimated photons, of incident intensity I_0, traversing a substance of thickness e and density ρ, has an emerging intensity I given by the following exponential absorption law,

$$I = I_0 \exp\left[-(\mu/\rho)\rho e\right] \qquad (15.1)$$

The quantity μ/ρ is the specific absorption coefficient of the material. The coefficient generally decreases piecewise with the energy of the photons and is independent of the physical state of the substance (solid, liquid or gas). Tables giving the values of this coefficient for different materials can be found in McMaster et al (1969), Storm and Israel (1970) and Stukenbroeker et al (1970). Figure 15.1 shows the mass absorption coefficient μ/ρ as a function of the incident beam energy for water.

15.2.2 Line Void Fraction Measurements

When a collimated beam is used the radiation is absorbed by the wall and the two phases in a series mode. At a given time the attenuation of the beam is given by the following relation,

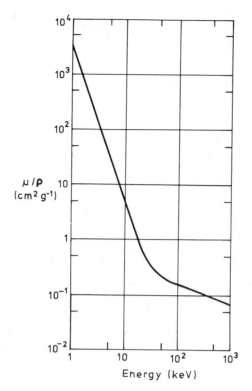

Figure 15.1: Mass Absorption Coefficient of Water as a Function of the Incident Beam Energy

$$I = I_o \exp(-\mu_p e_p) \exp\left[-\mu_L(1-R_{G1})d\right] \exp\left[-\mu_G R_{G1} d\right] \qquad (15.2)$$

where e_p is the total wall thickness, d the distance between the walls and μ_p, μ_G, μ_L the absorption coefficients of the wall, the gas and the liquid.

(i) For *low pressure gas-liquid flows at ambient temperature* (Galaup, 1975), the absorption by the gas can be considered negligible compared with that of the liquid. As a result, it is only necessary to measure intensities I_G and I_L corresponding respectively to the channel filled with gas and filled with liquid. The *instantaneous* line void fraction is then easily obtained, provided I_O is a constant,

$$R_{G1} = \frac{\text{Log } I/I_L}{\text{Log } I_G/I_L} \qquad (15.3)$$

(ii) For *high-pressure steam-water flows* we have,

$$\frac{\mu_L}{\rho_L} = \frac{\mu_G}{\rho_G} \qquad (15.4)$$

but the steam density is no longer negligible compared with that of the water. Consequently a calibration in conditions as close as possible to experimental conditions will be necessary (Martin, 1969, 1972; Reocreux, 1974).

15.2.3 Area Void-Fraction Measurements

The *instantaneous* area void fraction R_{G2} can be measured by means of two methods,

(i) In the *one-shot technique* (Nyer, 1969; Gardner et al., 1970; Charlety 1971), the radiation crosses the whole of the tube cross section containing the two-phase mixture. The beam height must be small compared with the tube diameter in order to restrict the measurement to a given cross section. Calibration with lucite mockups are generally necessary due to the lack of a specific absorption law.

(ii) The *multibeam gamma densitometer* has been used for a few years by laboratories involved in water reactor safety studies. Figure 15.2 shows a schematic of the three-beam densitometer used by Banerjee et al.(1978). A 25 Curie caesium 137 source provides three gamma ray beams which are directed through the pipe in the same cross section plane. The beams are attenuated according to the line void fractions and the measurement of the attenuation of each beam can be used to determine the area void fraction R_{G2}. This technique requires a model connecting the area void fraction, the flow pattern and the three measured line void fractions.

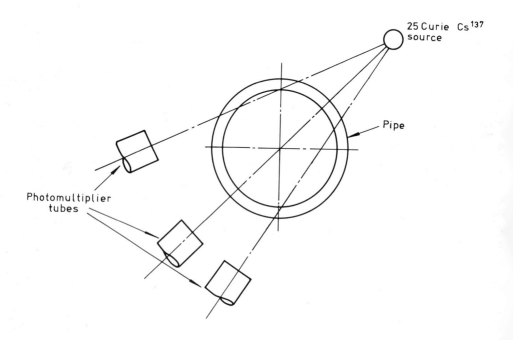

Figure 15.2: Three-Beam Gamma Densitometer (Banerjee et al, 1978)

The *time-averaged* area void fraction $\overline{R_{G2}}$ can be determined from a profile of time-averaged chordal void fraction $\overline{R_{G1}}$. In a pipe of circular cross section we have

$$\overline{R_{G2}} = \frac{1}{\pi R^2} \int_{y=-R}^{y=+R} 2\sqrt{R^2-y^2}\ \overline{R_{G1}}(y)\,dy \qquad (15.5)$$

where R is the pipe radius and y the distance from the axis of the pipe to the photon beam. Note that if the flow is axisymmetric the local void fraction α_G is a solution of the following Abel integral equation,

$$\int_{y}^{R} \frac{r\alpha_G(r)}{\sqrt{r^2-y^2}}\,dr = \overline{R_{G1}}(y)\ \sqrt{R^2-y^2} \qquad (15.6)$$

where r is the radial coordinate.

15.2.4 *Errors due to Fluctuating Voids*

In steady-state two-phase flow the emerging intensity I (in photons/s) is generally measured over a certain period θ (10 s to 1 min). Consequently the measurement gives actually the number N of emerging photons counted over the time interval θ,

$$N \overset{\Delta}{=} \int_{t-\theta/2}^{t+\theta/2} I\,dt \qquad \text{(in photons)} \qquad (15.7)$$

Since the average of an exponential is not equal to the exponential of the average, the measurement does not give in fact the void fraction $\overline{R_G}$ averaged over the time interval θ. This error can be considerable for highly fluctuating flows. The deviation can range from 0.05 for churn-flow up to 0.20 for slug-flow. However, the measurement of time-averaged void fractions in fluctuating flows can be handled by one of the following methods,

(i)   The averaged void fraction $\overline{R_G}$ can be determined from the probability density function of the instantaneous void fraction R_G obtained for a very short counting time (eg 0.05 s). This technique has been reported in a series of papers by Hancox et al (1972), Harms and Laratta (1973), and Laratta and Harms (1974).

(ii)   Levert and Helminski (1973) used a dual energy method based on the differential absorption of radiation energy.

(iii) Log amplifiers were used by Jones (1973) to linearize the instantaneous signal. The only requirement is that the response time of the amplifier should be smaller than the lower period of the physical fluctuations.

15.2.5 *Choice of Radiation*

A radiation is characterized by its *energy spectrum* and its *intensity.* The *contrast* is defined by the relationship,

$$c \triangleq \frac{I_G}{I_L} \qquad\qquad\qquad (15.8)$$

where I_G and I_L correspond respectively to the channel filled with gas and filled with liquid. For gas-liquid flow at low pressure,

$$c = \exp (\mu_L d) \qquad\qquad\qquad (15.9)$$

The greater the contrast, the higher the sensitivity of the measurement. To increase the contrast, the mass absorption coefficient μ_L/ρ_L must be increased and therefore the energy decreased (Fig. 15.1).

On the other hand, the *photon emission fluctuation* leads to the following statistical relationship,

$$\Delta N/N = (1/N)^{\frac{1}{2}} \qquad\qquad\qquad (15.10)$$

where N is the number of photons counted. For purposes of accuracy N must therefore be sufficiently high. As a result, either the counting time θ or the emerging intensity I must therefore be sufficiently high. Consequently, if the incident intensity I_0 is fixed, the mass absorption coefficient must be decreased, and therefore the energy increased.

Constraints due to the contrast and to the photon emission fluctuations have therefore opposite effects on the choice of the photon energy and of the source intensity. In this respect, X-rays have beam intensity 10^3 to 10^4 times higher than γ-rays.

The photon beam must be *monochromatic* for the exponential absorption law (Eq. 15.1) to be applicable. In this respect, γ-ray emission is more monochromatic than X-ray bremsstrahlung emission. However, filters or electronic discriminators can be used to select or detect only those photons with energy in a certain band.

The *stability* of the source in time is another important parameter of measurement accuracy. In this respect, γ-rays are to be preferred to X-rays due to the rather long half-lives of the principal sources of γ-rays. However, certain methods can be used to avoid the inconvenience of X-rays generator drifts and/or fluctuations (Nyer, 1969; du Bousquet, 1969; Jones, 1973; Lahey, 1977; Solesio, 1978).

15.2.6 *Radiation Detection*

The emerging intensity I is measured with a scintillator coupled to a photo-multiplier supplied with correctly stabilized high-voltage. Generally, a pre-time method is used in which the pulses are counted during a predetermined time. Xenon ionisation chambers seem to be more stable than photo-multipliers but further studies are needed to confirm this fact.

15.2.7 *Examples*

(i) *Steady-state steam-water flow at high pressure*
Martin (1969, 1972) measured the line void fraction at high
pressure (up to 140 bar) in a rectangular channel (50 x 2.8
mm) simulating a sub-channel of a nuclear reactor plate-type
fuel element. The test section consisted of two vertical
plates heated by Joulean effect and was enclosed in a thick
casing drilled with holes which allowed a horizontal X-ray
beam to pass through the two 0.5 mm thick lateral sides
(Fig. 15.3). This X-ray beam, 2 mm in height and 0.05 mm in
width, scanned the test cross section in a movement parallel
to the heated plates. The line void fraction $\overline{R_{G1}}$ of the
upward steam-water flow was measured every 0.025 mm with a
position accuracy of about 0.005 mm despite the severe
experimental conditions (ie fluid pressure : 140 bar; fluid
temperature : 335°C; ambient temperature : 50°C). Typical
distributions of the line void fraction $\overline{R_{G1}}$ are given in
Figure 15.3.

(ii) *Liquid film thickness measurements*
The thickness of a wavy liquid film flowing down an inclined
plane is measured by Solesio et al (1978) by means of an
X-ray absorption technique (Fig. 15.4). The film thickness

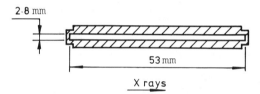

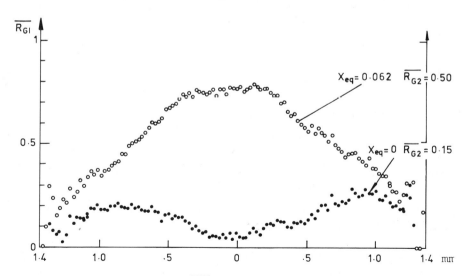

Figure 15.3: Line void fractions $\overline{R_{G1}}$ in steady-state steam-water flow at high
pressure (Martin, 1969, 1972). Pressure : 80 bar; heat flux
density : 110 W cm^{-2}; mass velocity : 220 g cm^{-2}s^{-1}; x_{eq} : equilibrium quality.

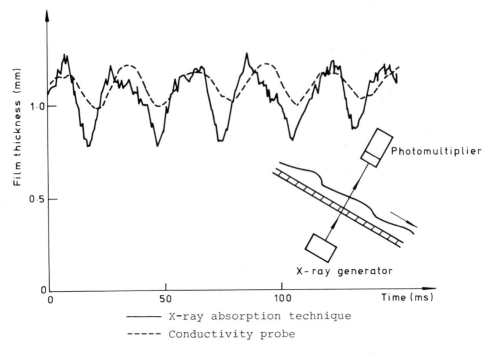

Figure 15.4: Instantaneous Liquid Height (Solesio et al, 1978)

whose averaged value is about one millimeter, is determined each millisecond with a precision finer than 0.050 mm. Figure 15.4 shows the film thickness versus time when 35 Hz waves are generated at the free surface of the liquid film. The continuous line represents the signal obtained by means of the X-ray absorption technique whereas the dashed line represents the signal obtained by means of a conductivity probe which measures the liquid conductance between two electrodes mounted flush with the wall.

15.3   Electromagnetic Flowmeters

15.3.1   *Use in Single-Phase Flow*

The theoretical and practical aspects of the use of electromagnetic flowmeters in single-phase flow are described in several papers or articles (Shercliff, 1962; Turner, 1968; Treenhaus, 1972; Bevir, 1973; Demagny, 1976). Whatever the conducting fluid used, the electromagnetic flowmeter gives a signal proportional to the area-averaged liquid velocity,

$$e = k \, B \, d \, \{w\} \tag{15.11}$$

where e is the induced emf, k a calibration constant, B the magnetic induction, d the tube diameter and $\{w\}$ the area-averaged fluid velocity in the cross section containing the electrodes.

The most commonly used flowmeter is the transverse flowmeter. The electrodes are placed diametrally opposite in a plane perpendicular to the plane formed by the magnetic induction and the flow axis. The magnetic induction is either constant or alternating. Constant induction flowmeters generally use a permanent magnet and their utilization is limited to liquid metals (Thatcher, 1972). Alternating induction flowmeters are used when the liquid is a poor conductor. In this case, the generation of a magnetic induction sufficiently high to generate a satisfactory induced emf would require too powerful direct currents. The use of an alternating field also avoids polarization of the electrodes in the case of electrolytes.

15.3.2 *Use in Steady-State Two-Phase Flow*

Although several attempts have been made to establish a complete theory on electromagnetic flowmeters in two-phase flow (Alad'yev et al, 1971; Fitremann, 1972, 1972a), their use is generally based on the assumption that they measure the area-averaged velocity of the continuous conducting phase in contact with the electrodes (Charlety, 1971). Hence, in dispersed annular flow, the electromagnetic flowmeter gives a signal proportional to the area-averaged velocity of the liquid film without taking into account the presence of liquid drops in the gas core.

Usually one flowmeter is placed in a single-phase zone of the flow (index ') while a second flowmeter is placed in the two phase zone under consideration (index "). We have,

$$e' = K'B'd \{w_L'\}_2 \tag{15.12}$$

$$e'' = K''B''d <w_L''>_2 \tag{15.13}$$

$$M'_L = \rho_L \{w_L'\}_2 A \tag{15.14}$$

$$M''_L = \rho_L <w_L''>_2 A(1-R_{G2}) \tag{15.15}$$

where M is the mass flowrate, ρ_L the liquid density assumed to be a constant, A the constant area of the tube cross section and R_{G2} the area fraction.

(i) *Two-component steady-state two-phase flow* (Heinemann et al, 1963; Hori et al, 1966; Milliot et al, 1967).

In that case,

$$M_L'' = M_L' \tag{15.16}$$

Hence,

$$R_{G2} = 1 - \frac{\{w_L'\}_2}{<w_L''>_2} = 1 - \frac{K''B''e'}{K'B'e''} \tag{15.17}$$

(ii) *One-component steady-state two-phase flow* (Heinemann, 1965; Lurie, 1965; Charlety, 1971)

In that case,

$$M_L'' = (1-x)M_L' \tag{15.18}$$

where x is the quality.

Hence,

$$R_{G2} = 1 - (1-x)\frac{\{w_L'\}_2}{<w_L''>_2} = 1 - (1-x)\frac{K''B''e'}{K'B'e''} \tag{15.19}$$

If x is not negligible compared to unity, it must either be measured or computed with a model to obtain the void fraction. In return, if the void fraction is known by another method, the quality can be determined.

15.4 Turbine Flowmeters

In *single-phase flow* one can define a flow coefficient C and a rotation Reynolds number N by the following relations,

$$C \triangleq \frac{Q}{nD^3} \tag{15.20}$$

$$N \triangleq \frac{nD^2}{\nu} \tag{15.21}$$

where Q is the volumetric flowrate, n the rotational frequency, D the pipe diameter and ν the kinematic viscosity of the fluid. Hochreiter has shown (1958) that the flow coefficient C is a function of the rotation Reynolds number N. However the flow coefficient appears to be a constant in a wide range of N and in this case the rotational frequency n is proportional to the volumetric flowrate. Introducing the mass velocity G and the fluid density ρ, we thus obtain,

$$n = K \frac{G}{\rho} \tag{15.22}$$

where K is a dimensional calibration constant.

With respect to *two-phase flow* the same type of formula as Eq. (15.22) has been looked for, relating the rotational frequency n, the total mass velocity G of the mixture, a two-phase density ρ and using the dimensional calibration constant K determined for single phase liquid flow.

The models encountered in the literature are based on an analysis of the different forces acting on the turbine blades.

There exist primarily two models: the first one was proposed
by Popper (1961, 1961a) and tested by Rouhani (1964) and the second
one was given by Aya (1975). These two models give results which
are practically equivalent.

The force balance written by Popper (1961) leads to the
following relation,

$$n = KA<w_L> \frac{(1-R_{G2})\rho_L + R_{G2}\rho_G\gamma^2}{(1-R_{G2})\rho_L + R_{G2}\rho_G\gamma}$$
(15.23)

where A is the pipe cross section area, $<w_L>$ the area-averaged
velocity of the liquid phase, R_{G2} the area void fraction, ρ_G and
ρ_L the gas and liquid densities and γ the slip ratio defined by,

$$\gamma \overset{\Delta}{=} \frac{<w_G>}{<w_L>}$$
(15.24)

where $<w_G>$ is the area-averaged velocity of the gas phase. If the
true quality is denoted by x, equation (15.23) becomes,

$$n = KA<w_L> \left[1 + x (\gamma-1)\right]$$
(15.25)

Comparing equations (15.22) and (15.25) and taking account of the
following equation,

$$(1-x) G = (1-R_{G2}) \rho_L <w_L>$$
(15.26)

we obtain an expression for the equivalent density,

$$\frac{1}{\rho} = \frac{(1-x)^2}{(1-R_{G2})\rho_L} + \frac{x^2}{R_{G2}\rho_G}$$
(15.27)

This equivalent density was also found by Rouhani (1964) who
started from a different expression of the forces acting on the
turbine blades. It agrees fairly well with the experimental
results obtained by Rouhani in steam water flows with the following
conditions:

. Pipe diameter : 6mm

. Pressure : 10 to 50 bar

. Mass velocity : 516 to 1850 $kgm^{-2}s^{-1}$

. Quality x : 0.0015 to 0.36

. Void fraction R_{G2} : 0.01 to 0.90

Frank et al (1977) used the same equivalent density but based
on a fictitious void fraction R_M determined as a function of the

quality and the pressure by a calibration on an argon-water loop. An example of a calibration curve is given in Figure 15.5. The tests carried out by Frank et al (1977) cover the following range,

- Pressure : 18 to 83 bar (equivalent to steam-water flows between 40 and 140 bar)

- Pipe diameter : 27 mm

- Quality : 0.03 to 0.75

- Void fraction : 0.30 to 0.99

- Mass flowrate : 0.10 to 0.80 kgs^{-1}

Aya (1975) used a different expression for the force balance on the turbine blades. He derives the following equation,

$$n = KA<w_L> \frac{\gamma + \sqrt{\dfrac{(1-R_{G2})\rho_L}{R_{G2}\rho_G}}}{1 + \sqrt{\dfrac{(1-R_{G2})\rho_L}{R_{G2}\rho_G}}} \tag{15.28}$$

The equivalent density to be used in equation (15.22) was then given by,

$$\frac{1}{\rho} = \frac{x}{R_{G2}\rho_G + \sqrt{R_{G2}(1-R_{G2})\rho_G\rho_L}} + \frac{1-x}{(1-R_{G2})\rho_L + \sqrt{R_{G2}(1-R_{G2})\rho_G\rho_L}} \tag{15.29}$$

Some recent experiments by Banerjee and Jolly (1978) have shown that this equation could be slightly better than Rouhani's equation (15.27) for the following conditions in steam-water flows,

- Pressure : 37 to 54 bar

- Pipe diameters : 38.1, 49.2 and 73.7 mm

- Liquid mass flowrate : 0 to 22.2 kgs^{-1}

- Steam mass flowrate : 0 to 3.9 kgs^{-1}

These models constitute exploratory investigations which have to be supplemented by a more rigorous and detailed analysis of the turbine flowmeter in steady-state and transient two-phase flow. A step in this direction has been made recently by Kamath and Lahey (1977) but more results are still sorely needed.

15.5 Variable Pressure Drop Meters

When the total mass flowrate is known, orifice plates and venturi tubes can be used in two-phase flow to measure the gas

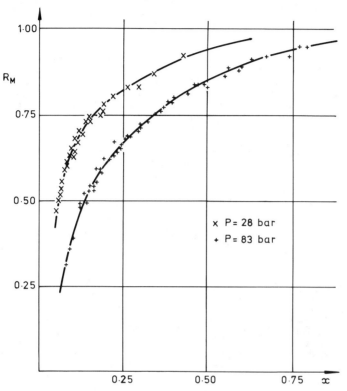

Figure 15.5: Calibration Curve used for the Determination of the Equivalent
Density in Equation (15.27) (Frank et al, 1977)

mass-quality which is defined as the ratio of the gas mass flowrate to the total mass flowrate (NEL, 1966).

Among several interesting studies by different authors,
Collins and Gacesa (1971) suggest the following correlation,

$$x = \frac{-R}{\left(\dfrac{D_1}{\varepsilon} - R\right)} + D_2 \frac{\alpha d^2}{\left(\dfrac{D_1}{\varepsilon} - R\right)} (10^8 y)^{\frac{1}{2}} + D_3 \left[\frac{\alpha d^2}{\left(\dfrac{D_1}{\varepsilon} - R\right)}(10^8 y)^{\frac{1}{2}}\right]^2$$

(15.30)

where

$$y \triangleq \frac{Hp_G}{M^2}$$

(15.31)

$$R \triangleq \frac{\rho_G}{\rho_L}$$

(15.32)

$$\alpha \overset{\Delta}{=} CEF \qquad\qquad\qquad (15.33)$$

The symbols used in Eqs. (15.30) to (15.33) have the following meaning : x, quality; H, differential water head in inches at 20°C; ρ_G and ρ_L, gas and liquid densities in lbs/ft³; M, total mass flowrate in lbs/hr; d, diameter of the orifice or of the throat in inches; ε, expansion factor; C, discharge coefficient; E, velocity of approach factor; F, thermal expansion factor.

The authors give the following values for *orifice plates* set in pipes of internal diameter greater than 50 mm,

$$D_1 = 1.275 \qquad D_2 = 3.066 \ 10^{-2} \qquad D_3 = 6.586 \ 10^{-4}$$

With *venturi tubes* placed in pipes of internal diameter greater than 65 mm, the authors find,

$$D_1 = 11.8 \qquad D_2 = 9.99 \ 10^{-2} \qquad D_3 = 0.278$$

Eq. (15.30) is valid for vertical upward steam-water flow, at 67 bar, with total mass flowrate between 1.9 and 12.6 kg/s and for steam quality between 0.05 and 0.90. Orifice plates and venturi tubes must comply with ASME standards (1959). The same correlation can be used for larger diameters, up to 200 mm, for lower pressures (1 bar), for higher mass flowrates (up to 64 kg/s). The influence of the diameter is important and the authors frequently recommend that the correlation should not be used for pipings with internal diameter less than 50 mm for orifice plates and less than 65 mm for venturi tubes.

Zanker (1968) used orifice plates and venturi tubes to measure the total mass flowrate in horizontal air-water flows, with void fractions less than 0.05. For tubes with internal diameter less than 5 cm, the total mass flowrate is equal to the mass flowrate calculated as if the water flowed alone, multiplied by the quantity,

$$\frac{\rho_G}{(1-x)\rho_G + 0.5x\rho_L}$$

In this method the discharge coefficient is equal to the value obtained in liquid single phase flow.

15.5.1 *Correlations Valid for Orifice Plates Only*

Kremlevskii and Dyudina (1972) used an orifice plate to measure the total mass flowrate, for steam-water flows at 1 to 4 bar and for quality higher than 0.7. The total mass flowrate is given by the following relation,

$$M = \alpha \ \varepsilon \ A_2 \ \sqrt{2\rho\Delta p} \ \left[1 + 0.56(1-x)\right] \qquad\qquad (15.34)$$

with,

$$\frac{1}{\rho} = \frac{x}{\rho_G} + \frac{1-x}{\rho_L} \tag{15.35}$$

In equation (15.34) A_2 is the cross section area of the orifice. The flow coefficient α and the expansion coefficient ε are calculated for a pure steam flow.

Smith et al (1977) evaluated the existing two-phase flow correlations for the flow of steam-water mixtures and two-component gas-liquid mixtures through ASME-code measuring orifices. They concluded that,

(i) For the flow of *steam-water* mixtures, James correlation (1965-1966) gives the best results. This correlation reads,

$$M = 0.61 \frac{YF}{\sqrt{1-\beta^4}} S_2 \sqrt{\frac{2\Delta p}{x^{1\cdot5}(1/\rho_G - 1/\rho_L) + 1/\rho_L}} \tag{15.36}$$

where M is the total mass flowrate, Y the gas expansion factor, F the orifice thermal expansion factor, β the ratio of orifice diameter to internal pipe diameter and S_2 the orifice cross section area. The correlation was tested by Smith et al (1977) for the parameter range given in Table 15.2. The expansion factor Y is calculated from the following equation,

$$Y = \sqrt{r^{2/\gamma}\left(\frac{\gamma}{\gamma-1}\right)\left[\frac{1-r^{(\gamma-1)/\gamma}}{1-r}\right]\left(\frac{1-\beta^4}{1-\beta^4 r^{2/\gamma}}\right)} \tag{15.37}$$

where r is the ratio of the differential pressure to the upstream pressure, γ the ratio of specific heats taken equal to 1.3 and β the ratio of the orifice diameter to the internal pipe diameter.

Table 15.2

Range of Parameters for which James Correlation was Tested (Steam-Water Flow)

Internal pipe diameter (mm)	63	200
β	0.500	0.707
Pressure (bar)	16.8	77.2
Quality	0.06	0.95
Mass velocity $(kgm^{-2}s^{-1})$	676	2514

(ii) For the flow of *two-component gas-liquid* mixtures, Murdock correlation (1962) gives the best results.

This correlation reads,

$$M = \frac{K_G Y_G FS_2 \sqrt{2 \Delta p \ \rho_G}}{x + 1.26 \ (1-x) \ \dfrac{K_G Y_G}{K_L} \sqrt{\dfrac{\rho_G}{\rho_L}}} \qquad (15.38)$$

where Y_G and Y_L are the gas and liquid expansion factors. The flow coefficients K_G and K_L are treated as unknown parameters so that this correlation requires an iteration scheme. The correlation was tested by Smith et al (1977) for the parameter range given in Table 15.3.

TABLE 15.3

Range of Parameters for which Murdock Correlation was Tested (Two-Component Two-Phase Flow)

Internal pipe diameter (mm)	75	98
β	0.26	0.43
Pressure (bar)	1	64
Quality	0.11	0.98
Mass velocity ($kgm^{-2}s^{-1}$)	46.5	1002

In a recent paper, *Chisholm* (1977) developed correlations for the pressure drop over sharp-edged orifices during the flow of incompressible two-phase mixtures. They read,

$$\frac{\Delta p}{\Delta p_L} = 1 + \frac{C}{X} + \frac{1}{X^2} \qquad (15.39)$$

$$X \triangleq \frac{1-x}{x} \left(\frac{\rho_G}{\rho_L}\right)^{\frac{1}{2}} \qquad (15.40)$$

$$C = \begin{cases} \left(\dfrac{\rho_L}{\rho_G}\right)^{\frac{1}{4}} + \left(\dfrac{\rho_G}{\rho_L}\right)^{\frac{1}{4}} & \text{for } X < 1 \qquad (15.41) \\[4mm] \left(\dfrac{\rho}{\rho_G}\right)^{\frac{1}{2}} + \left(\dfrac{\rho_G}{\rho}\right)^{\frac{1}{2}} & \text{for } X > 1 \qquad (15.42) \end{cases}$$

$$\frac{1}{\rho} \triangleq \frac{x}{\rho_G} + \frac{1-x}{\rho_L} \tag{15.43}$$

where Δp_L is the pressure change over the orifice when liquid flows alone. The proposed correlation agrees with the experimental results of Collins and Gacesa (1971) among others but disagrees with the data of James (1965-1966) and Murdock (1962).

15.5.2 *Correlations Valid for Venturis Only*

Harris (1967) suggested a formula for steam-water flows with quality less than 0.20, for pressures of 65.5 bar in horizontal tubes of internal diameter 105 mm, and for mass flowrates between 15.1 and 22.7 kg/s. The venturi tubes must comply with British standards. Harris gave the following formula,

$$10^2 x = 0.584 \frac{\Delta p - 17.0 M^2}{M^2} - 2.44 \ 10^{-3} \frac{(\Delta p - 17.0 M^2)^2}{M^2} \tag{15.44}$$

where the total mass flowrate is in 10^5 lbs/hour and Δp in inches of water.

Following the ideas developed by Collins and Gacesa (1971), Fouda (1975) and Fouda and Rhodes (1978) tried to correlate the pressure drop by using either a separated flow model or a homogeneous model. Although a better agreement was obtained by taking into account the hydrostatic pressure head and the diameter effect, much work remains to be done to generalize the results.

At high pressure the more extensive results seem to have been obtained by Frank et al (1977) who carried out an experimental program on an argon-water loop running between 18 and 83 bar. The results are fairly well correlated by the following equations,

$$\Delta p = K \frac{M^2}{\rho} \tag{15.45}$$

$$\frac{1}{\rho} = \frac{x^2}{R_v \rho_G} + \frac{(1-x)^2}{(1-R_v) \rho_L} \tag{15.46}$$

where K is a dimensional constant determined in single-phase liquid flow. The quantity R_v is a fictitious void fraction determined in terms of the quality x and the pressure by means of a calibration on an argon-water loop. An example of such a calibration curve is given in Figure 15.6.

The correlation was tested for the parameter range given in Table 15.4.

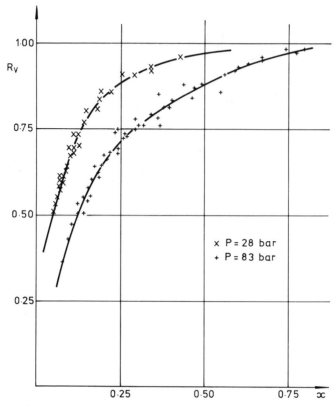

Figure 15.6: Calibration Curve used for the Venturi (Frank et al, 1977)

TABLE 15.4

Range of Parameters for which Eqs (15.45) and (15.46) were Tested

Argon-water pressure (bar)	18	83
Steam-water equivalent pressure (bar)	40	140
Quality	0.03	0.75
Void fraction	0.30	0.99
Mass flowrate (kg/s)	0.10	0.80

15.6 Local Measurements

Two-phase flow modelling efforts require among other information local instantaneous measurements of phase density functions, liquid and gas velocities, interfacial passage frequencies, and liquid and vapor temperatures and their statistical characteristics such as probability density functions and spectral densities. The purpose of these measurements is to obtain data regarding interfacial area densities, correlation coefficients between void and velocity, void and energy, etc, and to verify hypotheses regarding the shape of void, velocity and temperature profiles, their interrelations and their statistical variations. A detailed review of transient and statistical measurement techniques for two-phase flows, including local techniques, can be found in the paper by Jones and Delhaye (1976).

15.6.1 *Electrical Probes*

The first requirement to be met when using an electrical probe in two-phase flow is that one phase has a significantly different electrical conductivity from the other. Consequently, variations in conductance permit the measurement of the local void fraction and the arrival frequency of the bubbles at a given point in a continuous, conducting fluid. By using a double probe, a transit velocity can be measured but one has to be very careful when giving a physical significance to this velocity (Galaup, 1975; Delhaye and Achard, 1977).

Figure 15.7 shows the classical electrical diagram of a resistive probe while Figure 15.8 displays a typical probe geometry. Impedance changes due to the passage of bubbles at the tip of the probe produce a fluctuation in the output signal. One of the principal features which differentiate the electrical circuits shown in Figure 15.7 is the type of electrical supply. *Direct current supply* requires low voltages in order to reduce electrochemical phenomena on the sensor. Resultant electronics may become troublesome and sensors may still sustain alteration due to electrochemical deposits at low flows. When an *alternating current supply* is used, phase changes are detected by amplitude modulation of the alternating output signal. This technique has been used by several investigators to eliminate the electrochemical phenomena on the sensor. When high speed flows are investigated the required supply frequency can be very high (eg 1 MHz) and much trouble occurs with the electronics. Galaup (1975) used a supply frequency lower than the frequency of the physical phenomenon, eliminating electrochemical effects and providing pseudo DC operation in each half wave.

According to the way that the sensor is energized, the ideal output signal of a resistive probe is either a binary wave sequence, or a sequence of bursts of constant amplitude oscillations separated by zero voltage zones. Actually, the output signal is misshapen with respect to the ideal signal because of the interface deformations. The true signal is generally transformed into a binary sequence with the help of a trigger level. Galaup (1975) used a level adjustment based upon a comparison between the integrated void profile and the line void fraction obtained with a γ-ray absorption method.

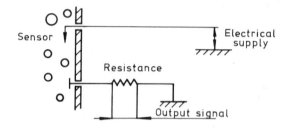

Figure 15.7: Electrical Diagram of a Resistive Probe

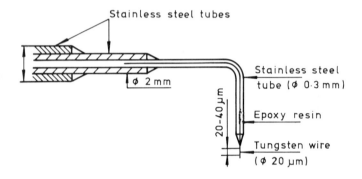

Figure 15.8: Miniature Probe Geometry (Lecroart and Porte, 1971)

The electrical probe described in Figure 15.8 does not work in dispersed droplet flow because of the impossibility of having a continuous electrical path in the liquid phase between the two electrodes. Reimann and John (1978) provided local void fraction data in steam-water flow at pressure up to 100 bar obtained with a high frequency probe. This probe is made of a coaxial wire energized by a 100 to 300 MHz current. A standing wave reflector senses the differential voltage between the forward and reflected waves which is a function of the nature of the medium surrounding the tip of the probe.

15.6.2 *Optical Probes*

An optical probe is sensitive to the change in the refractive index of the surrounding medium and is thus responsive to inter-facial passages enabling measurements of local void fraction and of interface passage frequencies to be obtained even in a non-conducting fluid. By using two sensors and a cross-correlation method, some information may be obtained on a transit velocity (Galaup, 1975).

A tiny optical sensor was proposed by Danel and Delhaye (1971) and developed by Galaup (1975). This probe consists of a single optical fiber, 40 µm in diameter. The overall configuration is shown in Figure 15.9. The active element of the probe is obtained

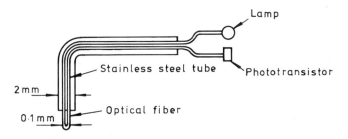

Figure 15.9: U-Shaped Fiber Optical Sensor (Danel and Delhaye, 1971)

by bending the fiber into a U-shape. The entire fiber, except the U-shaped bend, is protected inside a stainless steel tube, 2 mm in diameter. The active part of the probe, as shown in Figure 15.10, has a characteristic size of 0.1 mm.

Signal analysis is accomplished through an adjustable threshold which enables the signal to be transformed into a binary signal. Consequently, the local void fraction is a function of this threshold which is adjusted and then held fixed during a traverse in order to obtain agreement between the profile average and a γ-ray measurement of the line void fraction. Experimental results for void profiles in Freon two-phase flow with phase change and in air-water flow are given by Galaup (1975). Miniaturized U-shaped optical probes were used by Charlot et al (1978) to determine local void fraction profiles in a four rod bundle cooled by a vaporizing freon flow. ⌐Figure 15.11 shows some typical void profiles obtained in the central subchannel of the bundle.⌐

Single-fiber optical probes have been ⌐recently⌐ developed by Danel et al (1979). They enable void fractions to be measured in dispersed droplet flows. The light source is a He-Ne 5 mW laser connected to a single fiber via a beam-splitter (Fig. 15.12). The tip of the fiber is conical and its diameter is less than 40 µm.

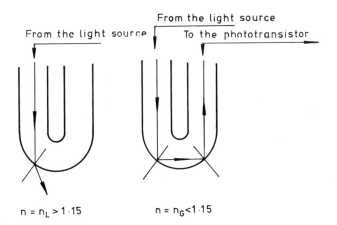

Figure 15.10: Active Part of the U-Shaped Fiber Optical Sensor (Danel and Delhaye, 1971)

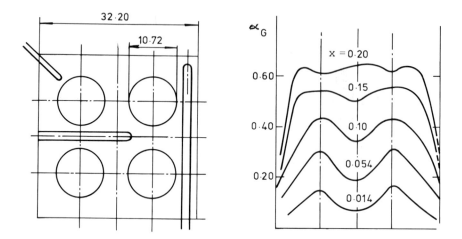

Figure 15.11: Local Void Fraction Profiles in the Central Subchannel of a Four
            Rod Bundle.   Freon-Freon Vapor Flow (23.3 bar; 80°C; 50 gcm^{-2}s^{-1})
            Charlot et al (1978)

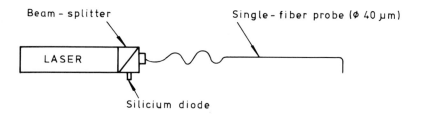

Figure 15.12: Single-Fiber Optical Probe (Danel et al, 1979)

The separation of the forward and backward light beams is ensured
by the beam-splitter and the backward light is detected by a
silicium diode.   The response time is of the order of 0.1 µs.

15.6.3   Thermal Anemometers

      It has been found that hot-wire or hot-film anemometry can be
used in two-component two-phase flow or in one-component two-phase
flow with phase change.   In the first case, eg an air-water flow,
it is possible to measure the local void fraction, the instantaneous
velocity and the turbulence intensity of the liquid phase.   In
the second case however, eg a steam-water flow, it has been so far
impossible to obtain consistent results on liquid velocity
measurements.   Nevertheless, in both cases, the anemometer gives
a signal which is characteristic of the flow pattern although the
signal is misshapen because of the interaction between the probe
and the pierced interfaces (Bremhorst and Gilmore, 1976; Remke,
1978).

It is evident that if the gas and liquid signals could be
separated, the turbulent structure of the liquid phase could be
obtained. To achieve this separation, Delhaye (1969) and Galaup
(1975) used the amplitude probability density function of the
output signal (Figure 15.13). In a first approximation, the local
void fractions were calculated as the ratio of the hatched area to
the total area. The separation line was set to ensure the
identity between the averaged value of the local void fraction and
the line void fraction measured by means of a γ-ray absorption
technique. The liquid time-averaged velocity and the liquid
turbulent intensity were calculated from the nonhatched area of
the amplitude histogram (Figure 15.13) and the calibration curve
of the probe immersed in the liquid. The same method was used by
Serizawa et al (1975), Herringe and Davis (1976) and Remke (1976)
for measuring the turbulent characteristics of air-water two-
phase flow in a pipe. A different signal processing was proposed
by Resch et al (1972, 1974, 1975) in a study of bubbly two-phase
flow in a hydraulic jump. The analog signal from the anemometer
is digitally analyzed according to a conditional sampling.
Another technique was proposed by Jones (1973) who used a discrim-
inator with a cutoff level depending on the local velocity.
Conditional sampling used in intermittent single phase flows have
provided new ideas for signal processing in two-phase flows. As
an example Kelada (1977) used a criterion based on the slope of
the signal.

For steam-water flow, Hsu et al (1963) noticed that the only
reference temperature is the saturation temperature. If water
velocity measurements are carried out, the probe temperature must
not exceed saturation temperature by more than $5^{\circ}C$ to avoid
nucleate boiling on the sensor. Conversely, if only a high
sensitivity to phase change is looked for, then the superheat
should range between $5^{\circ}C$ and $55^{\circ}C$ causing nucleate boiling to
occur on the probe. Signal processing for horizontal steam-water
flow is discussed by Abel and Resch (1978).

15.6.4 *Phase Indicating Microthermocouple*

Although the classical microthermocouple has contributed to
a large extent to the understanding of the local structure of two-
phase flow with change of phase, it has not provided any reliable
statistical information on the distribution of the temperature
between the liquid and the vapor phases.

The work done by Delhaye et al (1973) is based on the
possibility of separating the liquid temperature from the steam

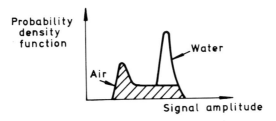

Figure 15.13: Typical Amplitude Histogram of Anemometer Signal (Delhaye, 1969)

temperature and of determining the local void fraction. An
insulated 20 μm thermocouple was used both as a temperature sensor
and as an electrical phase-indicator by means of a Kohlrausch
bridge which sensed the presence of conducting liquid between the
noninsulated junction and the ground. The phase signal is used to
route the thermocouple signal to two separate subgroups of a multichannel analyzer. As a result, separate histograms of liquid and
vapor temperatures can be obtained as shown in Figure 15.14 for
subcooled boiling.

The absence of a phase indicator does not preclude the
statistical analysis of the temperature signal but makes it less
precise. Recently Afgan and Jovic (1978) used a conditional
sampling to analyze the structure of the bubble boundary layer
occurring in forced convection local boiling.

15.6.5 Reliability of Local Measurements

Questions on reproducibility and the equivalence of one
method compared with another always arise. It is encouraging that
the results from three different instruments, optical probe,
thermal anemometer, and resistive probe all give comparable
results as shown by Galaup (1975) in Figure 15.15. Nevertheless
more work is needed on the cross-correlation technique to obtain
the gas velocity as well as the interface speed of displacement.

15.7 Conclusions

This review on measurement techniques in two-phase flow is
far from exhaustive. Some methods were not presented because
more work is needed to evaluate their reliability. Among others
we could mention the local isokinetic sampling (Schraub, 1969;
Delhaye et al, 1973a), the subchannel isokinetic sampling (Bayoumi,

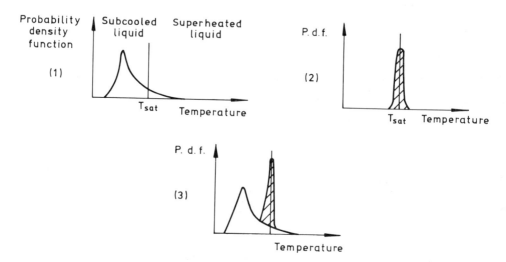

Figure 15.14: Subcooled Boiling Temperature Histogram (Delhaye et al, 1973)
         (1) Liquid Temperature, (2) Steam Temperature, (3) Unseparated
         Histogram

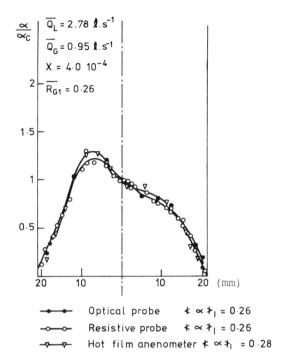

Figure 15.15: Comparison of results of optical, resistive and thermal probe
techniques in air-water flow (Galaup, 1975. $\overline{Q}$: volumetric flowrate; x : quality; $\overline{R_{G1}}$: time-averaged line void fraction measured
with γ-rays; α : local void fraction; α_c : centerline local void
fraction; $\{\alpha\}_1$: averaged void fraction measured on the profile

1976) or the Pitot tube technique (Delhaye et al, 1973a; Banerjee
and Nguyen, 1977). New techniques have been under development
for a few years. Except for the use of drag bodies (Aya, 1975)
and electrochemical probes (Burdokov et al, 1978; Surgenor, 1979;
Souhar, 1979) which have been known for a long time in single-
phase flow, most of these new techniques are highly sophisticated.
Good examples are given by the neutron scattering technique
(Rousseau et al, 1978; Rousseau and Riegel, 1978; Banerjee et al,
1978a, 1979) and by the pulsed neutron activation technique
 Kehler, 1978; Perez-Griffo et al, 1978). Laser techniques were
reviewed by Delhaye (1976) and seem to be promising for low or
high void fraction two-phase flows (Ohba et al, 1976, 1977, 1977a;
Mahalingam et al, 1976).

Apart from the photon attenuation technique which has been
studied rather extensively, most two-phase flow measuring
techniques need to be substantiated by (1) a more rigorous analysis,
(2) more systematic experimental studies. Furthermore, it has
been generally recognised that the validity of each technique
must be assessed according to the existing flow pattern and the
transient aspect of the flow.

Nomenclature

A   area

A   area

B   magnetic induction

c   contrast

C   flow coefficient (Eq. 15.20); Chisholm parameter; discharge coefficient

d   distance; diameter

D   diameter; Collins and Gacesa parameter

e   thickness; electromotive force

E   velocity of approach factor

F   orifice thermal expansion factor

G   mass velocity

H   differential water head

I   intensity

K   calibration constant

M   mass flowrate

n   rotational frequency

N   number of photons (Eq. 15.7); rotation Reynolds number (Eq. 15.21)

p   pressure

Q   volumetric flowrate

r   ratio of the differential pressure to the upstream pressure

R   area fraction; tube radius; density ratio (Eq. 15.32)

S   cross section area

w   axial component of the velocity vector

x   quality

X   Lockhart-Martinelli parameter

y   Collins and Gacesa parameter (Eq. 15.31)

Y   gas expansion factor

Greek

α local time fraction; flow coefficient (Eq. 15.33)

β ratio of orifice diameter to internal pipe diameter

γ slip ratio (Eq. 15.24); ratio of specific heats

ϵ expansion factor

θ time interval

μ linear absorption coefficient

ν kinematic viscosity

ρ density

Subscripts

c   centerline

G   gas

L   liquid

M   turbine

P   wall

o   incident

V   venturi

Operators

$\{\ \}_1$ line averaging operator

$\{\ \}_2$ area averaging operator over the pipe cross section

$< >$ area averaging operator over the area occupied by a given phase

References

Abel, R and Resch, F J, (1978) "A method for the analysis of hot-film anemometer signals in two-phase flows". *Int. J. Multiphase Flow,* **4**, 523-533.

Afgan, N and Jovic, L A, (1978) "Intermittent phenomena in the boiling two-phase boundary layer". *Int. J. Heat Mass Transfer,* **21**, 427-434.

Alad'yev, I T, Gavrilova, N D and Dodonov, L D, (1971) "Effective electrical conductivity of two-phase gas-liquid metal flows". *Heat Transfer - Soviet Research,* **13** (4), 21.

A.S.M.E., (1959) "Fluid meters - Their theory and application", *Report of the A.S.M.E. Research Committee on Fluid Meters,* A.S.M.E., New York, 5th Ed.

Aya, I, (1975) "A model to calculate mass flowrate and other quantities of two-phase flow in a pipe with a densitometer, a drag disk, and a turbine meter". *ORNL-TM-4759.*

Azzopardi, B J, (1977) "Measurement of drop sizes", *AERE-R8667.*

Banerjee, S and Nguyen, D M, (1977) "Mass velocity measurement in steam-water flow by Pitot tubes". *A.I.Ch.E.J.,* **23** (3), 385-387.

Banerjee, S, Heidrick, T R, Saltvold, J R and Flemons, R S, (1978) "Measurement of void fraction and mass velocity in transient two-phase flow". Transient Two-Phase Flow, *Proceedings of the CSNI Specialists Meeting,* August 3-4, 1976, Toronto. Banerjee, S and Weaver, K R, Eds, AECL, **2**, 789-832.

Banerjee, S and Jolly, O P, (1978) "Analysis of steady state steam-water mass velocity measurements". Interim report submitted to EG & G, INEL.

Banerjee, S, Chan, A M C, Ramanathan, N and Yuen, P S L, (1978a) "Fast neutron scattering and attenuation technique for measurement of void fractions and phase distribution in transient flow boiling". *Sixieme Congres International sur le Transfert de Chaleur,* **2**, Conseil National de Recherches du Canada, 351-355.

Banerjee, S, Yuen, P and Vandenbroek, M A, (1979) "Calibration of a fast neutron scattering technique for measurement of void fraction in rod bundles". *J. Heat Transfer*, **101**, 295-299.

Bayoumi, M A A, (1976) "Etude des repartitions de debit et d'enthalpie dans les sous-canaux d'une geometrie en grappe des reacteurs nucleaires en ecoulements monophasique et diphasique". These de docteur-ingenieur, Universite Scientifique et Medicale de Grenoble, Institut National Polytechnique de Grenoble.

Benedict, R P, (1969) "Fundamentals of temperature, pressure, and flow measurements". Wiley.

Bevir, M K, (1974) "Electromagnetic flow measurement". *Techniques de mesure dans les ecoulements,* Eyrolles, 408-425.

Bradshaw, P, (1971) "An introduction to turbulence and its measurement". Pergamon.

Bremhorst, K and Gilmore, D B, (1976) "Response of hot wire anemometer probes to a stream of air bubbles in a water flow". *J. Physics E : Scientific Instruments,* **9**, 347-352.

Brockett, G F and Johnson, R T, (1976) "Single-phase and two-phase flow measurement techniques for reactor safety studies", *EPRI NP-195.*

Burdukov, A P, Nakoryakov, C Ye, Kuvshinov, G G, Valukina, N V and Koz'Menko, B G, (1978) "Application of the electrodiffusion modelling method to investigation of hydrodynamics and transport processes in two-phase flow". *Heat Transfer - Soviet Research,* **10** (3), 116-122.

Charlety, P, (1971) "Ebullition du sodium en convection forcee". These de docteur-ingenieur, Faculte des Sciences, Universite de Grenoble.

Charlot, R, Oulmann, T and Ricque, R, (1978) "Boucle "Frenesie". Determination du taux de vide et du type d'ecoulement au moyen de sondes optiques dans une grappe a quatre barreaux chauffants". Compte Rendu d'Essais TT/SETRE/78-9-B/RC, TO,RR.

Chisholm, D, (1977) "Two-phase flow through sharp-edged orifices". *J. Mech. Engng Science,* **19** (3), 128-130.

Collins, D B and Gacesa, M, (1971) "Measurement of steam quality in two-phase upflow with venturimeters and orifice plates". *J. Basic Engng.,* 11-21.

Considine, D M, Ed, (1971) *"Encyclopedia of instrumentation and control".* McGraw Hill.

Danel, F and Delhaye, J M, (1971) Sonde optique pour mesure du taux de presence local en ecoulement diphasique". *Mesures-Regulation-Automatisme,* Aout-Septembre, 99-101.

Danel, F, Imbert, J L and Arnault, J, (1979) "Etude des ecoulements finement disperses, Compte rendu de fin d'etude d'une recherche financee par la DGRST" (Action Concertee : Mecanique).

Delhaye, J M, (1969) "Hot-film anemometry". *Two-Phase Flow Instrumentation,* LeTourneau, B W and Bergles, A E, Eds, ASME, 58-69.

Delhaye, J M, (1976) "Two-phase flow instrumentation and laser beam". *The*

accuracy of flow measurements by laser Doppler methods - Proceedings of the LDA Symposium, Copenhagen 1975, Buchhave, P, Delhaye, J M, Durst, F, George, W K, Refslund, K and Whitelaw, J H, Editors, Hemisphere.

Delhaye, J M, (1978) "Optical methods in two-phase flows". Proceedings of the Dynamic Flow Conference, Marseilles, Sept. 11-14, 321-343.

Delhaye, J M and Achard, J L, (1977) "On the use of averaging operators in two-phase flow modelling". Thermal and Hydraulic Aspects of Nuclear Reactor Safety, 1 : Light Water Reactors, Jones, O C and Bankoff, S G, Eds, ASME, 289-332.

Delhaye, J M, Galaup, J P, Reocreux, M and Ricque, R, (1973a) "Metrologie des ecoulements diphasiques ; Quelques procedes". C.E.A. R-4457.

Delhaye, J M, Semeria, R and Flamand, J C, (1973) "Void fraction, vapor and liquid temperatures : local measurement in two-phase flow using a micro-thermocouple". J. Heat Transfer, 365-370.

Demagny, J Ch, (1976) "Mesures de vitesses et de debits par la methode electromagnetique". Choix des materiels industriels, La Houille Blanche, No 5, 361-367.

Doebelin, E O, (1966) "Measurement systems. Application and Design". McGraw Hill.

Dowdell, R B, Ed, (1974) "Flow - Its measurement and control in science and industry". 1, I.S.A.

du Bousquet, J L, (1969) Etude du melange entre deux sous-canaux d'un element combustible nucleaire a grappe". These de docteur-ingenieur, Faculte des Sciences, Universite de Grenoble.

Fitremann, J M, (1972) "Theorie des velocimetres et debitmetres electromagnetiques en ecoulement diphasique". Comptes Rendus a l'Academie des Sciences, Paris, t. 274, Serie C, 440-443.

Fitremann, J M, (1972a) "La debitmetrie electromagnetique appliquee aux emulsions". Societe Hydrotechnique de France, Douziemes Journees de l'Hydraulique, Paris 1972, Question IV, Rapport 6.

Fouda, A E, (1975) "Two-phase flow behaviour in manifolds and networks". Ph.D Thesis, Dept. of Chem. Engng., Univ. of Waterloo, Canada.

Fouda, A E and Rhodes, (1978) "Total mass flow and quality measurement in gas-liquid flow". Two-Phase Transport and Reactor Safety, Veziroglu, T N and Kakac, S, Eds, Hemisphere, IV, 1331-1355.

Frank, R, Mazars, J and Ricque, R, (1977) "Determination of mass flowrate and quality using a turbine meter and a venturi". Heat and Fluid Flow in Water Reactor Safety, The Institution of Mech. Engrs. London, 63-68.

Galaup, J P, (1975) "Contribution a l'etude des methodes de mesure en ecoulement diphasique". These de docteur ingenieur, Universite Scientifique et Medicale de Grenoble, Institut National Polytechnique de Grenoble.

Gardner, R P, Bean, R H and Ferrel, J K, (1970) "On the gamma-ray one-shot-collimator measurement of two-phase flow void fraction". Nuclear Applications and Technology, 8, 88-94.

Hancox, W T, Forrest, C F and Harms, A A, (1972) "Void determination in two-phase systems employing neutron transmission". ASME Paper 72-HT-2.

Harms, A A and Laratta, F A, (1973) "The dynamic-bias in radiation interrogation of two-phase flow. *Int. Heat Mass Transfer*, **16**, 1459-1465.

Harris, D M, (1967) "Calibration of a steam quality meter for channel power measurement in the prototype S.G.H.W. reactor". Paper presented at *European Two-Phase Heat Transfer Meeting*, Bournmouth.

Heinemann, J B, (1965) "Forced convection boiling sodium studies at low pressure". *ANL* 7100, 189-194.

Heinemann, J B, Marchaterre, J F and Mehta, S, (1963) "Electromagnetic flow-meters for void fraction measurement in two-phase metal flow". *Rev. Sci. Instrum.*, **34** (4), 399.

Herringe, R A and Davis, M R, (1976) "Structural development of gas-liquid mixture flows". J. *Fluid Mechanics*, **73**, Part 1, 97-123.

Hewitt, G F, (1978) "Measurement of two-phase flow parameters". Academic Press.

Hewitt, G F and Whalley, P B, (1979) "Advanced optical instrumentation methods". *EPRI Workshop on Basic Two-Phase Flow Modelling in Reactor Safety and Performance*, Tampa, February 27-March 2.

Hochreiter, H M, (1958) "Dimensionless correlation of coefficients of turbine-type flowmeters". *Trans. ASME*, 1369.

Hori, M, Kobori, T and Ouchi, Y, (1966) "Method for measuring void fraction by electromagnetic flowmeters". JAERI 1111.

Hsu, Y Y, Simon, F F and Graham, R W, (1963) "Application of hot-wire anemometry for two-phase flow measurements such as void fraction and slip-velocity". *Multiphase Flow Symposium*, Lipstein, N J, Ed, ASME, 26-34.

James, R, (1965-66) "Metering of steam-water two-phase flow by sharp-edged orifices". *Proc. of the Institution of Mech. Engrs.*, **180**, Part 1, No 23, 549-566.

Jones, O C, (1973) "Statistical considerations in heterogeneous, two-phase flowing systems". Ph. D. Thesis, Rensselaer Polytechnic Institute, Troy, N.Y.,

Jones, O C and Delhaye, J M, (1976) "Transient and statistical measurement techniques for two-phase flows : a critical review". *Int. J. Multiphase Flow*, **3**, 89-116.

Jones, O C and Zuber, N, (1975) "The interrelation between void fraction fluctuations and flow patterns in two-phase flow". *Int. J. Multiphase Flow*, **2**, 273-306.

Kamath, P S and Lahey, R T, (1977) "A turbine-meter evaluation model for two-phase transients". *Topics in Two-Phase Heat Transfer and Flow*, Bankoff, S G, Ed, ASME, 219-225.

Katys, G P, (1964) "Continuous measurement of unsteady flow". Pergamon.

Kehler, P, (1978) "Two-phase flow measurement by pulsed neutron activation techniques". *Measurements in Polyphase Flow*. Stock, D E, Ed, ASME, 11-20.

Kelada, M I, (1977) "Application of anemometry in two-phase flow for local measurements and for the study of flow behavior in the entrance region of a long vertical pipe". M.S. Thesis, Univ. of Houston, Dept. of Chem. Engng.

Kremlevskii, P P and Dyudina, I A, (1972) "Measurements of moist vapor discharge by means of gauging diaphragms". *Measurement Techniques*, 15, 5, 741-744.

Lahey, R T, (1977) "Two-phase flow phenomena in nuclear reactor technology, Quarterly progress report, March 1, 1977-May 31, 1977". Contract at (49-24) - 0301 prepared for the USNRC by Dept. of Nuclear Engng., Rensselaer Polytechnic Institute, Troy, N.Y.

Laratta, F A and Harms, A A, (1974) "A reduced formula for the dynamic-bias in radiation interrogation of two-phase flow". *Int. J. Heat Mass Transfer*, 17, 464.

Lecroart, H and Porte, R, (1971) "Electrical probes for study of two-phase flow at high velocity". *International Symposium on Two-Phase Systems*, Haifa, Israel.

Leonchik, B I, Mayakin, V P and Lebedeva, P D, (1976) "Measurements in dispersed flows". AEC-tr-7600.

Letourneau, B W and Bergles, A E, Eds, (1969) "Two-phase flow instrumentation". ASME, Proceedings of a session held at the 11th National Heat Transfer Conf., Minneapolis, MI, 3-6 Aug.

Levert, F E and Helminski, E, (1973) "A dual-energy method for measuring void fractions in flowing medium". *Nuclear Technology*, 19, 58-60.

Lurie, H, (1965) "Sodium boiling heat transfer and hydrodynamics". ANL 7100, 549-571.

Mahalingam, R, Limaye, R S and Brink, J A, (1976) "Velocity measurements in two-phase bubble-flow regime with laser-Doppler anemometry". *A.I.Ch.E.J.*, 22 (6), 1152-1155.

Martin, R, (1969) "Mesure du taux de vide a haute-pression dans un canal chauffant". CEA-R 3781.

Martin, R, (1972) "Measurements of the local void fraction at high pressure in a heating channel". *Nuclear Science and Engineering*, 48, 125-138.

McMaster, W H, Kerr del Grande, N, Mallett, J H and Hubbell, J H, (1969) "Compilation of X-ray cross section". Report UCRL-50174, Sect. II, Rev. 1, Lawrence Radiation Lab.

Milliot, B, Lazarus, J and Navarre, J P, (1967) "Mesure de la fraction de vide dans un ecoulement double-phase NaK-Argon", EUR-3486 f.

Moore, M J, (1977) "A review of instrumentation for wet steam". *Two-Phase Steam Flow in Turbines and Separators*, Moore, M J and Sieverding, C H, Eds, Hemisphere - McGraw-Hill, 191-249.

Murdock, J W, (1962) "Two-phase flow measurements with orifices". *J. Basic Engineering*, 84, 419-432.

NEL, (1966) "Metering of two-phase mixtures" : Report of a meeting at NEL, 6th January 1965, NEL 217.

Nyer, M, (1969) "Etude des phenomenes thermiques et hydrauliques accompagnant une excursion rapide de puissance sur un canal chauffant" CEA-R 3497.

Ohba, K, Kishimoto, I and Ogasawara, M, (1976) "Simultaneous measurement of local liquid velocity and void fraction in bubbly flows using a gas laser - Part I, Principle and measuring procedure". *Technology reports of the Osaka University*, **26** (1328), 547-556.

Ohba, K, Kishimoto, I and Ogasawara, M, (1977) "Simultaneous measurement of local liquid velocity and void fraction in bubble flows using a gas laser - Part II, Local properties of turbulent bubbly flow". *Technology reports of the Osaka University*, **27** (1358), 229-238.

Ohba, K, Kishimoto, I and Ogasawara, M, (1977a) "Simultaneous measurement of local liquid velocity and void fraction in bubbly flows using a gas laser - Part III, Accuracy of measurement". *Technology reports of the Osaka University*, No 1383, 475-483.

Perez-Griffo, M, Block, R C and Lahey, R T, (1978) "N^{16} tagging for two-phase mass flow measurements". *ANS Trans*. **30**.

Popper, G F, (1961) "In-core instrumentation for the measurement of hydrodynamic parameters in water cooled reactors". Advanced course in in-core instrumentation for water-cooled reactors at Institutt for Atomenergi, Kjeller, Norway, **1**, Section V.IV.B, V.32-V.39.

Popper, G F, (1961a) "Proceedings of the power reactor in-core instrumentation meeting". Washington D.C., April 28-29, 1960. TID-7598, Instruments Reactor Technology, 37-47.

Reimann, J and John, H, (1978) "Measurements of the phase distribution in horizontal air-water and steam-water flow". *CSNI Specialists Meeting on Transient Two-Phase Flow*, June 12-14, 1978, Paris.

Remke, K, (1976) "Ein Beitrag zur Anwendung der Heissfilmanemometrie auf die Turbulenzmessung in Gas-Flussigkeitsstromungen". *ZAMM*, **56**, T 480 - T 483.

Remke, K, (1978) "Some remarks on the response of hot-wire and hot-film probes to passage through an air-water interface". *J. Physics E : Sci. Instrum*., **II**, 94-96.

Reocreux, M, (1974) "Contribution a l'etude des debits critiques en ecoulement diphasique eau-vapeur". These de doctorat es sciences, Universite Scientifique et Medicale de Grenoble.

Resch, F J, (1975) "Phase separation in turbulent two-phase flow". *Turbulence in Liquids*, Patterson, G K and Zakin, J L, Eds, Processing of the third symposium on turbulence in liquids, Sept. 1973, 243-249.

Resch, F J and Leutheusser, J H, (1972) "Le ressaut hydraulique ; mesures de turbulence dans la region diphasique". *La Houille Blanche*, No 4, 279-293.

Resch, F J, Leutheusser, J H and Alemu, S, (1974) "Bubbly two-phase flow in hydraulic jump". *J. Hydraulics Division*, ASCE, **100**, No HY1, Proc. Paper 10297, 137-149.

Richards, B E, Ed., (1977) "Measurement of unsteady fluid dynamic phenomena". Hemisphere - McGraw-Hill.

Rouhani, Z, (1964) "Application of the turbine type flowmeters in the measurement of steam quality and void". Symposium on in-core instrumentation, Oslo, June 1964, USAEC-CONF-640607.

Rousseau, J C, Czerny, J and Riegel, B, (1978) "Void fraction measurements during blowdown by neutron absorption or scattering methods". *Transient Two-Phase Flow*, Proceedings of the CSNI Specialists Meeting, August 3-4, 1976, Toronto, Banerjee, S and Weaver, K R, Eds, AECL, 2, 890-904.

Rousseau, J C and Riegel, B, (1978) "Super canon experiments". CSNI Specialists Meeting on Transient Two-Phase Flow, June 12-14, 1978, Paris.

Schraub, F A, (1969) "Isokinetic probes and other two-phase sampling devices : A survey". *Two-Phase Flow Instrumentation*, LeTourneau, B W and Bergles, A E, Eds, ASME, 47-57.

Serizawa, A, Kataoka, I and Michiyoski, I, (1975) "Turbulence structure of air-water bubbly flow-I. Measuring techniques". *Int. J. Multiphase Flow*, 2 (3), 221-223.

Shercliff, J A, (1962) "The theory of electromagnetic flow measurement". Cambridge University Press.

Smith, L T, Murdock, J W and Applebaum, R S, (1977) "An evaluation of existing two-phase flow correlations for use with ASME sharp edge metering orifices". *J. of Engng. Power*, **99** (3), 343-347.

Solesio, J N, (1978) "Mesure de l'epaisseur locale instantanee d'un film liquide ruisselant sur une paroi. Examen des methodes existantes. Mise en oeuvre d'une technique basee sur l'absorption de rayons X," CEA-R4925.

Solesio, J N, Flamand, J C and Delhaye, J M, (1978) "Liquid film thickness measurements by means of an X-ray absorption technique". *Topics in Two-Phase Heat Transfer and Flow*, Bankoff, S G, Ed, ASME, 193-198.

Souhar, M, (1979) "Etude du frottement parietal dans les ecoulements diphasiques en conduite verticale". Cas des regimes a bulles et a poches, These de docteur ingenieur, Institut National Polytechnique de Lorraine.

Storm, E and Israel, H I, (1970) "Photon cross sections from 1keV to 100 MeV for elements $Z = 1$ to $Z = 100$". *Nuclear Data Tables*, A7, 565-681.

Stukenbroeker, G L, Bonilla, C F and Peterson, R W, (1970) "The use of lead as a shielding material". *Nuclear Engineering and Design*, 13, I, 3-145.

Surgenor, B W, (1979) "Mass transfer and shear stress at the wall for cocurrent gas-liquid flows in a vertical tube". M.E. Thesis, McMaster University, Dept. of Engng. Physics.

Thatcher, G, (1972) "Electromagnetic flowmeters for liquid metals". *Modern Developments in Flow Measurement*, Clayton, C G, Ed., Peregrinus, 359-380.

Treenhaus, R, (1972) "Modern developments and new applications of magnetic flowmeters". *Modern Developments in Flow Measurement*, Clayton, C G, Ed., Peregrinus, 347-358.

Turner, G E, (1968) "Liquid metal flow measurement (sodium)". State-of-the-art study, LMEC - Memo-69-9.

Veteau, J M, (1979) "Mesure des aires interfaciales dans les ecoulements diphasiques". CEA-R-5005.

Zanker, K J, (1968) "The influence of air on the performance of differential pressure water flowmeters". Euromech Colloquium No 7, Grenoble.

Chapter 16

Advanced Optical Instrumentation

F. MAYINGER

16.1 Introduction

Optical measurement techniques have the advantage not to influence the process examined and to work inertialess, which makes it possible to investigate even ultra-fast phenomena. Photographic techniques instead of point by point measuring give information about a whole field and help better to understand physical effects, which consequently makes it easier to formulate wellposed correlations.

Interferometric methods (Hauf and Grigull (1970)) have been used for many years in heat and mass transfer. In 1949 Gabor (1949) invented a new optical recording technique, which is called holography. This new recording technique combined with the interferometry – thus called holographic interferometry – has many advantages, also for use in two-phase flow. It allows one, for example, to measure the temperature distribution at the interface around a growing or condensing vapour bubble.

High speed cinematography has also been used for many years in two-phase flow, and in recent years some progress was made with respect to ultra-short exposure times and to the arrangement of the optical system. Exposure times of 10^{-8}s with an interval of 10^{-6}s are possible today.

The X-ray technique gives information about density distribution also in fast changing fluiddynamic systems. The flash frequency can reach 10 kHz, however, the electronic processing of the X-ray picture is still rather slow (500/1000 Hz). Luminescence methods together with a stochastic analysis using cross- and autocorrelation, can be of help for determining velocities in two-phase mixtures under certain conditions.

16.2 Holographic Interferometry

The general theory of holography is so comprehensive, that for a detailed description one must refer to the literature by Kiemle and Ross (1969), Smith (1969), Caulfield and Sun Lu (1970) and Collier et al (1971). A simple and most commonly used arrangement for holographic interferometry is shown in Figure 16.1. A laser serves as the light source. By means of a beam splitter

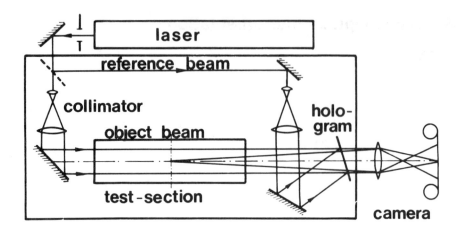

Figure 16.1: Holographic arrangement for the examination of transparent media.

the laser beam is divided into an object and reference beam. Both beams are then expanded to parallel waves by a telescope, which consists of a microscope objective and a collimating lens. The object wave passes through the test section in which the tempera- ture or the concentration field is to be examined, whereas the reference wave directly falls onto the photographic plate - called hologram - where from the picture later on can be registered by a camera.

In a first exposure the wave passing through the test section with constant temperature distribution is recorded. This wave may, however, already be distorted because of imperfections in the windows of the test-chamber or by distortions caused by pressure and liquid or gas flow. After recording of the comparison beam the phenomena to be investigated is started. For example the temperature field is established by heating the walls of a tube. Now the incoming wave receives a continuous additional phase shift due to the temperature changes. This resulting wave front, called the measuring beam, is recorded on the same plate. After processing, the hologram is repositioned and illuminated by the reference beam, as shown in Figure 16.2.

Now both object waves are reconstructed simultaneously; they will interfere. The interference picture can be observed or photographed.

An example of a hologram taken by this technique is shown in Figure 16.3. There the temperature distribution around a bubble growing at a heated wire of 0.4 mm diameter is shown. The bubble grows into the drift flux volume of a proceeding bubble, which left the wire a few milli-seconds ago.

By reconstructing a hologram, not only a virtual but also a real image can be obtained. This can be achieved when a parallel reference beam is used and the hologram is reconstructed with the conjugate reference wave. This can be done by inverting the

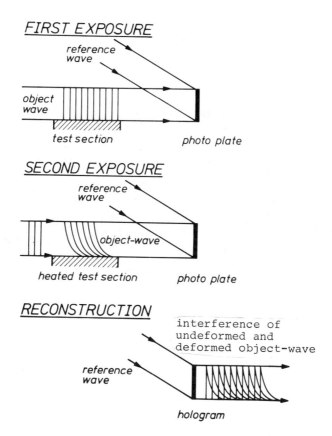

Figure 16.2: Double exposure technique.

direction of the reference wave or by simply turning the hologram itself around, as shown in Figure 16.4. The corresponding beams of the two waves recorded then form a real image of the interference pattern and intersect in the focusing plane. In this position the recording medium (e.g. a sheet of film) is placed. No additional lenses which might distort the image are necessary. The real image can also be easily examined by a microscope and thus very narrowly spaced fringes can be studied more easily than by conventional methods, which is of great benefit in nucleate boiling research.

With the technique described so far the investigated process cannot be continuously observed. This disadvantage can be overcome if the real time method is used, as illustrated in Figure 16.5.

After the first exposure, by which the comparison wave is recorded, the hologram is developed and fixed remaining at its place or repositioned accurately. The comparison wave is reconstructed continuously by illuminating the hologram with the reference wave. This reconstructed wave can now be superimposed onto the momentary object wave. If the object wave is not changed

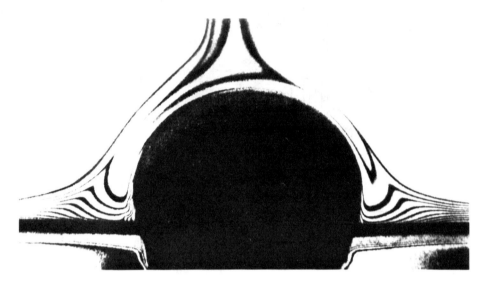

Figure 16.3: Temperature distribution around a "secondary bubble". Water,
p = 0.3 bar, q̇ - 30 W/cm², subcooling = 2K

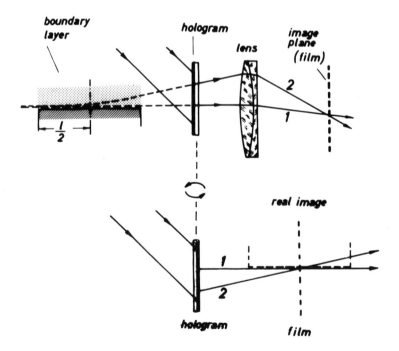

Fig ure 16.4: Arrangement for taking pictures of the virtual (above) and real
image (below)

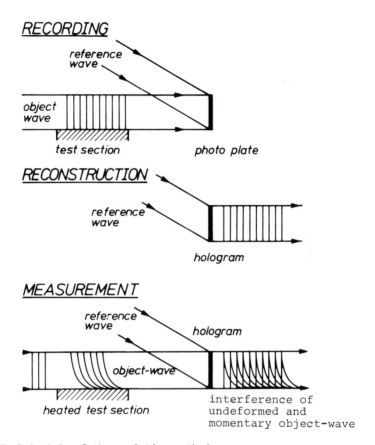

Figure 16.5: Principle of the real-time-method.

and the hologram precisely repositioned, no interference fringes
will be seen at first.

Now the heat or mass transfer process, which is to be examined,
can be started. The wave receives now an additional phase shift as
it passes through the newly imposed temperature field. Behind the
hologram both waves interfere with each other and the changes of
the interference pattern can be continuously observed or photographed on still or moving film (Mayinger and Panknin (1974)).

The evaluation of a holographic interferogram is very similar
to the evaluation of interference patterns recorded in a Mach-
Zehnder interferometer (Hauf and Grigull (1970)). Therefore only
the basic equations will be given. As illustrated in Figure 16.6,
the changes in optical path length between the two exposures can be
expressed in multiples S of a wave length λ. The length l of the
test section and the refractive index of the undistorted system
n_∞ and the refractive index of the distorted system n are required.

With the molar refractivity N and the molar mass M, and by

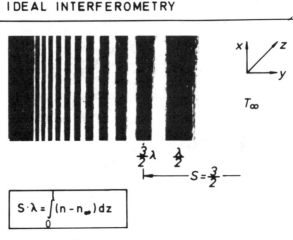

IDEAL INTERFEROMETRY

$$S \cdot \lambda = \int_0^l (n - n_\infty)\,dz$$

ASSUMPTIONS

- **Reference - condition**
  n_∞ = const. $\neq f(x,y,z)$

- **Measuring condition**
  Two - dimensional field of refraction index
  $n = n(x,y)$

- **Light beam runs straight in z-direction**

$$S(x,y) \cdot \lambda = l \cdot \left[n(x,y) - n_\infty \right] = \Delta n(x,y) \cdot l$$

Figure 16.6: Ideal interferometry.

using the Lorentz-Lorenz-formula for gases, the temperature field can be evaluated by the Gladstone-Dale-equation as shown in Figure 16.7. For liquids the integration is not so easy if the refractive index is a strong function of the temperature. However, for small temperature differences, as usually present in interferometric measurements, the first derivative of the refractive index with respect to time, can be assumed to be constant. Then the temperature field also can be evaluated easily as shown on the right side of Figure 16.7.

For many technical applications the local heat transfer coefficients are of special interest. In this case the temperature gradient at the wall is determined by the holographic interferometry and, assuming a laminar boundary layer next to the wall, the heat transfer coefficient can be obtained as demonstrated in Figure 16.8.

The evaluation of a holographic interferogram discussed up to

RELATIONS BETWEEN TEMPERATURE AND REFRACTION INDEX

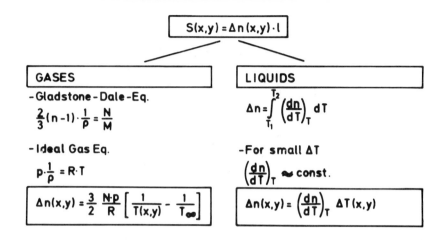

Figure 16.7: Relations between temperature and refraction index.

ENERGY BALANCE

$$q = \alpha \cdot \Delta T = -\lambda \left. \frac{dT}{dy} \right|_{wall}$$

$$\alpha = \frac{\lambda \cdot \left. \frac{dT}{dy} \right|_{wall}}{\Delta T}$$

$$Nu = \frac{\alpha \cdot L}{\lambda} = \frac{\left. \frac{dT}{dy} \right|_{wall} \cdot L}{\Delta T}$$

Figure 16.8: Energy balance.

now was for ideal conditions with a truly two-dimensional field
of the refractive index - i.e. constant value of n along the path
of a beam - and no deflection of the light travelling through
the temperature field. In reality - see Figure 16.9 - the laser
beam undergoes a deflection and - which is more important - in
many cases the temperature field is not two-dimensional but three-
dimensional, which means, that each laser beam is not travelling

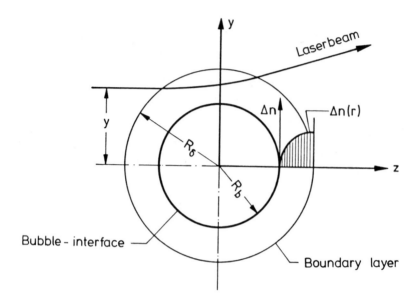

Figure 16.9: Abel-correction including light deflection.

the same distance in a certain temperature field. For spherical
and cylindrical geometries Abel (Hauf and Grigull (1970)) worked
out a correction method for taking in account non-homogeneous
temperature conditions along the travelling path of the laser
beam.

    If in addition the deflection has to be taken into account,
the evaluation procedure becomes much more complicated. Without
beam deflection, a point P in Figure 16.10 would form a picture
in the position A' on the holographic plate. Due to the deflect-
tion, the laser beam leaves the boundary of the assumed temperature
field at P' and forms a picture at B'. An evaluation of the interferogram needs a first estimation of the temperature field and in
a complicated iterative procedure (Nordmann (1980)), the temperature distribution and the heat transfer process can be reconstructed.

    In very thin boundary layers around a heated surface only a
few interferences fringes will occur. The evaluation then is not
very precise. The situation can be improved by superimposing a
finite fringefield after recording the reference hologram (see
Figures 16.2 and 16.5) by inclining slightly either one of the
mirrors in the holographic arrangement or the holographic plate
itself, as illustrated in Figure 16.11.

    An interferogram gained by this method is shown in Figure
16.12, which represents the temperature field around a condensing
bubble just grown by blowing vapour out of a nozzle into a subcooled liquid. The finite interference fringes in the region not
affected by the heat transfer around the bubble are straight lines
inclined under an angle of 45°. The deviation of each interference
fringe from this 45°-line is the temperature gradient, to a first
approximation, at the phase interface.

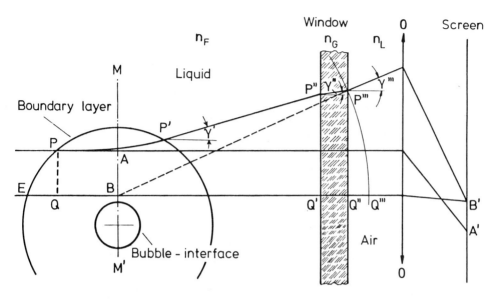

Figure 16.10: The way of the laser beam in the test chamber: illustration of the deflection caused by the bubble and the window.

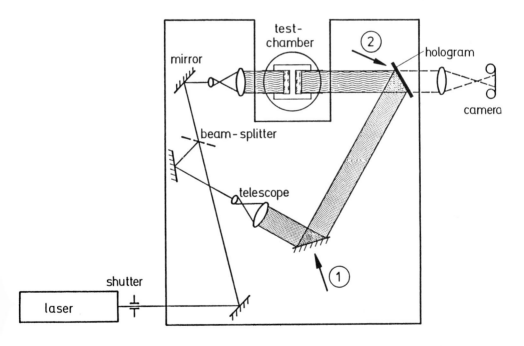

Figure 16.11: Holographic interferometer. The marked points show possibilities to adjust a fringe field.

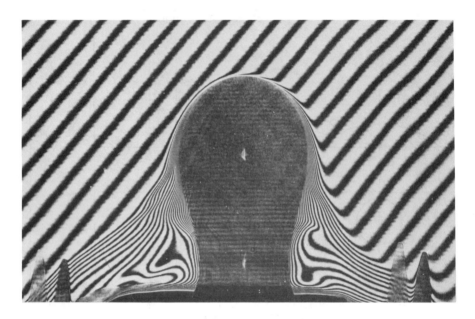

Figure 16.12: The temperature distribution around a vapour bubble growing in
              subcooled water; p = 4 bar, ΔT = 10 K. To get a better
              sensitivity at the top of the bubble, the fringe field was
              adjusted in a diagonal direction.

     By this method the heat transfer during the condensation of a
vapour bubble in a subcooled liquid, temporarily and locally can
be measured very precisely (Nordmann (1980)). Figure 16.13 gives
the distribution of the Nusselt number around the circumference
of a bubble condensing in a liquid which was subcooled by 16.8 K
at a system pressure of 4 bar.

     Thanks to the fact that this optical method works inertialess,
the extremely fast condensation process can be observed accurately
and interesting information can be gained of the dynamic behaviour
of the bubble and the interaction between heat transfer and inertia
effects, as illustrated in Figure 16.14.

16.3 High Speed Cinematography

     Cine pictures with high frequency can either be taken by a
rotating prism camera or by a high frequency flash equipment,
combined with a drum-camera. Those arrangements - flash equipment
and camera - can also be combined.

     A sketch of a rotating prism camera is represented in Figure
16.15. The film is continuously drawn from a spool and passes
the window of the lens without stopping periods, as in a usual
cine camera. To get well focused and sharp pictures, the film
movement has to be compensated by a rotating prism, as demonstrated
in Figure 16.16. Film movement and rotation of the prism have to

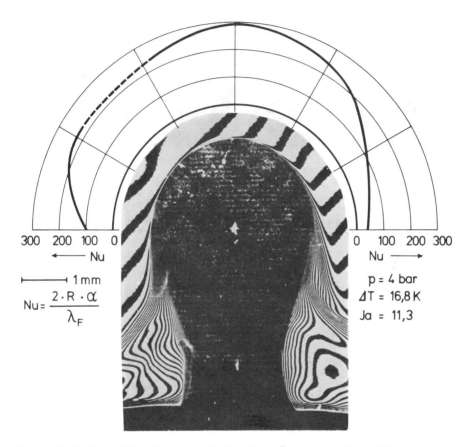

Figure 16.13: Nusselt number around the circumference of the bubble.

have to be exactly in phase.

    A principle diagram for combining high speed cine-cameras -
i.e. rotating prisma camera or drum camera - with the high-
frequency flash equipment is shown in Figure 16.17. Commercial
flash equipments have an illumination time of 10^{-6}s and make
possible exposure frequencies up to 100 kHz. The manipulation of
these flash units needs not only experience and experimental
skillfulness; usually it is very difficult, too, to find a suitable
triggering mechanism which makes it possible to catch exactly the
fluiddynamic procedure within the extremely short exposure time
available with this high speed technique. High speed cinematography is of great advantage in heat transfer measurements,
especially if fast changing interfacial areas and boundaries in
a two-phase system have to be analysed.

    An example of a high speed photograph is given in Figure
16.18, where the axial view method, as proposed by Hewitt and
described by Delhaye in Chapter 15, was used. Figure 16.18 shows
the phase distribution of an annular two-phase flow in a tube of
approximately 15 mm diameter.

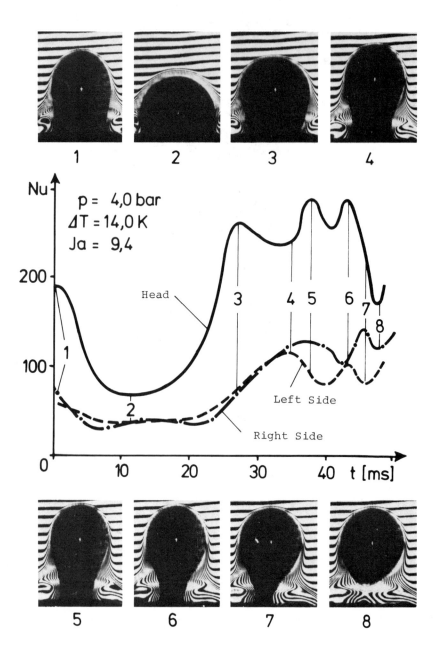

Figure 16.14: The heat transfer at the top and at the side of a growing bubble as a function of the time.

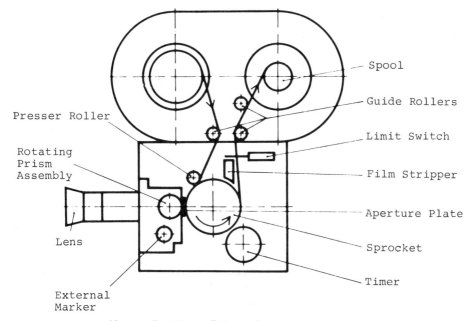

16mm-Rotating Prism Camera

Figure 16.15: The Hitachi high speed cine camera.

For higher velocities and faster changing phenomena a method with much shorter exposure times than described up to now is needed. Many years ago, Cranz and Schardin (1929) proposed an arrangement as illustrated in Figure 16.19 which works with a number of separate flash units arranged in series and ignited in short time distances. The flash time is approximately 10^{-8}s and the time distance can be as small as a few micro seconds. With a simple camera on one and the same photographic plate a series of pictures spacely separated can be registered.

With this method flow phenomena in two-phase mixtures up to velocities of more than 100 m/s can be recorded. An example for the photographic possibilities offered by this technique is shown in Figure 16.20, where a water particle formed in the nozzle of a venturi scrubber is photographed twice by a double exposure technique within a time difference of 10^{-5}s. The velocity of the particle is approximately 80 m/s; with this technique it is possible to observe how the surface of the particle is changed.

16.4 Luminescence Method and X-ray Technique

For determining the velocity also in two-phase flow the stochastic method, together with the auto- or cross-correlation is used. Signals for this technique may be temperature- or density fluctuations in the flow. If the stochastic distribution

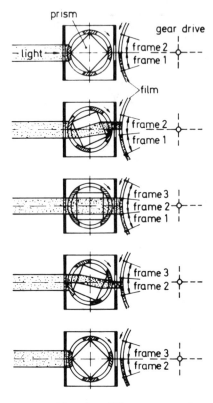

Figure 16.16: The compensation for film movement by a rotating prism.

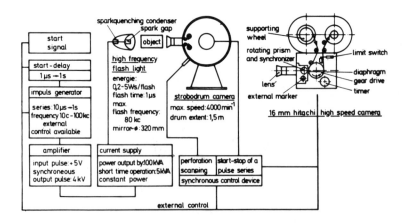

Figure 16.17: Principle diagram of the high-frequency flash equipment.

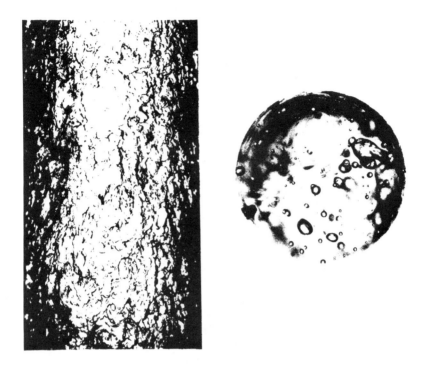

Figure 16.18: Droplets photographed by the axial view method; comparison of axial and radial view.

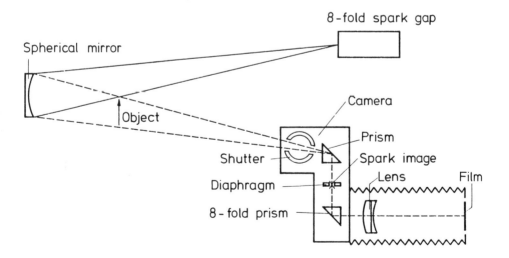

Figure 16.19: Cranz-Schardin multiple spark camera.

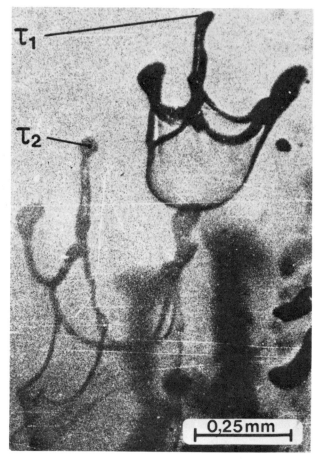

Figure 16.20: Water particle in venture scrubber, double exposure ($\Delta\tau = 10^{-5}$s)

of these fluctuations at the second sensor is not similar to the
situation at the first sensor, an evaluation is not possible, or
at least very erroneous. A simple pulse method as illustrated in
Figure 16.21, using the luminescence of certain liquids or crystals,
has a great advantage, because it offers a clear and well defined
signal and it can be even used without the stochastic procedure.
The luminescence is produced by a short ultra violet light flash
through a window into the flow, which then lasts for a few
seconds. In a second window, positioned in a certain distance
from the first one, it then can be observed how long it takes
until the luminescant fluid passes. From the distance and the
delay-time the velocity can be deduced.

Besides the γ attenuation-method, also the X-ray flash
technique can be used for measuring void distribution and flow
pattern in a gas-liquid mixture. An arrangement of an X-ray flash
unit is illustrated in Figure 16.22. While the X-ray flash unit

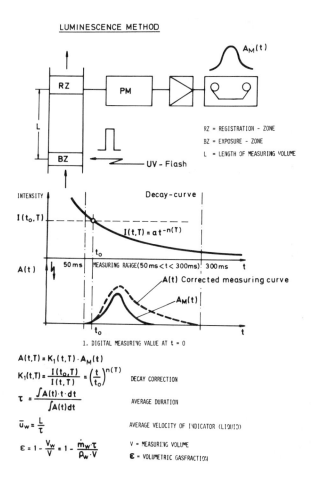

LUMINESCENCE METHOD

$$A(t,T) = K_1(t,T) \cdot A_M(t)$$

$$K_1(t,T) = \frac{I(t_0,T)}{I(t,T)} = \left(\frac{t}{t_0}\right)^{n(T)} \quad \text{DECAY CORRECTION}$$

$$\tau = \frac{\int A(t) \cdot t \cdot dt}{\int A(t) dt} \quad \text{AVERAGE DURATION}$$

$$\bar{u}_w = \frac{L}{\tau} \quad \text{AVERAGE VELOCITY OF INDICATOR (LIQUID)}$$

$$\varepsilon = 1 - \frac{V_w}{V} = 1 - \frac{\dot{m}_w \cdot \tau}{\rho_w \cdot V} \quad \begin{array}{l} V = \text{MEASURING VOLUME} \\ \varepsilon = \text{VOLUMETRIC GASFRACTION} \end{array}$$

Figure 16.21: Luminescence method.

itself is a rather reliable apparatus allowing flash frequencies up to 20 kHz, the image transformer plays still the weakest part in this system as it is only good for 500-1000 Hz. Without this image transformer an evaluation of the pictures produced by the X-ray is almost impossible for two-phase flow, because of the low contrast. The attenuation of the X-ray is similar to that of a γ-ray and the mathematical procedure is described by Delhaye in Chapter 15.

An easier evaluation of an X-ray picture is possible by a density-display system (Haertel, Stuttgart) which changes the different grades of darkness in the black and white picture to a coloured system. The arrangement of such a density display system is illustrated in Figure 16.23.

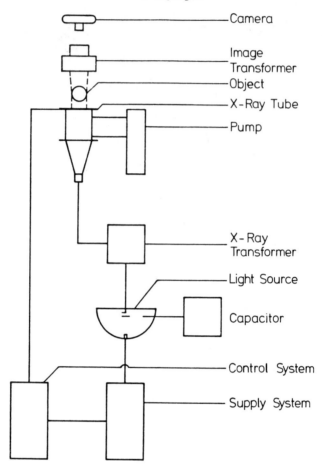

Figure 16.22: Arrangement of X-ray flash unit.

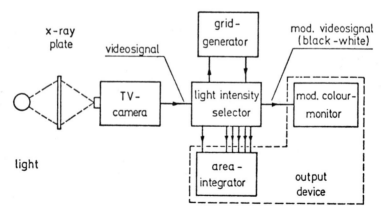

Figure 16.23: Density display system developed by Haertel.

16.5 Concluding Remarks

Optical measuring techniques have the advantage to give a visual impression of the phenomenon under study. This under certain conditions helps to get a better understanding and to formulate correlations physically in a more realistic way. However, one must not over-estimate the usefulness of these methods, as they give only qualitative information in many cases.

The layout and the design of an apparatus needs quantitative values of the heat- or mass transport or of the fluiddynamic conditions. Interferometric methods like the holographic interferometry offer both a qualitative overview of the thermohydraulic system and quantitative figures of the heat and mass transport. The use of this method, together with the high speed cinematography certainly opens a lot of interesting possibilities; however, it also needs a very skillful and patient experimentalist.

References

Caulfield, H J and Sun Lu, (1970) "The applications of holography", Wiley (Interscience), New York.

Collier, R J, Burckhardt, C B and Lin, L H, (1971) "Optical holography", Academic Press, New York.

Cranz, C and Schardin, H, (1929) Z. Phys. **56**, 147-183.

Gabor, D, (1948) "A new microscopic principle", *Nature*, 161, 777.

Gabor, D, (1949) "Microscopy by reconstructed wavefronts", *Proc. Roy. Soc.* A 197, 454.

Gabor, D, (1951) "Microscopy by reconstructed wavefronts II", *Proc. Phys. Soc.* B 64, 449.

Haertel, personal communication, Institut für Verfahrenstechnik und Dampfkessel-wesen, Universität Stuttgart.

Hauf, W and Grigull, U, (1970) "Optical methods in heat transfer", *Advances in heat transfer*, **6**, 133.

Kiemle, H and Roess, D, (1969) "Einfuehrung in die technik der holographie", Akademische Verlagsgesellschaft, Frankfurt/M.

Mayinger, F and Panknin, W, (1974) "Holography in heat and mass transfer", Heat Transfer 1974, **VI**, 28-43, *5th International Heat Transfer Conference*, Japan.

Nordmann, D, (1980) "Temperatur, druck und waermetransport in der umgebung kondensierender blasen", Diss. Universität Hannover.

Smith, H M, (1969) "Principles of holography", Wiley (Interscience), New York.

Varadi, C, Nabizadeh, H and Lordong, N (1977), "Entwicklung neuer messmethoden zur bestimmung der phasengeschwindigkeit in einer zweiphasenstroemung", Kerntechn. Ges. im Dt. Atomforum e.V. Experimentiertechnik auf dem Gebiet der Thermo- und Fluiddynamik, Teil 2.

Chapter 17

Introduction to Two Phase Flow Problems in the Process Industry

G. F. HEWITT

17.1 Introduction

At least 60% of all process industry heat exchange equipment involves two phase flow in one form or another. In this brief survey, my intention is to review the various types of equipment and to highlight some of the problems which occur in design and to which the attention of research workers must be directed. I will cover the following main areas:

(1) Pipeline systems

(2) Waste heat boilers

(3) Reboilers for distillation towers

(4) Cryogenic heat exchangers

(5) Fired heaters

(6) Condensers

(7) Direct contact heat transfer equipment

17.2 Pipelines

In previous chapters in this book, there has been extensive reference to two phase flow in tubes. One might have thought, therefore, that the situation on pipelines should, by now, be fully satisfactory. However, it must be admitted that the situation on the prediction of the performance of pipelines of the type now commonly used in off-shore operations (where two phase flow develops along the pipe as the pressure falls) is rather unsatisfactory. The following main problems occur:

(1) Most of the correlations for two phase flow have been developed for extremely small tubes (typically less than 3 inches diameter) whereas in the applications, pipe sizes up to, say, 36 inches are common. The normal correlations for hold-up and pressure drop are not valid for these large bore pipes and it is better to use flow modelling methods of the type described earlier in the course.

(2)   Very often, severe problems can occur if slug flow is present
in the pipeline. A number of circumstances can occur leading
to the generation of slug flow. For instance, if the oil/gas
mixture flows along a pipe on the sea-bed and then rises to
the platform level up a vertical pipe, the liquid may collect
upstream of the bend until it reaches a given level, at which
point it is swept up the vertical leg giving rise to mechanical
problems in the platform equipment.

In the future, one may expect to see much more sophisticated
approaches to the calculation of pipeline flows. Already, it is
proving possible to make some useful calculations on the slug
formation problem, using numerical fluid mechanics methods.

17.3 Waste Heat Boilers

Waste heat boilers function to cool, and recover heat from,
gases from various chemical processes. Another function is in
the rapid quenching of gases to "freeze" species generated in
chemical processes at high temperature. There are severe problems
in the design of waste heat boilers, which arise from the very
high temperature of the process gases, coupled often with severe
corrosion and other problems. A number of types of design have
been suggested including:

(1)   Vertical callandria

(2)   Bayonet type

(3)   U-type

(4)   Horizontal cross-flow type

I shall deal briefly with each of these forms; a more detailed
review is given, for instance, by Hinchley (1977).

Vertical Callandria Waste Heat Boilers

In this type, the hot gas is fed through the tubes of a
vertical callandria, water passing into the shell-side at the
bottom of the callandria and the generated steam water mixture
passing out at the top. Here, the main problem is associated with
the tube-sheet at the bottom of the callandria. The hot gas
impinges on this tube sheet and cooling has to be by water on the
shell-side which impinges on the tube sheet. However, the rate of
water ingress to the tube sheet is limited by the flooding
phenomenon (which I will describe in Chapter 20) and, often, the
tube sheet can become overheated. The problem can be aleviated
by using ferrules at the entrance of the tubes, though these
themselves are exposed to the hot gas, without appreciable
cooling, and it may be difficult to find suitable materials.

Bayonet Tube Waste Heat Boilers

In this form of boiler, the hot gas passes upwards over
bayonet tubes in which the water flows down in an inner tube and
then upwards in the annulus between this inner tube and the outer
tube which is in contact with the gas. Although these types of

boilers are widely used, they have problems due to flow separation effects leading to over-heating, problems of erosion and problems in the rather complex nature of the tube sheet design.

U-Tube Type Waste Heat Boilers

In this form of waste heat boiler, the hot gas is passed over a horizontal bundle of U-tubes mounted in a shell-and-tube heat exchanger. Water flows in the lower leg of each U-tube and steam-water mixtures are generated and passed out of the upper leg. Due to gravitational evaporation effects, the dryout heat fluxes are very much lower in horizontal tubes and, furthermore, the situation is complicated by the existence of the U-bend. Flow separation effects tend to occur downstream of the bend and give rise to premature dryout and overheating. These problems can be solved (though at the expense of much increased pressure drop on the tube side) by the use of inserted helical tapes in the tubes. The problems of U-tube waste heat boilers are discussed by Robertson (1972).

Horizontal Cross-Flow Boilers

Here, the gas flows on the tube side and boiling occurs in cross flow over the tubes. A typical design of unit is illustrated in Figure 17.1 There are still problems in these units of insulating the gas inlet and the example shown in Figure 17.1 uses refractory and insulating concrete. Ferrules can be inserted at the tube entrance as shown, but the advantage of the cross-flow design is that it is possible to sweep the tube sheet with the steam-water mixture. A more detailed discussion of this form of waste heat boiler is given by Hinchley (1977).

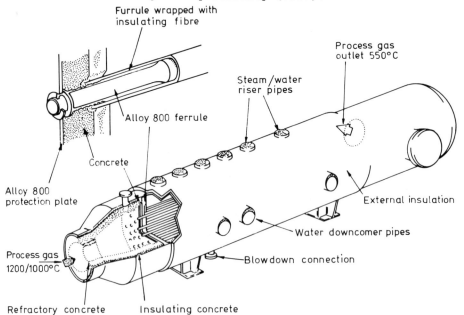

Figure 17.1: Typical horizontal cross-flow waste heat reboiler.

17.4 Reboilers

The function of a reboiler is to generate a vapour reflux at the bottom of a distillation tower. The main forms which have been used are the kettle reboiler, the horizontal thermosyphon reboiler and the vertical thermosyphon reboiler.

Horizontal Reboilers

A typical kettle reboiler installation is illustrated in Figure 17.2 and a typical horizontal thermosyphon reboiler in Figure 17.3. The main problems with horizontal boilers are as follows:

(1) Prediction of cross flow boiling heat transfer coefficients and dryout fluxes.

(2) Fouling of the heat transfer surfaces.

(3) Disentrainment of the liquid from the efflux vapour.

(4) Operation with multicomponent fluids.

Multicomponent boiling presents a number of special problems as is discussed in Chapter 11 by Mr Collier.

Vertical Thermosyphon Reboilers

A typical vertical thermosyphon installation is illustrated in Figure 17.4. The fluid circulates through the reboiler as a result of the heat generated by the difference between the density of the two phase fluid in the reboiler tubes and the liquid in the

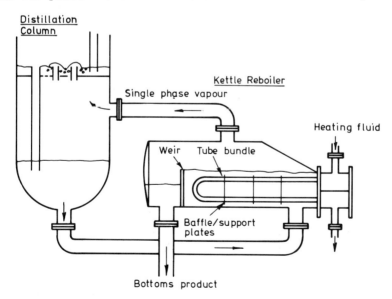

Figure 17.2: Kettle reboiler installation.

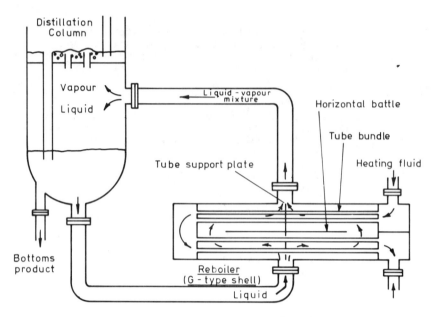

Figure 17.3: Horizontal thermosyphon boiler installation.

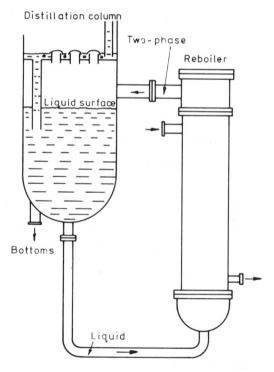

Figure 17.4: Vertical thermosyphon reboiler installation.

distillation tower and in the feed line. The main problems in this kind of installation are as follows:

(1) Calculation of circulation rate and heat transfer coefficients. This involves estimation of local pressure gradients and integration along the tubes of the heat transfer rate. An iterative calculation is required to establish a pressure balance around the circuit. It is normally best to do this using a computer program.

(2) Dryout. A number of different forms of dryout can occur in thermosyphon reboilers; a further discussion on dryout is given in Chapter 9.

(3) Instability. The various mechanisms of instability were reviewed in Chapter 13 by Professor Bergles; instability of the whole circulation loop and/or instability of flow between parallel channels in the reboiler may occur.

(4) Low pressure operation. Many of the normal correlations for two phase flow fail at sub-atmospheric pressures, where reboilers are often operated.

(5) Multicomponent boiling. Here again, there is a dearth of information on multi-component boiling in forced convection.

17.5 Cryogenic Heat Exchangers

The main problem in transferring heat between cryogenic fluids is that, to minimise compressor work, very low temperature differences are required. This leads to rather special types of heat exchangers like the Hampson coil type or the brazed aluminium, plate-fin type. The construction of the latter type is illustrated in Figures 17.5 and 17.6. The fins separating the plates may be in plain form or alternatively in herring-bone form or serrated form as illustrated in Figure 17.7.

The main problems with brazed aluminium (plate-fin) heat exchangers are as follows:

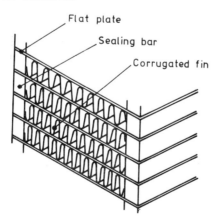

Figure 17.5: Construction of brazed aluminium (plate-fin) heat exchanger.

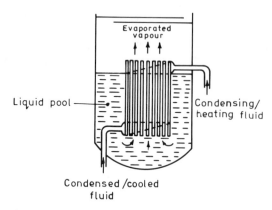

Figure 17.6: Use of brazed aluminium heat exchanger as a reboiler.

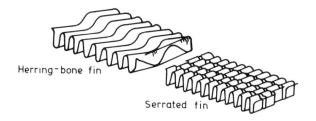

Figure 17.7: Alternative fin geometries for brazed aluminium (plate-fin) heat
           exchangers.

(1)   There is a dearth of data for this type of geometry in boiling
      and condensation duty. For this reason, Harwell has been
      carrying out an experimental programme over a number of years
      to obtain such data. A particular problem here is the very
      low temperature differences required and this gives a particular
      experimental challenge.

(2)   Flow distribution. It is important to obtain uniform flow
      distribution between the various sub-channels and special
      distributors have to be designed. A typical arrangement is
      shown in Figure 17.8.

(3)   Stability. Cryogenic exchangers have the normal two-phase
      flow stability problems some of which are particularly
      exaggerated due to the rather special physical properties of
      the fluids used.

(4)   Multi-component operation. A typical multi-component operation
      for a plate-fin heat exchanger would be in LNG plants. Again,
      the problems of simultaneous heat and mass transfer are
      severe.

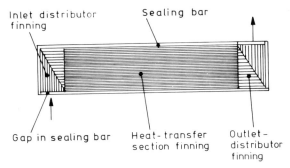

Figure 17.8: Flow distribution arrangement in a brazed aluminium (plate-fin) heat exchanger.

17.6 Fired Heaters

Fired heaters are used extensively in the chemical, petroleum and process industries. They take various forms including axi-symmetric heaters (with a burner placed centrally at the bottom) in which the fluid being heated passes through vertical tubes around the wall of the heater or, alternatively, in a helical coil. Another common type is the cabin heater in which a number of alternative arrangements of burners is possible and in which the process fluid may pass through horizontal tubes placed in various positions in the heater. Problems in the design of fired heaters occur on the radiant side where the prediction of combustion and heat flux distribution is complex, particularly for two-phase flow sprays. On the process side, the main problems are:

(1) Prediction of pressure drop and circulation rate (heat transfer coefficient is not normally a limiting factor in such heaters which are normally governed by the radiant flux).

(2) Stability.

(3) Distribution of flow between the various process side tubes.

(4) Dryout, leading to overheating and carbonisation of organic fluids. This is particularly severe in horizontal tubes and is complicated by the fact that the heat flux distribution around the tubes is non-uniform. For instance, a tube near the wall of a fired heater will receive heat mainly from one side.

17.7 Condensers

Condensers are used widely in the process industries for the condensation of water vapour and for the condensation of a great variety of single component and multi-component vapours. Below, I will consider very briefly the following main types:

(1) Baffled shell-and-tube condensers.

(2) "Surface" condensers.

(3)   Air-cooled condensers.

(4)   Direct contact condensers.

Shell-and-Tube Condensers

Practically all the common forms of baffled shell-and-tube heat exchangers are used for condensation duty. This includes one-pass (E) shells, dividing/combining (J) shells, two-pass (F) shells, double split (H) shells, triple-split shells and cross flow units.

The main problems in condensation in shell-and-tube condensers are as follows:

(1)   Prediction of heat transfer and pressure drop.

(2)   Flow pattern prediction.

(3)   Separation effects (eg in the use of shells for sub-cooling duty).

(4)   Mist (fog) formation.

(5)   Operation with multi-component mixtures.

The above problems are common to both shell-side and tube-side condensation. In tube side systems, there is a severe problem of flow distribution between the tubes, particularly in two-phase multi-pass operation. This should be avoided if possible.

"Surface" Condensers

These are commonly used in condensing the output from turbines, where the vapour pressure has to be as low as possible to maximise the turbine efficiency. Thus, the vapour is often led through "steam lanes" in order to minimise the pressure drop. The main problems are:

(1)   Heat transfer and pressure drop prediction with the specific context of complex three-dimensional flows.

(2)   Build-up of incondensible gases in various zones of the tube bundle. It is important to arrange vent lines at appropriate points and to design, as far as possible, for one-dimensional flow towards the vent positions.

Air-Cooled Condensers

These can operate in the forced draught (Figure 17.9) or induced draught (Figure 17.10) modes. The main problems arising are as follows:

(1)   Plenum aerodynamics and air flow distribution on the air-side.

(2)   In-tube condensation pressure drops and heat transfer coefficients.

(3)   Flooding limitations when operated in the reflux mode.

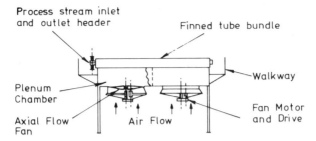

Figure 17.9: Air-cooled heat exchanger operating in the forced draught mode.

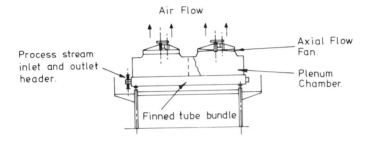

Figure 17.10: Air-cooled heat exchanger operating in the induced draught mode.

(4) Flow distribution problems in headers.

(5) Operation with multi-component mixtures.

Direct Contact Condensers

In this type of condenser, the product which is being condensed is recirculated through a cooler and sprayed into a vessel into which the vapour is fed. This kind of condenser is advantageous in situations where there is a large amount of inert gas present. It is usually advantageous to induce a swirl in the vessel by arranging the spray jets tangentially. Another form of direct contact condenser is the so-called "baffle column" type in which the cooled liquid splashes down a series of trays in counter-current flow to the vapour. This latter type has the advantage that it can handle dirty liquids which would block the nozzles in a spray condenser. It also has the advantage in that it operates in counter-current flow. However, the baffle column type has a relatively low interfacial area.

The design of direct contact condensers is at a relatively rudimentary stage. Important factors are the dropsize distribution produced by the spray nozzles in spray condensers and the extent of agitation produced by the splashing effects in baffle-column condensers.

17.8 Other Forms of Direct Contact Heat Exchanger

Cooling Towers

With the decreasing availability of suitable cooling water

sources, the installation and operation of cooling towers is becoming more and more common in the process industry. Cooling towers may be of the natural draught type (for which typical designs are shown in Figure 17.11) and of the forced draught type as illustrated in Figure 17.12.

There are many problems in the design and operation of cooling towers including environmental effects, prediction of air-flow distribution and gas-liquid interactions etc.

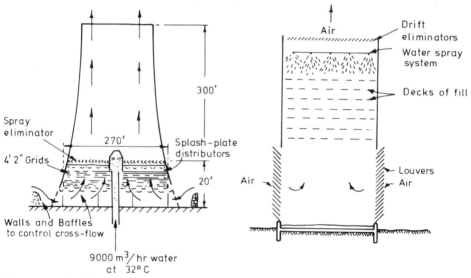

Figure 17.11: Natural draught cooling towers.

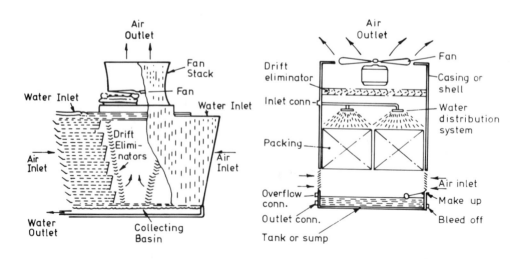

Figure 17.12: Forced draught cooling towers.

Submerged Combustion

Submerged combustion evaporators are particularly useful in the concentration of highly corrosive fluids. A typical unit is illustrated in Figure 17.13. The combustion gases pass down from the burner into the fluid and mix with the fluid transferring heat to it.

Two phase flow problems occurring in submerged combustion evaporators include problems of disentrainment, gas-liquid heat transfer and burner stability.

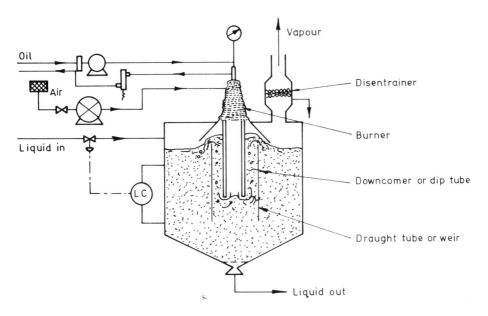

Figure 17.13: Typical submerged combustion unit.

References

Hinchley, P, (1977) "Waste heat boilers: problems and solutions". *Chem. Eng. Prog.* **73**, 90 and *Chem. Ing. Tech.* **49**, 553.

Robertson, J M, (1973) "Dryout in horizontal hair-pin waste-heat boiler tubes". *A.I.Ch.E. Symp. Ser.* **69** (131), 55.

Chapter 18

Multicomponent Boiling and Condensation

J. G. COLLIER

18.1 Introduction

So far, we have only considered the behaviour of pure single component fluids during evaporation and condensation.

However, in the petrochemical industries in particular, many processes involve the evaporation (and condensation) of binary (n = 2) and multi-component (n > 2) mixtures where n is the number of components. In the boiling of binary and multi-component mixtures, the heat transfer and the mass transfer processes are closely linked, with the evaporation rate usually being limited by the mass transfer processes. This is significantly different from single-component systems where interfacial mass transfer rates are normally very high.

In this chapter, we will first consider the differences to be expected in the basic physical processes when mixtures are evaporated and condensed. This will lead on to a review of the available information on pool boiling of mixtures. Finally, the problems of predicting heat transfer rates in forced convective evaporation and condensation of mixtures will be examined.

Whilst there is now a considerable body of published information on pool boiling of binary mixtures, very little has been published relating to forced convection evaporation of mixtures. The evaporation of multi-component mixtures is largely unexplored.

18.2 Elementary Phase Equilibria

18.2.1 *Definitions*

It may be useful to define some of the variables we will be using in this chapter:

- the *molar concentration* $\tilde{c}_i$ of component i is given by

$$\tilde{c}_i = \frac{\rho_i}{M_i}$$

where ρ_i is the density of component i and M_i is its molecular weight.

- the *total molar concentration* of the mixture is given by

$$\tilde{c} = \sum_{i = 1}^{n} \tilde{c}_i$$

- the *mol fraction* in the liquid $\tilde{x}_i$ or the vapour $\tilde{y}_i$ is defined by

$$\tilde{y}_i \ (\text{or} \ \tilde{x}_i) = \frac{\tilde{c}_i}{\tilde{c}}$$

- the *molar flux* $\dot{n}_i$ is the number of moles of component, i passing normally through unit area in unit time.

- the *diffusive molar flux* J_i is the flux of component i, *relative* to the *mixture total molar flux* $\dot{n}$.

- the *mixture total molar flux* $\dot{n}$ is defined as

$$\dot{n} = \sum_{i = 1}^{n} \dot{n}_i$$

18.2.2 *Binary Systems*

In a mixture of vapours the partial pressure, p_A, of component A is that pressure which would be exerted by A alone in a volume appropriate to the concentration of A in the mixture at the same temperature. Since $p = \sum p_A$, then for an ideal mixture the partial pressure, p_A, is proportional to the mol fraction of A in the vapour phase:

$$p_A = \tilde{y}_A \ p \qquad\qquad (18.1)$$

Two laws relate the partial pressure in the vapour phase to the concentration of component A in the liquid phase.

(a) *Henry's law* which states that the partial pressure p_A is directly proportional to the mol fraction $\tilde{x}_A$ of component A in the liquid phase

$$p_A = \text{constant} \ \tilde{x}_A \qquad\qquad (18.2)$$

This holds for dilute solutions only where $\tilde{x}_A$ is small.

(b) *Raoult's law* which states that the partial pressure p_A is related to the mol fraction $\tilde{x}_A$ and the vapour pressure of pure component A at the same temperature

$$p_A = p_A^{\ o} \ \tilde{x}_A \qquad\qquad (18.3)$$

where $p_A^{\ o}$ is the vapour pressure of pure component A at the same temperature. This holds for high values of $\tilde{x}_A$ only.

If a mixture follows Raoult's law it is said to be "ideal" and values of $\tilde{y}$ can be calculated for various values of $\tilde{x}$ from a knowledge of the vapour pressures of the two pure components at various temperatures.

Few mixtures behave in an "ideal" manner. If the ratio of the partial pressure p_A to the mol fraction in the liquid phase $\tilde{x}_A$ is defined as the "volatility" then a binary mixture of components A and B

$$\text{volatility of A} = \frac{p_A}{\tilde{x}_A}, \ \text{volatility of B} = \frac{p_B}{\tilde{x}_B}$$

The ratio of these two volatilities is known as the "relative volatility" α, given by:

$$\alpha = \frac{p_A \ \tilde{x}_B}{p_B \ \tilde{x}_A} \qquad\qquad (18.4)$$

Substituting equation (18.1):

$$\alpha = \frac{\tilde{y}_A \ \tilde{x}_B}{\tilde{y}_B \ \tilde{x}_A} \ \text{or} \ \frac{\tilde{y}_A}{\tilde{y}_B} = \alpha \ \frac{\tilde{x}_A}{\tilde{x}_B} \qquad\qquad (18.5)$$

Clearly, from equation (18.3) for an "ideal" mixture the relative volatility is just the ratio of the vapour pressures of the pure components A and B at the chosen temperature. Values of α vary somewhat with temperature and tables for different binary systems are available in chemical engineering handbooks.

Consider a container initially filled with a binary liquid mixture having a mol fraction $\tilde{x}_O$ of the more volatile component and held at constant pressure, p, and temperature, T_1 (Fig. 18.1(a)). What happens when the liquid mixture is heated at constant pressure can be shown conveniently on a *temperature-composition diagram* (Fig. 18.a(b)). In this diagram the mol fraction of the more volatil component in the liquid is plotted as the abscissa and the temperature at which the mixture boils as the ordinate.

The initial conditions in the container are denoted by point Q. The container is now heated at constant pressure. When the temperature reaches T_2 the liquid will start to boil as shown by point R. Some vapour is formed of composition $\tilde{y}_O$ as shown by point S. This initial vapour is rich in the more volatile component. If the experiment is repeated for the complete range of values of $\tilde{x}_O$ then a series of points such as R and S can be determined. The locus of points like R is a curve AXTRB which is known as the

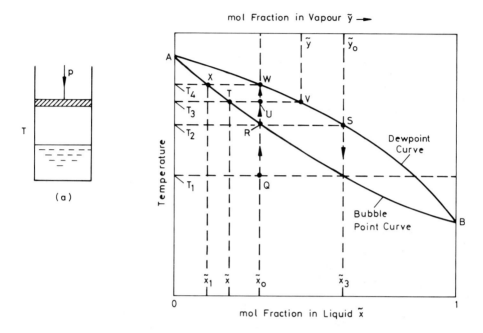

mol Fraction in Vapour $\tilde{y}$ ⟶

(a)

(b) Temperature–Composition Diagram

Figure 18.1: Elementary phase equilibria for binary systems.

"bubble point" or "boiling point" curve. The locus of points like S is a curve AWVSB which is known as the "dew point" curve.

If the container is heated further the composition of the liquid will change, because of the loss of the more volatile component to the vapour. The boiling point will therefore rise to some temperature T_3. Both the liquid and vapour phases become richer in the less volatile component with the liquid having a composition represented by point T and the vapour having a composition represented by point V. Since no material is lost from the container, the proportion of liquid to vapour must be given by:

$$\frac{L}{V} = \frac{UV}{TU} = \frac{\tilde{y} - \tilde{x}_o}{\tilde{x}_o - \tilde{x}} \tag{18.6}$$

The ratio $\tilde{y}/\tilde{x}$ at constant temperature is referred to as the *equilibrium ratio*, K

$$K = \tilde{y}/\tilde{x} \tag{18.7}$$

Combining equation (18.6) and (18.7) we have:

$$\tilde{x} = \frac{\tilde{x}_o \left(\frac{L}{V} + 1 \right)}{K + \frac{L}{V}}$$

(18.8)

On further heating to a temperature of T_4, all the liquid is vaporized to give a vapour of the same composition as the original liquid (point W). The last drop of liquid to disappear is very rich in the less volatile component ($\tilde{x}_1$) (point X). During this hypothetical experiment, the mass fraction vaporized, the specific volume and the heat added to the system to bring it to each temperature condition would also have been recorded. Note the analogy between this experiment and the once-through vaporization of a liquid mixture fed to the bottom of a tube.

In the design of evaporation equipment for liquid mixtures, thermodynamic equilibrium is usually assumed. However, clearly this must be an approximation since temperature and concentration differences must exist for evaporation to occur. Equilibrium is invariably assumed to exist at vapour-liquid interfaces, ie where the phases remain in intimate contact with one another. However, it is possible to conceive a situation where the vapour bubbles formed rise to the surface and into a vapour space such that there is no longer any close contact with the liquid phase. Further evaporation of the liquid must occur with a heavier (less volatile) liquid than originally designed for. There will be a corresponding rise in the boiling point, the effective temperature driving force for evaporation will be reduced and the available surface area may become insufficient to accomplish the required duty. Such a separation of the vapour from the liquid may occur to some extent, when evaporating a mixture on the shell side of a kettle reboiler, especially at low circulation rates. Alternatively, it could also occur within tubes where there is stratification or poor distribution of the flows.

Temperature-composition diagrams such as Figure (18.1B) are helpful in interpreting different physical situations. For example, consider the case where subcooled boiling of a binary liquid mixture of composition $\tilde{x}_o$ and temperature T_1 occurs (point Q). The vapour formed at the heat transfer surface will have a composition $\tilde{y}_o$. If the vapour bubble detaches from the surface, passes through the liquid and condenses, then the condensed liquid will have a composition $\tilde{x}_3$ corresponding to $\tilde{y}_o$. Thus the liquid adjacent to the heating surface will tend towards a composition $\tilde{x}_1$, whilst the liquid some way away from the heating surface will tend towards $\tilde{x}_3$. These composition changes will be accompanied by changes in physical properties of the liquid.

Some temperature-composition curves look like those shown in Figure (18.2). For these binary systems there are critical compositions $\tilde{x}_c$ where the vapour has the same composition as the liquid such that there is no change of composition upon evaporation. It will be seen later than for these special mixtures, called *azeotropes*, mass transfer limitations are absent and the mixture behaves when it is boiled or condensed like a pure component having mean mixture physical properties.

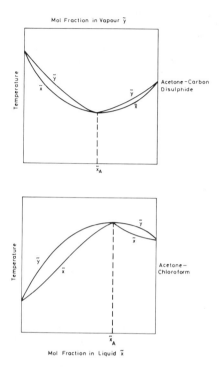

Figure 18.2: Temperature-composition diagrams for binary mixtures forming azeotropes

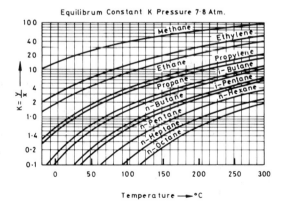

Figure 18.3: Vapour-liquid equilibrium data for hydrocarbons

18.2.3 Multi-Component Mixtures

The above concepts can readily be extended to multi-component mixtures. If Raoult's law applies, then equation (18.8) can be used with the appropriate $\tilde{x}_O$ and K values. Various values of L/V are assumed for a known liquid composition at a given pressure and temperature. The correct L/V ratio is that which satisfies the constraint:

$$\sum \tilde{x} = 1 \qquad\qquad (18.9)$$

The application of equation (18.8) to multi-component non-ideal mixtures is more difficult and involves a further constraint

$$\sum K\tilde{x} = 1 \qquad\qquad (18.10)$$

The calculations can be carried out in terms of the number of moles of each component in each phase instead of using mol fractions as above. K values have been measured for a wide range of hydrocarbons at various pressures and some values are shown in Figure (18.3). The methods used to measure or predict phase equilibrium data are, however, outside the scope of this lecture.

The evaporation of any binary or multi-component liquid can be expressed in terms of a curve of the equilibrium temperature (T) against an amount of heat (Q) - the conventional "cooling" or "condensation" curve. It is a unique curve provided the pressure (p) throughout the evaporator is constant. Another, and in some ways more convenient, way of expressing this curve is in terms of the temperature against the specific enthalpy ($\tilde{h}$) (Figure (18.4)). Transferring one curve to another is quite straightforward and can be done by:

$$\tilde{h} = \frac{Q - Q_r}{N} \qquad\qquad (18.11)$$

where Q_r is a reference value of Q which could conveniently be either the value at the lowest temperature reached in the system or the "bubble point" value at the lowest pressure, p; N is the total number of moles of fluid.

The condensation curve must always be provided to the heat exchanger designer. Typical information will be in a tabular form for a series of pressures spanning the range of interest for each mixture. Since the heat exchanger designer has to work with local conditions within the exchanger, the local pressure will always be known at any point in the calculation and it is, therefore, possible to allow for the effect of pressure on the equilibrium temperature by interpolation between tables at given pressures. For the sake of clarity, however, the effect of pressure is omitted from the rest of the discussion.

In addition to there being a unique T-$\tilde{h}$ relationship at each pressure level, so there is also a unique x-$\tilde{h}$ relationship where x is the quality or the vapour phase mass fraction. When the condensation curve (T-$\tilde{h}$) is specified the corresponding values of x must

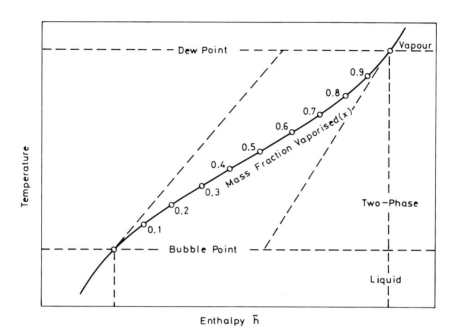

Figure 18.4: Temperature-enthalpy curve

also be given (Figure (18.4)).

It is possible to consider the change in enthalpy of a binary or multicomponent fluid with temperature as made up of three terms, viz.

$$d\tilde{h} = (1-x)\ dT\ \tilde{c}_{p_\ell} + x\ dT\ \tilde{c}_{pg} + \Delta\tilde{h}_v\ dx \qquad (18.12)$$

These are (i)   the sensible heat change in the liquid phase $(1-x)$

(ii)   the sensible heat change in the vapour phase (x)

(iii) the heat of vaporization of an amount of liquid (dx)

This concept differs from that for a single component system since, in that case, enthalpy changes are *either* sensible heat changes *or* latent heat changes (at constant temperature).

18.2.4   *Nucleation in a Binary System*

As with a single component system, vapour may be formed from a binary liquid mixture at either a planar (stable equilibrium) or curved interface (unstable equilibrium). The superheat $(T_g - T_{SAT})$ required to maintain a bubble embryo of radius r* in equilibrium is given by:

$$(T_g - T_{SAT}) = \frac{2\sigma}{\left(\dfrac{\partial p_{SAT}}{\partial T}\right) r^*} \qquad (18.13)$$

For mixtures, both $(\partial p_{SAT}/\partial T)$ and the surface tension σ depend on the composition as well as the temperature. The extended Clapeyron equation is given by Stein (1978) and by Malesinsky (1965).

$$\left(\frac{\partial p_{SAT}}{\partial T}\right) = \left(\frac{\partial p}{\partial T}\right)_{\tilde{x}} + \frac{p}{\tilde{R}T}\left(\frac{\partial \tilde{x}}{\partial T}\right)_p \quad (K-1) \quad \tilde{x} \left(\frac{\partial^2 g}{\partial \tilde{x}^2}\right)_{T,p} \qquad (18.14)$$

where K is the equilibrium constant and g is the Gibbs free energy.

The influence of this change in $(\partial p_{SAT}/\partial T)$ is usually small compared with changes in both the surface tension, σ, and the size of the active nucleation site on the heated surface, r^*.

The addition of relatively small quantities of a second component to water often greatly reduces the surface tension. Figure (18.5) shows the changes in bubble/dew point temperature and in surface tension (σ) for the two systems, ethanol/water and ethanol/benzene. This figure also shows the calculated superheat

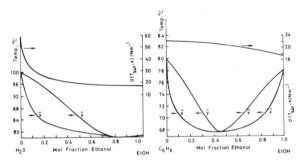

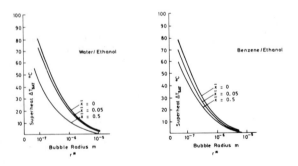

Figure 18.5: Nucleation in binary mixtures

values for these same two systems. For the ethanol/water system, the reduction in surface tension as the mol fraction of ethanol is increased causes a sharp drop in the superheat required to maintain a given size bubble embryo in equilibrium. Conversely, for the ethanol/benzene system the variation of surface tension and also of superheat with composition is small.

18.2.5 *Size of Nucleation Sites*

Before equation (18.13) can be used to estimate the superheat required of a mixture upon a heated surface it is necessary to know how the maximum size of nucleation site on the surface varies with composition. This maximum size is determined by the penetration of the liquid into pre-existing cracks and flaws in the surface.

This penetration is influenced both by changes in surface tension and in contact angle. Experimental evidence from Shock (1977) suggests that these changes in the maximum size of nucleation site with composition can be profound. Table (18.1) shows the variation of the computed size of nucleation site at the onset of boiling on a nickel plated surface for the enthanol/water system. On this evidence it would appear that changes in mixture composition, particularly when one of the components is water, may produce significant changes in the superheat required to initiate and sustain nucleation primarily as a result of changes in the contact angle and hence the maximum active nucleation site.

18.2.6 *Bubble growth in a Binary System*

Bubble growth in a single component system is limited very rapidly by the rate at which heat can diffuse to the interface to provide the latent heat of vaporization. In a binary liquid mixture, however, a limitation also occurs as a result of the liquid close to the bubble interface becoming depleted in the more volatile component (Scriven (1959)). For vaporization and bubble growth to continue the more volatile component must now diffuse from the bulk liquid through the depleted region. This is shown diagrammatically in Figure (18.6). The initial bulk liquid contains a mass fraction (x_O) of the more volatile liquid and it is superheated by an amount ΔT_{SAT} (to point E) above the boiling point corresponding to the initial liquid composition $T(x_O)$ (point D). At the bubble interface the mass concentration of the more volatile component in the liquid phase has fallen to x (point A) whilst the composition of the vapour within the bubble is y (point B). The corresponding rise in the saturation temperature *at the bubble wall* $(T(x)-T(x_O))$ is denoted by ΔT.

Van Stralen (1967) (1978) has given the theory for bubble growth in binary liquid mixtures. A mass balance at the interface for the more volatile component is

$$\rho_g (y-x) \dot{R} = \rho_L \, \delta \left(\frac{\partial x}{\partial r}\right)_{r \, = \, R} \tag{18.15}$$

If $(\partial x/\partial r)_{r=R}$ is approximated by assuming that the concentration falls from x_O to x across a diffusion shell of thickness z_M the bubble growth rate can be given as:

TABLE 18.1

Variation of Size of Active Nucleation Site with Composition for Boiling of Enthanol/Water on a Nickel Plated Surface

(Shock (1977))

p = 2.5 - 2.6 bar

Composition Mol Fraction EtOH	Wall Temperature at ONB	(T_g-T_{SAT}) at ONB	r_{MAX} x 10^6
$\tilde{x}$	^{o}C	^{o}C	m
0	138.2	9.9	1.05
0.058	143.5	26.6	0.23
0.058	143.0	26.7	0.24
0.058	146.5	29.3	0.20
0.058	145.4	28.6	0.21
0.197	145.2	36.8	0.095
0.197	145.0	36.7	0.095

$$\dot{R} = \left(\frac{\rho_L}{\rho_G}\right)\left(\frac{x_o-x}{y-x}\right)\left(\frac{\delta}{z_M}\right) = \left(\frac{\rho_L}{\rho_G}\right)\left(\frac{x_o-x}{y-x}\right)\frac{\delta}{(\frac{\pi}{3})^{\frac{1}{2}}(\delta t)^{\frac{1}{2}}} \qquad (18.16)$$

Van Stralen has termed $(x_o-x)/(y-x)$ the "*vaporized mass diffusion fraction*" (G_d).

It will also be seen from Figure (18.6) that the effective superheat conducting heat to the bubble wall is not ΔT_{SAT} but $(\Delta T_{SAT}-\Delta T)$. Thus the bubble growth rate is also given by:

$$\dot{R} = \left(\frac{\rho_L}{\rho_G}\right)\left[\frac{(\Delta T_{SAT}-\Delta T)\ c_{pL}}{\Delta h_v}\right]\left[\frac{\kappa_L}{(\frac{\pi}{3})^{\frac{1}{2}}(\kappa_L t)^{\frac{1}{2}}}\right] \qquad (18.17)$$

Equating equations (18.16) and (18.17), substituting for G_d and rearranging:

$$\frac{\Delta T}{G_d} = \left(\frac{\delta}{\kappa_L}\right)^{\frac{1}{2}}\left(\frac{\Delta h_v}{c_{pL}}\right)\left(\frac{\Delta T_{SAT}}{\Delta T} - 1\right) \qquad (18.18)$$

It may be assumed that $(\Delta T_{SAT}/\Delta T)$ and therefore $(\Delta T/G_d)$ are independent of the actual value of ΔT_{SAT}:

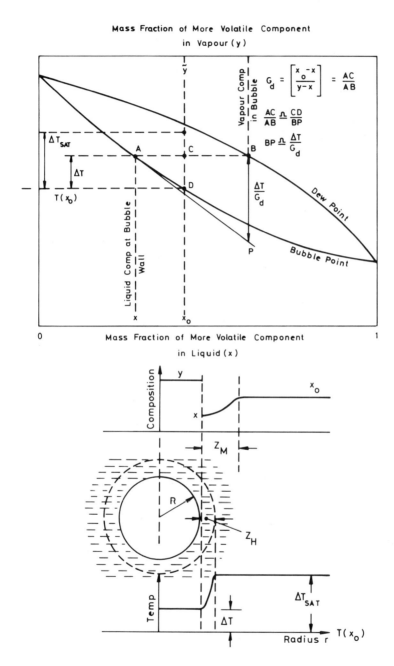

Figure 18.6: The growth of vapour bubbles in binary systems (Van Stralen (1967)).

$$\frac{\Delta T}{G_d}(x_o) = - (K-1) \times \left(\frac{\partial T}{\partial x}\right)_P \qquad (18.19)$$

It may be evaluated graphically from equilibrium data as shown in Figure (18.6).

The asymptotic bubble growth equation for binary fluid mixtures is obtained by substituting equation (18.18) in equation (18.17):

$$\dot{R} = \frac{\Delta T_{SAT}}{\left(\frac{\rho_G}{\rho_L}\right)\left[\frac{\Delta h_v}{c_{pL}} + \left(\frac{\kappa_L}{\delta}\right)^{\frac{1}{2}}\frac{\Delta T}{G_d}\right]}\left(\frac{3}{\pi}\frac{\kappa_L}{t}\right)^{\frac{1}{2}} \qquad (18.20)$$

It can be seen that a minimum value of $\dot{R}$ for a binary mixture corresponds to a maximum value of $(\Delta T/G_d)$. This was confirmed by Van Stralen using experimental observations.

The bubble growth rate in a binary system is sharply reduced from that for a single component system since the mass diffusivity, δ, of the more volatile component is an order of magnitude smaller than the thermal diffusivity (κ_L).

The much lower growth rate also results in much smaller bubbles at the point of departure from the surface.

A further important point is that the proportion of the heat transferred to the bubble whilst still attached to the heating surface is greatly reduced in the case of mixtures and as a result, the bubbles continue to grow significantly after their   r departure.

Zeugin et al (1975) have studied the behaviour of the micro-layer underneath a bubble formed from a binary liquid mixture and growing at a heated surface. As the bubble grows initially the surface temperature under the bubble falls sharply from the initial superheat to just above the boiling point of the liquid mixture. The surface temperature then recovers more slowly than in the case of pure liquids with the temperature rise following the increase in boiling point of the microlayer as the more volatile component is depleted. Evidently, vapour produced from evaporation of the microlayer will be progressively less rich in the more volatile component than that produced from the rest of the bubble wall. If the microlayer evaporates completely, then the composition of the vapour produced will be that of the original liquid.

18.3 Pool Boiling

The pool boiling curve is considerably altered when the fluid being evaporated is a binary mixture rather than a pure single component liquid. The principal changes are shown diagrammatically in Figure (18.7). Firstly, the onset of boiling is delayed to higher wall superheats as a result of the temperature

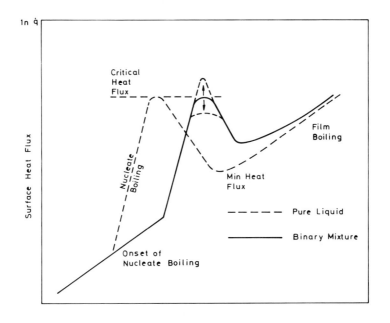

Figure 18.7: Schematic diagram showing principal changes to pool boiling curve
for binary mixtures

gradients set up in the pool to accommodate the corresponding
gradients in liquid composition. Heat transfer coefficients in
the nucleate boiling region are sharply reduced. The critical heat
flux may be increased or reduced depending on the extent of the
contribution from convection in the pool. The minimum heat flux
and the corresponding wall superheat are increased. Finally, heat
transfer rates in the film boiling region are also somewhat higher.

18.3.1 *Nucleate Boiling*

Provided the additive is not surface active even small amounts
of a second component cause considerable reductions in the heat
transfer rate under nucleate pool boiling conditions compared with
that measured for the pure liquid. The reason for this reduction
can be traced back to the influence of the second component upon the
bubble growth rate. The minimum bubble growth rate, the minimum
heat transfer coefficient and the occurrence of a maximum critical
heat flux all occur at the same liquid composition corresponding to
a maximum value of $|\tilde{y}-\tilde{x}|$. In the case of surface active agents
added to water, small measures (~25%) in boiling heat transfer
coefficients have been observed.

A good deal of useful experimental data for nucleate pool
boiling of binary liquids has been presented by Sternling and
Tichacek (1961). They used fourteen binary systems with components
ranging between water, light alcohols and heavy oils. All the

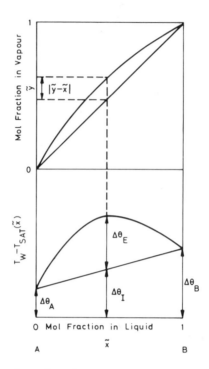

Figure 18.8: The evaluation of nucleate boiling heat transfer coefficients for binary systems (Stephan and Korner (1969))

mixtures had a wide boiling range, at least 90°C. For all the systems the heat transfer coefficient for a given heat flux was less than would be expected for an "ideal" single component fluid with the same physical properties.

Stephan and Korner (1969) have suggested a simple method whereby heat transfer coefficients in binary systems may be computed from data for the pure components. The method is illustrated in Figure (18.8). The upper part of this diagram represents an equilibrium diagram for the system A-B and shows the mol fraction of component B in the vapour phase ($\tilde{y}$) plotted against the mol fraction of component B in the liquid phase ($\tilde{x}$) for a constant pressure. The lower part of the diagram shows the difference in temperature between the heating surface T_W and the boiling point $T_{SAT}(\tilde{x})$ corresponding to a liquid composition ($\tilde{x}$). The curve of $[T_W - T_{SAT}(\tilde{x})]$ passes through a maximum corresponding to the maximum value of $|\tilde{y} - \tilde{x}|$. The value of $[T_W - T_{SAT}(\tilde{x})]$ may be expressed as:

$$\left[T_W - T_{SAT}(\tilde{x})\right] = \Delta\theta_I + \Delta\theta_E = \Delta\theta_I (1 + \Theta) \qquad (18.21)$$

$\Delta\theta_I$ is the "ideal" value of the temperature difference and is evaluated from the values of $\Delta\theta_{SAT}$ for the two pure components A and B calculated for the same pressure and heat flux as the mixture. These wall superheat values will be denoted $\Delta\theta_A$ and $\Delta\theta_B$ respectively. Then $\Delta\theta_I$ is given by

$$\Delta\theta_I = (1-\tilde{x})\ \Delta\theta_A + \tilde{x}\ \Delta\theta_B \tag{18.22}$$

Stephan and Korner found that θ in equation (18.21) could be correlated against the value of $|\tilde{y}-\tilde{x}|$. Thus

$$\theta = \Lambda\ |\tilde{y}-\tilde{x}| \tag{18.23}$$

Λ was found to vary with pressure and over the pressure range 1-10 bar the following expression was used:

$$\Lambda = \Lambda_0\ (0.88 + 0.12p) \tag{18.24}$$

where p is the system pressure in bar and Λ_0 is a constant dependent on the particular binary system studied. Values of Λ_0 are given in Table (18.2) for seventeen common systems. A value of Λ_0 of 1.53 was recommended when no experimental data for the system under consideration was available. This method is not reliable if one of the components is strongly surface active.

Happel and Stephan (1974) have reported the results of pool boiling experiments with benzene-toluene, ethanol-benzene and water-isobutanol mixtures over the entire range of compositions and at pressures 0.5, 1, and 2.0 bar. The results were presented in terms of heat transfer coefficient rather than temperature difference. If there was no influence of mass transfer on the boiling process the heat transfer coefficient ("ideal" value) at any liquid composition ($\tilde{x}$) to be related to the coefficients of the pure components at the same condition (α_A, α_B):

$$\alpha_I = \alpha_A\ (1-\tilde{x}) + \alpha_B\ \tilde{x} \tag{18.25}$$

This differs from equation (18.22) in that now

$$\frac{1}{\Delta\theta_I} = \frac{(1-\tilde{x})}{\Delta\theta_A} + \frac{\tilde{x}}{\Delta\theta_B} \tag{18.26}$$

Figure 18.9 shows results from benzene-toluene mixtures and confirms that the minimum coefficient does occur at the maximum value of $|\tilde{y}-\tilde{x}|$. For mixtures which form azeotropes (where the liquid and the vapour phases in equilibrium have the same composition; $|\tilde{y}-\tilde{x}| = 0$) such as ethanol-benzene, one might expect the heat transfer coefficient to approach the "ideal" value at the azeotropic composition and this also is confirmed. The ratio of the actual to the ideal coefficent was correlated as a function of $|\tilde{y}-\tilde{x}|$.

$$\frac{\alpha}{\alpha_I} = \frac{\Delta\theta_I}{\Delta\theta} = 1 - B\ \left[\tilde{y}-\tilde{x}\right]^n \tag{18.27}$$

where B depends on the mixture and the pressure. For benzene-toluene mixtures at 1 bar B = 1.5 and n = 1.4, for ethanol-benzene mixtures B = 1.25 and n = 1.0 and for water-isobutanol mixtures over their range of miscibility B = 0.9 and n = 0.7.

TABLE 18.2

Value of Λ_0 for use in Equation (18.23) and (18.24) for the Estimation of Nucleate Boiling Heat Transfer Coefficients for Binary Mixtures (Stephan and Korner)

System	Λ_0
Acetone-ethanol	0.75
Acetone-butanol	1.18
Acetone-water	1.40
Ethanol-benzene	0.42
Ethanol-cyclohexane	1.31
Ethanol-water	1.21
Benzene-toluene	1.44
Heptane-methyl cyclohexane	1.95
Isopopanol-water	2.04
Methanol-benzene	1.08
Methanol-amylalcohol	0.8
Methyl-ethyl ketone-toluene	1.32
Methyl-ethyl ketone-water	1.21
Propanol-water	3.29
Water-glycol	1.47
Water-glycerine	1.50
Water-pyridine	3.56

One of the few correlations to attempt quantitatively to relate the reduction in heat transfer coefficient to the reduction in bubble growth rate is that of Calus and Rice (1972). The correlation reduces to a form:

$$\alpha = \frac{\alpha_I}{\left[1 + |y-x| \left(\frac{\kappa_L}{\delta}\right)^{0.5}\right]^{0.7}} \tag{18.28}$$

where α_I is the "ideal" coefficient evaluated for the equivalent pure fluid using the modified Borishanskii-Minchenko correlations; $|y-x|$ is based on the mass rather than mol fractions. Whilst it is interesting to compare the form of this relationship with equation (18.21) and equation (18.27), it should be noted that $(\kappa_L/\delta)^{0.5}$ does vary considerably with changes in mixture composition.

To extend these concepts to multi-component mixtures (n>2), Stephan and Preußer (1978) have again made use of $\Delta\theta_I$, the "ideal"

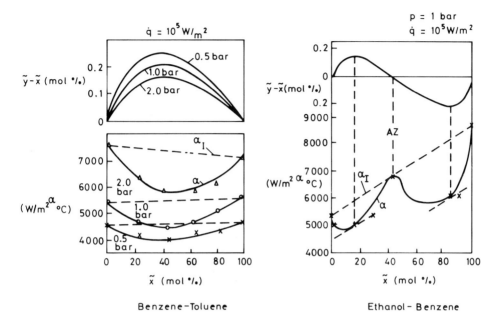

Figure 18.9: Variation of heat transfer coefficient for binary mixtures with composition and pressure (Happel and Stephan (1974)

value of the temperature difference, now defined as (cf equation (18.22)).

$$\Delta\theta_I = \sum_1^n \tilde{x}_i \, \Delta\theta_{SAT,i} \qquad (18.29)$$

also

$$\Theta = \sum_1^{n-1} \Lambda_i^{(n)} \, (\tilde{y}_i - \tilde{x}_i) \qquad (18.30)$$

For a ternary mixture (n = 3) we therefore obtain

$$\Theta = \Lambda_1^{(3)} \, (\tilde{y}_1 - \tilde{x}_1) + \Lambda_2^{(3)} \, (\tilde{y}_2 - \tilde{x}_2) \qquad (18.31)$$

Since a ternary mixture includes as limiting cases the three binary mixtures, for a binary mixture consisting of components 1 and 3 ($\tilde{y}_2 = \tilde{x}_2 = 0$) and for another mixture consisting only of components 2 and 3 ($\tilde{y}_1 = \tilde{x}_1 = 0$) then

$$\Lambda_1^{(3)} = \Lambda_{13} \text{ and } \Lambda_2^{(3)} = \Lambda_{23} \qquad (18.32)$$

In a similar manner than for a mixture of n components

$$\Lambda_i^{(n)} = \Lambda_{in}, \quad i = 1,2 \ldots (n-1) \tag{18.33}$$

In this way the unknown value of $\Lambda_i^{(n)}$ can be determined from the known values of Λ measured for the separate binary mixtures. In one example quoted by Stephan and Preußer, satisfactory agreement was achieved between this method and experimental results for an acetone/methanol/water ternary mixture.

18.3.2   Critical Heat Flux

Changes in the pool boiling critical heat flux occur when small amounts of a second component are added to a pure liquid. The critical heat flux may be higher or lower than the value for either component alone. As with the heat transfer coefficient in nucleate boiling, the change induced by the second component would appear to be large compared with the small changes in physical properties (surface tension; viscosity and density) attributed to the second component. It is, therefore, not possible to predict, a priori, the influence a second component might have on the critical heat flux simply by substituting the altered physical properties into an equation of the type proposed for pool boiling critical heat flux by Kutateladze or Zuber and Tribus.

The experimental and theoretical understanding of the critical heat flux phenomenon for binary mixtures has been increased very considerably by the work of Van Stralen (1966). The early investigations suggested the action of several conflicting mechanisms regarding such variables as the volatility, molecular weight of the second component and the geometry of the heating section. Some of this early work and the resulting explanations are briefly reviewed below.

Lowery and Westwater (1957) suggested that their observed increases in the critical heat flux for methanol were due to the second component providing particles which acted as nucleation sites within the superheated boundary layer. Van Stralen (1959) concluded that the increases he observed in the critical heat flux for water with the addition of a more volatile component were due to the formation of smaller bubbles. Dunskus and Westwater (1960) obtained data on the effects of various additives on the heat transfer characteristics of boiling isopropanol. They concluded that when the second component came from the same homologous series the magnitude of the increase was greater the higher the molecular weight. A decreased volatility of the second component also increased the critical heat flux. These authors suggested that the increases could be attributed to a depletion of the solvent with respect to the non-volatile second component at the bubble wall leaving the bubble surrounded by a fluid of high viscosity which tended to reduce coalescence between bubbles, promote foaming and also to reduce the motion of eddies.

An effect of geometry was observed by Kutateladze (1961) for boiling on a wire and a plate. In this instance, butanol was used to influence the critical heat flux for water at a pressure of 10 bar. For a wire heater the critical heat flux had a *maximum* at about 10% butanol of 2.5 times the value for pure water. For a plate heater the critical heat flux had a *minimum* at about 6%

butanol of 0.56 times the value for pure water. A similar
discrepancy was found by Owens (1964) when he compared his results
for the boiling on a stainless steel tube of water containing
various additives with the results obtained by Van Stralen (1959)
for similar binary systems using a thin wire heater. Also, Scarola
(1964) found that the addition of 2% of 1-pentanol to water produced
a *decrease* in critical heat flux in boiling from a 1.7mm diameter
horizontal stainless steel tube, except at very low or very high
subcooling. The explanation of these apparently diverse results may
lie in the fact that the critical heat flux is the sum of two terms,
one resulting from direct vaporisation at the surface and the second
resulting from the convection of hot liquid away from the surface.
The work of Costello et al (1965) and Kienhard and Keeling (1970)
shows that the second term is important even for pure liquids.
Reduction of bubble growth rate occurs in multicomponent mixtures
(van Stralen, 1966), reduces the first term and may therefore lead
to a *reduction* of the critical heat flux. However, the increase
in wall superheat which there occurs due to the delay in bubble crowding and the occurrence of CHF may be sufficient to so increase the
second term as to give an actual *increase* in the critical heat flux.

18.3.3 Transition Boiling

Little has been published on transition boiling of binary
mixtures. Happel and Stephan (1974) report that qualitatively the
composition of the mixture has a similar influence as in the
nucleate boiling region.

18.3.4 Minimum Heat Flux

The additional diffusional resistance in the liquid phase due
to the mass transfer of the more volatile component into the vapour
film at the heating surface results in the interface being at a
higher temperature (T_i) than the rest of the liquid (T_∞). A
significant proportion of the total heat flux passing through the
vapour film is therefore conducted from the interface into the bulk
liquid rather than forming vapour directly. As a consequence the
vapour film is thinner. Van Stralen et al (1972) studying film
boiling of water/2-butanone mixtures on a thin wire, observed that
direct vapour production at the heating surface corresponded to
only 53% of the applied heat flux compared with a figure of 95% for
pure fluids. The remaining component of the heat flux is convected
into the liquid phase and thence *indirectly* into the vapour bubbles.
Thus the minimum heat flux may be expected to be greater for a
binary mixture compared with an equivalent pure fluid.

Yue and Weber (1974) have examined the minimum heat flux for
horizontal cylinders.

18.3.5 Film Boiling

Yue and Weber (1973) have carried out a theoretical and
experimental study of film boiling of binary mixtures on a vertical
plate. The analysis assumes a two-phase boundary layer and involves
vapour formation by evaporation at an approximately planar interface with no vapour removal by bubble formation. The heat flux,
for a given wall superheat, is increased due to the heat convected
from the interface into the bulk liquid. Contrast this with the
situation in the nucleate boiling region where, for a given wall

superheat, heat fluxes are reduced by the addition of a second
component. A point is reached, however, when the heat flux is
sufficiently high to reduce the concentration of the more volatile
component at the interface to zero. The heat flux through the
liquid phase is then at its maximum and as the overall heat flux
is increased, this becomes a smaller fraction of the total. The
effect of the second phase is predicted to disappear at values of
wall superheat above that given by

$$(T_W - T_\infty) = \frac{3 \, \Delta h_v}{c_{pg}} \cdot Pr_g \qquad (18.34)$$

where T_∞ is the bulk liquid temperature remote from the vapour
film. Experimental results generally confirm the analytical
predictions. For acetone/cyclohexanol mixtures with a relative
volatility of up to 80, the simple Bromley relationship yields heat
fluxes as much as 30% too low. The contribution from radiation
must also be included.

18.4 Forced Convection Boiling

The published literature on forced convection vaporization of
mixtures is very limited. One of the earliest published studies is
that of McAdams et al (1940) who carried out experiments using a
four pass horizontal tube evaporator heated by steam. Each pass
had three separate steam jackets to allow the local heat flux to
be measured. The fluid was a benzene/oil mixture. Bulk fluid
temperatures were found to increase throughout the saturated boiling
length as the liquid became richer in oil. Thus, some of the heat
transferred to the liquid was retained in the form of sensible heat
to maintain the fluid at saturation conditions and was not available
for evaporation. Average boiling heat transfer coefficients were
calculated for each pass where boiling occurred in all three jackets.
At a given vapour mass quality the coefficient decreased as the oil
content of the feed increased.

A number of workers (Bonnet and Gerster (1951), Shellene et al
(1968)) have studied the performance of complete reboilers but such
studies cannot provide information on the local conditions in the
evaporating stream.

18.4.1 *Saturated Nucleate Boiling*

This will be influenced by the addition of a second component
in the same qualitative manner as nucleate pool boiling. The well-
known Chen correlation is formed by adding together a nucleate
boiling (α_{NcB}) and a convective heat transfer (α_c) contribution.

The nucleate boiling contribution may be modified in the
manner proposed by Stephan and Korner (1969) for pool boiling and
then becomes

$$\frac{1}{\alpha_{NcB}} = \left[\frac{(1-\tilde{x})}{S\alpha_A} + \frac{\tilde{x}}{S\alpha_B} \right] (1 + \Lambda \, |y-x|) \qquad (18.35)$$

where $\tilde{x}$ is the mol fraction of the more volatile component B in the liquid phase, $\tilde{y}$ is the mol fraction in the vapour phase, α_A and α_B are the nucleate pool boiling coefficients for the pure components A and B at the same heat flux and S is the "suppression factor" introduced by Chen to allow for temperature gradient effects.

The convective heat transfer coefficient (α_c) may be computed from the mixture physical properties in the manner for a pure fluid. However, this term also may be influenced by the second component, particularly at low qualities where bubbly flow occurs. The stripping of the more volatile component into the bubble cavity leaves around each bubble a shell of more viscous non-volatile liquid. This tends to inhibit bubble coalescence because of the increased time needed for drainage of the film between colliding bubbles. The consequence usually is the formation of a foam which will be stable to higher void fractions than is normal for pure liquids. In bubbly flow any increase in the convective heat transfer results primarily from the increased liquid velocity as a consequence of the increased liquid/vapour volumetric flow:

$$\frac{\alpha_{TP}}{\alpha_L} = \left(\frac{1}{1-\psi}\right)^{0.8} \tag{18.36}$$

The presence of vapour may also induce turbulence in an otherwise laminar liquid flow but, in any case, the increase over the single-phase heat transfer coefficient predicted by equation (18.36) is relatively small (<2). In some instances, the two-phase convective heat transfer coefficient (α_{TP}) may turn out to be *lower* than the single-phase value (α_L). This is because the correct fluid properties for the calculation of α_L in equation (18.36) are those of the more viscous non-volatile liquid surrounding the bubble cavity. The composition of this liquid may be estimated using the concept G_d *the vaporized mass diffusion fraction* introduced by Van Stralen and discussed under *Bubble Growth in a Binary System*.

18.4.2 Two-Phase Forced Convection Region

Further up the evaporator tube, annular flow is formed and this flow pattern will occupy most of the channel length. The liquid is displaced to the heating wall and it is necessary to evaluate mass transfer effects, firstly within the liquid film, and, subsequently, within the vapour core. Shock (1976) has considered these effects in some detail.

Consider first the liquid film, Figure 18.10. *If the molar concentration of component A is represented by* $\tilde{c}$ then the conservation equation in the liquid film is given by:

$$\left[u_x \frac{\partial \tilde{c}}{\partial x} + u_y \frac{\partial \tilde{c}}{\partial y} + u_z \frac{\partial \tilde{c}}{\partial z}\right] = \delta_{EFF} \left(\frac{\partial^2 \tilde{c}}{\partial x^2} + \frac{\partial^2 \tilde{c}}{\partial y^2} + \frac{\partial^2 \tilde{c}}{\partial z^2}\right) \tag{18.37}$$

There is no net convection in the x and y direction and assuming that concentration profile in the y direction is "fully developed":

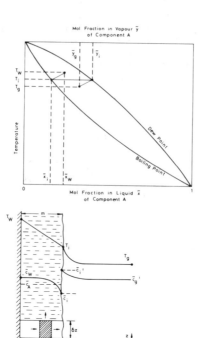

Figure 18.10: Mass transfer from a thin liquid film

$$u_z \frac{\partial \tilde{c}}{\partial z} = \delta_{EFF} \left(\frac{\partial^2 c}{\partial y^2} \right) \qquad (18.38)$$

This equation must be integrated using an expression for the mass diffusivity, δ_{EFF}:

$$J = - (\delta + \epsilon_M) \frac{\partial \tilde{c}}{\partial y} = - \delta_{EFF} \frac{\partial \tilde{c}}{\partial y} \qquad (18.39)$$

For a laminar film $\delta_{EFF} = \delta$ and

$$\int_o^m u_z \frac{\partial \tilde{c}}{\partial z} \, dy = \delta \int_o^m \frac{\partial^2 \tilde{c}}{\partial y^2} \, dy \qquad (18.40)$$

Define a mean bulk molar concentration $\tilde{c}_A$ at any location z such that:

$$\tilde{c}_A = \frac{1}{Q} \int_o^m u_z \tilde{c} \, dy \qquad (18.41)$$

where Q is the volumetric flowrate per unit wetted perimeter:

$$Q = \int_{o}^{m} u_z \, dy \qquad (18.42)$$

also

$$Q \, \frac{\partial \tilde{c}_A}{\partial z} = - J_i \qquad (18.43)$$

where J_i is the molar flux at the interface.

$$\frac{\partial \tilde{c}}{\partial z} = \frac{\partial \tilde{c}_A}{\partial z} = - \frac{J_i}{Q} \qquad (18.44)$$

Substituting for $\frac{\partial \tilde{c}}{\partial z}$ in equation (18.38) leads to:

$$\frac{\partial^2 \tilde{c}}{\partial y^2} = - \frac{J_i \, u_z}{Q \delta} \qquad (18.45)$$

or

$$\frac{\partial \tilde{c}}{\partial y} = - \frac{J_i}{Q \delta} \int_{o}^{y} u_z \, dy \qquad (18.46)$$

Define a molar mass transfer coefficient k_L:

$$k_L = \frac{J_i}{(\tilde{c}_A - \tilde{c}_i)} \qquad (18.47)$$

(a) *for a laminar falling film*

The Sherwood number is given by

$$Sh_L = \left[\frac{k_L m}{\delta} \right] = \frac{70}{17} \qquad (18.48)$$

or combining it with the relevant expression for film thickness to give an explicit relationship for the mass transfer coefficient (k_L):

$$m = \left[\frac{3Q \, \mu_L}{(\rho_L - \rho_G) g_n} \right]^{1/3} , \quad k_L = \frac{70}{17} \left[\frac{(\rho_L - \rho_G) g_n \, \delta^3}{3Q \, \mu_L} \right]^{1/3} \qquad (18.49)$$

By analogy with the situation for heat transfer it seems probable that the actual mass transfer coefficient will be higher than that given by equation (18.49) due to the presence of waves.

(b) *for a laminar film with constant shear stress across the film*
     (as in upwards annular flow at high interfacial shear)

The Sherwood number is given by:

$$Sh_L = \left[\frac{k_L m}{\delta}\right] = 5 \qquad (18.50)$$

or combining it with the relevant expression for film thickness to give an explicit relationship for the mass transfer coefficient (k_L):

$$m = \left[\frac{2Q \mu_L}{\tau_i}\right]^{1/2} , \quad k_L = 5 \left[\frac{\tau_i \delta^2}{2Q \mu_L}\right]^{1/2} \qquad (18.51)$$

where τ_i is the interfacial shear stress. If the analogy with the heat transfer situation applies mass transfer coefficients in practice may be lower than estimated from equation (18.52) by about 30%.

(c) *for a turbulent film*

Turbulence within the liquid film will enhance the mass transfer coefficient for a given film thickness. Equation (18.39) applies where ε_m is the "eddy mass diffusivity". Expressions for turbulent mass transfer in liquid films can be developed by an analogous procedure to that used to establish the hydrodynamic and heat transfer characteristics of the film. Shock (1976) assumed that ε_m for use in equation (18.39) was identical to the "eddy viscosity" (ε) and that the expression given by Deissler for ε could be used to establish the relationship between a dimensionless concentration, c^+, and a dimensionless distance, y^+. c^+ is defined as:

$$c^+ = \frac{u^*}{J_i} \; (\tilde{c} - \tilde{c}_i) \qquad (18.52)$$

where u^* is the friction velocity $(= \sqrt{(\tau/\rho_L)})$.

Mass transfer coefficients were deduced from the dimensionless concentration profile evaluated at the appropriate value of the dimensionless film thickness (y_i^+). The molar mass transfer coefficient k_L is defined as:

$$k_L = \frac{J_i}{(\tilde{c}_A - \tilde{c}_i)} = \frac{u^*}{c_b^+} \qquad (18.53)$$

where c_b^+ is the bulk value of c^+ evaluated from the profiles of c^+ and y^+. A liquid film Sherwood number can also be defined as:

$$Sh_L = \frac{k_L m}{\delta} = \frac{Sc \; y_i^+}{c_b^+}$$

For the case of constant shear stress across the film Shock (1976) has produced tables of Sh_L against the film Reynolds number Re_L

with Schmidt number (Sc) as parameter. The following equation fits
the results of Shock acceptably well:

$$Sh_L = 5 - 0.11\ Sc + 0.01\ Re_L\ Sc \qquad (18.54)$$

18.4.3 *Critical Heat Flux in Forced Convection Boiling*

The critical heat flux (CHF) for binary and multi-component
mixtures in forced convection boiling is best considered in terms
of the separate mechanisms of *departure from nucleate boiling* (DNB)
which occurs under subcooled and low quality conditions, and
dryout which occurs at higher vapour qualities.

.4.3.1 Departure from Nucleate Boiling (DNB). Data for the
critical heat flux of binary organic mixtures for subcooled and
saturated forced convection boiling have been given by a number
of workers (Table (18.3)). Maxima in the CHF versus composition
curves, corresponding to maximum values of $|\tilde{y}-\tilde{x}|$ are seen in these
results also (Figure 18.11) although not so marked as for the pool
boiling case. The CHF for the binary mixture increases with both
increased velocity and increased subcooling in a similar manner
to that for pure liquids. The value of the critical heat flux
for a binary mixture, $q_{cr,m}$ may be expressed by analogy of the
treatment by Stephan and Körner for pool boiling heat transfer,
as the sum of two terms; $\dot{q}_{cr,I}$, the "ideal" value of the critical
heat flux evaluated from the values of $\dot{q}_{cr}$ for the two pure
components A and B calculated at the same pressure, velocity and
subcooling as the mixture, and $\dot{q}_{cr,E}$ an additional heat flux
which allows for the enhancement in CHF brought about by mass
transfer effects, etc.

$$\dot{q}_{cr,m} = \dot{q}_{cr,I} + \dot{q}_{cr,E} = \dot{q}_{cr,I}\ (1 + \chi) \qquad (18.55)$$

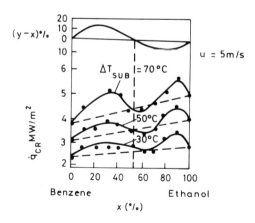

Figure 18.11: Critical heat flux for forced convection boiling of benzene-ethanol
mixtures (Tolubinskiy and Matorin)

TABLE 18.3

Experimental Studies of DNB for Binary Mixtures

Source, Ref, Date	Fluids	Geometry	Experimental Ranges				
			Pressure bar	Subcooling °C	Velocity m/s	Critical Heat Flux (MW/m²)	
Tolubinskiy & Matorin (1973)	ethanol-water acetone-water ethanol-benzene ethylene-glycol-water	tube 4 mm i.d. 60 mm long	3.3-13.2	10-110	2.5-10	2-13	
Andrews, Hooper and Butt (1968)	acetone-toluene benzene-toluene	annulus 6.35 mm i.d. 20.9 mm o.d. 76mm long	1	20-50	1.09-4.36	1-4.5	
Sturman, Abramov, Checheta (1968)	mono-iso-propyl di-phenyl/benzene	annulus 10 mm i.d. 16 mm o.d. 110 mm long	2	25-70	4-12	1-4.5	
Naboichenko, Kiryutin and Gribov (1965)	mono-iso-propyl di-phenyl/benzene	annulus 6 mm i.d. 10 mm o.d. 80 mm long	2.94-16.37	25-125	4-8	1-6	
Carne, M (1963)	benzene-toluene acetone-toluene	Internally heated annulus 6.35 mm i.d. 19.05 mm o.d. 76.2 mm long	2.05	20-80	2.18-4.36	0.5-3.0	

where

$$\dot{q}_{cr,I} = \left[\dot{q}_{cr,A} (1-\tilde{x}) + \dot{q}_{cr,B} \tilde{x} \right] \qquad (18.56)$$

Tolubinskiy and Matorin (1973) and Sterman et al (1968) found that χ in equation (.55) could be expressed:

$$\chi = fn \left\{ |\tilde{y}-\tilde{x}|, \left[\dot{m}D/\mu_L\right]_A, \left[T_{SAT,B}/(T_{SAT,m}-T_{SAT,B})\right] \right\} \qquad (18.57)$$

The Reynolds number $\left[\dot{m}D/\mu_L\right]_A$ is based on the properties of the less volatile component since this component becomes concentrated at the heated surface.

Tolubinskiy and Matorin (1973) suggest the following expression for χ:

$$\chi = 1.5(|\tilde{y}-\tilde{x}|)^{1.8} + 6.8(|\tilde{y}-\tilde{x}|) \left[(T_{SAT,m}-T_{SAT,B})/T_{SAT,B}\right] \qquad (18.58)$$

This expression correlated CHF data for ethanol-water, acetone-water, ethanol-benzene and water-ethylene glycol mixtures within $\pm$ 20%. For mixtures forming an azeotrope, Tolubinskiy and Matorin recommend that $T_{SAT,m}$ be set equal to $T_{SAT,B}$ at the azeotropic composition, so that $\dot{q}_{cr,m} = \dot{q}_{cr,I}$. This recommendation is confirmed by examination of the results for benzene-ethanol mixtures shown in Figure (18.11).

In the experiments of Sterman et al (1968) the values of χ varied from 0 to 0.8 and an alternative expression for χ was given

$$\chi = 3.2\times10^5 \frac{(|\tilde{y}-\tilde{x}|)^3}{\left[\dot{m}D/\mu_L\right]_A} + 6.9 \frac{(|\tilde{y}-\tilde{x}|)^{1.5}}{\left[\dot{m}D/\mu_L\right]_A^{0.4}} \left[\frac{T_{SAT,B}}{T_{SAT,m}-T_{SAT,B}}\right] \qquad (18.59)$$

Bergles and Scarola (1966) observed a reduction in $\dot{q}_{cr}$ in forced convective subcooled boiling of water/2.2% 1-pentanol mixtures, at low subcooling. They suggested that this was due to a reduction of subcooled void fraction and, hence, velocity.

18.4.3.2 Dryout. For higher vapour qualities, associated with the annular flow pattern dryout is likely to be determined by the point at which the flowrate of the liquid film on the heated surface approaches zero. No experimental data or theoretical models have been published relating to dryout in binary or multi-component mixtures. However, it is possible to recommend the method proposed by Hewitt and his co-workers. On the evidence of Shock (1976) it is apparent that the flowrates within the liquid film and of entrained liquid within the vapour core are dictated primarily by hydrodynamic effects and that mass transfer effects on the distribution of the phases will be small. In carrying out the step-by-step integration of the equations given by Hewitt it is recommended that equilibrium be assumed between the liquid and vapour at each axial location. If it is considered important the small departures from the equilibrium situation can be intro-duced using the equatios given by Shock (1976). For positive mixtures, breakdown of the liquid film into rivulets may occur some-

what earlier than for a pure liquid due to surface tension and temperature gradient effect. Breakdown may be delayed for negative
mixtures.

18.5 Simplified Treatment for Binary and Multi-Component Evaporation and Condensation

Consider a simple single-pass evaporator heated by a fluid
stream in counter-current flow. The subscripts H and e stand for
the heating and evaporating streams respectively:

$$\frac{d\dot{Q}}{dz} = \dot{M}_e \frac{dh_e}{dz} = \dot{M}_H \frac{dh_H}{dz} = UA\ (T_H - T_e) \tag{18.60}$$

In equation (18.60) A is the "effective" heat exchange area per
unit length (perimeter) allowing for any secondary surface which
may be present. U is the overall heat transfer coefficient and T_H
and T_e are respectively the local temperature of the heating system
and the equilibrium (*bubble point*) temperature of the evaporating
stream.

If the local values of the enthalpies h_H and h_e are known, then
the temperatures T_H and T_e are also known, together with the mass
fraction vaporized (x_e). This, together with the flowrates,
provides sufficient information to calculate the overall coefficient
U. Thus equation (18.60) can be progressively integrated along the
evaporator starting from known conditions at the evaporator inlet
$(z = 0)$.

Define Γ as the flowrate per unit perimeter

$$\Gamma_e = \frac{\dot{M}_e}{A} \tag{18.61}$$

Likewise, the product $U(T_H - T_e)$ represents the heat flux $(\dot{q}_W)$ passing
through the wall separating the two streams:

$$\dot{q}_W = U(T_H - T_e) \tag{18.62}$$

Substituting equation (18.61) and (18.62) into equation (18.60):

$$\dot{q}_W\ dz = \Gamma_e\ dh_e \tag{18.63}$$

Consider Figure (18.12) which shows a section of an evaporator heated
by a fluid stream in counter-current flow separated from the
vaporizing stream by a wall. The heating stream is flowing downwards and the evaporating stream upwards. It is assumed that the
flow pattern on the evaporating side is *annular flow* and the
evaporating liquid is displaced to the wall as a thin liquid film.

Consider the situation at a position z from the evaporator
inlet and over a length dz. There is a change of evaporating fluid

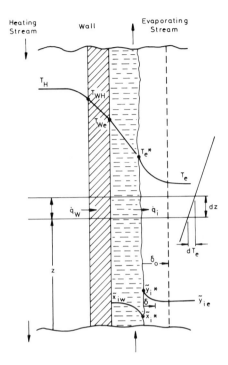

Figure 18.12: Evaporation of a multi-component fluid in annular flow

bubble point temperature over this length so that

$$dh_e = (1-x_e)dT_e c_{pL} + x_e dT_e c_{pG} + dx_e \Delta h_v \qquad (18.12)$$

Now the heat flux through the vapour-liquid interface, $\dot{q}_i$, is less than $\dot{q}_W$ by the amount of the flux absorbed in heating the liquid flow $\Gamma_e(1-x_e)$ through dT_e. Therefore

$$\dot{q}_i = \dot{q}_W - \Gamma_e(1-x_e)\left(\frac{dT_e}{dz}\right) c_{pL} \qquad (18.64)$$

Of course, in pure single component vaporization (or condensation) $\dot{q}_W \simeq \dot{q}_i$ because there is no axial temperature gradient of any consequence.

The interfacial heat flux $\dot{q}_i$ is, in turn, made up of two terms; an evaporation flux ($\dot{q}_v$) at the interface and a sensible heat flux ($\dot{q}_G$) required to heat the vapour phase through dT_e.

18.5.1 *Interfacial Mass Transfer during Evaporation and Condensation of Binary and Multi-Component Mixtures*

At the interface mass transfer and heat transfer processes

occur in parallel and the resistances to mass and heat transfer must be considered together. To induce these parallel mass and heat transfer processes there must be parallel concentration and temperature driving forces from the interface to the bulk vapour phase. Shock (1976) has shown that mass transfer resistances in the liquid phase are negligible by comparison and therefore it is assumed that the resistance to heat and mass transfer occurs in the vapour phase across a boundary layer of thickness δ_o.

From the definitions given earlier we see that the basic transport equation for the molar flux of component i from the interface is:

$$\dot{n}_i = \dot{n} \, \tilde{y}_{ie} + J_i \tag{18.65}$$

where $\tilde{y}_{ie}$ is the mol fraction of component i in the bulk vapour.

These equations were firset set down by Colburn and Drew (1937) for condensation in a binary system and have recently been reiterated by Price and Bell (1974).

The *diffusive molar flux* J_i can be evaluated from

$$J_i = k_{i,eff} \, (\tilde{y}_i{}^* - \tilde{y}_{ie}) \tag{18.66}$$

where $k_{i,eff}$ is an effective mass transfer coefficient. The sensible heat flux ($\dot{q}_G$) from the interface to the vapour core is the sum of the sensible heat flux due to the bulk flow of the vapour from the interface and the heat conducted through the boundary layer:

$$\dot{q}_G = (T_e{}^* - T) \sum_{i=1}^{n} \dot{n}_i \, \tilde{c}_{pi} + \alpha_g \frac{dT}{d\eta} \tag{18.67}$$

Integrating this equation from $\eta = 0$, $T = T_e{}^*$ to $\eta = 1$, $T = T_e$ gives:

$$\dot{q}_G = \frac{a}{1 - e^{-a}} \left[\alpha_g (T_e{}^* - T_e) \right] = \alpha_g{}' \, (T_e{}^* - T_e) \tag{18.68}$$

where

$$a = \frac{1}{\alpha_g} \sum_{i=1}^{n} \dot{n}_i \, \tilde{c}_{pi}$$

The evaporative heat flux required to form the molar vapour flux $\dot{n}$ is:

$$\dot{q}_V = \dot{n} \, \Delta \tilde{h}_v \tag{18.69}$$

where $\Delta \tilde{h}_v$ is the molar latent heat of vaporization.

Thus, the interfacial heat flux is given by:

$$\dot{q}_i = \dot{q}_G + \dot{q}_V$$

$$= \frac{a}{1-e^{-a}} \left[\alpha_g (T_e^* - T_e) \right] + \dot{n} \, \Delta \tilde{h}_v \qquad (18.70)$$

In general the solution of such sets of equations is complicated by the fact that the conditions at the interface are not known and must be guessed. An iteration is then necessary to achieve a solution. A *simplifying assumption* is to say that for situations where the thickness of the mass transfer and heat transfer boundary layers in the vapour phase are equal:

$$Le = \frac{Sc}{Pr} = \frac{\kappa}{\delta} \simeq 1 \qquad (18.71)$$

then the resistance to heat transfer and the resistance to mass transfer will also be about equal. Thus, if we evaluate the resistance to the transfer of the sensible heat component $\dot{q}_G$ in equation (18.68) then it is argued that there will be sufficient mass transfer driving force to provide $\dot{q}_V$ as well. Thus the overall heat transfer coefficient on the evaporating side will be described by the equation:

$$(T_{We} - T_e) = (T_{We} - T^*) + (T^* - T_e)$$

Now, if $\dot{q}_i \simeq \dot{q}_W$ (but see equation (18.64)) then the overall local evaporating side coefficient, α_e, can be given by:

$$\frac{\dot{q}_W}{\alpha_e} = \frac{\dot{q}_W}{\alpha_{fe}} + \frac{\dot{q}_G}{\alpha_g}, \qquad (18.72)$$

where α_{fe} is the coefficient for the liquid film and α_g' is the "modified" single-phase vapour heat transfer coefficient (defined by equation (18.68)). Dividing both sides of equation (18.67) by $\dot{q}_W$ gives:

$$\frac{1}{\alpha_e} = \frac{1}{\alpha_{fe}} + \frac{(\dot{q}_G/\dot{q}_W)}{\alpha_g'} \qquad (18.73)$$

The term $(\dot{q}_G/\dot{q}_W)$ was denoted Z by Bell and Chaly (1972) and is given by:

$$\left(\frac{\dot{q}_G}{\dot{q}_W}\right) = \mathcal{Z} = \left[x_e \left(\frac{dT_e}{dh_e}\right) c_{pG}\right] \tag{18.74}$$

This term is easily computed from the known phase equilibrium information and the approximation can be used for both binary and multi-component systems. Although the liquid/vapour interface is rough, it is recommended that the smooth tube value of α_g be used in equation (18.68) to ensure adequate mass transfer driving force.

The more rigorous approach given earlier (equation (18.70)) is strictly only correct for binary mixtures and can be solved using the methods described by Price and Bell (1974). The extension of these methods to multi-component systems, primarily in relation to condensation is being pursued by Standart (1974) and Krishna (1976a) (1976b) (1976c), although the interactions between the diffusive fluxes of the various components makes the solution complex. Krishna and Standart (1976a) have pointed out that equation (18.66) is incorrect for multicomponent evaporation and condensation. The correct relationship is:

$$J_i = \sum_{j=i}^{n} k_{ij} (\tilde{y}_j{}^* - \tilde{y}_{je}); \quad i = 1 \ldots n \tag{18.75}$$

Methods of calculating the matrix k_{ij} of mass transfer coefficients have been discussed by Krishna and Standart (1976b) (1976c). The energy balance becomes:

$$\dot{q}_i = \frac{a}{1-e^{-a}} \left[\alpha_g (T_e{}^* - T_e)\right] + \sum_{i=1}^{n} \dot{n}_i (\Delta\tilde{h}_v)_i \tag{18.76}$$

Using vapour-liquid equilibrium data equation (18.65) and equation (18.75) can be solved to provide $T_e{}^*$ and $\dot{n}_i$. The heat balance relationships equation (18.68) and (18.76) are then used to calculate the bubble point at some point further down the evaporator. The vapour phase composition is calculated from:

$$\frac{1}{A} \frac{d\dot{N}_i}{dz} = \dot{n}_i$$

from which the vapour phase mol fractions $\tilde{y}_{ie}$ are readily calculated. No comparisons have been published between this method and the more simplified method given earlier for the case of multi-component evaporation. Some comparisons are, however, available for the analogous condensation situation and have been given by Price and Bell (1972) and by Krishna and Standart (1976b).

Nomenclature

A	- area or effective area per unit length	(m² or m)
a	- term defined below equation (18.68)	(-)
B	- constant in equation (18.29)	(-)
$\tilde{c}_p$	- molar specific heat	(J/k mol °K)
$\tilde{c}$	- molar concentration	(k mol/m³)
C^+	- dimensionless concentration	(-)
G_d	- "vaporized mass diffusion fraction"	(-)
g	- Gibbs free energy	J/k mol
h	- specific enthalpy	J/kg
$\tilde{h}$	- molar enthalpy	J/k mol
Δh_v	- latent heat of vaporization	J/kg
$\Delta \tilde{h}_v$	- molar latent heat of vaporization	J/k mol
J	- diffusive molar flux	(k mol/m²s)
k	- mass transfer coefficient	(m/s)
K	- equilibrium ratio	(-)
L/V	- liquid/vapour ratio	(-)
$\dot{m}$	- mass flux	(kg/m²s)
m	- film thickness	(m)
M	- molecular weight	(k mol/kg)
N	- number of moles	(k mol)
$\dot{n}$	- molar flux	(k mol/m²s)
n	- number of components	(-)
n	- exponent	(-)
p	- pressure	(N/m²)
Q	- volumetric flow per unit perimeter	(m²/s)
Q	- amount of heat	(W)
$\dot{q}$	- heat flux	(W/m²)
r	- radius	(m)
r*	- radius of bubble embryo	(m)
$\tilde{R}$	- universal gas constant	(Nm/k mol)
$\dot{R}$	- velocity of bubble wall radius	(m/s)
S	- Chen suppression factor	(-)
ΔT	- temperature difference	(°K)
T	- temperature	(°K)
t	- time	(s)
U	- overall heat transfer coefficient	(W/m²°C)
u	- velocity	(m/s)
x,y,x	- co-ordinates	(-)

z_m	– diffusion shell thickness	(m)
z	– parameter defined by equation (18.74)	(–)
u*	– friction velocity	(m/s)
$\tilde{x}$	– mol fraction in liquid phase	(–)
x	– mass fraction in liquid phase	(–)
	– vapour mass quality	(–)
$\tilde{y}$	– mol fraction in vapour phase	(–)
y	– mass fraction in vapour phase	(–)

Greek

ψ	– void fraction	(–)
Γ	– flowrate per unit perimeter	(kg/ms)
κ	– thermal diffusivity	(m^2/s)
ρ	– density	(kg/m^3)
α	– relative volatility	(–)
	– heat transfer coefficient	$(W/m^2\,{}^OC)$
σ	– surface tension	(N/m)
μ	– viscosity	(Ns/m^2)
δ	– diffusivity of mass	(m^2/s)
$\Delta\theta$	– temperature difference	$({}^OK)$
Θ	– constant in equation (18.23)	(–)
Λ	– constant in equation (18.25)	(–)
τ	– shear stress	(N/m^2)
χ	– constant in equation (18.55)	(–)

Subscripts

A	– component A
B	– component B
C	– critical (azeotropic) composition
cr	– critical
e	– relating to evaporating stream
EFF	– effective
fe	– liquid film
g, G	– vapour or gas
H	– heating stream
I	– ideal
i	– component i
	– interface
l, L	– liquid
NcB	– nucleate boiling

o	–	initial condition
r	–	reference value
SAT	–	saturation
TP	–	two-phase
V	–	vapour or evaporation
W	–	wall
x,y,x	–	coordinates
∞	–	bulk

Superscripts

o	–	relating to pure component
'	–	modified as indicated

Dimensionless Numbers

Pr	–	Prandtl number
Sh	–	Sherwood number
Sc	–	Schmidt number
Re	–	Reynolds number

References

Andrews, D G, Hooper, F C and Butt, P, (1968) "Velocity, subcooling and surface effects in the departure from nucleate boiling of organic binaries". *Can. J. of Chem. Engng.*, **46**, 194-199.

Bell, K J and Ghaly, M A, (1972) "An approximate generalized design method for multi-component/partial condensers". *Am. Inst. Chem. Engrs. Symp. Series.* **69** 131, 72-79.

Bergles, A E and Scarola, L S, (1966) "Effect of a volative additive on the control heat flux for surface boiling of water in tubes". *Chem. Engng. Sci.* 21 721-723.

Bonnet, W E and Gerster, J A, (1951) "Boiling coefficients of heat transfer - C_4 hydrocarbon/furfural mixtures inside vertical tubes". *Chem. Eng. Prog.* **77** (3), 151-158.

Calus, W F and Rice, P, (1972) "Pool boiling - binary liquid mixtures". *Chem. Engng. Science*, **7**, 1687-1697.

Carne, M, (1963) "Studies of the critical heat flux for some binary mixtures and their components". *Can. J. of Chem. Engng.* 235-240.

Colburn, A P and Drew, T B, (1937) "The condensation of mixed vapours". *Trans. Am. Inst. Chem. Engrs.* **33**, 197-215.

Costello, C P, Bock, C O and Nichols, C C, (1965) "A study of induced convective effects on pool boiling burnout". *Chem. Engng. Prog. Symp. Series.* **61**, 271-280.

Dunskus, T and Westwater, J W, (1960) "The effect of trace additions on the heat transfer to boiling isopropanol". *A.I.Ch.E. - ASME Heat Transfer Conf. Buffalo, N.Y. A.I.Ch.E.* Preprint No. 30.

Happel, O and Stephan, K, (1974) "Heat transfer from nucleate boiling to the beginning of film boiling in binary mixtures". Paper B7.8 presented at the *5th Int. Heat Transfer Conf. Tokyo.*

Krishna, R and Standart, G L, (1976a) "Determination of interfacial mass and energy transfer rates for multi-component vapour-liquid systems". *Lett. Heat Mass Transfer.* **3**, No 2, 173-182.

Krishna, R, Panchal, C B, Webb, D R and Coward, I, (1976b) "An Ackerman-Colburn and Drew type analysis for condensation of multi-component mixtures". *Lett. Heat Mass Transfer,* **3**, 163-172.

Krishna, R and Standart, G L, (1976c) "A multi-component film model incorporating a general matrix method of solution to the Maxwell-Stefan equation". *A.I.Ch.E.J.* **22**, 383.

Kutateladze, S S, (1961) "Boiling heat transfer". *Int. J. Heat Mass Transfer.* **1**, 31-45.

Lienhard, J H and Keeling, K B, (1970) "An induced-convection effect upon the peak boiling heat flux". *J. Heat Transfer.* **92**, 1-5.

Lowery, A J Jr and Westwater, J W, (1957) "Heat transfer to boiling methanol - effect of added agents". *Ind. Engng. Chem.* **49**, 1445-1448.

Malesinsky, W, (1965) "Azeotrophy and other theoretical problems of vapour-liquid equilibrium". *Interscience, Wiley, London.*

McAdams, W H et al, (1943) "Vaporization inside horizontal tubes - II. benzene-oil mixtures". *Trans ASME,* 193-200.

Naboichenko, K V, Kiryutin, A A and Grihov, B S, (1965) "A study of critical heat flux with forced flow of monosiopropyldeiphenyl-benzene mixture". *Teploenergetika,* **12** (11), 81-86. *Thermal Engng.* **12** (11), 107-114.

Owens, W L Jr, (1964) "An analytical and experimental study of pool boiling with particular reference to additions". *UKAEA Report, Winfrith, AEEW-180.*

Price, B C and Bell, K J, (1974) "Design of binary vapor condensers using the Colburn-Drew equations". *A.I.Ch.E. Symp. Series.* **70**, No 138, 163-171.

Scarola, L S, (1964) "Effect of additives on the critical heat flux in nucleate boiling". S.M.Thesis in *Mech. Engng,* MIT.

Scriven, L E, (1959) "On the dynamics of phase growth". *Chem. Engng. Science.* **10** (1/2), 1-13.

Shellene, K R et al, (1968) "Experimental study of a vertical thermosyphon reboiler". *Chem. Eng. Prog. Symp. Series.* **64** (82), 102-113.

Shock, R A W, (1976) "Evaporation of binary mixtures in upward annular flow". *Int. J. Multiphase Flow.* **2**, 411-433.

Shock, R A W, (1977) "Nucleate boiling in binary mixtures". *Int. J. Heat Mass Transfer.* **20**, 701-709.

Standart, G L, Krishna, R and Cullinan, H T, (1974) "Multi-component isothermal mass transfer". Paper presented at the *A.I.Ch.E. Conf, Tulsa, Oklahoma.* See also *Ph.D. Thesis* by Krishna, R. "Interphase transport of mass and energy in multi-component systems". UMIST (1975).

Stein, H N.   Chapter 17 in SJD Van Stralen and R Cole (1978) "Boiling phenomena". *Hemisphere Publishing Corp., Washington.*

Stephan, K and Körner, M, (1969) "Calculation of heat transfer in evaporating binary liquid mixtures". *Chemie-Ingenieur Technik.* **41**, No 7, 409-417.

Stephan, K and Preußer, P, (1978) "Heat transfer in natural convection boiling of polynary mixtures". *6th Int. Heat Transfer Conf.* I, 187-192, Toronto.

Sterman, L, Abramov, A and Checketa, G, (1968) "Investigation of boiling crisis at forced motion of high temperature organic heat carriers and mixtures". Paper E2 presented at *Int. Symp. on Research into Co-current Gas-Liquid Flow.* **2**, University of Waterloo, Ontario, Canada.

Sternling, C V and Tickacek, L J, (1961) "Heat transfer coefficients for boiling mixtures". *Chem. Engng. Science.* **16**, No 4, 297-337.

Tolubinskiy, V I and Matorin, P S, (1973) "Forced convection boiling heat transfer crisis with binary mixtures". *Heat transfer - Soviet Research,* **5**, No 2, 98-101.

Van Stralen, S J D, (1959) "Heat transfer to boiling binary liquid mixtures". Pt. I and Pt. II. *Brit. Chem. Engng.* **4**, No 1, 8-17 and **4**, No 2, 78-82.

Van Stralen, S J D, (1966) "The mechanisms of nucleate boiling in pure liquids and in binary mixtures". Pt. I and Pt. II. *Int. J. Heat Mass Transfer.* **9**, 995-1046.

Van Stralen, S J D, (1967) "Bubble growth rates in boiling binary mixtures". *Brit. Chem. Engng.* **12**, No 3, 390-394.

Van Stralen, S J D, Joosen, C J J and Sluyter, W M, (1972) "Film boiling of water and an aqueous binary mixture". *Int. J. Heat Mass Transfer.* **15**, 2427-2445.

Van Stralen, S J D and Zijl, W, (1978) "Fundamental developments in bubble dynamics". *6th Int. Heat Transfer Conf. Toronto.*

Yue, P L and Weber, M E, (1973) "Film boiling of saturated binary mixtures". *Int. J. Heat Mass Transfer.* **16**, 1877-1887.

Yue, P L and Weber, M E, (1974) "Minimum film boiling flux of binary mixtures". *Trans. Inst. Chem. Engng.* **52**, 217-221.

Zeugin, L, Donovan, J and Mesler, R, (1975) "A study of microlayer evaporation for three binary mixtures during nucleate boiling". *Chem. Engng. Science.* **30**, 679-683.

Chapter 19

Introduction to Two-Phase
Flow Problems
in the Power Industry

J. G. COLLIER

19.1 Introduction

This chapter is concerned with two-phase flow problems in central station generating plant and begins by reviewing the various types of fossil, nuclear and renewable energy equipment in which two-phase flow occurs. The chapter then goes on to consider specific two-phase flow processes which are relevant to power plant equipment and where special difficulties may arise. These include

- dryout in fossil-fired boilers

- evaporation inside horizontal tubes

- two-phase pressure drop

- sub-channel analysis

- fouling and corrosion

19.2 Fossil-Fired Equipment

Boilers

A modern fossil station may be gas, oil or coal-fired. The steam generator consists of the following elements (Figure 19.1):

(1) a water-cooled furnace

(2) a large superheater (with radiant superheater surface)

(3) a reheater (common but not essential)

(4) a convective heat transfer section (large or small) depending on design conditions)

(5) an economizer or air heater or both.

The furnace, superheater, convection region and economiser are arranged to form an integral unit. The walls of the furnace are formed by a gas-tight welded membrane in which each boiler tube has an integrally formed fin which is welded to its neighbour using

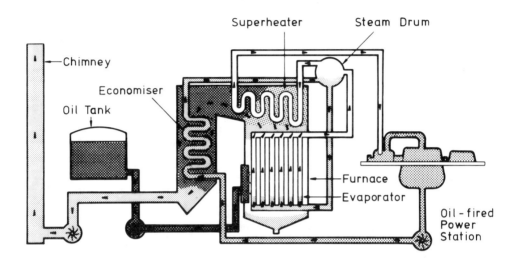

Figure 19.1   A modern fossil-fired boiler

semi-automatic machines in the factory and erected in the form of
large panels. The heat transfer situation is complicated by the
fact that the heating is from one side only and because there are
horizontal, inclined and curved tube lengths within the boiler
configuration. Other problems encountered with this type of plant
include instabilities (both parallel channel and full loop
instabilities) and also separation and two-phase flow effects in
the steam drums.

     One of the more recent developments in fossil-fired steam
plant is the concept of fluidised bed combustion (Thurlow (1978)).
In this concept, fuel, usually coal, is fed into a turbulent bed
of inert particles usually coal-ash. This bed of ash is kept
turbulent by flowing up through it a stream of air evenly distributed over the whole base of the bed and is kept at a temperature
below that at which the particles begin to soften or sinter, ie
generally at a bed temperature of 800-850°C which is 250-300°C
lower than the typical temperatures leaving conventional fossil
fuel-fired furnaces. A further consideration is the high heat
transfer coefficients obtained within the bed at $\sim$ 300 W/m^2 $^{\circ}$K or
60 Btu/hr^{2}ft $^{\circ}$F. As a result, the heat can be extracted using
less surface area than in a convectional furnace. Also, more of
the heat transfer surface can be used since the entire circumference
of the tube is available for heating. Generally, about 45-50% of
the total boiler heat absorption occurs within the bed. One of the

major problems is, however, that the boiler tubes must be horizontal.

It is possible to operate a fluidised combustion unit under
pressure giving higher heat release rates and the possibility of
increasing thermal efficiency by combining with a gas turbine
cycle (Figure 19.2).

Condensers

A diagramatic cross-sectional arrangement of a modern
condenser is shown in Figure 19.3 (Modern Power Station Practice
(1963)). The shell is a large construction and cannot be
transported as a complete unit from the manufacturer to site. The
shell is bolted and seal-welded to the turbine exhaust flange.
The condenser tubes are supported from the shell by suitably spaced
mild steel or brass plates to ensure that there is no danger of
tube vibration. These plates also support the condenser shell
against the external pressure existing under normal operating
conditions. Under working conditions the tube plates are subjected
to the pressure difference between the cooling water and vacuum.
The axial load is taken off the tubes themselves by the support
stays.

For thermodynamic reasons, power station condensers must
operate at the lowest possible pressure and very high vapour

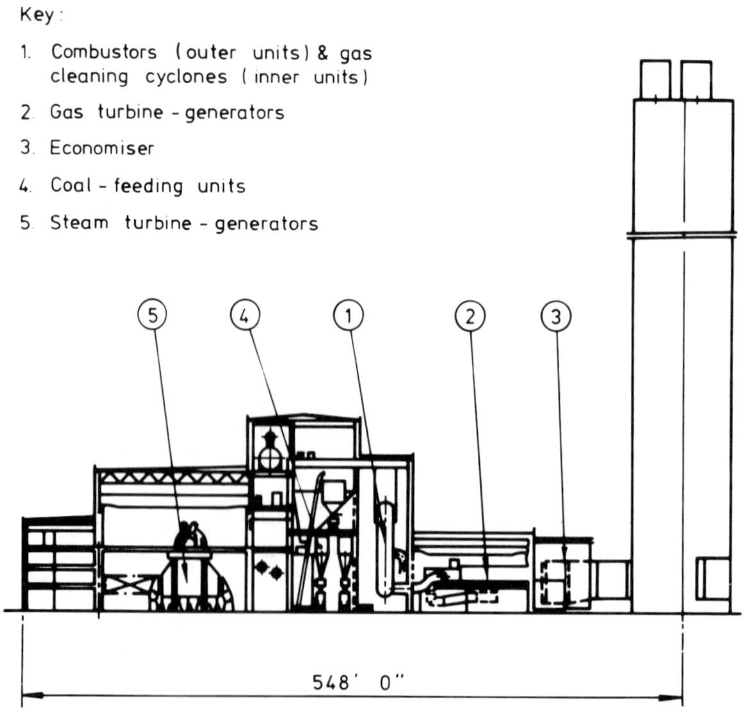

Key:

1. Combustors (outer units) & gas
   cleaning cyclones (inner units)

2. Gas turbine - generators

3. Economiser

4. Coal - feeding units

5. Steam turbine - generators

Figure 19.2: Conceptual design of a 2000 ME pressurised fluid bed fired power
station (courtesy of British Columbia Hydro & Power Authority)

velocities result. It is usual to allow wide "vapour lanes" in order to allow easy access to the tube array. The main problem with such condensers is the build-up of incondensible gases and this can often result in gross under-performance from such units. To solve this it is necessary to vent the incondensible gases at likely points of build-up and the best designs encourage one-dimensional flow from the inlet to the point of venting.

Most central power stations utilise lakes, cooling ponds, rivers or coastal waters for once-through cooling of the condenser. In the UK, however, evaporative natural draught cooling towers are used as an alternative. In many locations in the Middle East, for example, the unavailability of make-up water may make dry cooling towers the only available cooling method (Leung and Moore (1971)). There are two basic types of closed cooling water system. One is the GEA (Gesellschaft fuer Luftkondensation) system; and the other, the Heller system. These systems are illustrated in Figure 19.4.

The GEA system utilises an extended surface air-cooled condenser in which the turbine exhaust steam is condensed directly. A typical layout for such a unit will involve a bundle of finned tubes through which the condensing steam is passed. Air is drawn over the tube bank to remove the condensation heat load. Large ducts are needed to convey the exhaust steam to the tube bank. Such units also raise problems of the calculation of heat transfer coefficients and pressure drop for condensation inside tubes, particularly with incondensible gases present. Another important problem in such units is that of flow distribution in the inlet and outlet manifolds. This is a particularly severe problem when the unit has more than one tube side pass.

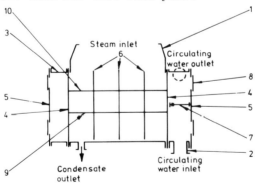

1 Condenser shell
2 Inlet end water box
3 Return end water box
4 Tubeplates
5 End covers
6 Tube support plates
7 Inlet water box division plate
8 Inspection doors
9 Tubes
10 Steam space stays

Figure 19.3: Diagrammatic arrangement of modern power station condenser (Modern Power Station Practice, Central Electricity Generating Board, London, UK)

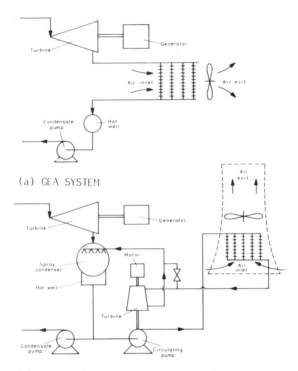

(a) GEA SYSTEM

(b) HELLER (RUGELEY) SYSTEM

Figure 19.4: Two systems using dry cooling towers

The Heller system utilises a direct contact or jet condenser
in lieu of the conventional tubed unit. Circulating water, which
must be high quality condensate, is sprayed into the jet condenser
where it mixes with and absorbs the heat from the exhaust steam.
Large circulating pumps recycle most of the heated condensate to
the dry-type cooling tower and return the remaining condensate to
the feedwater cycle.

In the direct contact spray condenser, the resistance to heat
transfer may be in the liquid phase (droplets), in the vapour
phase (due to incondensibles), or both. If the vapour phase
resistance is negligible, then the design can be based on the
liquid phase resistance. A heat balance will provide the inlet
and outlet temperatures of the spray flow. In most cases this
flow is large compared with the amount of water condensed and hence
the droplet radius varies very little as it passes through the
steam phase.

The transit time, t, required to give the design temperature
rise of the spray droplets is given by the following equation
derived by Brown (1951):

$$\left[\frac{T_{\ell 2}-T_{\ell 1}}{T_g-T_{\ell 1}}\right] = 1 - \frac{6}{\pi^2}\sum_{n=1}^{\infty}\frac{1}{n^2}\exp\left\{-\frac{n^2\pi^2\alpha_T t}{r_d}\right\} \tag{19.1}$$

where $T_{\ell 1}$ and $T_{\ell 2}$ are the initial and final temperatures of the droplet respectively, T_g is the steam temperature, r_d is the droplet radius and α_T is the thermal diffusivity of the water droplet. Curves derived from equation (19.1) are shown in Figure 19.5. For a given size of droplet, the transit time required to achieve the desired duty can be easily determined from this graph. The droplet size in the above calculation is obtained from the manufacturer's data for spray nozzles.

19.3 Alternative Energy Sources

Following the oil crisis in 1973-74 there has been an upsurge in the research and development into alternative energy sources. Many of these concepts, such as solar, geothermal, and ocean-thermal energy conversion (OTEC) use a Rankine power cycle and thus involve two-phase gas-liquid flows. In the case of OTEC, for example, very small temperature driving forces are available for evaporation and condensation and means of enhancing the heat transfer coefficients may have to be used. An accurate knowledge of in-tube evaporating and condensing coefficients will be essential for this concept.

In the case of geothermal energy (Axtmann and Peck (1976)) optimisation of the flow in a well is an important aspect. Geothermal water reaches the surface with approximately the same energy content as in the reservoir. Flashing in the high temperature ($T > 100^\circ C$) bore holes begins at about two-thirds the way of the surface. The fluid in the upper part of the well then has a lower density and thus hydrostatic pressure in the reservoir drives the two-phase flow out. The flow rate is determined by a balance between the well-head pressure, the hydrostatic pressure drop and the frictional resistance compared to the reservoir pressure. These aspects have led to a large requirement for information on two-phase flow pressure gradients (and hence void fraction) for both straight pipe lengths and also for the various fittings and metering devices which occur in such systems.

Important limitations occur on the use of geothermal pipelines (James (1967)) due to undesirable flow patterns. For example, in pipelines with slug flow, there is a tendency for the pipe to vibrate within its supports. Another consequence of slug flow is that the slugs (largely liquid filled lengths) tend to grow as the

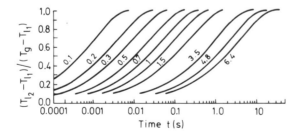

Figure 19.5: Time of heating of water droplets (thermal diffusivity $\alpha_T = 1.68 \times 10^{-7} m^2/s$). The numbers on the curves are the droplet diameter in mm (after Brown 1951)

flow proceeds along the pipeline. Large cyclone separators are
used to separate out the water which emerges from the bore hole
with the steam.

Various types of solar power plant have been conceived.
France has built at La Paz (Mexico) a 30 kWe power plant using
plane collectors and a demonstration plant at Odeillo (CNRS,
Southern France). The French designs of plant use as the heat
transfer medium, passing through the collectors, a mixture of
sodium and potassium nitrate salts which are heated to between 400
and 500°C. This salt then transfers its heat to high pressure
boiling water in what amounts to a waste heat boiler. Other solar
electric schemes make use of liquid metals as the heat transfer
medium.

19.4 Nuclear Power Plant

Two-phase flow problems figure prominently in the design of
nuclear power plants. Nuclear power plants in common use around
the world can be classified into four main types:

- the Light-Water Reactor (LWR)

- the Heavy-Water Reactor (HWR)

- the Gas-Cooled Reactor (GCR)

- the Liquid Metal Cooled Reactor (LMFBR)

Two types of light water reactor are under construction around
the world, viz.

- *the Pressurised Water Reactor* (PWR) where the primary circuit
  is cooled by light water at a pressure of 2300 psia. Hot
  water from the core is transferred via a series of two or
  more coolant loops to a number of steam generators. Each
  steam generator consists of a bundle of tubes in the form
  of an inverted U, through which the primary coolant passes.
  Boiling takes place on the shell side (outside of the tubes)
  at a pressure of around 1000 psia.

- *the Boiling Water Reactor* (BWR) where boiling occurs
  directly within the core of the reactor and the water/steam
  mixture passes to separators where the steam is dried before
  going direct to the turbine. The separated water is returned
  to the core with the aid of a series of recirculating jet
  pumps.

Both the PWR and BWR employ a saturated steam cycle. Two-
phase problems in respect of the LWR relate mainly to the loss-of-
coolant accident (LOCA) (Solbrig (1976)) which will be discussed
in a later chapter. Other two-phase flow problems relate to the
behaviour of the coolant within the interstices of the core and
to the performance of the steam generators on the PWR.

The gas-cooled reactor and the liquid-metal cooled reactor
use either a high pressure gas (CO_2 or He) or sodium respectively
to transfer heat from the reactor core to the steam generator.
Because higher temperatures can be used with these types of

reactor, higher steam pressures (up to 2700 psi) and superheat/
reheat cycles are used. Once-through steam generators are often
employed in these designs. Details of some designs of gas-cooled
reactor and liquid-metal cooled reactor steam generators compared
with conventional fossil-fired boilers are given in Table 19.1.
Figure 19.6 shows an isometric view of the inverted U-tube design
of evaporator used on the UK Prototype Fast Reactor (PFR).

19.5 Two-Phase Heat Transfer and Fluid Flow Processes of Relevance in Power Plant Design and Operation

Dryout

A fossil-fired boiler tube is heated only over half the tube
circumference. Conduction in the tube wall, however, ensures that
some heat is transferred to the insulated half of the tube. It is
important to establish to what extent this asymmetric heating
around the circumference influences the dryout condition. Alekseev
et al (1964) have examined the situation for a 10 mm id tube.
Three ratios of the maximum heat flux (ϕ_{MAX}) to the average heat
flux ($\bar{\phi}$) were used, namely 1.12, 1.28 and 1.50. Allowance was
made for the circumferential conduction in the tube wall when
calculating the heat flux profile. The experimental results for

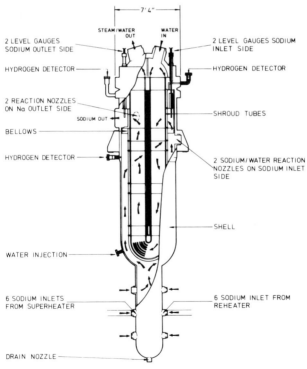

Figure 19.6: Cross section view of evaporator work on UK Prototype Fast Reactor
          (PFR)

TABLE 19.1

Main Design Parameters for Steam Generators of
Gas-Cooled Reactors and Liquid Metal-Cooled Reactors

	MAGNOX		AGR		HTR	LMFBR	
	Trawsfynydd	Oldbury	Hinkley 'B'	Hartlepool	—	PFR	—
Type of Boiler	Drum assisted recirculation	Once-through	Once-through	Once-through	Once-through	Forced recirculation	Once-through or forced recirculation
Steam pressure bar (psia)	65 (965) / 20.6 (315)	96 (1400) / 52 (750)	165 (2400)	165 (2400)	165 (2400)	165 (2400)	165 (2400)
Tube geometry	Serpentine (finned)	Serpentine (finned)	Serpentine (finned)	Helical (finned)	Helical	U-tube	U-tube helical
Material — Evaporator	Carbon steel	Carbon steel	Mild steel followed by 9 Cr - 1 Mo	Mild steel followed by 9 Cr - 1 Mo	as per AGR	$2\frac{1}{4}$Cr-1Mo stabilised	Incoloy 800
Material — Primary Superheater	Carbon steel	Carbon steel	9 Cr - 1 Mo	9 Cr - 1 Mo		Austenitic stainless steel	Incoloy 800
Material — Secondary	Carbon steel	Carbon steel	Austenitic stainless	Austenitic stainless		Austenitic stainless steel	
Tube Dimensions — Evaporator o.d. mm	56	38	28.6	18 (fin base)	20-30	25	15-25
thickness mm	7	5	3.25	3	3-5	2.3	2-4
Tube Dimensions — Superheater o.d. mm	—	38	38	20	25-35	16	15-25
thickness mm	—	5	4.1	3	~ 2.5	2	2-4

TABLE 19.1 (cont'd)

	MAGNOX		AGR		HTR	IMFBR	
	Trawsfynydd	Oldbury	Hinkley 'B'	Hartlepool	–	PFR	–
Type of Boiler	Drum assisted recirculation	Once-through	Once-through	Once-through	Once-through	Forced recirculation	Once-through or forced recirculation
Steam pressure bar (psia)	65 (965) 20.6 (315)	96 (1400) 52 (750)	165 (2400)	165 (2400)	165 (2400)	165 (2400)	165 (2400)
Tube geometry	Serpentine (finned)	Serpentine (finned)	Serpentine (finned)	Helical (finned)	Helical	U-tube	U-tube helical
Heat flux kw/m² inlet	25	54 19	50	140	160	630	R 700 OT 150
Heat flux kw/m² outlet	63	190 110	137	320	380	110	~700 ~110
Heat flux kw/m² peak			137	320	~380	~700	~110 ~700
Mass Velocity kg/m²s Evaporator at full load	575	410 540	1200	1650	2030	2580	~3000

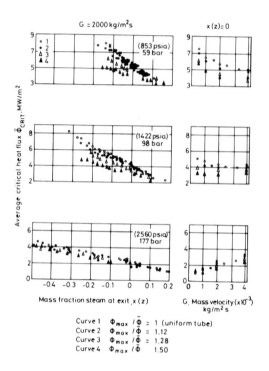

Curve 1 $\Phi_{max} / \bar{\Phi}$ = 1 (uniform tube)
Curve 2 $\Phi_{max} / \bar{\Phi}$ = 1.12
Curve 3 $\Phi_{max} / \bar{\Phi}$ = 1.28
Curve 4 $\Phi_{max} / \bar{\Phi}$ 1.50

Figure 19.7: The influence of non-uniform heating around the tube circumference. (Alekseev (1964))

each tube are shown in Figure 19.7 for three different pressures, together with the results of a uniform tube of the same dimensions. It is seen that as the circumferential heat flux profile becomes more asymmetric, so the average critical heat flux, $\bar{\Phi}_{CRIT}$ (evaluated from the total power to the tube) falls. At high sub-coolings the value of $(\phi_{MAX})_{CRIT}$ for each of the asymmetric tubes approximately corresponds to the uniform tube value, ϕ_{CRIT}. As the exit steam quality increases, so the average heat flux approaches the value for a uniform tube. These experiments confirm the "local" nature of the critical condition at high subcoolings and the tendency to an "overall power" condition in the higher steam quality regions. It is recommended that the "overall power" hypothesis be used to calculate the critical power which can be applied to a boiler tube with an asymmetric heating. The power calculated should be reduced by 10% to cover the worst condition for cases with $(\phi_{MAX}/\bar{\phi})$ between 1.0 and 1.5 and by 20% for cases with $(\phi_{MAX}/\bar{\phi})$ between 1.5 and 2.0.

In a once-through boiler the temperature rise at dryout is more important than the precise location of the dryout point. Figure 19.8 shows the wall temperature variation for a once-through boiler as determined from the measurements of Schmidt (1959). These diagrams show clearly the characteristic sharp increase in wall temperature. Analysis of the extensive programme of work

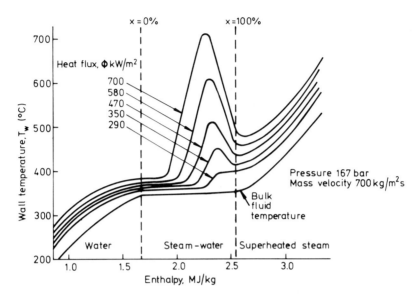

Figure 19.8: Tube wall temperature in the liquid deficient region. (Schmidt (1959))

carried out by Herkenrath (1967), together with that of other workers, has shown that the temperature rise at dryout ($\Delta T_{MAX} = T_{W(MAX)} - T_{SAT}$) for a uniformly heated vertical tube can be expressed by the simple relationship:

$$\Delta T_{MAX} = C\{\phi/G\}^{2.5} \qquad (19.2)$$

The influence of tube diameter on ΔT_{MAX} is relatively small in the range 5-20 mm. Equation (19.2) is plotted in Figure 19.9 and this shows the strong dependence on system pressure. For example, with a constant mass velocity and heat flux the value of ΔT_{MAX} (and therefore C) decreases by a factor 4.5 as the pressure is increased in the range 140-190 bar. It might be expected that this decrease would extend all the way up to the critical pressure. This, in fact, is not the case, and above 210 bar the value of ΔT_{MAX} rises sharply. This singularity in the temperature rise at the critical heat flux has been investigated in detail by Bahr (1969). The explanation of the effect stems from the fact that as the pressure rises above 210 bar the mass quality corresponding to the critical heat flux falls with increasing speed from a positive value to a negative value; the mechanism of the critical heat flux changes from dryout to departure from nucleate boiling and the increase in ΔT_{MAX} results from the substitution of the relatively high heat transfer coefficients in the liquid deficient region with the much lower coefficients in the subcooled film boiling region.

Horizontal and Inclined Tubes

Evaporation and condensation within inclined and horizontal

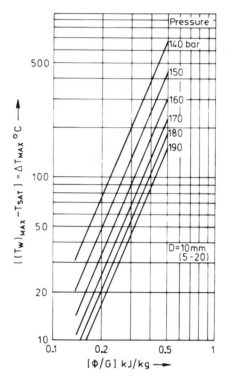

Figure 19.9: The influence of pressure on the temperature rise at dry-out for a uniformly heated vertical tube cooled by high pressure water.

tubes is encountered frequently. Stratification of the flow occurs so that the majority of the liquid flows along the bottom of the tube. To a degree this is beneficial in the case of condensation but can cause problems during evaporation in that the liquid film at the top of the tube may dry out (Figure 19.10).

Dryout in a horizontal tube differs from that in a vertical tube in two ways:

(1) *Stratification* of the flow can occur at low velocities for both low quality and subcooled conditions. Such conditions can lead to overheating at steam boiler tubes at quite modest heat fluxes.

(2) *Dryout* of the tube at high vapour qualities occurs over a relatively long tube length, starting at the top of the tube where the film thickness and flowrate are lowest and ending up with the final evaporation of the rivulet running along the bottom of the tube. Under these conditions the vapour flow in the upper part of the tube may become superheated before dryout occurs at the base of the tube.

Styrikovich and Miropolskii (1950) reported the effects of stratification of a high pressure steam-water mixture in a horizontal pipe. These caused wide temperature differences between

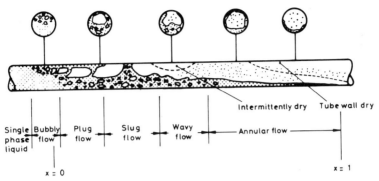

Figure 19.10: Flow patterns in a horizontal tube evaporator (Collier (1972))

the top and bottom of the boiler tube. Experiments were carried out on a single 7.5 m, 56 mm id tube at pressure between 10 bar and 220 bar with heat fluxes in the range 22-135 kW/m^2 and inlet velocities between 0.24 and 1 m/s. It was found that there was a critical two-phase velocity, j, below which stratification occurred and above which it did not (Figure 19.11). Using an alcohol-water analogue for steam-water flow Gardner and Kubie (1976) established the following expression for the critical velocity, j:

$$j^2 = 6.6 \frac{\{\sigma \, g \, \cos \alpha \, (\rho_f - \rho_g)\}^{0.5}}{\rho_f \, f \, K^2} \tag{19.3}$$

α is the angle over inclination of the tube to the horizontal, f is the friction factor and K is an empirical constant found to be 2.17. Substituting for f and K leads to the following equation for j:

$$j^{1.8} = 30.43 \frac{D^{0.2} \, \{\sigma \, g \, (\rho_f - \rho_g)\}^{0.5}}{\rho_f^{0.8} \, \mu_f^{0.2}} \tag{19.4}$$

Table 19.2 gives the values of j calculated from equation (19.4) for steam-water flows over a range of pressures and tube diameters. Excellent agreement is seen with the data in Figure 19.11 and equation (19.4) is recommended as giving the minimum single-or two-phase velocity below which stratification will occur in a horizontal tube.

Stratification may also occur in vertical bends and in helical coils. For the case of a vertical bend, Bailey (1977) showed that the gravitational force maintaining stratified flow is supplemented by centrifugal forces so that g cos α in equation (19.3) is replaced by (g + j^2/r). This neglects the secondary flow known to occur in both bends and coils.

Another problem relates to the estimation of heat transfer coefficients in horizontal tubes. The Chen correlation is only valid for situations where all surfaces of the tube remain wetted. More recently, Shah (1976) has proposed an alternative correlation which has the advantage that it can be applied to partially

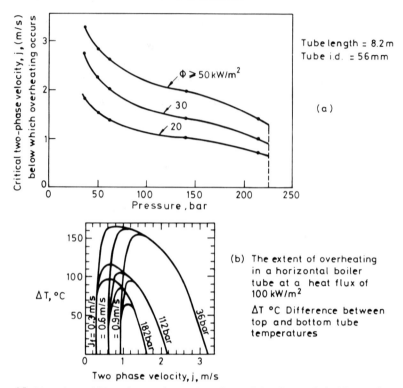

Tube length = 8.2 m
Tube i.d. = 56 mm

(a)

(b) The extent of overheating
in a horizontal boiler
tube at a heat flux of
100 kW/m²

ΔT °C Difference between
top and bottom tube
temperatures

Figure 19.11:   Stratification and overheating of horizontal boiler tubes.
(Stryikovich and Miropolskii (1950))

TABLE 19.2

Values of j (m/s) to prevent stratification for
for steam/water flow in horizontal tubes

p (bar) D (mm)	33.5	64.2	112.9	146	165	187
20	2.71	2.47	2.11	1.82	1.63	1.37
40	2.92	2.67	2.28	1.97	1.77	1.48
60	3.06	2.80	2.38	2.06	1.85	1.55

stratified flows in horizontal channels. The correlation is in
the form of a graphical chart (Figure 19.12). The ordinate is the
ratio (h_{TP}/h_L) and the abscissa is C_O where:

$$C_O = \left(\frac{1-x}{x}\right)^{0.8} \left(\frac{\rho_g}{\rho_f}\right)^{0.5}$$

Two other parameters are used:

$$Bo - \text{the boiling number } \left(\frac{\phi}{G\, i_{fg}}\right)$$

$$\text{and } Fr_f - \text{the Froude number } \left(\frac{G^2}{\rho_f{}^2\, g\, D}\right)$$

Figure 19.12 illustrates the use of the chart. For vertical tubes the value of Fr_f is ignored and the line AB is used; for horizontal tubes different lines (according to the value Fr_f) are employed. When the correlation is used for a horizontal tube the value of h_{TP} is the mean coefficient over the tube perimeter. The Shah correlation agrees well with experimental data for refrigerants but for water should only be used under conditions where the heat flux is a dependent variable, ie where heating is from a primary fluid.

19.6 Two-Phase Pressure Drop and Void Fraction ("Best Buys")

The evaluation of the pressure drop across an evaporating or condensing channel or in a geothermal pipeline involves integrating the local pressure gradients along the channel length. This local pressure gradient is made up of three separate components; a frictional term, an accelerational term and a static heat term.

$$-\left(\frac{dp}{dz}\right)_{TP} = -\left(\frac{dp}{dz}\ F\right) - \left(\frac{dp}{dz}\ A\right) - \left(\frac{dp}{dz}\ z\right) \qquad (19.5)$$

Using the separated flow model these three components are respectively given by:

$$-\left(\frac{dp}{dz}\ F\right) = -\left(\frac{dp}{dz}\ F\right)_{LO} \phi_{LO}{}^2 \qquad (19.6)$$

where $\phi_{LO}{}^2$ is known as the two-phase frictional multiplier and $-\left(\frac{dp}{dz}\right)F_{LO}$ is the frictional pressure gradient calculated from the Fanning equation for the total flow (water and steam) assumed to be water:

$$-\left(\frac{dp}{dz}\ A\right) = G^2 \frac{d}{dz}\left[\frac{x^2\ v_G}{\alpha} + \frac{(1-x)^2\ v_L}{(1-\alpha)}\right] \qquad (19.7)$$

where G is the total mass velocity, x is the quality and v_G and v_L are the steam and water specific volumes and α is the void fraction.

$$-\left(\frac{dp}{dz}\ z\right) = g\,\sin\theta\left[\frac{\alpha}{v_G} + \frac{(1-\alpha)}{v_L}\right] \qquad (19.8)$$

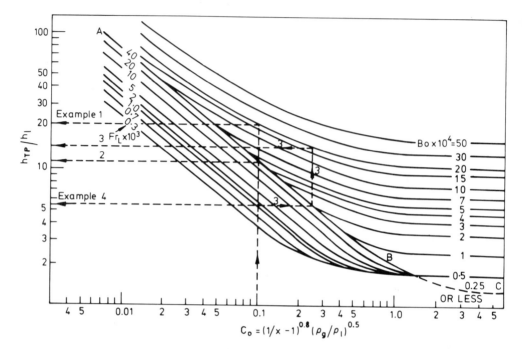

Figure 19.12: The correlation for two-phase heat transfer coefficient given by
          Shah (note that Bo, Co and Fr_L are defined in the text and that
          the curve AB relates only to vertical tubes and the curves labled
          $Fr_L \times 10^3$ to horizontal tubes). (Shah (1976)).

where g is the acceleration due to gravity, and θ is the angle of
inclination of the tube to the horizontal.

        To evaluate the local pressure gradient, expressions are
required for the functions ϕ_{LO}^2 and α. A very large number of
correlations have been proposed for these functions and a summary
of the better known correlations for ϕ_{LO}^2 is given in Table 19.3.
Recently, a number of workers have carried out systematic
comparisons between these various correlations and data banks
containing large numbers of experimental pressure drop measurements for steam-water mixtures. A summary of these various
comparisons is given in Table 19.4. It can be seen that the various
studies agree that the most accurate correlations for ϕ_{LO}^2 are
those of Baroczy, Chisholm and CISE but that in each case the
standard deviation of errors about the mean is 30-35%.

        To evaluate the changes in momentum (or kinetic energy)
and also the mean density of a two-phase flow, it is necessary to
be able to establish the local void fraction or fraction of the
flow cross section occupied by steam. Once again a large number
of correlations have been proposed for the evaluation of void
fraction (α). Sometimes the correlation is expressed in terms of
the slip ratio (S) which is related to the void fraction by the
identity:

TABLE 19.3

Two-Phase Pressure Drop Correlations for Steam-Water Mixtures

	Correlation
Homogeneous (Collier (1972)	$\phi_{LO}^2 = \left[1 + x\left(\frac{\rho_L - \rho_G}{\rho_G}\right)\right]\left[1 + x\left(\frac{\mu_L - \mu_G}{\mu_G}\right)\right]^{-1/4}$
Baroczy (1966)	$\phi_{LO}^2 = \Omega\phi_{LO}^2$ (G = 1356 kg/m²s) where ϕ_{LO}^2(G = 1356) = $f\left(\left[(\mu_L/\mu_G)^{0.2}\ (\rho_G/\rho_L)\right], x\right)$ $\Omega = f\left(\left[(\mu_L/\mu_G)^{0.2}\ (\rho_G/\rho_L)\right], x\right)$
Chisholm (1973)	$\phi_{LO}^2 = 1 + (\Gamma^2 - 1)\left[Bx^{(2-n)/2}(1-x)^{(2-n)/2} + x^{2-n}\right]$ where $\Gamma = \left[(dp/dz)_{GO}/(dp/dz)_{LO}\right]^{\frac{1}{2}}$ and B = f(G, Γ)
CISE Lombardi (1972)	$\left(\frac{dp}{dz}\right)_{TP} = \left[\frac{K\ G^n\ \bar{v}^{-0.86}\ \sigma^{0.4}}{D^{1.2}}\right]$ where $\bar{v} = \left[x/\rho_G + (1-x)/\rho_L\right]$ K = f (geometry) n = f (geometry)

TABLE 19.3 (cont'd)

Correlation

Martinelli-Nelson
(1948)

$$\phi_{LO}^2 = fn(P,x)$$

Smith-Macbeth
(1976)

$$\phi_{LO}^2 = \left[\left\{ e(1-x) + \left(\frac{\rho_L}{\rho_G} \right) \times \left(e(1-x) + x \right) \right\}^{\frac{1}{2}} + (1-e)(1-x) \right]^2 \quad \text{where } e = 0.4$$

TABLE 19.4

Two-Phase Pressure Drop Data for Steam-Water Mixtures

Data Bank	ESDU (1976)			Friedel (1977)			Idsinga (1975)			Harwell (1977)		
Flow Direction	Upflow, downflow and horizontal			Upflow only			Upflow and horizontal			Upflow and horizontal		
Correlation	n	e	σ	n	e	σ	n	e	σ	n	e	σ
Homogeneous Collier (1972)	1709	-13.0	34.2	2705	-19.9	42.0	2238	-26.0	22.8	4313	-23.1	34.6
Baroczy (1966)	1447	4.2	30.5	2705	-11.6	36.7	2238	- 8.8	29.7	4313	- 2.2	30.8
Chisholm (1973)	1536	19.0	36.0	2705	- 3.8	36.0	2238	0.5	40.5	4313	13.9	34.4
CISE Lombardi (1972)	-	-	-	2705	16.3	28.0	2225	22.6	28.9	-	-	-
Martinelli-Nelson (1948)	1422	16.3	36.6	-	-	-	2238	47.8	43.7	-	-	-
Smith-Macbeth (1976)	-	-	-	-	-	-	-	-	-	4313	-16.6	24.1

n = number of data points analysed

e = mean error (%) = $(\Delta p_{cal} - \Delta p_{exp}) \times 100/\Delta p\ exp$

σ = standard deviation of errors about the mean (%)

$$S = \left(\frac{x}{1-x}\right)\left(\frac{\rho_G}{\rho_L}\right)\left(\frac{1-\alpha}{\alpha}\right) \tag{19.9}$$

A summary of the better known correlations for S or α is given in Table 19.5. Similarly, systematic comparisons have been carried out between these various void fraction correlations and data banks containing large numbers of experimental measurements of either void fraction or fluid density measurements for steam-water mixtures. A summary of these comparisons is given in Table 19.6. It can be seen that the various studies agreed that the most accurate void fraction correlations are those of Smith, CISE and Chisholm, with the latter having the added advantage of great simplicity. Again, the standard deviation of error on the mean density is about 20-30%.

19.7 Sub-Channel Analysis for Nuclear Reactor Fuel Elements

Rod bundles, particulars which form nuclear reactor fuel elements, differ greatly in geometry. They may be housed in square, hexagonal or round outer channels. The flow space between the rods is made up of a series of interconnected "sub-channels" which are usually delineated by lines joining the rod centres. Two methods exist for estimating the local conditions for axial two-phase flows within rod bundles, Weisman and Bowring (1975) viz.

- "mixed flow" models in which the duct is assumed to behave one-dimensionally and the equivalent diameter of a simple duct is calculated using the total flow cross-sectional area and the total wetted perimeter of the rods and housing;

- "sub-channel" models in which the duct is divided into a series of closely connected sub-channels and the conservation equations for mass, momentum and energy are solved simultaneously for each sub-channel to establish the local flow, enthalpy, void fraction and pressure drop. These calculations are complex and require computer codes, for example, HAMBO, COBRA and THINC.

The cross-section of the cluster is divided into a number of hydraulically interconnected sub-channels (Figure 19.13). The size of these sub-channels is arbitrarily selected, usually by drawing sub-channel boundaries at the minimum rod-to-rod and the rod-to-wall spacings. Sometimes it is advantageous to define a sub-channel to be associated with a particular rod. This is particularly useful in the case of analyses involving annular flow as will be described in the following chapter. Once the sub-channel grid has been established, then the appropriate two-phase conservation equations can be written down and solved for the individual sub-channel dependent variables. The equations can be written on a nodal basis, or, alternatively, in a finite difference form. Figure 19.13 shows the control volume for two adjacent sub-channels, i and j, and the equations for sub-channel i assuming steady state.

The continuity equation stages that the change in mass flow $(\partial M_i/\partial x)$ sub-channel i is equal to the sum of the cross flows in

TABLE 19.5

Void Fraction Correlations for Steam-Water Mixtures

	Correlation
Lockhart & Martinelli	$\alpha = f(X)$ where $X = \left[\left(\dfrac{dp}{dz}\right)_L \Big/ \left(\dfrac{dp}{dz}\right)_G\right]^{\frac{1}{2}}$
Hughmark (1962)	$\alpha = K\beta$ where $K = f\left[Re,\, Fr,\, (1-\beta)\right]$
Smith (1969)	$S = e + (1-e)\sqrt{\dfrac{\rho_L/\rho_G + e\,(1/x - 1)}{1 + e\,(1/x - 1)}}$ where $e = 0.4$
CISE Premoli (1970)	$S = f(G,D,\rho_L,\rho_G,\mu_L,\sigma,\beta) + 1$
Chisholm (1973)	$S = \left[x\left(\dfrac{\rho_L}{\rho_G}\right) + (1-x)\right]^{\frac{1}{2}}$
Thom (1964)	$S = f\left(\dfrac{\rho_L}{\rho_G}\right)$
Bankoff-Jones (1961)	$S = \left[\dfrac{1 - \alpha}{A - \alpha + (1-A)\alpha^B}\right]$ where $A,B = f(P)$
Bryce (1977)	$S = \left[\dfrac{1 - \alpha}{A - \alpha + (1-A)\alpha^B}\right]$ where $A = f(P,G,x,\rho_G,\rho_L)$ $B = f(P,\rho_G,\rho_L)$

TABLE 19.6

Void Fraction Data for Steam-Water Mixtures

Data Bank Correlation	Analysis of Mean Density									Analysis of Slip Ratio								
	Friedel (1977)			ESDU (1977)			Bryce (1977)			Friedel (1977)			ESDU (1977)			Bryce (1977)		
	n	e	σ	n	e	σ	n	e	σ	n	e	σ	n	e	σ	n	e	σ
Lockhart & Martinelli (1949)	–	–	–	–	–	–	–	–	–	–	–	–	598	-57.6	50.3	–	–	–
Hughmark (1962)	484	-10.8	33.0	–	–	–	–	–	–	–	–	–	598	-9.1	29.2	–	–	–
Smith (1969)	484	0.5	26.8	–	–	–	639	8.6	31.5	–	–	–	–	–	–	639	18.0	77.8
CISE Premoli (1970)	484	9.3	35.0	–	–	–	639	-1.4	22.7	–	–	–	598	-23.7	27.2	639	-1.2	68.6
Chisholm (1973)	484	-0.4	26.0	–	–	–	–	–	–	–	–	–	598	-14.5	30.8	–	–	–
Thom (1964)	484	7.4	36.5	–	–	–	639	43.3	61.7	–	–	–	–	–	–	639	132.6	200.0
Bankoff-Jones (1961)	–	–	–	–	–	–	639	9.32	31.6	–	–	–	–	–	–	639	34.5	137.6
Bryce (1977)	–	–	–	–	–	–	639	0.1	20.7	–	–	–	–	–	–	639	6.1	86.9

n = number of data points analysed

e = mean error (%) = (cal-exp)×100/exp

σ = standard deviation of error about mean (%)

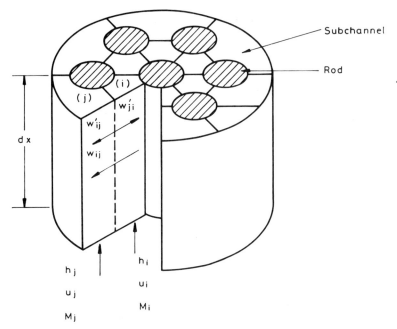

Figure 19.13: Typical definitions and conservation equations used in sub-channel analyses.

CONTINUITY EQUATION

$$\frac{\partial M_i}{\partial x} = - \sum_{j}^{N} w_{ij}; \quad j = 1,2,3\ldots N \tag{1}$$

ENERGY EQUATION

$$M_i \frac{\partial h_i}{\partial x} = q_i P_i - \sum_{j}^{N} (h^* - h_i) w_{ij} - \sum_{j}^{N} (h_i - h_j) w_{ij}' \tag{2}$$

P_i is the heated perimeter for sub-channel i; h^* is the effective cross-flow enthalpy.

AXIAL MOMENTUM EQUATION

$$\frac{\partial p}{\partial x} = - (\dot{m}_i)^2 \left[\frac{f_i \phi^2 LO}{2 D_i \rho_i} + \frac{K_i}{2 \rho_i} + \frac{\partial}{\partial x} \left(\frac{1}{\rho_i} \right) \right] - g \rho_i \cos\theta$$

$$- f_t \sum_{j}^{N} (u_i - u_j) \frac{w_{ij}'}{A_i} + \sum_{j}^{N} (2 u_i - u^*) \frac{w_{ij}}{A_i} \tag{3}$$

u^* is the effective cross-flow velocity

TRANSVERSE MOMENTUM EQUATION

$$\frac{\partial (u^* w_{ij})}{\partial x} = \frac{s}{\ell} (p_i - p_j) - F_{ij} \tag{4}$$

and out of the sub-channel. The energy equation equates the change
in enthalpy ($\partial h/\partial x$) to the heat added, ϕ_i, and the two energy
flows transferred by diversion cross flow and by mixing respectively. The axial momentum equation equates the pressure gradient
($\partial p/\partial x$) to the friction, grid, acceleration and static head terms,
together with drag terms, resulting from cross flow and mixing.
The transverse momentum balance relates the effective cross flow
velocity, u^*, to the pressure gradient between the sub-channels
(p_i-p_j), the distance between sub-channel centres (ℓ), the gap
between the rods (s) and the frictional or drag losses between
sub-channels (F_{ij}). Cross flow and mixing terms can be evaluated
by assuming that the two-phase flow behaves like a single-phase
flow with the appropriate density and viscosity (the so-called
"homogeneous flow" assumption). However, it has been observed
experimentally that cross flow and mixing in two-phase flows are
flow-pattern dependent and this fact is usually taken into
account in modern sub-channel codes.

This set of equations must be solved for boundary conditions
of zero lateral pressure differential at the channel inlet and
exit. Two methods of solution have been developed (a) "step-by-
step", or "marching" solutions, and (b) "simultaneous" or
"relaxation" methods. The former are satisfactory provided the
coupling between the sub-channels is weak. However, when strong
coupling exists between sub-channels, as might be caused by a
partial blockage inducing recirculation or asymmetric power
distribution, then the latter method must be employed. Examples
of codes which use the relaxation technique are COBRA IV, SABRE,
and THINC IV.

Having established the local mass flowrate and enthalpy
in each sub-channel, then the critical heat flux is evaluated
from an empirical correlation developed specially for use with
a sub-channel analysis. Correlations of this latter type are
the W-3 correlation, the WSC correlation of Bowring and the B&W
correlation. Typically, such correlations when used with the
appropriate sub-channel code can predict the critical heat fluxes
in PWR and BWR rod clusters with a RMS error or 5-10%. Table
19.7 reproduced from Weisman and Bowring (1975) summarises the
performance of these various correlations in relation to PWR and
BWR conditions respectively.

19.8 Fouling and Corrosion

So far the discussion of convective boiling heat transfer has
been limited to the condition where the heat transfer surface is
clean. This, however, is rarely the case in operating power
plant and it is necessary to discuss briefly the incidence of
deposition and corrosion, particularly on steam generating
surfaces. With a once-through evaporator all dissolved salts and
suspended matter entering with the liquid feed to the evaporator
must either:

(a)   remain in the evaporator to form a deposit,

(b)   be mechanically entrained in the outflow of steam,

(c)   be actually dissolved in the steam outflow.

TABLE 19.7

Performance of Critical Heat Flux Correlations
(adapted from Weisman & Bowring (1975))

(i) for PWR Conditions

Correlation (used in conjunction with HAMBO)	1.19m long ferrule grid 9-rod cluster		1.83m long ferrule grid 9-rod cluster		2.13m long mixing vane 25-rod cluster	
	e	σ	e	σ	e	σ
W-3	-7.2	8.8	-1.9	5.6	-12.4	8.3
WSC-2	6.7	6.7	0.3	5.7	2.3	9.8
B&W-2	5.2	8.4	-2.1	2.4	-6.2	4.4

e = mean error $(\dot{q}_{CR,CAL} - \dot{q}_{CR,EXP}) \times 100/\dot{q}_{CR,EXP}$

σ = std deviation of errors about the mean (%)

(ii) for BWR Conditions

Analysis	Correlation	Bundles with uniform heat flux e (%)	Bundles with non-uniform heat flux e (%)	Long bundles e (%)
Mixed flow	Barnett	9.86	9.36	21.33 12.79
Mixed flow	Macbeth	12.48	16.73	12.96 10.5
COBRA Sub-channel	Becker	7.67	13.34	4.7 5.2
CISE Rod centred Sub-channel	Gaspari	7.9	9.68	14.6 8.7
CISE Rod centred Sub-channel	Hewitt	8.1	10.3	4.5 4.17

A comprehensive review of solubility of various salts in steam was published by Styrikovich and Martynova. For sub-critical systems the partitioning of the salt between the water and steam phases is required. The partition (or distribution) coefficient denoted by K_D is:

$$K_D = \frac{C_G}{C_L} = fn\left(\frac{\rho_G}{\rho_L}\right) = \left(\frac{\rho_G}{\rho_L}\right)^m \qquad (19.10)$$

where C_G and C_L are the concentrations of salt in the steam and water phases respectively. The partition coefficient is found to be a simple power law of the ratio of the phase densities, the value of m varying from salt to salt.

"On-load corrosion" is a particularly rapid and severe attack of boiler tubes in zones where steam is raised. As a result of intensive research into this problem of the past few years a considerable amount is known about the reasons for this type of corrosion. In particular, work carried out by Masterson, Castle and Mann (1969) has elucidated three mechanisms by which corrosive salts may be concentrated in the boiler. The first mechanism is that of dryout, either complete as in a once-through boiler, or partial as may occur due to stratification in horizontal tubes of waste-heat boilers. A second concentration mechanism is that occurring in crevices in the heating surface, particularly where the heating is applied asymmetrically. The third mechanism is closely associated with the build-up of porous deposits on the heat transfer surfaces. Liquid is drawn into the deposit between the micron-size magnetite particles by a "wicking" effect whilst steam is liberated into tunnels in the deposit. Very high concentration factors in excess of 10^4 have been measured for the liquid held up within the deposit.

Mann (1975) has developed a model for the behaviour of aggressive agents such as sodium chloride or sodium hydroxide in a once-through boiler. His results show that the equilibrium indicated by equation (19.10) is not achieved at dryout and that concentrated solutions form as rivulets which penetrate a short distance beyond the dryout point. Mann postulated that the level of solute concentration in the liquid film is determined primarily by the temperature of the tube wall and the pressure in the system. Finally, a point is reached where sodium chloride will exceed its solubility limit in the liquid phase and be precipitated. However, the concentration of sodium hydroxide can continue to very high levels.

Nomenclature

C	- concentration of salt	(kg/m^3)
C	- constant in equation (19.2)	
D	- tube diameter	(m)
e	- mean error	(-)
F_{ij}	- drag losses between subchannels	(-)
f	- friction factor	(-)

G   - mass velocity $(kg/m^2 s)$

g   - acceleration due to gravity (m/s^2)

h   - enthalpy (J/kg)

i_{fg} - latent heat of vaporisation (J/kg)

j   - superficial velocity (m/s)

K_D - partition coefficient $(-)$

K   - empirical constant (= 2.17) in equation (19.3)

ℓ - distance between subchannels (m)

M_i - mass flow in subchannel i (kg/s)

m   - index in equation (19.10)

n   - number of points

P   - pressure (N/m^2)

r_d - droplet radius (m)

s   - gap between rods (m)

S   - slip ratio $(-)$

t   - transit time for droplet (s)

T   - temperature (^oK)

$T_{\ell 1}$ - initial droplet temperature (^oK)

$T_{\ell 2}$ - final droplet temperature (^oK)

T_{SAT} - saturation temperature (^oK)

T_g - steam temperature (^oK)

T_{MAX} - maximum wall temperature (^oK)

ΔT_{MAX} - difference between maximum wall temperature
          (T_{MAX}) and saturation temperature (T_{SAT}) $(^oK$

          $(= T_{MAX} - T_{SAT})$

u   - velocity (m/s)

v   - specific volume (m^3/kg)

x   - vapour quality $(-)$

$\left(\dfrac{dp}{dz}\right)_{TP}$ - two-phase pressure gradient (N/m^3)

$\left(\dfrac{dp}{dz}\right)_F$ - frictional pressure gradient (N/m^3)

$\left(\dfrac{dp}{dz}\right)_A$ - accelerational pressure gradient (N/m^3)

$\left(\dfrac{dp}{dz}\right)_z$ - static head pressure gradient (N/m^3)

$\left(\dfrac{dp}{dz}F\right)_{LO}$ - frictional pressure gradient assuming <u>total</u>
          flow as liquid (N/m^3)

Greek

α	- void fraction	(-)
α_T	- thermal diffusivity of water droplet	(m²/s)
θ, α	- angle of inclination of tube to horizontal	(º)
ϕ	- heat flux	(W/m²)
ϕ_{MAX}	- maximum heat flux (of a heat flux profile)	(W/m²)
$\bar{\phi}$	- average heat flux (of a heat flux profile)	(W/m²)
σ	- surface tension	(N/m)
σ	- std deviation	(-)
ρ	- density	(kg/m³)
μ	- viscosity	(kg/ms)
ϕ_{LO}^2	- two-phase friction multiplier	(-)

Subscripts

CRIT - critical

L,f - liquid

G,g - gas

i   - pertaining to subchannel i

j   - pertaining to subchannel j

Dimensionless Numbers

C_O - $\{(1-x)/x\}^{0.8} \{\rho_g/\rho_f\}^{0.5}$

B_O - Boiling number (= $\phi/G\ i_{fg}$)

Fr_f - Froude number (= $G^2/\rho_f^2\ g\ D$)

References

Alekseev, G V, et al, (1964) "Burnout heat fluxes under forced flow". *Paper presented at USSR at 3rd Int. Conf. on Peaceful uses of Atomic Energy, Geneva,* A/CONF 28/P/327a.

Axtmann, R C and Peck, L B, (1976) "Geothermal chemical engineering". *AIChEJ* **22** (5), 817-828.

Bagley, R, (1965) "The application of heat transfer due to the design of once-through boiler furnaces". *Inst. of Mech. Engrs. Symp. on "Boiling Heat Transfer in Steam Generating Units and Heat Exchangers", Manchester.*

Bahr, A, et al, (1969) "Anomale Druckabhangegkeit der Warmeubertragung im Zwerphasengebeit bei Annaherung an der kritischen Druck". *Brennstoff-Warme-Kraft,* **21** (2), 631-633.

Bailey, N A, (1977) "Dryout in the bend of a vertical U-tube evaporator". *Personal Communication.*

Baroczy, C J, (1966) "A systematic correlation of two-phase pressure drop". *Chem. Engng. Prog. Symp.* Series **62** (64), pp 323.

Brown, G, (1951) "Heat transmission by condensation of steam on a spray of water drops". *Inst. Mech. Engrs. Proc. Discussion on Heat Transfer,* 49-52.

Bryce, W M, (1977) "A new flow dependent slip correlation which gives hyperbolic steam-water mixture flow equations". *AEEW-R1099.*

Chisholm, D, (1973) "Pressure gradients due to friction during flow of evaporating two-phase mixtures in smooth tubes and channels". *Int. J. Heat Mass Transfer,* **16**, 347-358.

Chisholm, D, (1973) "Research note: void fraction during two-phase flow". *J. Mech. Engng. Sci.* **15** (3), 235-236.

Collier, J G, (1972) "Convective boiling and condensation". Published by *McGraw Hill Book Co. (UK) Ltd.*

ESDU, (1976) "The frictional component of pressure gradient for two-phase gas or vapour/liquid flow through straight pipes". *Engineering Sciences Data Unit (ESDU) London.*

ESDU, (1977) "The gravitational component of pressure gradient for two-phase gas or vapour/liquid flow through straight pipes". *Engineering Sciences Data Unit (ESDU) London.*

Friedel, L, (1977) "Mean void fraction and friction pressure drop: Comparison of some correlations with experimental data". *European Two-Phase Flow Group Meeting, Grenoble,* Paper A7.

Gardner, G C and Kubie, J, (1976) "Flow of two liquids in sloping tubes: an analogue of high pressure steam and water". *Int. J. Multiphase Flow,* **2**, 435-451.

Herkenrath, H, et al, (1967) "Heat transfer in water with forced circulation in 140-250 bar pressure range". *EUR 3658d.*

Hughmark, G A, (1962) "Hold-up in gas-liquid flow". *Chemical Engineering Progress.* **58** (4), 62-65.

Idsinga, W, (1975) "An assessment of two-phase pressure drop correlations for steam-water systems". *M.Sc. Thesis, MIT.*

James, C R, (1967) "Pipeline transmission of steam-water mixtures for geothermal power". *New Zealand Engineering,* **22** (6), 221-228.

Jones, A B, (1961) "Hydrodynamic stability of a boiling channel". *KAPL-2170.*

Lockhart, R W and Martinelli, R C, (1949) "Proposed correlation of data for isothermal two-phase, two component flow in pipes". *Chemical Engineering Progress,* **45** (1), 39-48.

Lombardi, C and Peddrochi, E, (1972) "A pressure drop correlation in two-phase flow". *Energia Nucleare,* **19** (2),

Leung, P and Moore, R E, (1971) "Thermal cycle arrangements for power plants employing dry cooling towers". *Journal of Engineering for Power,* 257-264.

Macbeth, R V, (1976) Private communication quoted in paper by Brittain, I. and Fayers, F J. "A review of UK developments in thermal-hydraulic methods for loss of coolant accidents". *Paper presented at CSNI Meeting on Transient Two-Phase Flow, Toronto.*

Martinelli, R C and Nelson, D B, (1948) "Prediction of pressure drop during forced circulation boiling of water". *Trans.* ASME **70**, p695.

Mann, G M W, (1975) "Distribution of sodium chloride and sodium hydroxide between steam and water dryout in an experimental once-through boiler". *Chem. Engng. Sci.* **30** (2), 249-260.

Masterson, H G, Castle, J E and Mann, G M W, (1969) "Waterside corrosion of power station boiler tubes". *Chemistry and Industry,* 1261-1266.

"Modern power station practice" (1963, Published by *Central Electricity Generating Board.* **1-7.**

Premoli, A, Di Francesco, D and Prima, A, (1970) "An empirical correlation for evaluating two-phase mixture density under adiabatic conditions". *European Two Phase Flow Group Meeting,* Paper B9, Milan.

Schmidt, K R, (1959) Warmetechnische Untersuchungen an hock belasteten kesselheizflacken. Mitteilungen der Vereinigung der grosskessel - bezitzer. 391-401.

Shah, M, (1976) "A new correlation for heat transfer during boiling flow through tubes". *ASHRAE Trans.* **82** (2). 66-86.

Smith, S L, (1969) "Void fractions in two phase flow. A correlation based on an equal velocity head model". *Proc. Inst. Mech. Engrs.* **184**, Pt 1, (36), 647-664.

Solbrig, C W, (1976) "Experimental loss-of-coolant accident". *Proceedings of NATO Advanced Study Institute, Instanbul,* **II**, 913-968. Published by *Hemisphere Publishing Corp.*

Styrikovich, M A and Miropolskii, Z L, (1950) *Dok. Akad. Nank. SSSR.* **71** (2).

Thom, J R S, (1964) "Prediction of pressure drop during forced circulation boiling of water". *Int. J. Heat Mass Transfer,* **7**, 709-724.

Thurlow, G G, (1978) "The combustion of coal in fluidised beds". *Inst. of Mech. Engrs. (London) Proceedings,* **192** (15).

Ward, J A, (1977) Private communication.

Weisman, J and Bowring, R W, (1975) "Methods for detailed thermal and hydraulic analyses of water-cooled reactors". *Nucl. Sci. and Engng.* **57**, 255-276.

Chapter 20

Design of Boiling
and Condensing Plant

G. F. HEWITT

20.1 Introduction

In this chapter, I shall first consider the overall design
problems for a boiler or a condenser, with particular emphasis on
those cases where the fluid is multi-component and does not have a
fixed boiling or condensing temperature. This raises some
interesting and difficult problems in the devising of overall
design approaches from the point of view of economy and accuracy.

Having dealt with the overall approach to design, I will then
discuss a number of aspects of design in two-phase systems which,
in my view, give particular difficulty or are of special interest.
This includes:

(1) Condensation on the shell side of shell and tube heat
     exchangers.

(2) Boiling in cross-flow (as would be found in shell and tube
     boilers).

(3) Flow distribution problems in headers and manifolds.

(4) Two-phase flow and heat transfer in coiled tubes.

(5) Fouling in two-phase flow systems.

(6) Tube vibration problems.

(7) The problem of flooding with particular reference to reflux
     condensers.

20.2 Overall Design Approaches

Figure 20.1 illustrates the temperature profile in heat
transfer from one fluid to another, through a heat transfer surface.
It is usual to define an overall heat transfer coefficient, U,
according to the equation:

$$\frac{1}{U} = R_1 + R_2 + \frac{x}{k_W} + \frac{1}{h_1} + \frac{1}{h_2} \qquad (20.1)$$

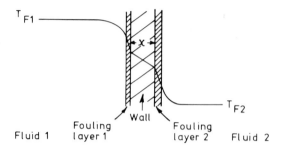

Figure 20.1: Temperature profile for fluid-to-fluid heat transfer through a heat
transfer surface.

where R_1 and R_2 are fouling layer resistances and h_1 and h_2 the
heat transfer coefficients for the respective sides and x and k_W
are the thickness and the thermal conductivity of the wall.

In designing heat exchangers, a number of approaches have been
made in calculating the size of the heat exchanger from a knowledge
of the U values and the temperature differences applicable along
the exchanger, from one fluid to the other. These methods vary in
the extent to which simplifying assumptions are made and I shall
discuss some of the more common approaches.

Logarithmic mean temperature difference (LMTD) approach.

In this approach, the usual assumptions are:

(a)   The value of the overall heat transfer coefficient, U, is
      constant along the exchanger.

(b)   The fluid has a constant temperature or a constant specific
      heat.

(c)   The exchanger is operating in pure counter-current flow
      (though corrections can be made for multi-pass operations).

The LMTD method can be used for the approximate design of
pure fluid boilers and condensers where the fluid on the other
side is single phase and has a constant heat transfer coefficient
and specific heat. Clearly, from what has been said previously in
this course, the boiling or condensing side coefficient will vary
along the exchanger. However, in some practical applications, the
exchanger is governed mainly by the heat transfer resistance on the
single phase flow side and it is of sufficient accuracy to assume
a constant coefficient for the boiling or condensing. The
variation of temperature with length for these cases is illustrated
in Figure 20.2. The overall amount of heat transferred in the
exchanger (Q) is related to the mean value of the overall heat
transfer coefficient ($\bar{U}$) and the logarithmic mean temperature
difference (LMTD) as follows:

$$Q = \bar{U}. \ A. \ LMTD \hspace{4cm} (20.2)$$

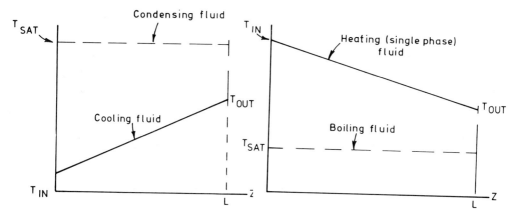

Figure 20.2: Temperature/length profiles for condensing and boiling of pure
fluids with single phase cooling or heating fluids.

where the LMTD is defined for the condensing case as:

$$LMTD = \frac{T_{out} - T_{in}}{\ell n\left[(T_{SAT} - T_{out})/(T_{SAT} - T_{in})\right]}$$

(20.3)

and for the boiling case by:

$$LMTD = \frac{T_{in} - T_{out}}{\ell n\left[(T_{in} - T_{SAT})/(T_{out} - T_{SAT})\right]}$$

(20.4)

where the various temperatures are defined in Figure 20.2 (note
that T_{SAT} is the saturation temperature).

Integral methods for multi-component fluids

In the most general case, a heat exchanger may operate with
multi-component (ie variable condensation and boiling temperature)
fluids with a condensing fluid on one side and the boiling fluid
on the other. This case is illustrated in Figure 20.3; two distinct
forms of behaviour can be identified:

(a) Heat transfer with maintenance of thermodynamic equilibrium
     between the vapour and liquid phases of the respective fluids.
     This case can never actually occur in practice though it is
     sometimes not too inaccurate to assume it.

(b) The vapour and liquid phases of the respective fluids are in
     a state of non-equilibrium as illustrated in Figure 20.3.
     The condensing vapour, for instance, is super-heated and
     the liquid subcooled. Also, the compositions of the vapour
     and liquid are different from their equilibrium value.

Many multi-component systems are so complex that it is not
possible to calculate details of the state of equilibrium of the

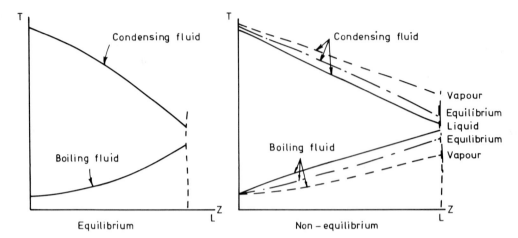

Figure 20.3: Heat exchange between multi-component condensing and boiling fluids.

system. Thus, it is common design practice to assume thermodynamic
equilibrium along the heat exchanger. For this case, and for a
counter-current flow, the area of heat exchange surface required
may be calculated by integration of the equation:

$$dA = \frac{dq}{U\Delta T} \qquad\qquad (20.5)$$

where dq is the amount of heat transferred in area dA where the
operating overall coefficient is U and the operating temperature
driving force is ΔT. Thus, the total exchanger area is given by:

$$A = \int_{0}^{Q} \frac{dq}{U\Delta T} \qquad\qquad (20.6)$$

and the procedure is illustrated schematically in Figure 20.4.

    In multi-pass operation, the situation is obviously more
complicated since the overall heat transfer coefficients and,
indeed, the areas per pass can vary from one pass to another.
However, a simplication can be made for the case where the heat
transfer coefficient for the two passes is identical and where the
pass areas are the same. Since two phase heat transfer depends
on temperature driving force, it may seem strange that this
assumption is made. However, the variation of the shell-side
coefficient from pass to pass, at a given position in the shell,
may be relatively small in comparison with the overall heat
transfer resistances and the assumption is often made. For this
case, the temperature profiles can be expressed in terms of the
shell-side enthalpy as shown in Figure 20.5.

Point-wise calculation methods for heat exchangers

    For many practical applications, the assumptions of equal
areas in passes and of cross-sectionally uniform overall

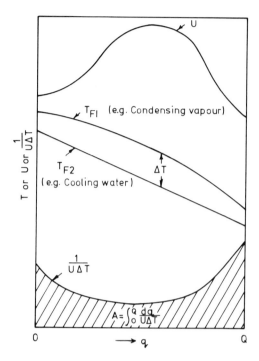

Figure 20.4: Integral design method for single pass, counter-current, equilibrium heat exchange with variable U.

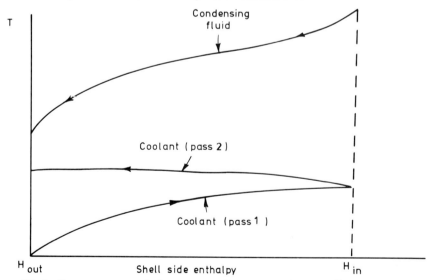

Figure 20.5: Temperature-enthalpy curves for the condensing fluid and the coolant in a two-pass condenser in which U is constant over the cross-section and the pass areas are equal.

coefficients are invalid and it is necessary to adopt a more
"hair-shirted" approach and to calculate point values of coefficients for each pass. We have found it convenient to do such
calculations on a baffle space to baffle space basis calculating
the changes in temperature which occur in each of the passes for
each baffle space. Incidentally, this allows an easy option for
variation of baffle space along the shell to maximise performance
for a given heat exchanger size. The procedure is to guess the
inter-pass temperatures at one end of the exchanger and to carry
out step-wise calculations along the shell. The inter-pass
temperatures at the other end of the shell must be identical for
each pass (obviously) and an iteration is carried out using new
values at the starting end until a match is obtained. This principle
is illustrated in Figure 20.6. Point calculation methods are
certainly very useful in shell and tube condenser design but they
are probably too expensive for every-day design optimisation where
literally thousands of options have to be investigated. A combination of the simpler, integral, methods with the point methods
seems to offer the best solution.

20.3   Aspects of Shell-and-Tube Condenser Design

Shell side flow patterns

Two-phase flow patterns in cross-flow over tube bundles have
been investigated by Grant and Murray (1972) (1974) and the types
of flow which occur are illustrated in Figure 20.7. As will be
seen, the flow patterns have some similarity to those found in
single tubes, although, naturally, the detailed interactions of the
phases with the tube bundle are highly complex.

In shell and tube exchangers, additional effects occur. For

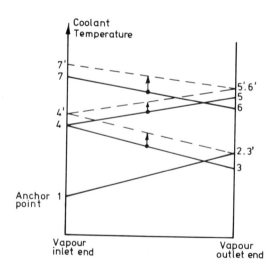

Figure 20.6: Point-wise iterative calculation of pass temperatures in a multi-
      pass shell-and-tube condenser.

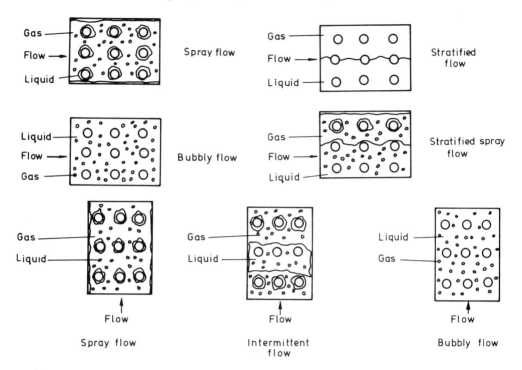

Figure 20.7: Two-phase flow patterns in cross-flow over a tube bundle.

instance, in two phase flow around a baffle, a gas "bubble" tends to form on the down-stream side of the baffle.

Effects of inundation and cross flow on heat transfer coefficient

Figure 20.8 shows the idealised and real situations occurring in multiple tube condensation. The left-hand sketch in Figure 20.8 represents the original Nusselt concept for which:

$$\frac{\bar{h}_N}{h_1} = N^{-\frac{1}{4}}$$
(20.7)

where $\bar{h}_N$ is the mean heat transfer coefficient for a vertical row of N tubes and h_1 is the coefficient (calculated from the Nusselt equation) for the first tube in the row. In practice, the heat transfer coefficient decreases much less rapidly with the number of tubes in the row and an equation of the form:

$$\frac{\bar{h}_N}{h_1} = N^{-1}/6$$
(20.8)

is more appropriate (as has already been pointed out in Chapter 11, pp. 346-347). In practice, condensers operate with a finite cross-flow velocity and this completely changes the effects of inundation. In many condensers, the vapour flows horizontally at right angles

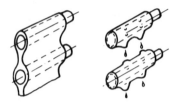

Figure 20.8: Idealised and actual flow in condensing over a vertical row of
            tubes.

to the gravitational flow of the condensate. As might be expected,
the behaviour is extremely complex and has been the object of
experimental studies at the Winfrith (UK) and Harwell (UK)
laboratories, and elsewhere. Figure 20.9 shows some of the results
obtained for steam-water flows. At low approach velocities, the
condensation coefficient is enhanced but, for high velocities,
there is a reduction in heat transfer coefficient with increasing
approach velocity. This surprising result is probably accounted
for by entrapment of liquid on the leeward side of the tube and a
also by pressure variations around the tube. The results are not
unique to water and have been repeated with a number or organic
fluids.

Liquid subcooling layers in shell-and-tube condensers

     Many condensers in the process industry are designed for
simultaneous condensing and liquid subcooling duty. This is

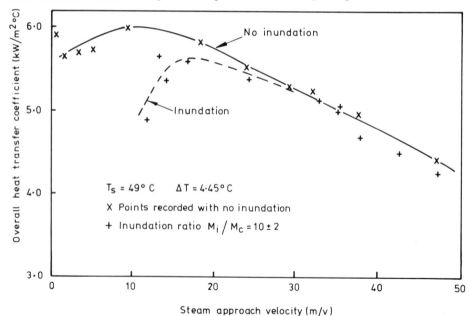

Figure 20.9: Effect of vapour approach velocity on condensation coefficient in
            horizontal cross-flow over a tube bundle.

achieved by adjusting the liquid level at the shell outlet and a simplified approach is to assume that this level is constant throughout the shell. However, because the flow of condensate is increasing along the shell and because the pressure is changing, the level may be by no means constant. In Figure 20.10, I have shown some calculations done using a point-wise method built into an HTFS computer program. As will be seen, the level varies considerably along the shell and the introduction of weirs into the baffles does not eliminate this variation. In general, it is probably preferable to carry out the subcooling in a separate shell, though one recognises that this is sometimes extremely inconvenient.

Fogging effects in condensers

Fog formation in condensers occurs when the temperature of the vapour falls sufficiently below the saturation temperature to allow the formation of small ("fog") droplets. Consider the situation illustrated in Figure 20.11 for the condensation of a vapour from a mixture of the vapour and incondensable gas. Condition G represents the bulk condition of the mixture and condition S represents the interface between the mixture and the condensate, with line SC representing the saturation condition. Note that point G corresponds to a slight superheat. The temperature gradient in the region of the vapour/gas phase immediately adjacent to the interface may be of the form GJS or GKS. In the former case, the vapour will be superheated right up to the interface, but in the latter case the vapour temperature is below the saturation line and fog formation can take place. Calculations show that the effect of fog formation is not too great on the calculation of the performance of a condenser, though the presence of a considerable amount of fog in the outlet vapour/gas mixture can be troublesome in operation.

20.4 Boiling in Cross Flow

Cross flow boiling occurs in a variety of situations, including kettle and horizontal thermosyphon reboilers. Here, I

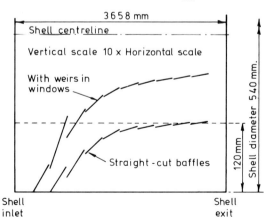

Figure 20.10: Calculated variation of liquid level in a shell-and-tube condenser with a liquid pool for subcooling.

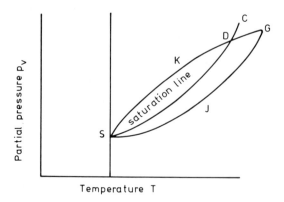

Figure 20.11: Temperature gradients adjacent to the interface in condensation;
            formation of fog.

shall consider briefly two aspects, namely dryout and distribution
of heat transfer coefficeint.

Dryout in cross flow boiling

In cross-flow boiling, a variety of situations can occur.   To
explain these, it is simpler to consider a vertical tube as an
analogy to the channel between two adjacent vertical rows of tubes
in cross flow.   The various types of dryout for this analogy are
illustrated in Figure 20.12.   At one extreme, the liquid in the
tube forms a static pool and the dryout (or critical heat flux)
condition occurs in the normal vapour-blanketing mode.   In the
second example (b) there is again zero circulation but the rate
at which liquid may enter the tube is limited by the "flooding"
phenomenon which restricts the rate at which liquid can flow down
the tube in counterflow to the upwards vapour flux (I shall return
to this phenomenon in the context of reflux condensers below).

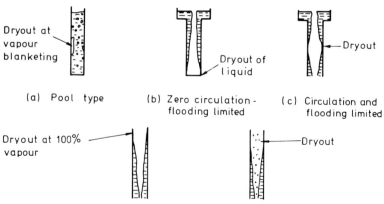

Figure 20.12: Illustration of dryout mechanisms cross flow boiling by analogy
            to the case of the single tube.

In case (c) the lower part of the tube is wetted by a limited
upward circulating flow and the upper part by a counter-current
down flow (which is limited by flooding). Dryout occurs at an
intermediate point as illustrated. As the circulation rate is
increased, then more of the tube may be wetted and further limiting
point is where all of the circulating liquid entering the bottom
of the tube is evaporated by the exit (d). Finally, it is possible
to entrain some of the circulating liquid in the form of droplets
and this particular form of dryout is discussed in detail in
Chapter 9.

Harwell and NEL have been conducting a series of experiments
to investigate dryout and flow mechanisms in cross flow boiling.
These have been done with vertical "slices" both of idealised
cross-flow and of slices representing the reboiler geometry. Some
of the results are illustrated in Figure 20.13 from which it will
be seen that the dryout power for an ideal cross flow bundle
increases with the number of heated rows and with the mass flux,
but that the dryout power per row continuously decreases as the
number of heated rows is increased.

Heat transfer coefficients in kettle reboilers

In designing kettle reboilers, it is often assumed that the
heat transfer coefficient is uniform over the cross section of the
reboiler. Recent experiments carried out for HTFS at Heriot Watt
University show that very considerable variations occur with the
heat transfer coefficient increasing with bundle height. The
situation is rather complex as illustrated in Figure 20.14.

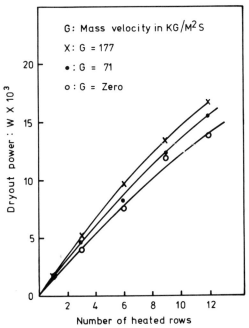

Figure 20.13: Dryout power data for the evaporation of Freon 113 in an idealised
       tube bundle.

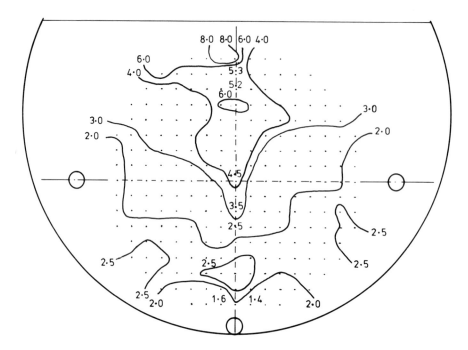

Figure 20.14: Contours of heat transfer coefficient in evaporation in a kettle
reboiler.

Much more needs to be known about the detailed phenomena in
cross flow boiling and this seems to be a rich area for further
research.

20.5 Flow Distribution in Headers and Manifolds

In practical applications of two phase flow, it is necessary
to split two phase flow streams into one or more parts. A
typical case here would be the second pass distribution system in
an air-cooled heat exchanger. Typically, the distributor might
consist of a series of T-junctions and, obviously, it is helpful to
consider first the case of two phase flow division at a single T-
junction. Results obtained for this case are illustrated in
Figure 20.15 and it will be seen that, usually, there is a strong
tendency for the liquid phase to pass along the header, with the
gas phase separating preferentially at the junction.

Even in single phase flow, there is a strong possibility of
maldistribution in multi-tube systems. This point is illustrated
in Figure 20.16. Intuitively, it might have been expected that
the distribution between the four tubes would have been better for
the case where the flow in inlet and outlet headers is in the same
direction (asymmetric flow). However, a considerable maldistribution can occur as shown in Figure 20.16. The reason for this is
that there is pressure recovery in the inlet header, at each

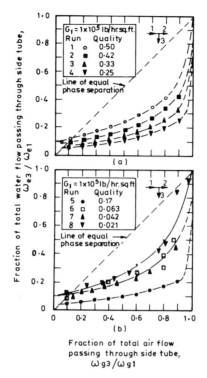

Figure 20.15: Flow separation at a single T-junction in air-water flow (plane of junction- horizontal).

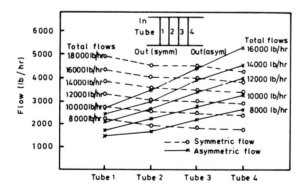

Figure 20.16: Single phase flow distribution between four tubes with symmetric and asymmetric flow in the header tubes.

junction, due to the change in flow direction. Pressure builds up along the header. Similarly, in the oulet header the pressure falls as the entering flow is accelerated thus, the fourth tube has the highest pressure in the inlet header and the lowest pressure

in the outlet header and, thus, carries the greatest flow, contrary
to what might have intuitively been expected.

For two-phase flow in the same multi-tube system, the
separation effects mentioned above for single T-junctions lead to
a preferential separation of the gas flow. However, the liquid
must come out somewhere and, indeed, the last tube carries a very
high liquid flow as might be expected. Figure 20.7 shows typical
results obtained with this system.

There is no satisfactory general method for the design of
header and manifold systems in two-phase flow and my recommendation
is that two-phase flow division should be avoided wherever
possible.

20.6 **Flow in Coiled Tubes**

In many industrial applications of two-phase, the flow takes
place in a coiled tube. Coils are used in a number of forms of
package boiler and are also used in cryogenic evaporation plant and
in fired heaters.

The flow in coiled tubes is somewhat complex. Intuitively,
it might be expected that liquid phase would always flow on the
outside of the bend but, because of slip between the two phases,
the centrifugal force on the gas phase can, in fact, be higher than
that on the liquid phase in some circumstances. This leads to
flow of the liquid phase on the inside of the bend. In a typical
situation, droplets are torn from this inner film, are accelerated
and deposited on the outside of the bend and then returned in a
liquid film to the inside of the bend. There is thus a very
efficient wetting process and this tends to inhibit the dryout
(burnout) phenomenon. A comparison of data for dryout in coiled
and straight tubes respectively is given in Figure 20.18. As
will be seen, there is a very large increase in the quality at
dryout for a given heat flux level, in the case of the coil. Note

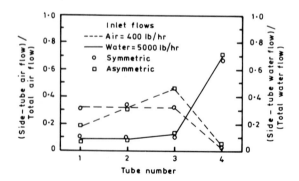

Figure 20.17: Two-phase flow distribution between four tubes illustrating
              liquid/gas separation effects.

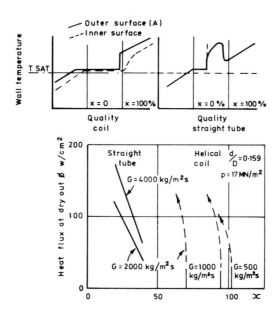

Figure 20.18: Dryout data for high pressure steam-water flow in a coiled tube (data obtained at AEEW, Winfrith).

that the temperature increase associated with dryout occurs first on the outer surface of the coil.

Confirmation that there is a higher film flow on the inside of the bend in a coil has been obtained in air-water experiments by Dr Whalley at Harwell. Some of his results are illustrated in Figure 20.19.

20.7 Fouling in Boiling Systems

A key consideration in all heat transfer situations is the formation of fouling layers on the heat transfer surfaces. Often, fouling resistances are dominant in the heat transfer process. An important feature in boiling systems is that resistances obtained in boiling may be very much lower than those obtained in single phase flow for the same fouling layer. The reason for this is the so-called "wick boiling" mechanism investigated, for instance, by Macbeth (1971). A close examination of the fouling layers formed on boiling surfaces shows that they often have small "chimneys" in them through which the vapour escapes. Liquid enters the region around the chimney by capillary wicking through the adjacent porous deposit. This is a very efficient mechanism of boiling and the surface temperature differences are relatively small. However, if trace quantities of a dissolved salt are present in the liquid, then this salt is deposited in the pororus fouling layer and prevents seepage to the chimneys. The temperature of the surface then increases rapidly due to the failure of the wick boiling mechanism. Results illustrating this effect are shown in Figure 20.20.

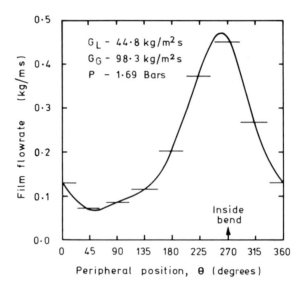

Figure 20.19: Data obtained by Dr P B Whalley, Harwell for film flow rate
distribution as a function of peripheral position in a coiled
tube.

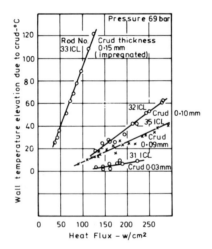

Figure 20.20: Wall temperature elevation as a function of heat flux for salt-
impregnated and non-impregnated fouling layers.

One of the most important influences of fouling layers in heat
exchanger design is that on the pressure drop in the system. The
fouling layer is usually rough and increases the pressure drop.
However, the effect of surface roughness on pressure drop is often
much less in two-phase flow than it is in single phase flow. Thus,
the pressure drop multiplier is often smaller for rough channels
than for smooth channels. This phenomenon was discussed by Shires
(1972) and Figure 20.21 illustrates the sort of effect that can

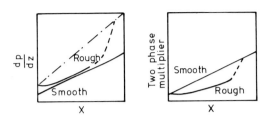

Figure 20.21: Effect of surface roughness on two-phase pressure drop and two-phase friction multiplier.

occur. The reason for the smaller influence of roughness in two-phase flow is that the pressure drop is dominated (as we have seen earlier in Chapter 5) by the interfacial roughness and, thus, the effect of the surface roughness is not so important.

Tube Vibration in Two Phase Flow Systems

One of the most important problems in heat exchangers nowadays is that of tube vibration. As design methods have improved, there has been a tendency to move to higher fluid velocities in heat exchangers and this has led to an increasing number of vibration problems. The mechanisms of tube vibration include:

(1) Vortex shedding: Here, as the flow passes over the tubes, vortices are shed on the downstream side with a resulting oscillating pressure force on the tube. For single tubes, this force is 180° out of phase between the top and the bottom of the tube and this oscillating force leads to vibration if the vortex shedding frequency is identical to the tube natural frequency. In tube bundles, the picture is more complex and it seems possible that, in a number of situations, the pressure oscillations at the top and bottom of the tube are in phase rather than out of phase. Thus, for the pitch-to-diameter ratios normally encountered in heat exchangers, the vortex shedding mechanism is unlikely to be a dominant one.

(2) Turbulent buffeting: The flow within the interstices of a heat exchanger tube bundle is highly complex and turbulence levels are very high. If the dominant frequency of this turbulence is similar to the natural frequency of the tubes, then vibration can occur.

(3) Fluid-elastic instability: This type of excitation mechanism is probably the most important in normal heat exchangers and occurs as a result of changes in fluid flow patterns as a result of tube movements in vibration. This type of vibration mechanism is often calculated using the equation of Connors (1970):

$$V_g/fd = K\sqrt{m\delta/\rho d^2} \tag{20.9}$$

where V_g is the velocity in the gaps between the tubes, f the natural frequency, d the tube diameter, m the virtual mass per unit length of the tube, δ the logrithmic decrement for damping, ρ the fluid density and K a constant.

Very little information exists on tube vibration in two-phase flow systems. Some limited experiments are reported by Pettigrew and Gorman (1977) which seem to indicate that the Connors equation can be used to predict two-phase flow systems using the homogeneous model for calculating V_g and ρ.

20.8   Flooding

The flooding phenomenon is important, for instance, in tubular reflux condensers where the condensate flows down the tubes, counter-current to the upward rising vapour. At a certain vapour velocity (the "flooding velocity") the counter-current flow regime ceases and the liquid phase starts to be carried upwards through the outlet of the tube. A variety of correlations has been used to predict the flooding transition, for example that of Wallis (1961) who expresses the flooding condition in terms of dimensionless gas and liquid velocities as follows:

$$V_G^{*\frac{1}{2}} + V_L^{*\frac{1}{2}} = C \qquad\qquad (20.10)$$

where C is a constant having a value typically in the range 0.75-1 and V_G^* and V_L^* are defined as follows:

$$V_G^* = V_G \rho_G^{\frac{1}{2}} \left[g d_o (\rho_L - \rho_G) \right]^{-\frac{1}{2}} \qquad\qquad (20.11)$$

$$V_L^* = V_L \rho_L^{\frac{1}{2}} \left[g d_o (\rho_L - \rho_G) \right]^{-\frac{1}{2}} \qquad\qquad (20.12)$$

where ρ_G and ρ_L are the gas and liquid densities, g the acceleration due to gravity and d_o the tube diameter.

A review of flooding problems is given, for instance, by Hewitt and Hall-Taylor (1970) but I would like to make a few points here on design:

(1)   Effect of length.   Earlier data at Harwell tended to show a significant effect of length. This data was obtained in adiabatic air-water flow in which the water was injected and removed smoothly from the tube surface. More recent tests at Harwell, in which flooding was studied in a sharp-ended tube entering an exit volume into which the gas was fed, have shown that the effect of the tube length was small in this case.

(2)   Effect of cutting of tube end. The flooding velocity can be considerably increased by cutting the tube end at an angle.

(3)   The flooding rate varies with the angle of inclination of the tube. Typically, the highest flooding rate is obtained

for a tube inclination of the order of 60° from the horizontal.

More data is required on flooding, particularly on the effect of tube diameter and fluid physical properties. The flooding rate is decreased as surface tension decreases and the effect of liquid viscosity is somewhat complex.

Nomenclature

A	– Total surface area of exchanger	(m^2)
dA	– Increment of surface area corresponding to increment dq of heat transferred	(m^2)
d	– Tube outside diameter	(m)
d_o	– Tube inside diameter	(m)
f	– Tube natural frequency	$(^1/s)$
g	– Acceleration due to gravity	(m/s^2)
h_1	– Heat transfer coefficient on one side of exchanger	(J/m^2sK)
	Heat transfer coefficient on first (uppermost) tube in vertical row	(J/m^2sK)
h_2	– Heat transfer coefficient on one side of exchanger	(J/m^2sK)
$\bar{h}_N$	– mean heat transfer coefficient for a vertical row of N tubes	(J/m^2sK)
k_W	– Thermal conductivity of wall material	(J/msK)
K	– Constant in Connors equation (equation 20.9)	$(-)$
LMTD	– Logarithmic mean temperature difference (equations 20.3 and 20.4)	(K)
m	– Virtual mass of tube per unit length	(kg/m)
N	– Number of tubes in vertical row	$(-)$
dq	– Incremental amount of heat transferred in area dA	(J)
Q	– Total heat transferred in exchanger	(J)
R_1, R_2	– Fouling resistances on respective sides of heat exchanger	(m^2sK/J)
T_{in}	– Inlet temperature of heating or cooling fluid	(K)
T_{out}	– Outlet temperature of heating or cooling fluid	(K)
T_{SAT}	– Saturation temperature	(K)
U	– Local overall heat transfer coefficient	(J/m^2sK)
$\bar{U}$	– Mean overall coefficient for exchanger	(J/m^2sK)
V_g	– Maximum velocity between tube gaps	(m/s)
V_L	– Superficial velocity of liquid	(m/s)
V_G	– Superficial velocity of gas	(m/s)
V_L^*	– Dimensionless superficial liquid velocity (equation 20.11)	$(-)$

V_G^* - Dimensionless superficial gas velocity   (-)
          (equation 14.12).

x   - Wall thickness   (m)

Greek

δ - Logarithmic decrement $(^1/m)$
ρ - Fluid density (kg/m^3)
ρ_L - Liquid density (kg/m^3)
ρ_G - Gas density (kg/m^3)

References

Connors, H J, (1970) "Fluid elastic vibration of tube arrays excited by cross-flow". *Paper presented at winter annual meeting, ASME, New York.*

Grant, I D R and Murray, I, (1972) "Pressure drop on the shell-side of segmentally baffled shell and tube heat exchanger with vertical two-phase flow". *NEL-500.*

Grant, I D R and Murray, I, (1974) "Pressure drop on the shell-side of a segmentally baffled shell and tube heat exchanger with horizontal two phase flow". *NEL-560.*

Hewitt, G F and Hall-Taylor, N S, (1970) "Annular two phase flow". *Pergamon Press*, London.

Macbeth, R V, (1971) "An investigation into the effect of "crud" deposits on surface temperature, dryout and pressure drop, with forced convection boiling of water at 69 bar in an annular test section". *AEEW-R705.*

Pettigrew, M J and Gorman, D J, (1977) "Experimental studies on flow-induced vibration to support steam generator design. Part III: Vibration of small tube bundles in liquid and two-phase cross-flow". *AECL-R5804.*

Wallis, G B, (1961) "Flooding velocities for air and water in vertical tubes". *AEEW-R123.*

Chapter 21

Industrial Applications
of Heat Transfer Augmentation

A. E. BERGLES

21.1 Introduction

The material presented in Chapter 12 indicates that many
techniques have been shown in laboratory tests to have considerable
potential for augmenting heat transfer in practical situations.
A number of the techniques have made the transition to full-scale
industrial heat exchange equipment. Many factors enter into the
decision to use an augmentation technique, either in the original
equipment or as a retrofit. In addition to thermal-hydraulic
performance, the operational aspects and cost must be considered.
Since these factors are numerous and sometimes difficult to quanti-
tize, the designer or field engineer must make a decision based
on the particular system and requirements. This chapter reviews
considerations pertinent to the applications of augmentation in
the power and process industries.

21.2 Available Techniques

The techniques which are available have been described in
Chapter 12 and previous survey papers. For convenience, the world
technical literature (excluding patents and manufacturer's literature) represented in Figure 12.1 has been organized according to
augmentation technique and mode of heat transfer. The distribution
of the citations is shown in Table 21.1. Over one-third of the
studies cited have application to boiling or condensing. Many of
the other studies are, however, still of interest to the present
discussion as augmentation of the single-phase side of a two-fluid,
phase-change heat exchanger will often be considered. Actual
citations can be obtained from the report by Bergles et al (1979a)
or the series by Bergles and Webb (1978). Updating of this
bibliography, which is stored in a computerized information
retrieval system, continues in the Iowa State University Heat
Transfer Laboratory.

The patent literature is also important because granted
patents are generally indicative of industrial interest in commercialization of a technique. Patents can be used to identify
variations of techniques now in commercial use or available for
commercial use, to obtain descriptions of major manufacturing
methods for current applications and to discover possible applications of techniques not now commercially applied. It must be

A E Bergles

Table 21.1

Classification of augmentation techniques.
(Bergles et al (1979a))

	Single-Phase Natural Convection	Single-Phase Forced Convection	Pool Boiling	Forced-Convection Boiling	Condensation	Mass Transfer
Passive Techniques (No External Power Required)						
Treated Surfaces	-	-	73	12	32	-
Rough Surfaces	4	282	21	43	37	22
Extended Surfaces	*	269	34	33	87	15
Displaced Enhancement Devices	-	36	1	9	3	5
Swirl Flow Devices	-	89	-	67	9	5
Coiled Tubes	-	118	-	44	3	5
Surface Tension Devices	-	-	11	1	-	-
Additives for Liquids	3	10	43	23	-	3
Additives for Gases	-	155	-	-	5	2
Active Techniques (External Power Required)						
Mechanical Aids	4	34	19	1	17	10
Surface Vibration	44	21	9	2	6	9
Fluid Vibration	40	89	15	4	1	32
Electric or Magnetic Fields	37	30	23	7	13	8
Injection or Suction	3	17	7	+	6	2
Compound Enhancement (Two or More Techniques)	+	32	2	4	4	2

- Not applicable   * Not considered in this survey   + No citations located

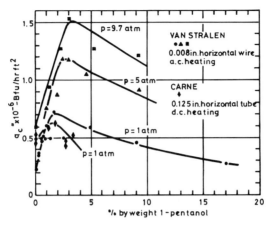

Figure 21.1: Effect of addition of 1-pentanol on CHF for saturated pool boiling
of water (Bergles (1969)).

recognized, however, that most patents stress legal rather than
technical aspects. Of the approximately 500 patents now being
reviewed at Iowa State University and Pennsylvania State
University (Bergles et al (1979b)), only about 25% disclose
actual heat transfer data.

    Product information supplied by manufacturers is direct
evidence of commercialization of augmentation techniques. The
situation is very dynamic, as new products are continually being
introduced while others are being discontinued. The major international marketing activity is observed in Germany, Japan, Sweden,
United Kingdom, and the United States. Several hundred manufacturers are involved with products ranging from tube inserts to
complete heat exchange systems which include augmented heat
exchangers.

21.3 Suitability of Basic Performance Data

    Caution must always be exercised when extracting heat
transfer and pressure drop data from the published literature and
applying these data to a different set of conditions. For
instance, the performance of special boiling surfaces for a given
fluid and pressure level can be very sensitive to geometry, eg
groove pitch and pore diameter in the case of Thermoexcel-E (Arai
et al (1977)). The performance and optimum geometry are expected
to be different as the fluid and operating pressure change.
The situation can be even more serious than this, as illustrated
by an example in the category of additives.

    As shown in Figure 21.1, large increases in critical heat
flux (CHF) can be realized when a small percentage of 1-pentanol
is added to water. However, when the data for the two studies
with heaters of different diameter are compared, it is evident
that the increase in CHF is much less with the larger heater.
Furthermore, even with small-diameter heaters, a substantial
decrease in CHF is observed in subcooled pool boiling and in CHF

for forced convection subcooled boiling, except at very high subcooling (Bergles (1969)).

Another concern is the question of scale-up. The example
given above is one illustration; but more typically data will be
obtained for boiling or condensing outside single horizontal tubes
of the correct diameter with the intent of applying the data to a
large bundle of tubes. Katz et al (1955) and Nakajima and Shiozawa
(1975), for example, found that boiling coefficients on the upper
finned tubes in a bundle are higher than coefficients on the lower
tubes, due to bubble-enhanced circulation. Similar results are
reported by Arai et al (1977) for a bundle of Thermoexcel tubes.
It is quite probable that with certain types of augmented tubes
it is sufficient to use the special tubes only in the lower rows
since the augmented circulation in the upper rows is so high that
augmentation is not effective there. In condensers, there may be
a decrease in enhancement in the lower rows due to buildup of
condensate (Withers and Young (1971)).

Scale-up problems also occur when applying electric fields,
vibrations, and sound fields or pulsations. It is usually not too
difficult to influence boiling or condensing from small test
sections; however, industrial size heat exchangers with large
surface areas and flow path lengths are less tractable.

Finally, one should not be hesitant to critically examine
the accuracy of the experimental data. Phase-change heat transfer
processes are low thermal resistance processes in general; hence,
wall temperatures and fluid (or saturation) temperatures must be
determined with precision if the temperature difference or the
heat transfer coefficient is to be accurately established. This
situation is compounded when the higher coefficients of augmented
systems are considered. The boiling curves for most special
boiling surfaces (see, for example, Figure 12.2) have wall superheats of the order of $1^\circ C$. It is difficult to imagine that such
low ΔT_S's can be accurately established when the usual uncertainties in measuring the wall temperature and the saturation temperature (pressure) are considered.

21.4 Thermal-Hydraulic Performance Evaluations

For two-fluid heat exchangers, the ratio of thermal resistances
between the two fluid streams is of primary importance in determining whether augmentation will be of benefit. Augmentation should
be considered for the stream which has the larger thermal resistance. If the thermal resistances of both streams are approximately
equal, augmentation of both streams may be considered.

Methodologies have been suggested for assessing the thermal-
hydraulic performance of augmented surfaces (Webb and Eckert (1972),
Bergles et al (1972, 1974)); however, the examples considered
involved single-phase flows. The extension of these criteria to
two-phase heat transfer is complicated by the dependence of the
heat transfer coefficient on the local temperature difference and/
or the quality. Royal and Bergles (1978) and Luu and Bergles
(1979) have evaluated in-tube condensation by two parameters:

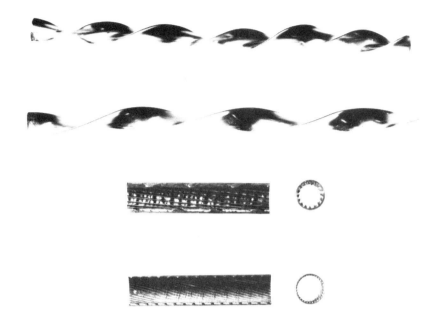

Figure 21.2: Twisted-tape inserts and inner-fin tubes used to augment in-tube
condensation. Top to bottom: Tubes 2,3,4,5. For reference,
bottom tube is 15.88 mm o.d. (Luu and Bergles (1979)).

$$R_h = \left| \frac{A_a}{A_o} \right|_{q,w} = \left| \frac{\bar{h}_o}{\bar{h}_a} \right|_{q,w} \tag{21.1}$$

$$R_{\Delta p} = \left| \frac{\Delta p_a}{\Delta p_o} \right|_{q,w} \tag{21.2}$$

The first parameter represents the area (length) of a condenser
relative to the smooth tube (same nominal diameter) unit for equal
heat duty, flow rate, and similar temperature or pressure level.
The ratio is given simply by the ratio of the heat transfer
coefficients, assuming, of course, that the condensing side
resistance controls heat transfer. The second parameter represents
the pressure drop of the augmented condenser relative to the
smooth tube unit, subject to the same constraints. In calculating
this ratio, the various components of pressure drop must be
considered. The results from this evaluation of data for the tubes
with twisted-tape inserts and internally finned tubes (Figure 21.1)
are shown in Figures 21.3 and 21.4. This analysis suggests that
Tube 5 has a clear advantage over the other tubes.

While the preceding type of assessment is useful for screening

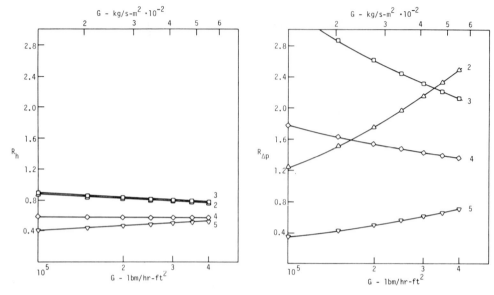

Figure 21.3: Condenser size reduction index vs. mass velocity, P_{in} = 6.55 bar (Luu and Bergles (1979)).

Figure 21.4: Condenser pressure drop index vs. mass velocity, P_{in} = 6.55 bar (Luu and Bergles (1979)).

surfaces, in general, the thermal resistances of both sides of a two-fluid heat exchanger must be accounted for. Consider the example of the use of augmented tubes in an Ocean Thermal Energy Conversion (OTEC) evaporator. As described by Bergles and Jensen (1977) the parameter R_5 from Bergles et al (1972) is useful in this particular application:

$$R_5 = \left|\frac{A_a}{A_o}\right|_{q,P} = \left|\frac{\bar{h}_o}{\bar{h}_a}\right|_{q,P} \qquad (21.3)$$

This parameter denotes the ratio of the area of the augmented heat exchanger to that of the normal heat exchanger (same nominal diameter tubes), subject to the constraints of constant heat duty and pumping power (sea water side only). As shown in Figure 21.5, large reduction in heat transfer area are possible when the inside surface of the tube is modified to increase h for the single-phase sea water (dashed curves). Further reductions in area result from augmentation of the evaporating ammonia (assumed here to be a 200% increase in h). The area reduction depends on the number of tubes (or frontal area); the length decreases faster than the frontal area increases.

Several measures of performance have been proposed for sub-cooled boiling CHF. Gambill et al (1961) reported that the CHF's for twisted-tape inserts (Figure 12.4) were approximately twice those for straight flow at the same test-section pumping power.

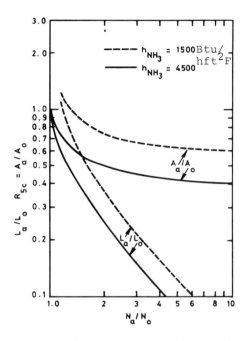

Figure 21.5: Performance evaluation of an OTEC heat exchanger with repeated-rib
roughness on the interior (sea-water) and a 200% evaporating
ammonia augmentation (Bergles and Jensen (1977)).

Megerlin et al (1974) suggested, in view of the fact that the
cooling passage volume is important when cooling high power electronic devices, that the maximum volumetric heat transfer rate be
compared with the pumping power per unit volume.

It must be pointed out that these thermal-hydraulic comparisons consider only the basic heat exchanger. A full analysis of
the effect of the augmentation may require consideration of the
entire system if flow rates or temperature levels change as a
result of the augmentation.

21.5 Manufacturing Considerations and Codes

There are several important questions which must be raised
when the thermal-hydraulic analysis is favorable and the decision
is made to use augmentation in a heat exchanger. In most cases,
the heat exchanger materials are specified for the application;
however, the augmentation technique (tube configuration, insert,
etc.) may be limited to the use of a particular material in its
manufacture (Gilbert (1975)).

Codes and standards must be checked to see if use of the
augmentation technique is allowed. For instance, certain fabrica-
tion procedures which deform the tube wall reduce the strength
below the minimum specified by the pressure vessel code. The
bending strength may also be affected, as in the case of some

spirally fluted tubes; this might necessitate extra support plates
to prevent vibration and fretting corrosion. Differences in
national standards also complicate production. It appears to be
cheaper, for example, to purchase tubing in Germany, ship it to
the U.S. where a special boiling surface is applied, and return
the tubes to Germany rather than fabricate the tubing in the U.S.
to German standards.

By modifying the rod-spacer configuration or the shroud, the
CHF can be increased for nuclear fuel element rod clusters (Tong
et al (1967), Ryabov et al (1977)). However, strict nuclear codes
usually prohibit use of such devices in actual reactors.

21.6 Operational Considerations

21.6.1 *Operating History*

Boiling performance is occasionally quite sensitive to
operating history. This was demosntrated in the pioneering experiments of Jakob and Fritz (1931) shown in Figure 7.12. The
mechanically roughened surface ("roughness screen") exhibited
increasingly higher wall superheats as the surface was boiled and
soaked. The spread in wall superheat is greater than for a surface
with roughness typical of normal surface finishing ("sand blast").

Large temperature overshoots which lead to boiling curve
"hysteresis" are observed with highly wetting fluids, as noted in
Chapter 7. This problem may be reduced with special boiling
surfaces, due to increased probability of doubly re-entrant
cavities; however, reports of hysteresis with special surfaces are
beginning to appear. Lewis and Sather (1978) reported erratic
heat transfer behavior in a 279-tube, "High Flux" flooded evaporator intended for OTEC applications. This was attributed to
deactivation of nucleation sites by the flooding of the sites with
liquid ammonia. The lowest value of the overall heat transfer
coefficient, U_O, was estimated at 600 Btu/hr ft^2 F (3407 W/m^{2}K).
The excellent heat transfer behavior of the High Flux surface was
restored by drying out the bundle prior to filling and initiation
of heating. After "soaking" in liquid ammonia, a five-day run was
initiated to determine the reactivation period. As shown in Figure
21.6, U_O increased slowly with time and reached a steady value of
785 Btu/hr ft^2 F after about 100 hours of operation.

Subsequent tests at Argonne National Laboratory were run with
a sprayed bundle, Figure 21.7, (Hillis et al (1979)) and disclosed
similar deactivation/activation behavior. However it was also
found that the heat transfer coefficient can be decreased by dryout
if done when the surface is already fully activated, as shown in
Figure 21.8. This was attributed to difficulty in rewetting the
dry heated surface by the thin falling film of ammonia. Brief
mention of hysteresis effects with the Hitachi Thermoexcel tubing
is given by Torii et al (1978).

The above-mentioned studies of the Linde surface are for
multiple tubes where the boiling heat transfer coefficients must
be derived from the overall coefficients. A fundamental study of
single-tube behavior, with the flooded-tube boiling coefficients

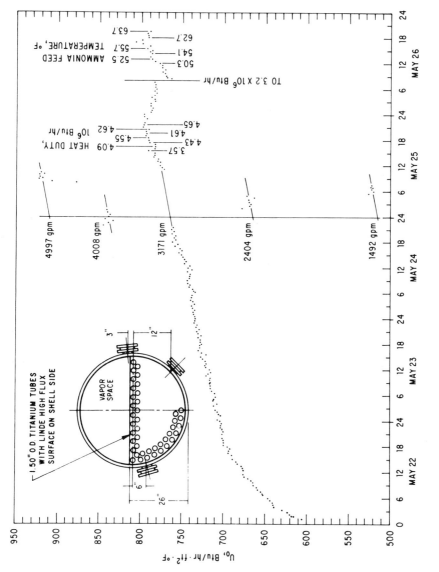

Figure 21.6: Variation of overall heat transfer coefficient for High Flux flooded-bundle evaporator with time (Lewis and Sather (1978)).

A E Bergles

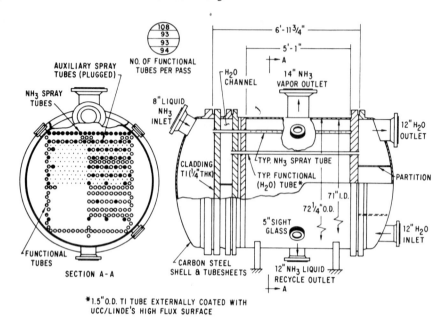

Figure 21.7: Diagram of Union Carbide sprayed-bundle evaporator (Hillis et al (1979)).

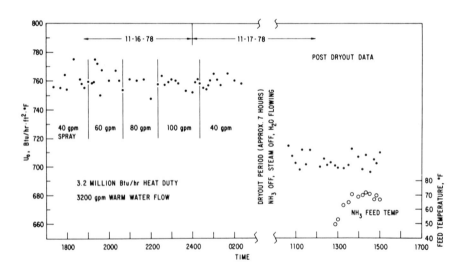

Figure 21.8: Variation of overall heat transfer coefficients for High Flux sprayed bundle evaporator with time and procedure (Hillis et al (1979)).

obtained directly, was carried out by Chyu (1979). As shown in
Figures 21.9 and 21.10, the temperature overshoots with the High
Flux surface area are quite dramatic. Explanations are offered as to
the effect of surface ageing (preboiling or drying), pool subcooling at start of test run, and rate of heat flux change on the
boiling curves. These data confirm that special operating procedures are necessary to insure stable, low ΔT boiling of special
boiling surfaces.

21.6.2 *Fouling and Corrosion*

In general, special boiling surfaces should be used with
clean fluids, as concentrations of impurities or corrosion products
could plug up the small pores. Few data are available; however,
laboratory experiments, eg Chyu (1979), suggest that there is no
difficulty in boiling distilled water for extended periods with
High Flux surfaces. On the other hand, the rather grim descriptions
of boiler "crud" presented by Macbeth (1977) suggest that many
enhancement techniques will not be effective for long in industrial
boiler environments. It is interesting to note, though, that some
investigators have reported enhanced heat transfer with porous
magnitite deposits, apparently due to increased nucleation sites
and "wick boiling".

Special boiling surfaces have potential wide application in
refrigeration. Here, the fluid is inherently very clean; however,
oil is frequently present. Arai et al (1977) report a rather
substantial reduction in the boiling coefficient with addition of
lubricating oil (up to 3.4% by weight) to R-12 if there is no
foaming. However, if the tubes are immersed in the foaming region,
the nearly-oil-free coefficient is restored.

Care must be taken that the augmentation device itself does
not lead to accelerated fouling or corrosion. For instance, twisted-
tape inserts could promote deposits at the tube-tape junction. To
avoid such depsoits, Pai and Pasint (1965) advocated a gap between
the wall and the tape used to improve a once-through steam generator.

Certain augmentation techniques may be considered under fouling conditions if cleaning is possible. This requirement may
restrict the type of enhancement device. For example, tube inserts
or large surface extensions would preclude mechanical cleaning of
evaporator tubes.

21.6.3 *Reliability*

If an augmented exchanger is to be successful, the augmentation must perform reliably. The active techniques are particularly
suspect from this viewpoint as loss of auxiliary power or failure
of a piece of equipment could wipe out the augmentation. For
instance, the loss of power, failure of the field generator, or
breakdown of the fluid would obviate the beneficial effects of
electrohydrodynamic augmentation.

The possibility of accelerated fouling or corrosion is a
major concern in industrial plants which often inhibits use of
augmentation techniques. Although this concern may be unfounded,
there is reluctance to take the chance that heat exchanger failure

A E Bergles

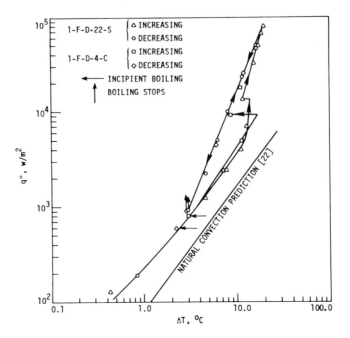

Figure 21.9: Saturated pool boiling of R-113 on a smooth copper surface, Chyu (1979).

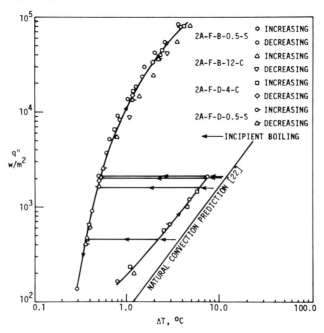

Figure 21.10: Saturated pool boiling of R-113 on a High Flux surface, Chyu (1979).

could shut down the entire plant, at a financial loss which would far exceed the original savings in the heat exchanger.

21.6.4 *Safety and Environmental Considerations*

It is apparent that there are objections to use of certain augmentation techniques because of the potential hazards. Electric fields are an example; the heat transfer coefficient for film condensation of R-113 can be increased several times by the application of a transverse electric field (Velkoff and Miller (1965)); however it is not apparent how 30,000-40,000 volts can be safely provided in a practical condenser.

In the area of acoustic augmentation, sonic fields result in some enhancement, eg a 60% increase in the heat transfer coefficient for condensation of isopropanol vapor flowing downward in a vertical tube when subjected to a siren providing sound at 50-330 Hz (Mathewson and Smith (1963)). The intensity required, however, was 186 db! The precautions which would be required to shield personnel would probably cost more than the original equipment savings.

21.7 Economics

Basic cost goals must be established early. Augmentation usually will not be attractive unless the augmented heat exchanger offers a cost advantage relative to a conventional heat exchanger. A frequent complication is that reliable manufacturing cost data are not available for the desired surface geometry and material. For instance, manufacturers of enhanced tubing expect a lower unit price if the market expands, but generally have very little experience to back up the expectation. In general, the costing should be applied to the entire system and done on a life cycle basis.

As an example of partial cost analysis consider the use of augmented tubing in an aromatics reboiler. Table 21.2 lists the manufacturer's analysis of this application. The point is that the High Flux boiling surface on the inside of the tube together with a fluted exterior condensing surface increases the overall coefficient dramatically. The driving temperature difference is reduced accordingly, with the result that a large annual saving in fuel costs for the heating steam is expected. The increased tubing cost is not cited, but a reasonable payback period is expected.

21.8 Examples of Applications of Augmentation

After sounding many warnings about possible complications in applying augmentation techniques to boiling or condensing systems, it is appropriate to cite some examples of actual or proposed commercial useage. Only representative applications will be given, in rather sketchy form - in keeping with the way the performance and cost are usually reported.

21.8.1 *Chemical Process Industry*

Wolf et al (1975) report the use of Linde High Flux surfaces

Table 21.1

Example of energy savings with High Flux
tubing in aromatics reboiler application.
(Linde Division, Union Carbide Corp.)

	BARE TUBE DESIGN	HIGH FLUX DESIGN
Heat Duty, Btu/hr	10×10^6	10×10^6
Boiling Fluid	BTX Aromatic	BTX Aromatic
Condensing Fluid	Steam	Steam
Steam Pressure, psig	400	125
Steam Condensing Temperature, °F	448	352
Boiling Temperature, °F	325	325
Mean Temperature Difference, °F	123	27
Exchanger Area, ft²	6000	6000
Exchanger Size (diameter x length)	52in.x240in.	52in.x240in.
Overall Heat Transfer Coefficient, Btu/hrft²F	95	435
Steam Cost, $/1000 lb	2.50	1.50
Energy Cost, $/yr	1,924,320	1,030,600
Energy Savings, $/yr		893,000

in distillation, and note how rising energy costs can be mitigated by better cycle arrangement and use of augmented heat exchangers. Many other examples of the use of this particular surface in distillation and gas liquifaction have been cited by the manufacturer. (See also Section 21.7.)

21.8.2 *Power Industry*

Graham and Aerni (1970) discuss the potential for dropwise condensation in surface condensers, particularly for marine applications. A surface area reduction of 30% and a total weight reduction of 15% is typically predicted. The more recent review by Tanasawa (1978) should be consulted for an assessment of the practical difficulties, including removal of non-condensables and maintenance of the promoter, which must be surmounted before the high heat transfer coefficients of dropwise condensation can be realized in practical equipment.

21.8.3 *Refrigeration Industry*

Arai et al (1977) report use of enhanced boiling (Thermoexcel-E) and condensing surfaces (Thermoexcel-C), along with repeated ribs on the interior surfaces of the tubing in both the evaporator and condenser. A 28% reduction in the heat exchanger length was reported.

Boling et al (1953) suggested a large reduction in refrigerant evaporator size by heavily finning the evaporator tubes (Dunham-Bush "Inner Fin").

21.8.4 *Desalination*

Much development work in enhanced heat transfer tubing has been motivated by desalting applications. There is a large potential market due to the large surface areas required. A rather detailed economic analysis by Cox et al (1970) was performed for Long Tube Vertical (LTV) and Horizontal Tube Multiple Effect (HTME) plants. Internally grooved tubes (with some effects using these tubes with knurled exterior as well) were shown to reduce the capital cost by about 10%. An important screening study of enhanced tubes for LTV systems was reported by Alexander and Hoffman (1971). Increases of over 200% in the overall heat transfer coefficient were observed with several tubes.

In Multistage Flash (MSF) plants, only condensing surface is required. Withers and Young (1971) analyzed a steam condenser, typical of the low temperature end of the heat recovery section of a horizontal MSF plant, fabricated from corrugated ("Korodense") tubes. The combination of increased condensing coefficient and increased internal water coefficient led to a surface area reduction of about 30%, subject to constraints of constant heat duty, fixed pressure drop, and fixed coolant flow rate. Rather than comparing smooth and enhanced tubes in a series of full-scale plant designs, Newson (1976) elected to use the more common single-tube analysis. A commercial "roped" tube was chosen with tubeside (single-phase brine coolant) and shell-side (condensing pure water) enhancements of 1.63 and 1.67, respectively. Fixed are heat duty, temperatures, and N_a. As enhanced tube velocity varies, L_a and N_o vary. The smooth tubes were 18.1 mm i.d., while the enhanced tubes were 24.5 mm i.d. As shown in Figure 21.11, in the region of 1 m/s water velocity in the enhanced tube, the smooth tube cost is 40-50% more than the enhanced tube. Process and economic data are given in the figure. It is noted that here the enhanced tubing premium is only 7 1/2%. The most important factor becomes the operating cost or pumping power; thus, the comparison is very sensitive to water velocity.

The comprehensive survey of Thomas (1973) gives interesting speculations as to maximum heat transfer coefficients which might be realized with enhanced surfaces for desalination. The upper limit is suggested as about 6000 Btu/hr ft^2 F (34000 W/m^{2}K) which was at that time, and still is, about 3 times the state-of-the-art coefficient. Fouling is a major problem, as even a small fouling factor can drastically reduce the coefficients predicted on the basis of clean surface data.

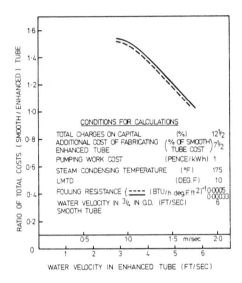

Figure 21.11: Cost ratio for tubing in typical MSF condenser (Newson (1976)).

21.8.5 Electronics Industry

Liquid cooling with boiling or evaporation is occasionally
required to dissipate very large heat generation rates produced
by electrical or electronic equipment. An established application
of extended surfaces is found in the "Vapotron" series of cooling
arrangements for high power communications tubes. As described
by Beurtheret (1970), the fins are shaped so as to maximize the
average boiling heat transfer coefficient for either pool boiling
or forced convection boiling.

21.8.6 Status of Application

Table 21.3 summarizes the present and potential use of
special geometries and other passive augmentation techniques
according to mode of heat transfer, as suggested in a recent
evaluation (Bergles et al (1979b)). The first part of the table
considers augmenting the inner and outer surfaces of circular tubes
while the second part considers plate-type and plate-fin heat
exchangers. A coding system is employed to define the degree to
which an enhancement technique has found application. A dash
entry indicates not applicable or not intended to augment that heat
transfer mode. The third column lists typical materials used for
commercial manufacture. The last column indicates the authors'
assessment of the performance potential, independent of whether a
commercial manufacturing technique exists. It is evident that
only a limited number of the available enhancement techniques are
commercially utilized. In only a few cases are techniques
considered to be in common use. Notably, some manufacturers of a
given type of enhanced surface have not been able to market their
product in significant quantities.

Table 21.3

Application of augmentation techniques (Bergles et al (1979b)).

	Commercial Availability	Forced Convection	Boiling	Condensation	Typical Material	Performance Potential
			Mode			
Inside Tubes						
Metal Coatings	Yes	—	2	—	Al, Cu, St	Hi
Integral Fins	Yes	2	3	4	Al, Cu	Hi
Flutes	Yes	4	4	4	Al, Cu	Mod
Integral Roughness	Yes	2	3	4	Cu, St	Hi
Wire Coil Inserts	Yes	3	4	4	Any	Mod
Displaced Promoters	Yes	2	4	4	Any	Mod (Lam)
Twisted Tape Inserts	Yes	2	3	4	Any	Mod
Outside Circular Tubes						
Coatings						
Metal	Yes	—	2	4	Al, Cu, St	Hi (Boil)
Non-Metal	No	—	4	4	"Teflon"	Mod
Roughness (Integral)	Yes	3	2	4	Al, Cu	Hi (Boil)
Roughness (Attached)	Yes	3	4	—	Any	Mod (For Conv)
Axial Fins	Yes	1	4	4	Al, St	Hi (For Conv)
Transverse Fins						
Gases	Yes	1	—	—	Al, Cu, St	Hi
Liquids/Two-Phase	Yes	1	1	1	Any	Hi
Flutes						
Integral	Yes	—	—	2	Al, Cu	Hi
Non-Integral	Yes	—	—	4	Any	Hi
Plate-Fin Heat Exchanger						
Metal Coatings	Yes	—	3	—	Al	Hi
Surface Roughness	Yes	4	3	4	Al	Hi (Boil)
Configured or Interrupted Fins	Yes	1	2	2	Al, St	Hi
Flutes	No	—	—	4	Al	Mod
Plate Type Heat Exchanger						
Metal Coatings	No	—	4	—	St	Lo
Surface Roughness	No	4	4	4	St	Lo
Configured Channel	Yes	1	3	3	St	Hi (For Conv)

Use Code

1. Common Use
2. Limited Use
3. Some Special Cases
4. Essentially No Use

21.9 *Factors Affecting Commercial Development*

It is appropriate to view augmentation techniques as "second generation" heat transfer technology. Such new technology normally requires several phases of development for successful commercialization. These phases include:

(1) Basic performance data for heat transfer and pressure drop, if applicable, must be obtained. General correlations must be developed to predict heat transfer and pressure drop as a function of the geometrical characteristics.

(2) Design methods must be developed to facilitate selection of the optimum surface geometry for the various techniques and

particular applications.

(3)   Manufacturing technology and cost of manufacture must be
        available for the desired surface geometry and material.

(4)   Pilot plant tests of the proposed surface are required to
        permit a complete economic evaluation and to establish the
        design, including long-term fouling and corrosion character-
        istics.

Some augmentation techniques have completed all steps
required for commercialization. Table 21.4 attempts to summarize
the status of technology development for several of the more
important enhancement techniques applied to circular tubes.
Commercialization represents the ultimate stage of development;
however, even commercial products usually require additional
development work.

The further development of any technique will depend on two
factors: (1) its performance potential, and (2) manufacturing
cost. Presently, cost-effective manufacturing technology is
probably the most significant barrier to commercial application of
high performance enhancement techniques. Researchers appear
intent on identifying surface geometries which offer the highest
performance. It is suggested that more communication between
researchers and manufacturing engineers is needed to identify the
most cost-effective concepts. An example of a recent research
workshop intended to promote this type of communication is
described by Junkhan et al (1979).

Table 21.4

*Status of augmentation technology development
(Bergles et al (1979b)).*

Inside Tubes	Single Phase Forced Convection	Boiling	Conden-sation
Inside Tubes			
Coatings	—	3, 5	—
Roughness	2, 3, 5	1	1
Internal fins	1, 2, 3, 5	1, 5	1
Flutes	5	5	—
Insert devices	5	1, 5	1, 5
Outside Tubes			
Metal coatings	—	. 3, 5	—
Non-metal coatings	—	1	4
Roughness	1, 2	1, 2, 5	1
Extended surface	5	5	5
Flutes	—	—	5

Code: 1. Basic performance data
        2. Design methods
        3. Manufacturing technology
        4. Heat exchanger application
        5. Commercialization

Nomenclature

A   - heat transfer area (nominal basis

G   - mass velocity

$\bar{h}$ - average heat transfer coefficient for complete condensation

L   - length of tube bundle

N   - number of tubes in tube bundle

p   - pressure

Δp - pressure drop

q'' - heat flux

q''_{cr} - critical heat flux

R_h - performance ratio defined in Eq. 21.1

$R_{\Delta p}$ - performance ratio defined in Eq. 21.1

R_{5c} - performance ratio defined in Eq. 21.3

ΔT - wall-minus-saturation temperature difference

U_o - overall heat transfer coefficient

Subscripts

a   - refers to augmented tubes

in   - inlet condition

o   - refers to normal smooth tubes; referred to outside diameter

P   - evaluated at constant pumping power

q   - evaluated at constant heat futy

w   - evaluated at constant mass flow rate

References

Alexander, L G, and Hoffman, H W, (1971) "Performance characteristics of corrugated tubes for vertical tube evaporators". ASME Paper No. 71-HT-30.

Arai, N, Fukushima, T, Arai, A, Nakajima, T, Fujie, K and Nakayama, Y, (1977) "Heat transfer tubes enhancing boiling and condensation in heat exchanger of a refrigerating machine". *ASHRAE Trans.*, **83**, Part 2, 58-70.

Bergles, A E, (1969) "Survey and evaluation of techniques to augment convective heat and mass transfer". in *Progress in Heat and Mass Transfer*, U Grigull and E Hahne (Editors), **1**, 331-424, Pergamon, Oxford.

Bergles, A E, Blumenkrantz, A R and Taborek, J, (1972) "Performance evaluation criteria for enhanced heat transfer surfaces". AIChE Preprint 9 for 13th National Heat Transfer Conference, Denver, Colorado, August.

Bergles, A E, Bunn, R L and Junkhan, G H, (1974) "Extended performance evaluation criteria for enhanced heat transfer surfaces". *Letters in Heat and Mass Transfer*, **1**, 113-120.

Bergles, A E and Jensen, M K, (1977) "Enhanced single-phase heat transfer for OTEC systems". *Proceedings Fourth Annual Conference on Ocean Thermal Energy Conversion*, University of New Orleans, March, VI-41 - VI-54.

Bergles, A E and Webb, R L, (1978) "Bibliography on augmentation of convective heat and mass transfer - Part 1". *Previews of Heat and Mass Transfer*, **4**(2), 61-73.

Bergles, A E, Webb, R L, Junkhan, G H, and Jensen, M K, (1979a) "Bibliography on augmentation of convective heat and mass transfer". HTL-19, ISU-ERI-Ames-79206, Iowa State University, May.

Bergles, A E, Webb, R L and Junkhan, G H, (1979b) "Energy conservation via heat transfer enhancement". *Energy*, **4**, 193-200.

Beurtheret, C A, (1970) "Transfer de flux superieur a 1 kw/cm^2 par double changement de phase entre une paroi non isotherme et un liquid en convection forcee." *Heat Transfer 1970*, Proceedings of the Fourth International Heat Transfer Conference, **V**, Paper B4.2, Elsevier, Amsterdam.

Chyu, M-C, (1979) "Boiling heat transfer from a structured surface". MS Thesis in Mechanical Engineering, Iowa State University.

Cox, R B, Matta, G A, Pascale, A S and Stromberg, K G, (1970) "Second report on horizontal tube multiple effect process pilot plant test and designs". U.S. Office of Saline Water Research and Development Progress Report No. 592.

Gambill, W R, Bundy, R D, and Wansbrough, R W, (1961) "Heat transfer, burnout, and pressure drop for water in swirl flow tubes with internal twisted tapes". *Chem. Eng. Prog. Symp. Ser.*, **57**, No. 32, 127-137.

Gilbert, P T, (1975) "Selection of materials for heat exchangers". Paper presented at Sixth International Congress on Metallic Corrosion, Sydney, Australia, December.

Graham, C and Aerni, E F, (1970) "Dropwise condensation - a heat transfer process for the 70's". Naval Ship Systems Command 7th Annual Technical Symposium.

Hillis, D L, Lorenz, J J, Yung, D T and Sather, N F, (1979) "OTEC performance tests of the Union Carbide sprayed-bundle evaporator". ANL/OTEC-PS-3, Argonne National Laboratory, May.

Jakob, M and Fritz, W, (1931) "Versuche über den verdampfungsvorgang". *Forschung auf dem Gebiete des Ingenieurwesens*, **2**, 435-447.

Junkhan, G H, Bergles, A E and Webb, R L, (1979) "Research workshop on energy conservation through enhanced heat transfer". HTL-21, ISU-ERI-Ames-80063, Iowa State University, October.

Katz, D L, Myers, J E, Young, E H, and Balekjian, G, (1955) "Boiling outside finned tubes". *Petroleum Refiner*, **34**, 113-116.

Lewis, L G and Sather, N F, (1978) "OTEC performance tests of the Union Carbide flooded-bundle evaporator." ANL/OTEC-PS-1, Argonne National Laboratory.

Linde Division, Union Carbide Corp., Tonawanda, New York. "Technical information High Flux tubing," September 1977.

Luu, M and Bergles, A E, (1979) "Experimental study of the augmentation of in-tube condensation of R-113". *ASHRAE Trans.*, **85**, Part 2, 132-145.

Macbeth, R V, (1977) "Fouling in boiling water systems". in *Two-Phase Flow and Heat Transfer*, D Butterworth and G F Hewitt (Editors), Oxford University Press, 323-342.

Mathewson, W F and Smith, J C, (1963) "Effect of sonic pulsation on forced convective heat transfer to air and on film condensation of isopropanol". *Chemical Engineering Progress Symposium Series*, **59**(41), 173-179.

Megerlin, F E, Murphy, R W and Bergles, A E, (1974) "Augmentation of heat transfer in tubes by means of mesh and brush inserts". *J. Heat Transfer*, **96**, 145-151.

Nakajima, K and Shiozawa, A, (1975) "An experimental study on the performance of a flooded type evaporator". *Heat Transfer - Japanese Research*, **4**(4), 49-66.

Newson, I H, (1976) "Enhanced heat transfer condenser tubing for advanced multistage flash distillation plants". *Proceedings of the Fifth International Symposium on Fresh Water from the Sea*, **2**, 107-115.

Pai, R H and Pasint, D, (1965) "Research at Foster Wheeler advances once-through boiler design". *Electric Light and Power*, January, 66-70.

Royal, J H and Bergles, A E, (1978) "Pressure drop and performance evaluation of augmented in-tube condensation". *Heat Transfer 1978*, Proceedings of the Sixth International Heat Transfer Conference, **2**, 459-464.

Ryabov, A N, Kamen'shchikov, F T, Filipov, V N, Chalykh, A F, Yuagy, T, Stolyarov, Y V, Blagovestova, T I, Mandrazhitskiy, V M, and Yemel'yanov, A I, (1977) "Boiling crisis and pressure drop in rod bundles with heat transfer enhancement devices". *Heat Transfer-Sov. Res.*, **9**, No. 1, 112-121.

Tanasawa, I, (1978) "Dropwise condensation - the way to practical applications". *Heat Transfer 1978*, Proceedings of the Sixth International Heat Transfer Conference, **6**, Hemisphere Publishing Corp., 393-405.

Thomas, D G, (1973) "Prospects for further improvements in enhanced heat transfer surfaces". Desalination, **12**, 189-215.

Tong, L S, Steer, R W, Wenzel, A H, Bogaardt, M, and Spigt, C L, (1967) "Critical heat flux of a heater rod in the center of smooth and rough sleeves, and in line-contact with an unheated wall". ASME Paper No. 67-WA/HT-29.

Torii, T, Hirasawa, S, Kuwahara, H, Yanagida, T and Fujie, K, (1978) "The use of heat exchangers with Thermoexcels' tubing in Ocean Thermal Energy Power Plants". ASME Paper No. 78-WA/HT-65.

Velkoff, H R and Miller, J H, (1965) "Condensation of vapor on a vertical plate with a transverse electric field". *J. Heat Transfer*, **87**, 197-201.

Webb, R L and Eckert, E R G, (1972) "Application of rough surfaces to heat exchanger design". *Int. J. Heat Mass Transfer*, **15**, 1647-1658.

Withers, J G and Young, E H, (1971) "Steam condensing on vertical rows of horizontal corrugated and plain tubes". *Industrial and Engineering Chemistry, Process Design and Development*, **10**(1), 19-30.

Wolf, C W, Weiler, D W and Ragi, E G, (1974) "Energy costs prompt improved distillation". *Oil and Gas Journal*, September 1.

Chapter 22

Nuclear Plant Safety

A. E. BERGLES

22.1 Introduction

Nuclear power reactors are designed to provide a reliable safe source of electrical energy. Plant and public safety concerns have led to the formulation of accident scenarios and the development of safety systems to arrest the progress of potentially hazardous transient thermal-hydraulic processes. In light water reactors (LWR's) in the US and other countries, the worst conceivable accident or design basis accident (DBA) provides the design criteria for the safety systems. This hypothetical accident is the double-ended break of the recirculation suction line in a boiling water reactor (BWR) and the double-ended rupture of the primary coolant inlet line for a pressurized water reactor (PWR).

A loss-of-coolant accident (LOCA) would result in rapid reactor shutdown; however, an emergency core cooling system (ECCS) must be provided to remove the decay heat and stored energy so as to limit the peak clad temperature (PCT). Many complex thermal-hydraulic phenomena have been observed during the course of laboratory experiments which model LOCA's. The study of these transient two-phase flow phenomena has dominated worldwide research in two-phase heat transfer during the past decade. See, for example, the heat transfer review of Tong (1978). In this chapter, the basic accident postulates will be outlined briefly, the thermal-hydraulic events will be identified, and the fundamental processes and their transport mechanisms will be discussed. The discussion, organized according to typical examples of both reactor types, will conclude with comments on thermal-hydraulic codes which have been developed to analyze the hypothetical accidents.

Reactor safety technology includes, of course, design for steady-state operation and normal operational transients. Discussions of these expected conditions are included in the reviews by Lahey (1977) and Hsu and Sullivan (1977) for the BWR and PWR, respectively.

22.2 BWR Description

In a BWR, energy release from nuclear fission of the fuel results in steam being produced directly at about 70 bar by vaporization of water in the core. As shown in Figure 22.1,

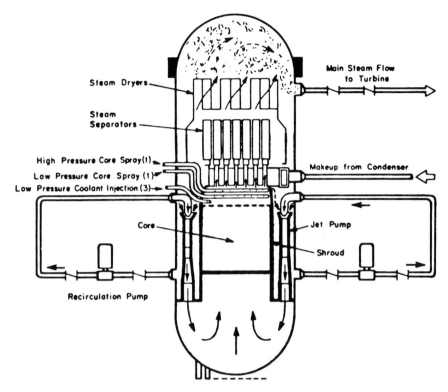

Figure 22.1: Typical US BWR system. (Adapted from Brockett et al, 1972).

feedwater from the condenser enters the pressure vessel and is injected into the pool through a ring sparger. Mixed condensate and recirculation flow from the separators is forced into the lower plenum by jet pumps driven by coolant taken from the down-comer. Partial vaporization occurs in the core, with the dried steam going to the turbine and the water falling back to the pool. The recirculation ratio is approximately 7:1. The ECCS consists of the core sprays (1) and coolant injection to the lower plenum (3).

22.3 BWR Accident

DBA

The reactor is assumed to be operating at 102% of rated power when a double-ended "guillotine" rupture instantaneously occurs in one of the two recirculation pump suction lines, as illustrated in Figure 22.2. There is a simultaneous loss of off-site power for the recirculation pump and other vital components. The reactor is scrammed so that within 1 second the power is reduced to decay heat level, 2-4% of rated power (core voiding would insure this in any case). The ECCS is initiated within 30 seconds after the break by low water level or high dry-well pressure. As shown in Figure 22.3, the emergency coolant is

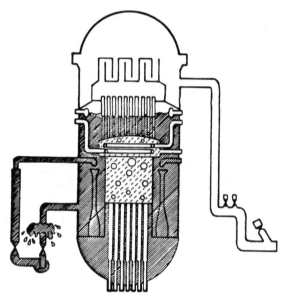

Figure 22.2: Hypothetical BWR LOCA event at time of initiation.   (Lahey, 1979a)

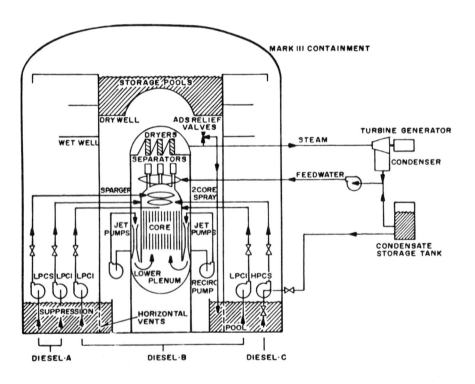

Figure 22.3: BWR/6 ECCS schematic.   (Lahey, 1979a).

supplied from the suppression pool or condensate storage tank by pumps driven by emergency diesel power. The hypothetical LOCA event is divided into "blowdown", "refill", and "reflood" phases.

Thermal-Hydraulic Events

Blowdown. Two-phase critical flow occurs after the rupture, first with steam-water flow and then with steam as the water level in the downcomer falls below the jet pump suction. Reasonable core flow is maintained, until the jet pump suctions are uncovered, by pump coast-down in the intact loop and natural circulation.

During the flow decay, it is possible that parallel channel instability occurs. This could take the form of a flow excursion where a channel or fuel assembly suffers sharply reduced flow or where flow oscillations are observed among interacting channels, eg, U-tube oscillations. The oscillations might also occur during the reflooding phase. The effect of the oscillation is felt as oscillation of the clad temperature.

After uncovering of the jet pump suction, the pool level drops rapidly and the break flow changes to steam. The pressure vessel depressurization rate then increases and there is vigorous flashing of the residual water in the lower plenum. A two-phase mixture is then forced up through the core and diffusers.

As a result of the low flow rate and reduced pressure, the fuel rods experience dryout or boiling transition within 5 seconds after rupture. The dryout point propagates along the rods, from top to bottom, aided, of course, by the drop in water level. A mist or dispersed flow two-phase heat transfer occurs above the dryout point. As a result of the low heat transfer coefficient, the rod temperature rises sharply. If the rod temperature is high enough, an exothermic chemical reaction between the rod and the coolant may occur; this augments the rod temperature rise. Rewetting of the rods may occur during lower plenum flashing; however, the rod temperature rise is only momentarily arrested.

Refill. Emergency core coolant (EEC) injection is necessary if the PCT on the hottest fuel rod is to remain below 1200°C, the DBA upper limit. As an example, the General Electric BWR/6 ECCS shown in Figure 22.3 has four independent subsystems:

High Pressure Core Spray System (HPCS) which sprays water through ring headers onto the top of the fuel assemblies.

Automatic Depressurization System (ADS) employs relief valves to vent steam to the suppression pool so that additional cooling can be provided by low pressure injection systems.

Low Pressure Core Spray System (LPCS) is also introduced through a semi-circular "showerhead" arrangement at the top of the core.

Low Pressure Coolant Injection System (LPCI) delivers a large volume of water to the lower plenum, through a core bypass, once the vessel is depressurized.

Bottom flooding through the LPCI is the main line of defense;

however, the spray systems assist with lower plenum refilling and prevent unacceptably high fuel rod temperatures until flooding is completed.

Spray distribution is affected by steam entrainment, steam up-draft, direct-contact condensation of vapor, and interaction of sprays from adjacent nozzles. If the steam flow through the core is large, reflood is delayed by the counter-current-flow-limited (CCFL) condition. Water accumulates in the upper header until at least some channels are available for downflow. If the bundle is not penetrated, water may bypass the "steambound" core entirely. The overflow from the separator goes to the downcomer and from there out the break. Even if the top sprays are ineffective, the LPCI should flood the lower plenum. The only difficulty that might arise is if LPCI coolant bypasses the steambound core by flowing up the jet pump diffuser tail pipes and out the break.

Reflood. As the water level rises up into the core, the fuel rods are quenched. Normal temperature levels can be achieved when nucleate boiling occurs. The fuel rods experience large temperature gradients as the quench front progresses. Liquid is entrained by vapor generated below the liquid level as well as by agitation at the pool interface. The two-phase mixture preceding the quench front provides some "precursory cooling" of the rods. The reflooding of the core typically takes 200 seconds for the DBA.

If inadequate emergency coolant reaches the core, fuel rod melting could occur. The fuel-coolant interaction (FCI) might then lead to a vapor explosion, ie a very rapid release of energy due to spontaneous nucleation.

Fuel Rod Temperature

Attention focuses on fuel rod behavior during the LOCA transient. The temperature responses of upper and lower cladding are shown schematically in Figure 22.4. Core flow decreases (1), but is momentarily restored due to lower plenum flashing (5). The pressure decay (2) is accelerated once the recirculation pump suction is uncovered (4). Cladding temperature increases sharply at dryout (3). Rewetting occurs at the lower portions of the fuel elements (6) and dryout does not occur again until the lower plenum flashing has subsided (7). The rate of temperature rise is arrested as the spray is initiated, (8), (8'). The cladding temperature turnarounds, (9), (9'), result from flooding at the bottom of the core. The increased rate of temperature rise of the upper cladding prior to flooding is due to release of energy from the metal-water reaction. An understanding of the basic thermal-hydraulic mechanisms involved is required if the temperature history is to be accurately estimated.

Containment Integrity

Relief valve line clearing. As a relif valve is actuated, a liquid plug and a compressed air plug must be vented to the suppression pool. When the bubble enters the pool it expands rapidly and then contracts after the overexpansion. Containment walls and baffles are stressed by the compression and rarefaction waves.

Pressure suppression and chugging. Steam released to the drywell

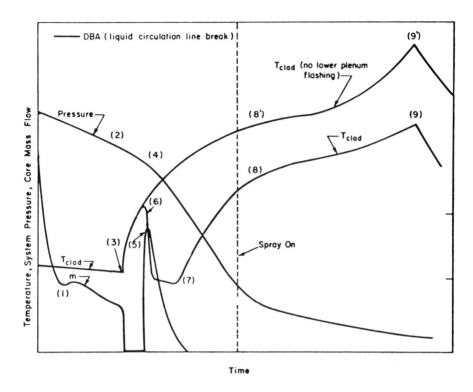

Figure 22.4: Generalized loss-of-coolant behavior for large breaks in a BWR.
(Adapted from Brockett et al, 1972).

must be quenched in the suppression pool if excessive pressures
are to be avoided in either the wet well or the dry well. The
condensation process becomes less efficient as the steam flow rate
and the pool temperature are raised. This sporadic, noisy,
direct-contact condensation ("chugging") can damage the containment
structure.

Key Mechanisms and Analytical Techniques

Critical flow. Critical flow is normally calculated using the
slip model of Moody; however, there is evidence that the critical
flow rates thus calculated are high and that a homogeneous model
(HEM) is more appropriate (Lahey, 1979a). A recent review of
critical flow, with particular emphasis on nuclear reactor applications, is given by Bouré (1977).

Transient analysis. Transient analysis of the flow is carried out
by nodalizing the system, applying the mass, momentum, and energy
equations to each node and solving the system of equations with
the aid of a digital computer. Accurate estimates of the mass
inventory are required if key events such as uncovering of the
jet pump suction and filling of the lower plenum are to be
properly identified. Simple drift-flux models assuming thermodynamic

equilibrium have been found effective in predicting phase separation; however, exact analytical solutions can only be obtained
for relatively simple situations (Lahey, 1979a and 1979b). The
fundamental work in this area includes studies of transient flow
regimes (Taitel et al, 1978, Prassinos and Liao, 1979) and
pressure drops in piping components under transient flow conditions
(Weisman et al, 1976).

Flow instability. Flow instability has been identified as a
possible complicating factor during flow coastdown (Lahey, 1979a)
or bottom reflood (Yadigaroglu, 1978a). Further information on
the flow cxcursions or oscillations which might be encountered
can be found in Chapter 13.

Flashing. Flashing is of considerable importance in the piping
as the system pressure level drops and the fluid moves through a
pressure gradient to the break. For instance, the upstream
quality is necessary input to any critical flow model. Jones
(1979) treats the problem of flashing in flowing liquids. For
rapid depressurization, the analysis is hampered by lack of
information on the nonequilibrium vapor generation. For a
depressurizing pool, eg the lower plenum during blowdown, rather
accurate estimates of the pool height or "level swell" can be
obtained by a one-dimensional drift-flux model with equilibrium
vapor generation (Lahey, 1979b). Model tests showed that
rewetting of the rods can occur during the lower plenum flashing
surge (Muralidharan, 1976). These tests further established that
after the surge the vapor generated by lower plenum flashing
(and by heat transfer from the pressure vessel wall) can be
effective in holding up liquid in the core itself due to CCFL.

Dryout. Dryout or critical heat flux (CHF) condition under
transient conditions is the subject of considerable debate. For
BWR LOCA the accepted procedure is to use local instantaneous
flow, pressure, and enthalpy together with a conventional CHF
correlation, and compare the calculated CHF with the heat flux
estimated for the rod. Even if quasi-steady behavior can be
assured, the validity of available correlations at reduced flow
and pressure can be questioned.

Ishigai et al (1974) observed "premature dryout" when steam
flow was reduced in a low pressure steam-water (mixed-inlet)
system. On the other hand, Cumo et al (1978) observed large
increases in dryout heat fluxes during rapid coastdown of R-12
flow (scaled to BWR and PWR pressures) in a conventional subcooled-
inlet system. These results translated to increases in the "time
to CHF" as shown in Figure 22.5. These data are not directly
applicable to BWR blowdown due to differences in geometry, fixed
pressure level, and uncertainties in fluid-to-fluid modeling.
However, since the BWR flow coastdowns are approximately $t_{0.5} = 2s$,
some delay in dryout would be expected due to the transient.
Belda and Mayinger (1977) confirm that the dryout delay during
blowdown (pressure and flow decay of R-12, scaled to BWR) may be
quite large (~10 s) because of the time required to evaporate the
liquid film of an annular flow. In spite of this evidence, it
appears that the conservative approach using steady-state CHF
correlations will be retained. In the area of correlation, Bowring
(1977) described a new Macbeth-type correlation for rod bundles

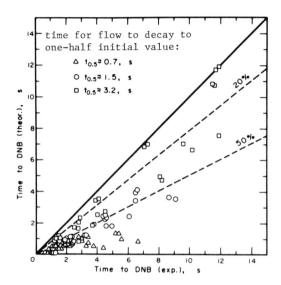

Figure 22.5: Times to DNB computed on the basis of steady state CHF correlation compared to the experimental ones. (Cumo et al, 1978).

which includes data at low pressures and mass velocities. The correlation has been adapted to a reactor blowdown code.

Post dryout. Subsequent to dryout, it is conservatively assumed that heat transfer is governed by stable film boiling. In addition to the discussion in Chapter 10, comprehensive reviews of post-dryout heat transfer are given by Groeneveld and Gardiner (1977) and Mayinger and Langner (1978). The low heat transfer coefficients are set to zero when the core is uncovered. The probable rewetting during lower plenum flashing is ignored. Radiation to the bundle channel is an important heat transfer mechanism. A metal-water reaction must be considered for the zircaloy cladding and the steam. This involves specification of a suitable rate equation. The zirconium-oxide film thus developed is usually included in solving for the temperature distribution in the fuel pin.

Spray cone angle. The two core spray systems have nozzles which are sized in an atmospheric air facility to provide a specified flow to each fuel bundle. This sizing is subject to some uncertainty as actual use, if a LOCA event occurs, will be in a pressurized steam environment. The reduction of the cone angle due to entrainment of stagnant steam has been studied by Block and Rothe (1975). Other factors which influence the cone angle are the upward velocity of the steam (Sandoz and Sun, 1976), augmented transverse steam velocity due to condensation on the cold water droplets, and the strong interaction with adjacent nozzles. Accurate sizing obviously requires testing of multiple nozzles in an appropriate steam environment.

CCFL. If steam comes from the core at a high velocity, the liquid has difficulty entering the core. The criterion, essentially a

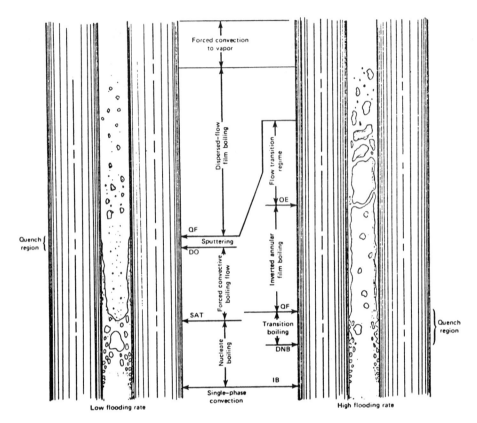

Figure 22.6: Flow and heat transfer regimes observed during reflooding
            (Yadigaroglu, 1978b).

balance between fluid-dynamic drag and buoyancy, for the onset of
this condition in simple systems is well established from "flooding" experiments (Wallis, 1969). It seems improbable that the
spray water could accumulate in the upper plenum without "breaking
through" at least some of the channels and reflooding the lower
plenum (Letzring, 1977).

Reflooding and rewetting. Complex two-phase flow and heat
transfer phenomena occur as the spray and/or lower plenum liquid
level enters the core. Essentially following the boiling curve
in reverse, a rod segment sequentially experiences cooling by
steam, dispersed-flow film boiling, film boiling, transition
boiling, nucleate boiling, and finally by single-phase water.
This succession of heat transfer regimes is schematically shown
in Figure 22.6. It should be noted that a realistic picture of
spray cooling could also involve an annular film of liquid
advancing from the *top* of the core, as described in Chapter 10.10.

   Figure 22.6 suggests that conventional annular flow occurs at
low flooding rates and an inverted annular flow is established at

high flooding rates. The reflooding rate for a BWR is controlled
by the balance between the driving head and the pressure in the
upper plenum. The necessary head comes from the LPCI or from a
column of water in shunted channels. These two pictures
correspond to forced-convection vaporization dryout (DO) and
departure from nucleate boiling (DNB), respectively. In both
cases, downstream of the quench front (QF) the rods are cooled
by droplets ("precursory cooling"). With top spray cooling,
however, the downward progression of both droplets and film may
be retarded or even halted (CCFL) by the rising vapor. Thus, top
spray cooling generally has poor heat transfer downstream of the
quench front. The rods eventually cool below a rewetting temperature and are "quenched".

A number of BWR reflooding experiments are mentioned in the
reviews by Dix and Anderson (1977) and Yadigaroglu (1978b).
While a large amount of data has been generated, correlation of
these data is difficult since local flow conditions are generally
not recorded. Accurate modeling of the rewetting phenomenon is
required. Elias and Yadigaroglu (1978) survey the models, noting
that two-dimensional (axial and radial) analyses are necessary
for high flow rates. The rewetting temperature and the heat
transfer coefficient behind the quench front are chosen to fit
the experimental data.

Meltdown/vapor explosion. Thermal fuel-coolant interactions are
very low-probability accidents for LWR plants. Fuel-coolant
interactions involve a number of possible processes: coolant
vaporization, fuel fragmentation, generation of pressure pulses
due to rapid coolant vaporization, cooling of the fuel without
fragmentation via film boiling. The more severe interactions
termed "large-scale events" involve energy transfer from the hot
liquid to the cold liquid on a time scale which is less than
the characteristic time of pressure relief for the system of
interest (of the order of milliseconds). Excellent surveys of
recent work in this area have been reported by Bankoff (1978,
1979).

Chugging. Chugging is a quasi-periodic phenomenon which may
occur when steam is injected into a pool of subcooled water. In
addition to the "chugs" generated by direct-contact condensation
in the pool, random condensation events are observed within the
downcomer pipes. The phenomenon is only observed at pool temperatures greater than about 70°C, a condition which might be attained,
for example, if the relief valve is struck open. With certain
relief valve geometries or vent configurations, chugging can lead
to large forces on the containment walls. Marks and Andeen (1979)
reported experimental data which clearly demonstrate the irregular
behavior. Sargis et al (1979) used a statistical analysis and
found calculated steam pressures and wall pressure histories to
be in qualitative agreement with data from their small-scale
experiment.

Variations in Reactor Types

While all BWR's manufactured in Europe are produced according
to GE licensee agreements, there are significant differences in
circulating systems and ECCS supply which lead to different
calculated DBA behavior. The older AEG BWR (eg Würgassen) is

essentially the same as the US BWR described above. However, the
current German (Kraftwerk Union) version of the BWR utilizes an
integral mechanical pump instead of a jet pump. There are no
large recirculating lines which might rupture. The DBA condition
specifies rupture of one makeup water line or a steam line, with
the result that calculated blowdown time is several hundred
seconds as compared to the calculated depressurization time of
about 30 seconds for a US BWR. The ECCS is primarily introduced
in a flooding mode. The makeup line "small break" condition is
less severe and a steam line break in general results in much
lower PCT than calculated for the US BWR.

The current Swedish (ASES-ATOM) version also has internal
mechanical pumps so that the calculated depressurization time for
the DBA is relatively long.

Computer Codes

The postulated accident is divided into sections and analyzed
by different codes. Typically, eight different types of computer
codes are used in BWR safety analysis. Five of these involve
core thermal-hydraulic considerations as shown in Figure 22.7;
the others are: containment codes, piping codes, and fission
product release codes. The thermal and hydraulic loop code is
solved iteratively with the nuclear and fuel rod codes. The
results from this calculation are transmitted to the core thermal
and hydraulic code which is also solved iteratively with the
nuclear and fuel rod codes. Weak coupling between portions of
the accident is implied here. For instance, the loop code computes
the flow rate and enthalpy of the fluid entering the core and the
core average cladding temperature as a function of time. This
information is used as input to the core code which calculates
fluid flow and heat transfer in the core in greater detail; the
output of primary interest is then the maximum cladding temperature as a function of time.

Code inputs, above all the assumptions used for the accident
initiation, are not necessarily realistic, but are taken as a
combination of extreme conditions which represent a "worst case".

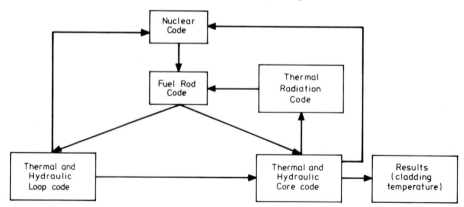

Figure 22.7: Coupling of BWR safety analysis codes (Ybarrando et al, 1972).

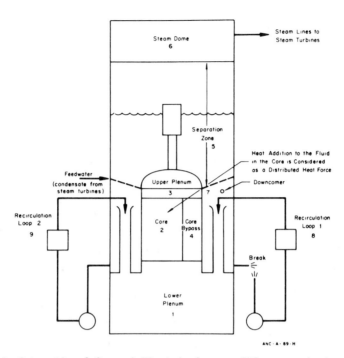

Figure 22.8: Schematic of General Electric Company BWR power plant nodalization for a large break (Slifer, 1971).

Some of these assumptions have been noted previously; however, other sources should be consulted for a full account, eg Ybarrondo et al (1972).

Nodalization of the system for loop code calculations is shown in Figure 22.8. The nodes or control volumes for the two thermal-hydraulic codes are separated by flow paths. Mass and energy conservation equations are applied to each control volume and the momentum equation is used to calculate flow rates through the flow paths. As discussed above, appropriate thermal-hydraulic input is required for the boundary conditions (critical flow, CHF, CCFL, etc.). The initial conditions for the transient solution come from previous steady-state conditions. Finite-difference techniques are used to solve the resulting set of simultaneous partial differential equations. Instantaneous coolant flows, pressures, and enthalpies are determined throughout the system. Further details of the "Calculated LOCA" are given by Ybarrondo et al (1972). The final result, of course, is BWR plants are licensed on the basis that ECCS protection is deemed adequate for the DBA, that is, calculations indicate that the PCT will not be exceeded in the event of a LOCA.

While there are numerous safety issues associated with BWR's, there is less concern about the possibility of system failure during LOCA than in the case of PWR's. This is due to the generally more severe operating conditions of the PWR as well as

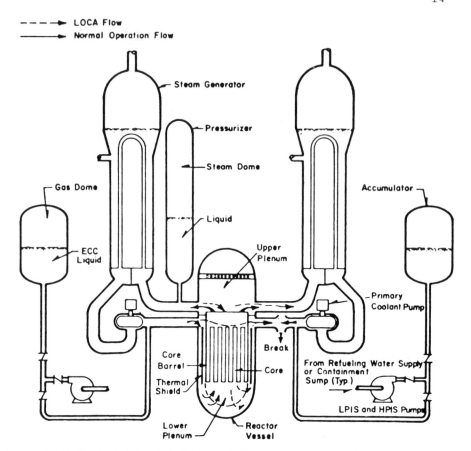

- - - - → LOCA Flow
———→ Normal Operation Flow

Figure 22.9: Typical US multiple loop PWR system. (Adapted from Brockett et
           al, 1972).

the fact that considerably more PWR's than BWR's are in operation.
Accordingly, the bulk of LWR safety research has focussed on PWR's.

22.4 PWR Description

     In a PWR, the primary coolant system is pressurized such
that no net vaporization of the coolant occurs. Energy released
by nuclear fission of the fuel increases the primary cooling water
temperature, with only localized subcooled boiling. The primary
water is circulated to steam generators where vaporization of
secondary cooling water takes place. Reactors have from two to
four primary loops with the loops consisting of either one pump,
a cold leg and a hot leg (Westinghouse) or two pumps, two cold
legs, and one hot leg (Babcock & Wilcox, Combustion Engineering).
The secondary loop is similar to a conventional power cycle.
Primary water is returned to the reactor core to complete that
cycle. The ECCS is injected as indicated in Figure 22.9. The
primary loop pressure is approximately 155 bar. In addition to

the higher pressure, PWR's have higher volumetric power density than BWR's.

22.5 PWR Accident

The hypothetical LOCA is initiated due to an instantaneous guillotine break of the primary cold leg piping, as shown in Figure 22.9. The reactor is assumed to have been operating at a slight over power condition with the peak core power density at the maximum allowable value. Off-site power is lost on initiation of the accident. The reactor is scrammed; however, core voiding provides sufficient negative reactivity for shutdown (to decay heat power level). The steam generators are promptly isolated on the secondary side by closing the turbine steam supply valves and the feed water valves.

Thermal-Hydraulic Events

The course of expected events during LOCA is divided into blowdown, refill, and reflood periods, as in the case of the BWR; however, the details of these processes are considerably different.

Subcooled blowdown. Water loss through the break results in rapid depressurization of the reactor vessel (from about 150 bar) to the vapor pressure of the coolant (about 100 bar). This period is within the range of 5-50 ms.

Saturated blowdown. Depressurization is less severe due to critical flow at the break. The continued heat input and flashing creates a large void in the core. The coolant flow initially decreases: however, the flow rate and direction in the core depend on the break conditions and the hydraulic resistance of components in the broken loop. The decrease of coolant flow together with the generation of void causes CHF. A possible scenario is stagnation or reversal after about 1 second due to flow being maintained by the pump coastdown in unbroken loops counteracting the tendency for flow reversal due to flow out the break. The lower plenum decompression proceeds more rapidly than that of the upper plenum; this then induces a positive flow through the core which reduces the rate of rise of the fuel pin temperature. Loss of the forced circulation shortly results in establishment of reverse flow due to the lower hydraulic resistance of the path from the lower plenum to the break. This reverse flow further controls the fuel pin temperature excursion.

In response to low pressure, the accumulator ECCS is initiated during the latter stage of blowdown, about 10 seconds into the accident. The high pressure accumulator injects water into the cold legs with the intent of filling the lower plenum with liquid and flooding the core. Due to resistance from the reverse flow (or simply the upward flow of steam generated in the lower plenum, CCFL) and the horizontal momentum of water, some ECC bypasses the downcomer and flows out the break ("accumulator bypass"). The mixing of this ECC with the steam flowing in the cold legs may lead to flow and pressure oscillations. Heat storage in the wall affects both the expelled primary coolant and the incoming ECC. Due to these complications, it is unlikely that much of the accumulator water directly cools the core. As the pressure drops further and the primary system water inventory (original coolant and accumulator ECC) is depleted, the core flow rate decreases. The cladding temperature again rises.

Refill. High pressure injection system (HPIS) and low pressure injection system (LPIS) pumps are brought into action after about 30 seconds so that the lower plenum is refilled and the core refloods. The core quench begins at about 200 seconds.

Reflood. The flooding process proceeds as shown in Figure 22.6. The steam produced from quenching the lower portions of the fuel rods cools the upper portions of the rods and limits the PCT. The transient is terminated when the vessel is refilled; however, continual addition of water is required to replace that lost through boiloff.

There is some resistance to flooding as a pressure gradient is established by steam flowing up the core and out the break through the hot leg.

Fuel Rod Temperatures

Conditions within the core and the cladding temperature history are illustrated schematically in Figure 22.10. The core

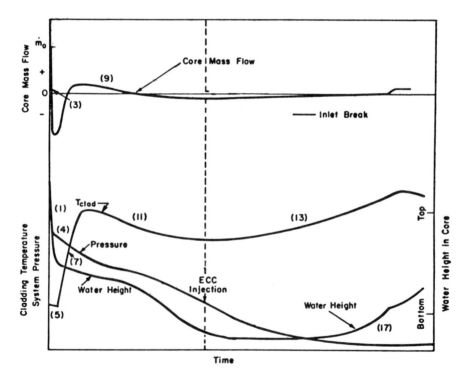

Figure 22.10: Generalized loss-of-coolant behavior for large pipe breaks in a PWR. (Adapted from Brockett et al, 1972).

flow decreases rapidly (3) and reverses before resuming positive flow (9), after which the reverse flow again occurs. The flow rate (absolute) gradually decreases until reflooding (17). The water level (all liquid equivalent) behaves accordingly. The subcooled blowdown rapidly reduces the pressure (1); the depressurization rate is less during saturated blowdown (4). The cladding temperture rises sharply at CHF (5) and then is turned around by both the positive flow and the negative flow (11). The temperature turnaround is a function not only of the improved heat transfer coefficients but also of the reduction of heat flux due to power decay and removal of stored energy in the fuel rods. The clad temperature rises again as the core is uncovered (13) until the reflood level reaches the bottom of the core (17). The fuel pin temperature rise may be accelerated prior to reflood due to the zirconium-steam reaction.

Key Mechanisms and Analytical Techniques

Most of the items covered under BWR mechanisms have application to PWR LOCA analysis. The same general format will be used, but the discussion will center on those items which are of major concern in present analyses.

Decompression and break flow. The subcooled decompression is controlled by pressure wave propagation through the system. The period is short as the fluid path from the break to the lower plenum is relatively unobstructed and the fluid is essentially liquid. "Water hammer" poses a potential hazard to the structure. The rate of void formation is important in establishing when the discharge flow becomes "choked". There is concern that present critical flow models are inadequate because of lack of consideration of non-equilibrium effects and two-dimensional effects.

Transient Analysis. Flows in the system are space- and time-dependent, with sudden changes in flow pattern and direction being common. Reactor applications and the current worldwide emphasis on the dynamic aspects of two-phase flows in general has directed a reevaluation of the basic equations. Practically all of the early research was conducted using the classical "three-equation" models of two-phase flows, considering the mixture mass, momentum, and energy conservation equations (Yadigaroglu and Lahey, 1976). Although this approach is generally acceptable in steady-state situations, it became evident that up to "six-equation" models of two-phase flow must be employed to correctly describe transient situations (Ishii, 1975). In the six-equation approach one considers separately the conservation equations of each phase. This leads to the need for considering the detailed interactions between phases, ie the mass, momentum, and energy exchanges between the phases. However, the phase interaction laws required for coupling the two sets of conservation equations at the interfaces are generally not well established.

In many situations that are under study today the state of one of the two phases may significantly depart from thermodynamic equilibrium. These situations include the critical flow condition, rapid depressurization, subcooled boiling, dispersed flow film boiling (post-dryout heat transfer), and direct contact condensation. Thus, a considerable effort has been devoted to the study of non-equilibrium two-phase flows (Jones and Saha, (1977).

Another fairly new aspect of two-phase flow modeling is the consideration of two- and three-dimensional effects. Phenomena such as those occurring in the cores and vessels (including steam generators) of LWR's during postulated accidents, as well as during steady state, are clearly multi-dimensional and should be modeled as such. It is thus necessary not only to consider the two phases separately, but also to write the conservation equations in two and three dimensions. Weisman (1979) comments in more detail on modifications being introduced into contemporary codes.

Flow instability. During the later stages of blowdown, the subcooled ECC may interact with steam flowing in the cold legs to produce flow and pressure oscillations or rapid transients. Oscillations have been observed in scale model tests (1/30 - 1/3 PWR scale) of the ECC injection region. Rothe et al (1977) report data taken in a transparent 1/20-scale cold leg model. A one-dimensional analysis was presented which identifies the governing dimensionless parameters, predicts flow regime maps for various types of water plug behavior, and predicts the amplitude, frequency, and history of the unsteady flow and pressure. The predictions are compared with these data and with data from other scale models. Flow regimes are predicted closely, and frequency and pressure amplitude are generally predicted to within 25% of the measured values.

The reflooding rate of the core is controlled by the balance between the driving head of the ECC and the back pressure in the upper plenum. Frequently this sets up a "U-tube manometer" between the downcomer and the quenched rods. The oscillations are probably due to the feedbacks between flooding rate, vapor generation rate, and back pressure caused by the mixture leaving this core. As pointed out by Yadigaroglu (1978a), this instability is thus more of a density wave instability than a manometer oscillation. Oscillations of the flow rate can occur which are especially violent during the initial phases of the transient; these oscillations produce fluctuations in the fuel pin surface temperature. Observed frequencies are approximately 0.3 Hz. A comprehensive survey of work in this area, particularly at Westinghouse and Kraftwerk Union is presented by Hochreiter and Riedle (1977).

DNB/Dryout. The CHF condition can either be of the DNB or dryout type. As in the case of BWR's, correlations developed from experiments under steady conditions are used together with local instantaneous parameters. PWR behavior is more complex, however, since CHF could occur during flow stagnation or flow reversal.

Experiments directed specifically at determining critical heat flux or "time to critical heat flux" for LOCA conditions have been performed at many laboratories around the world. The experiments in the US range from the large, highly complex Loss of Fluid Test and Semiscale experiments at the Idaho National Engineering Laboratory, which stresses acquisition of data under conditions approximating those in actual core, to the small experiments at Massachusetts Institute of Technology, directed at obtaining approximate or limiting solutions based on physical understanding. Griffith et al (1977) discuss the latter studies and propose a method of estimating the "minimum time to CHF".

This assumes flow stagnation in the hottest region of the core. A standard pool boiling CHF correlation, modified for geometry, was converted to a void fraction (or quality) basis. The minimum time to CHF is then the time required to remove the subcooling in the stagnant mass plus the time required to reach the critical void. For a flowing situation, at low velocity, the CHF-void correlation is used directly (Bjornard and Griffith, 1977).

Snider (1979) recently described experimental and analytical results for the Semiscale Mod-1 electrically heated core simultator. The calculated in-core hydraulic behavior was used to obtain a better understanding of observed critical heat flux conditions during the early stage of a simulated LOCA initiated by a large cold leg break. The critical condition occurred during the instantaneous flow stagnation associated with flow reversal; however, it did not occur in the region of the prolonged flow stagnation. Rapid voiding of certain flow channels prior to flow reversal was considered to be responsible for the critical condition. Temporary rewetting was observed during flow reversal. Additional transient CHF work is noted by Bergles (1979).

CCFL. CCFL is an important issue in PWR lower plenum refill. Block and Schrock (1977) note that the countercurrent flow in PWR downcomers is significantly different from the simple air-water experiments used to derive flooding correlations due to: more complex geometry, influence of the liquid inlet flow rate, condensation effects, and the periodic effects mentioned earlier. The conclusion is that accurate correlation requires data in a model which approaches the order of magnitude of the size of the actual pressure vessel.

Reflooding and rewetting. The core is quenched as indicated in Section 22.3, except that the front proceeds only from the bottom. The tracking of the quench front is complicated by the reflood oscillations noted above.

Steam Generator Safety Issues

There do not appear to be specific problems in PWR steam generators which require analysis during LOCA. The PWR steam generators seem to be free from deletereous oscillatory behavior of the type studied in connection with liquid metal fast breeder reactor steam generators. Tube "denting", brought about by dry-out and buildup of corrosive deposits in the tube - support plate gaps is a subject of much study, eg Thomas et al (1978). Condensation-induced water hammer, caused by loss of normal feedwater flow and collapse of the resultant steam volume by cold auxiliary water, can cause severe mechanical damage (Jones et al 1979).

Variations in Reactor Types

The recent PWR designs in Germany (eg Biblis) employs both cold leg and hot leg injections. The accumulator injection is activated at a somewhat higher pressure (20 bar) than US systems. The higher pressure increases the probability that the injected coolant will reach the lower plenum. The hot leg injection is also more desirable due to the higher gravity head for lower

plenum refill. It must be emphasized that reactor cooling configurations are continually modified in response to the results
of safety studies. For example, hot leg injection and even direct
lower plenum injection are mentioned by the US Nuclear Regulatory
Commission (1978).

Computer Codes

The types of thermal-hydraulic codes as well as the coupling
between the codes are similar to those of the BWR (Figure 22.7)
except that the thermal radition code is deleted (due to the open
lattice fuel arrangement).

Current production versions of the computer codes used for
PWR licensing and licensing audit calculations include RELAP4-
MOD1 and RELAP4-MOD3. Solbrig and Barnum have discussed the
basic codes (Solbrig and Barnum, 1976), input modeling (Barnum
and Solbrig, 1976a), and typical results (Barnum and Solbrig,
1976b). Among the limitations noted are assumption of homogeneous
flow, one-dimensional flow (although attempts are made to correct
for multidimensional effects), and use of steady-state correlations.

Newer codes address these limitations. The RELAP5 code is an
advanced, one-dimensional, code based on a nonhomogeneous, non-
equilibrium model (Ransom et al, 1978). Five field equations
characterize the two-fluid model. The TRAC-P1A code (Pryor and
Sicilian, 1978) emobides the current state of the art and features
a nonhomogeneous, multidimensional fluid dynamics treatment;
nonequilibrium thermodynamics; and a flow-regime-dependent
constitutive equation package to describe the basic physical
phenomena that occur under accident conditions. A six-equation,
two-fluid formulation is used in the reactor vessel.

The basic approach of the worldwide LWR safety research
programs is to develop computer codes to analyze LOCA and other
abnormal events. These codes are to be based on physical understanding, yet these codes have become so large and complex that
few people understand all of the models employed or the numerical
techniques. Experimental programs are thus required to provide
information to code developers ("separate effects tests") or to
verify the code ("integrated system tests"). A description of the
coordinated program which has been launched in the US, with the
cooperation of laboratories around the world, is given in a report
by the US Nuclear Regulatory Commission (1978).

As the incident at Three Mile Island (Nuclear News, 1979) has
indicated, however, more safety research is needed in non-LOCA or
"small break events. The computer programs which have been
developed for the DBA-LOCA should, however, prove valuable for
simulating plant behavior for small breaks under a variety of
operator responses and equipment malfunctions.

References

Bankoff, S G, (1978) "Vapor explosions: a critical review". Heat Transfer
1978, Proceedings of the Sixth International Heat Transfer Conference, 6, 355-
360.

Bankoff, S G, (1979) "Vapour explosions". Preprint proceedings of the Japan-US Seminar on Two-Phase Flow Dynamics, Kobe, Japan, July 31-August 3, 227-248.

Barnum, D J and Solbrig, C W, (1976a) "The RELAP4 computer code: Part 2. Engineering design of the input model". *Nuclear Safety,* 17, 299-311.

Barnum, D J and Solbrig, C W, (1976b) "The RELAP4 computer code: Part 3. LOCA analysis of a typical PWR plant". *Nuclear Safety,* 17, 422-436.

Belda, W and Mayinger, F, (1977) "Dryout delay during blowdown". *AIChE Symposium Series,* 73, No. 164, 21-29.

Bergles, A E, (1979) "Burnout in boiling heat transfer. Part III: High-Quality Forced-Convection Systems". *Nuclear Safety,* 20, 671-689.

Bnornard, T A and Griffith, P, (1977) "PWR blowdown heat transfer". in *Symposium on the Thermal and Hydraulic Aspects of Nuclear Reactor Safety,* 1: *Light Water Reactors,* O C Jones, Jr. and S G Bankoff (Editors), ASME, New York, 17-39.

Block, J A and Rothe, P H, (1975) "Aerodynamic behavior of liquid sprays". Creare report TN-206.

Block, J A and Schrock, V E, (1977) "Emergency cooling water delivery to the core inlet of PWR's during LOCA". in *Symposium on the Thermal and Hydraulic Aspects of Nuclear Reactor Safety,* 1: *Light Water Reactors,* O C Jones, Jr. and S G Bankoff (Editors), ASME, New York, 109-132.

Bouré, J A, (1977) "The critical flow phenomenon with reference to two-phase flow and nuclear systems". in *Symposium on the Thermal and Hydraulic Aspects of Nuclear Reactor Safety,* 1: *Light Water Reactors,* O C Jones, Jr. and S G Bankoff (Editors), ASME, New York, 195-216.

Bowring, R W, (1977) "A new mixed flow cluster dryout correlation for pressure in the range 0.6-15.5 MN/m^2 (90-2250 psia) - for use in a transient blowdown code". Proceedings of the Conference on Heat and Fluid Flow in Water Reactor Safety, Manchester, United Kingdom, September 13-15.

Brockett, G F, Shumway, R W, Zane, J O and Griebe, R W, (1972) "Loss of coolant: control of consequences by emergency core boiling". Proceedings of the 1972 International Conference on Nuclear Solutions to World Energy Problems, Washington, D C, November 13-17.

Cumo, M, Fabrizi, F and Palazzi, G, (1978) "Transient critical heat flux in loss-of-flow-accidents (L.O.F.A.)". *International Journal of Multiphase Flow,* 4, 497-509.

Dix, G E and Andersen, J G M, (1977) "Spray cooling heat transfer for a BWR fuel bundle", in *Symposium on the Thermal and Hydraulic Aspects of Nuclear Reactor Safety,* 1: *Light Water Reactors,* O C Jones, Jr. and S G Bankoff (Editors), ASME, New York, 217-248.

Elias, E and Yadigaroglu, G, (1978) "The reflooding phase of the LOCA in PWRs. Part II: Rewetting and liquid entrainment". *Nuclear Safety,* 19, 160-175.

Griffith, P, Pearson, J F and Lepkowski, R J, (1977) "Critical heat flux during a loss-of-coolant accident". *Nuclear Safety,* 18, 298-305.

Groeneveld, D C and Gardiner, S R M, (1977) "Post-CHF heat transfer under forced convective conditions". in Symposium on the Thermal and Hydraulic Aspects of Nuclear Reactor Safety, 1: Light Water Reactors, O C Jones, Jr. and S G Bankoff (Editors), ASME, New York, 43-74.

Hochreiter, L E and Riedle, K, (1977) "Reflood heat transfer and hydraulics in pressurized water reactors". in Symposium on the Thermal and Hydraulic Aspects of Nuclear Reactor Safety, 1: Light Water Reactors, O C Jones, Jr. and S G Bankoff (Editors), ASME, New York, 75-107.

Hsu, Y Y and Sullivan, H, (1977) "Thermal-hydraulic aspects of PWR safety research". in Symposium on the Thermal and Hydraulic Aspects of Nuclear Reactor Safety, 1: Light Water Reactors, O C Jones, Jr. and S G Bankoff (Editors), ASME, New York, 1-16.

Ishii, M, (1975) "Thermo-fluid dynamic theory of two-phase flow". Monograph No. 22, Collection de la Direction Des Etudes et Recherches D'Electricité de France, Eryolles, Paris.

Ishigai, S, Nakanishi, S, Yamauchi, S and Masuda, T, (1974) "Effect of transient flow on premature dryout in tube". Heat Transfer 1974, Proceedings of the Fifth International Heat Transfer Conference, 4, 300-304.

Jones, O C, Jr. and Saha, P, (1977) "Non-equilibrium aspects of water reactor safety". in Thermal and Hydraulic Aspects of Water Reactor Safety, 1: Light Water Reactors, O C Jones, Jr. and S G Bankoff (Editors), ASME, New York, 249-288.

Jones, O C, Jr., (1979) "Flashing inception in flowing liquids". Paper presented at 18th National Heat Transfer Conference, San Diego, California.

Jones, O C, Jr., Saha, P, Wu, B J C and Ginsberg, T, (1979) "Condensation induced water hammer in steam generators". Preprint Proceedings of the Japan-US Seminar on Two-Phase Flow Dynamics, Kobe, Japan, July 31-August 3, 227-248.

Lahey, R T, Jr., (1977) "The status of boiling water nuclear reactor safety technology". in Symposium on the Thermal and Hydraulic Aspects of Nuclear Reactor Safety, 1: Light Water Reactors, O C Jones, Jr. and S G Bankoff (Editors), ASME, New York, 151-171.

Lahey R T, Jr., (1979a) "Current boiling water nuclear reactor LOCA issues". Preprint Proceedings of the Japan-US Seminar on Two-Phase Flow Dynamics, Kobe, Japan, July 31-August 3, 389-414.

Lahey, R T, Jr., (1979b) "Transient analysis of two-phase systems". Preprint Proceedings of the Japan-US Seminar on Two-Phase Flow Dynamics, Kobe, Japan, July 31-August 3, 175-196.

Letzring, W J, Editor, (1977) "BWR blowdown/emergency core cooling program prelminary facility description report for the BD/ECC1A test phase". GEAP-23592, General Electric Company, December.

Marks, J S and Andeen, G B, (1979) "Chugging and condensation oscillation". ASME paper presented at 18th National Heat Transfer Conference, San Diego, California, August.

Mayinger, F and Langner, H, (1978) "Post-dryout heat transfer". Heat Transfer 1978, Proceedings of the Sixth International Heat Transfer Conference, 6, Hemisphere Publishing Corp., Washington, 181-198.

Muralidharan, R, Editor, (1976) "BWR blowdown heat transfer program final report". GEAP-2124, General Electric Company.

Nuclear News, (1979) "The ordeal at Three Mile Island". April 6.

Prassinos, P G and Liao, C K, (1979) "An investigation of two-phase flow regimes in LOFT piping during loss of coolant experiments". NUREG/CR-0606, Idaho National Engineering Laboratory.

Pryor, R J and Sicilian, J M, (1978) "TRAC code overview and status". Presented at Sixth Water Reactor Safety Research Meeting, Gaithersburg, Maryland, November.

Ransom, V H, Wagner, R J, Trapp, J A, Carlson, K E, Kiser, D M and Kuo, H H, (1978) "RELAP5-Models, code structure, and applications". Presented at Sixth Water Reactor Safety Research Meeting, Gaithersburg, Maryland, November.

Rothe, P M, Wallis, G B and Block, J A, (1977) "Cold leg ECC flow oscillations". in *Symposium on the Thermal and Hydraulic Aspects of Nuclear Reactor Safety*, **1**: *Light Water Reactors*, O C Jones, Jr. and S G Bankoff (Editors), ASME, New York, 133-150.

Sandoz, S A and Sun, K H, (1976) "Modeling environmental effects on nozzle spray distribution". ASME Paper No. 76-WA/FE-39.

Sargis, D A, Masiello, P J and Stuhmiller, J H, (1979) "A probabilistic model for predicting steam chugging phenomena". ASME paper presented at 18th National Heat Transfer Conference, San Diego, California, August.

Slifer, B C, (1971) "Loss-of-coolant accident and emergency core cooling models for General Electric boiling water reactors". NEDO-10329, General Electric Company, April.

Snider, D M, (1979) "The thermal-hydraulic phenomena resulting in early critical heat flux and rewet in the Semiscale core". *Journal of Heat Transfer*, **101**, 43-47.

Solbrig, C W and Barnum, D J, (1976) "The RELAP computer code: Part 1. Application to nuclear power plant analysis". *Nuclear Safety*, **17**, 194-204.

Taitel, Y, Lee, N and Dukler, A E, (1978) "Transient gas-liquid flow in horizontal pipes: modeling the flow pattern transitions". *AIChE Journal*, **24**, 920-934.

Thomas, J M, Cipolla, R C, Ranjan, G V and Derbalian, G, (1978) "Mechanical analysis of steam generator tube denting". ASME Paper No. 78-NE-5.

Tong, L S, (1978) "Heat transfer in reactor safety". *Heat Transfer 1978*, Proceedings of the Sixth International Heat Transfer Conference, **6**, 285-309.

US Nuclear Regulatory Commission, (1978) "Water reactor safety research program, a description of current and planned research". NUREG-0006, February.

Wallis, G B, (1969) "One-dimensional two-phase flow". McGraw-Hill, New York.

Weisman, J, Ake, T and Knott, R, (1976) "Two-phase pressure drop across abrupt area changes in oscillatory flow". *Nuclear Science and Engineering*, **61**, 297-309.

Weisman, J, (1979) "Two-phase flow problems in the loss-of-coolant accident in pressurized water reactors". Preprint Proceedings of the Japan-US Seminar on Two-Phase Flow Dynamics, Kobe, Japan, July 31-August 3, 415-439.

Yadigaroglu, G and Lahey, R T, Jr., (1976) "On the various forms of the conservation equations in two-phase flow". *International Journal of Multiphase Flow*, **2**, 447-494.

Yadigaroglu, G, (1978a) "Two-phase flow instabilities and propagation phenomena". Von Karman Institute for Fluid Dynamics Lecture Series 1978-5, April.

Yadigaroglu, G, (1978b) "The reflooding phase of the LOCA in PWRs. Part 1: Core heat transfer and fluid flow". *Nuclear Safety*, **19**, 20-36.

Ybarrondo, L J, Solbrig, C W and Isbin, H S, (1972) "The 'calculated' loss-of-coolant accident: a review". AIChE Monograph Series, **68**, No. 7.

Chapter 23

Chemical Plant Safety

F. MAYINGER

23.1 Safety Problems in Chemical Engineering

Safety problems arising from unit operations in chemical
engineering depend on the mechanical, thermal or chemical process
to be performed. Especially in thermal and chemical processes
high pressures have to be mastered, and with exothermic reactions
the conditions of the heat balance have to be carefully observed
to avoid an unallowable temperature increase or a pressure
excursion.

Chemical reactions can be controlled by mass transport, heat
transfer or in a catalytic way. Heat transfer and reducing mass
flow rate of the reacting components may be sometimes too slow to
avoid an excursion of the reaction, and then a pressure relief by
vent valves can be an additional measure to keep the system within
safe borders. If the reacting components are liquids, flashing may
occur after opening the vent valves and a two-phase flow mixture
leaves the nozzle. Under certain conditions - e.g. with poisoning
or explosive components - it is not possible to blow the two-phase
mixture into the free atmosphere. Pressure suppression systems or
containment vessels have then to be used. Pressure suppression
can be performed by injecting the two-phase mixture into a cold
liquid, where the gas phase is condensed and absorbed.

For the safe layout of a plant or an apparatus under these
venting conditions, several two-phase flow phenomena have to be
well understood. For pure substances the following phenomena have
to be taken in account:

- flashing

- thermodynamic non-equilibrium between liquid and gas

- phase separation and entrainment in the vessel

- critical mass flow rate in the nozzle

- pressure drop under high velocities and

- chugging phenomena during condensing in a pressure suppression system.

In a multi-component system, where some substances are more
volatile than the others, additional questions arise, like

- how the flashing process differs from that for single
  components,

- does the more volatile component stay in solution or is
  there time enough for destillation,

- may the more volatile component readily form nuclei for
  boiling and flashing.

Finally, with reacting components, the reaction kinetics and
the heat and mass transfer between the liquid and the gas phase has
to be known.

For pure substances, some two-phase flow phenomena related to
these safety problems are known from nuclear engineering; however,
there the experiments and theoretical deliberations mainly were
made for water substance. In some cases also tests were done with
refrigerants. For organic liquids, however, very little is known
about flashing, critical two-phase flow and quenching during fast
transients.

The situation becomes even worse for safety studies. One can
perhaps start from the assumption that during fast depressurizations
there is not enough time to allow the more volatile components to
go out of solution. Up to what extent these components, however,
may accelerate the flashing process by forming nuclei has to be
studied for each system separately. For constant pressure, usually
the reaction kinetics are well known. The question is, whether
they behave the same under fast transients. For calculating the
heat and mass transfer under accident conditions, the present
status of knowledge only allows one to start from steady state
experience.

23.2 General Safety Strategy

Like in all technical disciplines, also in chemical engineering
the safety strategy follows the principle which may be called
defence in depth. This is accomplished in the manufacturing of the
components and by repeated tests, up to actual operation.

Already the mechanical layout of components relevant for
safety is made in a way that in addition to careful and detailed
calculations, large safety margins are observed. Great care is taken
for material selection, manufacturing and material tests. The
probability for a component failure is minimized.

Safety systems, for example, emergency coolers or relief
valves, are usually provided in a redundant manner to improve the
availability of the safety systems in case one subsystem should
fail, thereby decreasing the probability of a catastrophic failure.

The engineer very often is in the position that he has to lay
out an apparatus or an entire plant without knowing all details
of physical phenomena to be expected during operation. In safety
deliberations a way out of the arising problems sometimes seems
to be offered by performing the calculations with conservative
assumptions and worst case initial conditions. "Conservative" means
that the worst physical behaviour is assumed with respect to
safety effects.

A certain assumption may be conservative in one way and may not be in another. For example, in laying out a safety relief valve in a conservative way, flow conditions have to be assumed in the nozzle, which give minimum critical flow rate. On the other hand, calculating the depressurization in a vessel after a break of a pipeline, maximum critical flow rates have to be taken in account, to predict the forces acting onto the internals due to pressure pulses in a pessimistic way. So one easily can run out of a logical correlation system. Finally, one always can argue what conservatism is conservative enough.

Basing safety calculations on conservative assumptions, certainly gives preliminary results. Doing the calculations with best-estimate initial and boundary conditions requires a detailed and well based knowledge of the physical phenomena, which may necessitate long and expensive research. In the end, an optimal layout of safety systems has to be based on a good and fundamental understanding of the physical phenomena.

23.3 Flashing and Phase Separation

Pressure waves in a liquid near saturation conditions may cause stable or unstable behaviour. A sudden pressure increase always stabilizes the thermodynamic state, because the saturated liquid or vapour will be changed to subcooled liquid or to super- heated vapour (see Figs. 23.1 and 23.2). Depressurization, however, causes unstable conditions, i.e. the saturated liquid will become superheated and the saturated vapour subcooled.

The question now is, how fast this unstable situation can be stabilized again by boiling or condensing. For safety delibera- tions, mainly pressurized liquids are of interest and therefore the boiling delay and the maximum thermodynamic non-equilibrium has to be known to calculate the flashing phenomena correctly.

Observations of the pressure and temperature behaviour during a depressurization - also called blowdown - measured in a vessel originally filled with a refrigerant, partially in liquid and partially in gaseous phase under saturated conditions, is shown in Fig. 23.3. After opening a relief valve, the pressure decreases rapidly for a short period; however, it then increases again. After reaching a maximum, the pressure is reduced more slowly, but continuously.

The temperature in the liquid and in the gas phase shows at the beginning a remarkable deviation from saturation conditions, and not until the pressure reaches the maximum value can thermo- dynamic equilibrium between the phases be observed. In the period B - C, there is an evaporation delay in the mixture, and the temperature of the liquid stays almost at the value of the initial conditions. At the point C bubble growth starts in the mixture due to flash evaporation. During the period C - D the flashing mixture moves upwards and the superheating of the liquid slowly decreases. At point D the flashing boundary reaches the outlet nozzle at the top of the vessel, which originally was not completely filled with liquid, and now instead of single phase vapour a two-phase mixture flows through the nozzle. The maximum volumetric flow rate out of the nozzle is decreased by this

F Mayinger

High pressure wave

subcooled liquid

stable

superheated vapour

Low pressure wave

superheated liquid

unstable

subcooled vapour

Figure 23.1: Stable and unstable behaviour with pressure waves.

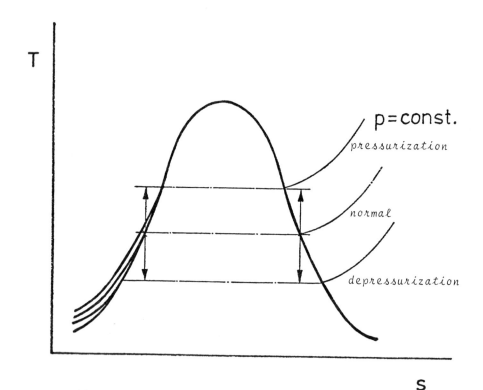

Figure 23.2: Pressure waves and stability.

effect. In the period D - F, due to non-equilibrium conditions the flash-evaporated volume prevails over the out-streaming volume, which allows the pressure to rise again. At the point F the evaporated volume by flashing is equal to the out-streaming volume, which can be deduced from the maximum in the pressure curve. In the period D - F the thermodynamic non-equilibrium is rapidly reduced due to the evaporation by flashing. The period

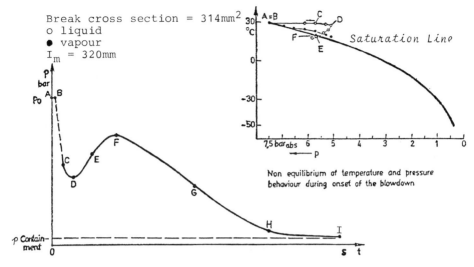

Break cross section = 314mm²
o liquid
● vapour
I_m = 320mm

Non equilibrium of temperature and pressure
behaviour during onset of the blowdown

A : initial conditions

B-C : evaporation delay in the mixture
      onset of vapour decompression

        At point C : onset of bubble growth in
                     the mixture due to flash-
                     evaporation

C-D : Flashing mixture moves upwards
        At point D : Flashing boundary reaches
                     the outlet pipe at the top of
                     the vessel.
                     Onset of two-phase outflow.

D-F : Due to the non-equilibrium effects the flash-
      evaporated volume prevails over the outstreaming
      volume
        At point F : Flash evaporated volume =
                     outstreaming volume

F-H : Continuous pressure decrease due to homogeneous
      outflow
        At point H : Mixture level drops below the
                     altitude of the outlet pipe

I: Pressure equilibrium is reached

Figure 23.3: Thermodynamic non-equilibrium during depressurization.

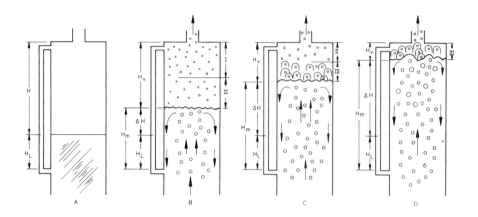

Figure 23.4: Schematic description of hydrodynamic effects with phase
             separation.

F - H shows continuous pressure decrease with homogeneous outflow
and at the point H the mixture level drops below the position of
the outlet nozzle. Finally at I, pressure equilibrium with the
atmosphere is reached.

        Schematically, the hydrodynamic effects during depressurization
and blowdown are illustrated in Figure 23.4. Besides the thermodynamic non-equilibrium the phase separation, characterized by the
upward movement of the vapour and a fall back of the liquid, plays
an important role in defining the thermodynamic and fluiddynamic
conditions at the outlet. As long as the mixture level or swell
level is far away from the outlet nozzle, the carry-over of
droplets also influences the quality at the outlet.

        Measuring the void fraction over the height of the vessel
during the blowdown, gives the conditions shown in Figure 23.5.
By carrying-over liquid droplets, the void fraction in the upper
part of the vessel decreases, whereas originally only vapour was
present. In the lower parts of the vessel - originally filled
with liquid - the void fraction increases continuously with time.

        There are several models in the literature for predicting
void fraction and phase separation in a pool, when vapour is
blown through from below and is rising in the liquid (see Figure
23.6). The references in this figure are discussed in detail by
Viecenz (1980); the citations are either in the present bibliography or are given by Viecenz. Most of these models are for
steady state conditions.

        Viecenz (1980) conducted a comprehensive experimental
programme and developed a simple correlation for the mean void
fraction in the vessel. As shown in Figure 23.7, the correlation
(equation 14.22) predicts not only his own experiments but also
measurements by Behringer (1972), Margulova (1953) and Wilson
(1961).

        Viecenz (1980) observed two regions where the phase separation

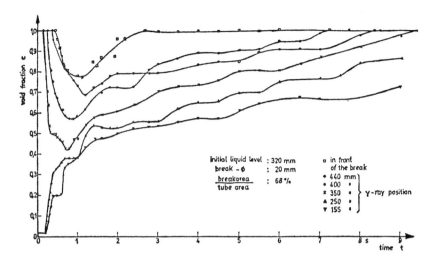

Figure 23.5: Transient void-fraction distribution at different heights of the

Authors	Medium	Geometry	Empirical relation for average void fraction in mixture		flow pattern boundary	Pressure range (bar)
Wilson (1961)	H_2O- H_2O steam	d_{cont} 480mm 100mm Sintered Plate Own Data	$\langle \bar{\epsilon} \rangle = C \cdot (Fr)_a^a (We)_a^{0.1} \left[\frac{\rho}{\rho' - \rho''} \right]^{0.17}$	C=0.68 a=0.31	$\frac{u_o''}{\left[g \frac{\sigma}{g(\rho' - \rho'')} \right]^{\frac{1}{4}}} > 2$	20÷41
				C=0.88 a=0.20	$\frac{u_o''}{\left[\sqrt{\frac{\sigma}{g(\rho' - \rho'')}} \right]^{\frac{1}{4}}} > 2$	1.07 190
Sterman Dementiev Leplin	H_2O- H_2O steam	d_{cont} 51mm 69mm 85mm 200mm Boiler Sintered plate Behringer data /23/ Margulova data /19/	$\langle \bar{\epsilon} \rangle = C \cdot (Fr)_a^a (We)_a^{0.25} \left[\frac{\rho''}{\rho' - \rho''} \right]^{0.17}$	C=0.88 a=0.40	$\frac{u_o''}{\left[g \frac{\sigma}{g(\rho' - \rho'')} \right]^{\frac{1}{4}}} < 3.7$	1.07÷190
				C=1.90 a=0.17	$\frac{u_o''}{\left[g \frac{\sigma}{g(\rho' - \rho'')} \right]^{\frac{1}{4}}} > 3.7$	
Margulova (1953)	H_2O- H_2O steam	d_{cont} 200mm Sintered plate Own data	$\langle \bar{\epsilon} \rangle = (0.576 + 0.00414 p(at\ m)) \cdot (u_a'')^{0.75}$		none	91÷190
Kurbatov	H_2O- H_2O steam	d_{cont} 51-200mm Boiler/ sintered plate Behrinter data /23/ Margulova data /19/	$\langle \bar{\epsilon} \rangle = 0.67\ (Fr)^{1/3} (We)^{1/6} \left[\frac{\rho'}{\rho' - \rho''} \right]^{1/3} \left[\frac{v'}{v''} \right]^{2/9}$		none	1.07÷190
Labuncov		$d_{cont} = 17 \div 748mm$	$\langle \bar{\epsilon} \rangle \left(1 + \frac{u_{BLS}}{u_o''} \right)^{-1}$ $u_{BLS} = u_{BL} \cdot \psi_L$ $u_{BL} = 1.5 \left(\frac{g \cdot \sigma \cdot (\rho' - \rho'')}{\rho'^2} \right)^{\frac{1}{4}}$ $\psi_L = 1.4 (\frac{\rho'}{\rho''})^{1/5} (1 - \frac{\rho'}{\rho''})^5$		Ba > 500	1÷196
Mersmann /22/	Air H_2O H_2O-Hg Toluol H_2O		$\langle \bar{\epsilon} \rangle (1 - \langle \bar{\epsilon} \rangle)^n = 0.14 \cdot u_o'' \left(\frac{\rho'^2}{g \sigma(\rho' - \rho'')} \right)^{\frac{1}{4}} \left(\frac{\rho'}{\rho' - \rho''} \right)^{\frac{1}{4}} \left(\frac{d^3 \rho^2}{\pi (\rho' - \rho'') g} (\frac{\rho'}{\rho''})^{5/3} \right)^{1/24}$		n=1 $(\frac{\rho'}{\rho''})$ from plot	

with $Fr = \dfrac{\sqrt{\frac{\sigma}{g(\rho' - \rho'')}}}{d_{cont}}$ $We = $ $Bo = g \cdot d_{cont}^2 \frac{(\rho' - \rho'')}{\sigma}$

Figure 23.6: Equations for phase separation. Refer to p. 452 for symbols.

behaves somewhat differently and seems to be influenced by the Froude-number (see Figure 23.8). The correlation for predicting the void fraction, therefore, must be used with two different sets of the empirical factor C and the exponent n, depending whether the Froude-number is greater or smaller than 3. Also the behaviour

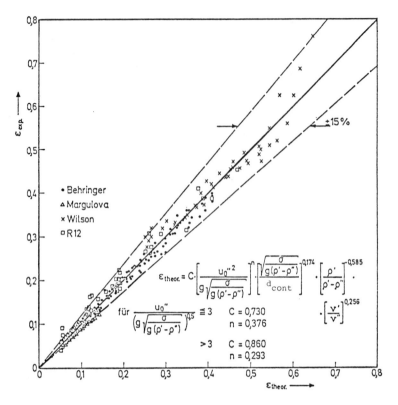

Figure 23.7: Comparison of experimental data measured at the IfV (Freon 12) and data of Margulova, Wilson and Behringer with empirical equation.

of the drift velocities undergoes a change at a Froude-number of approximately 3. As illustrated in Figure 23.9, also the drift velocities can be correlated by a similar equation like the void fraction. In addition, however, the diameter and the height of the vessel as well as the mean bubble diameter have to be taken into account in the correlation. In the Froude-number u_0'' is the superficial velocity of the vapour in the vessel if no liquid at all would be present.

For calculating the phase separation and the depressurization during blowdown, the correlation by Viecenz (1980) in a somewhat different way can be used, too; however, there has to be performed an iterative calculation procedure between the predictions of the phase separation, the critical mass flow rate in the nozzle and the depressurization.

If the pressurized system is not a geometrically simple vessel but consists of a complicated arrangement of apparatus, pipes, pumps and valves, predicting the blowdown becomes much more complicated because additional phenomena occur. The pump may - intentionally or accidentally - not be stopped and then its head ratio can influence the conditions in the system. The effect of

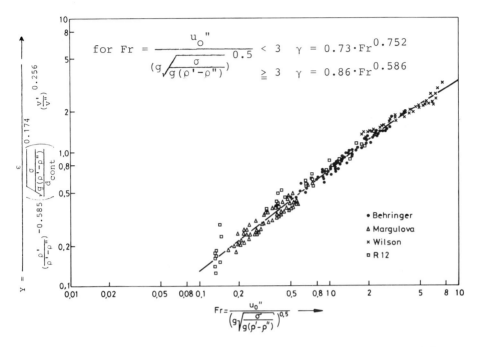

Figure 23.8: Dependence of void fraction function on Froude number, (Viecenz (1980).

void fraction on the homologous head ratio measured in water (Olson (1974)) is illustrated in Figure 23.10. It can be seen that the head ratio at a void fraction of approximately 0.2 starts to decrease rapidly, however, the pump still can transport a certain amount of mass.

In a complicated arrangement of different apparatus and other components, the flow at the nozzle is not continuous and the flow pattern may change suddenly, as illustrated in Figure 23.11. This is due to the fact that the different apparatus, depending on the pipeline connections, are not blown out and evacuated simultaneously, and therefore periods of high void may suddenly change with almost pure liquid flow at the nozzle or at the break. So the flow pattern varys between annular-, dispersed- and pure gas flow.

In the pipes connecting the various apparatus, liquid carried in the vapour flow may be de-entrained which also influences the thermo- and fluiddynamic conditions at the nozzle. However, also the opposite phenomenon occurs, namely that liquid flowing in form of an annulus at the wall, is entrained by the high shear stresses resulting from the high vapour velocity. Measurements by Langner (1978) demonstrated, that the rate of entrainment becomes larger with increasing acceleration i.e. with a faster depressurization and a shorter blowdown (see Figure 23.12).

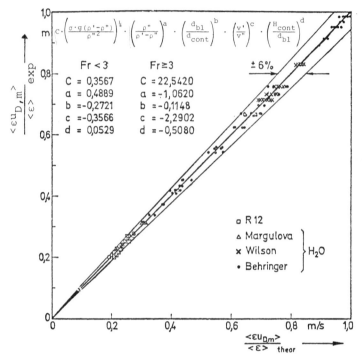

Figure 23.9: Comparison of experimental drift velocities with an empirical
equation for data measurement at IfV (Freon 12) and data by
Margulova, Wilson and Behringer.

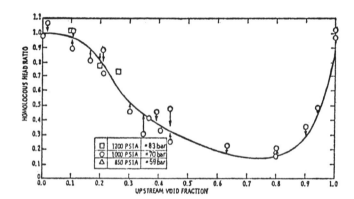

Figure 23.10: Effect of void fraction on homologous head ratio for rated
speed and near rated flow (Specific speed N_s = 1200).

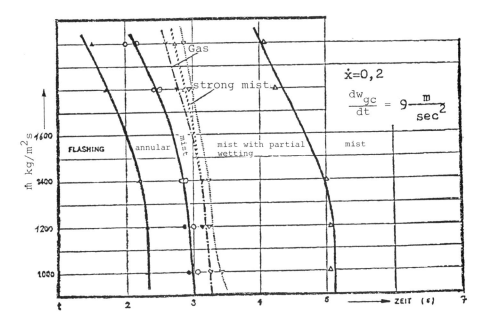

Figure 23.11: Flow patterns during blowdown.

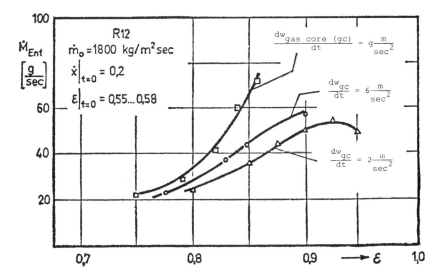

Figure 23.12: Entrainment during blowdown.

23.4 Critical Flow

From a certain pressure ratio on in single phase gas flow, the velocity in the nozzle remains constant at the value of the sonic velocity, which is also called Laval-velocity. This maximum velocity can easily be deduced from the laws of thermodynamics. In two-phase gas-liquid mixtures the situation is more complicated. It's not fully correct to speak about a sonic velocity because the travelling velocity of small pressure disturbances - which is the sonic velocity - is different in the liquid and in the gas phase. One usually speaks in two-phase flow about a critical mass flow rate. The velocity in each phase, the quality and the density distribution influences the critical mass flow rate strongly. At critical and super critical pressure ratios there is a strong and sudden depressurization in a short nozzle and thermodynamic non-equilibrium may also play an important role.

The flow conditions in a nozzle with critical flow are difficult to measure, and therefore theories describing the flow phenomena and predicting the critical flow rate contain a number of assumptions. Three classes of assumptions are demonstrated in Figure 23.13. The first assumption - homogeneous and thermal equilibrium flow - implies that the thermodynamic and the fluid-dynamic state is completely equalized between the phases. The other extreme is the assumption of frozen flow without any momentum and heat exchange between the phases. Between these two situations any condition is conceivable, for example, homogeneous flow. Complete thermal non-equilibrium or slip models, are presented by Moody (1965), Henry (1970) or Fauske (1963). It would go beyond the frame of this chapter to discuss in detail critical flow slip models. Here only a few of the important physical phenomena influencing the critical flow shall be discussed.

Let us first consider a short nozzle with a small L/D ratio. In case of a very fast depressurization and with saturated or slightly subcooled liquid at the entrance of the nozzle, the

ASSUMPTIONS
FOR CRITICAL BLOWDOWN

1. Homogeneous and thermal equilibrium

2. Homogeneous and complete thermal

   non-equilibrium

3. Slip models:
   Moody
   Henry
   Fauske

Figure 23.13: Assumptions for critical blowdown.

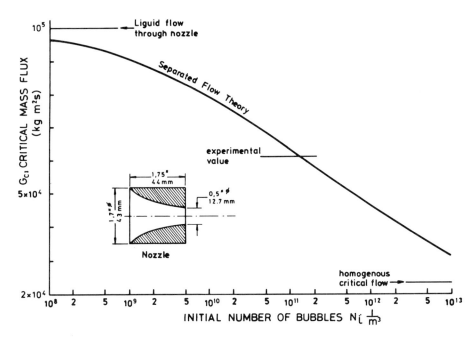

Figure 23.14: Effect of nucleation site density on prediction of short nozzle
         data of Sozzi and Sutherland (1975).

number of the nuclei being present in the liquid, plays an
important role in flashing by expansion and thereby in the
critical flow rate. Figure 23.14 shows that assuming pure liquid
phase passing through the nozzle the highest mass flow rate could
be gained. With increasing initial number of the bubbles or
nuclei per unit of volume the critical flow rate decreases. The
minimum value would be predicted by a theory assuming homogeneous
and equilibrium flow.

     From this flashing behaviour it is easily understood that the
length of the nozzle influences the critical mass flow rate
strongly, too, because with increasing nozzle length the depressurization along the flow path is less rapid and there is more time
available for flashing. Boiling and nucleation is less effective
with increasing subcooling at the nozzle inlet as demonstrated in
Figure 23.15 and therefore the critical flow rate increases with
smaller subcooling and higher quality (Henry (1979)).

     With a two-phase mixture already being present at the inlet
of the nozzle, the nucleation problem is ruled out; however, still
one can not expect that the rapid expansion through the nozzle
follows a fully thermal equilibrium path. The phases have
different densities and the pressure gradient will tend to accelerate the lighter vapour phase more than the liquid. The resulting
temperature, free energy and velocity differences cause interfacial
transfer of heat, mass and momentum. These interface processes
determine the thermodynamic and fluiddynamic paths followed by each
phase in the expansion.

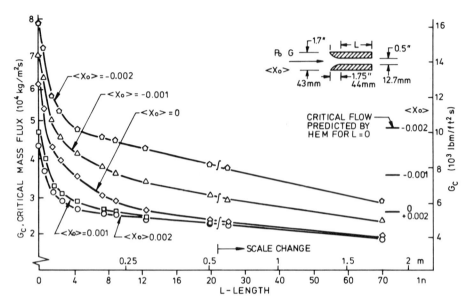

Figure 23.15: Effect of flow length on critical two-phase flow through 0.5
inch diameter tubes, Sozzi and Sutherland (1975).

If there is a long and narrow pipe between the pressure
vessel undergoing the blowdown and the nozzle, the situation for
critical flow rate can change completely and other phenomena not
discussed up to now may govern the flow conditions. Friction
may now play an important role and its influence may become so
large that the flow rate at the nozzle is not determined by the
critical flow conditions there, but by the resistance in the pipe
between the pressure vessel and the outlet to the atmosphere.

The mass flow rate in the nozzle is not only important for
predicting the depressurization and the loss of liquid in the
vessel, but also for calculating the momentum forces acting on
the discharge geometry. If long pipes are not well fastened the
momentum of the outstream jet may push them out of their holding
structure and the resulting whipping effect may destroy other
pipelines and escalate the accident.

23.5 Pressure Suppression

Poisoning gases or gas-liquid mixtures must not be blown out
into the atmosphere. With explosive components it must be
guaranteed that by mixing with air a rapid rarefication takes
place that no clouds would be formed which can later on explode.

The expansion, distribution and the mixing of a vertical
free jet in the air are functions of the jet momentum and the phase
distribution at the nozzle outlet, of the flashing effects in the
jet and of the buoyancy forces between the mixture and the air.
With increasing momentum of the jet at the nozzle outlet, the jet

diameter is smaller along its path and it has a stronger penetration into the air. The mass mixed with air and thus forming an explosive system is proportional to the third power of the nozzle diameter and inversely proportional to the 1.5th power of the jet outlet momentum (Gärtner (1978)). With increasing distance from the nozzle, buoyancy forces in the jet gain more influence. With hydrocarbons one has to be aware of the fact, that the density of their gases is usually higher than that of air and therefore if the momentum in the jet is used up, the mixture starts to flow down again. At this point the concentration of hydrocarbons in the mixture should be so low that ignition can be avoided.

Another possibility to prevent additional impacts from the accident to the environment is the quenching of the mixture flowing out from the vessel in a pool of cold and not burnable liquid where the gas is condensed. Direct contact condensation, as demonstrated in Figure 23.16, may cause serious instabilities with respect to pressure and to the liquid level. Only at low vapour fluxes is there a smooth interface between both phases, and no chugging occurs.

With steam blown into a water pool under certain flow rates and at low pressures, very strong condensation pulses, as demonstrated in Figure 23.17, were observed (Chan et al (1977/8)). There is a close and dynamic coupling between the compressible volume in the vessel and in the pipe where the vapour comes from and the liquid pool where it is condensed. These pressure pulses act on the walls of the pressure suppression chamber and may cause some damage there. To calculate the local forces on the wall from the measured pressure pulses is not quite simple because the response of the wall has to be known, especially with respect to its flexibility. Highest forces are predicted assuming a

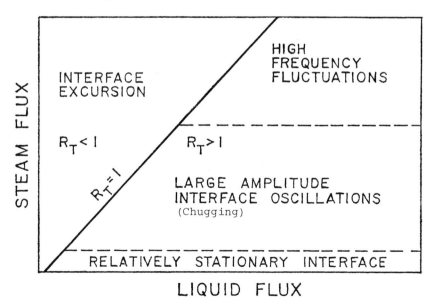

Figure 23.16: "Universal" regime map for direct contact condensation.

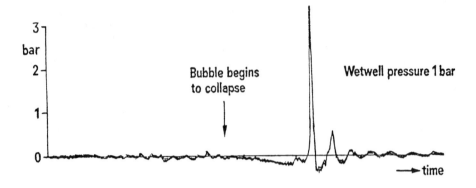

Figure 23.17: Pressure suppression system strong condensation pulses in a stiff and narrow tank at atmospheric and at elevated pressure.

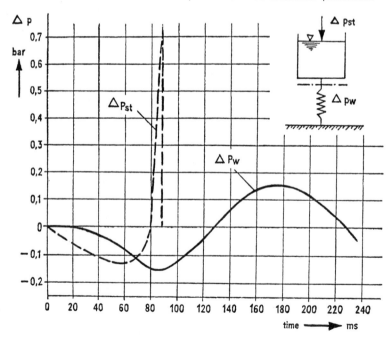

Figure 23.18: Bubble collapse: Calculated pressure transient p_{st} for a rigid tank and transformation p_w on a flexible wall.

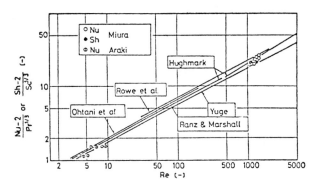

Figure 23.19: Relationship between Re and $(Nu-2)/Pr^{1/3}$ or $(Sh-2)Sc^{1/3}$
Comparison by Miura et al. (1977).

rigid wall (see Figure 23.18). The condensation pulses become
smaller with increasing pressure in the pressure suppression
system.

23.6 Heat and Mass Transfer

For predicting chemical reactions during a blowdown, and also
for estimating the probability of explosive clouds formation, the
heat and mass transfer between the phases should be known.
Unfortunately our knowledge in this field - especially under fast
transient conditions - is very poor (Mayinger (1978)).

Information is available in the literature about interfacial
transfer in annular flow and in dispersed flow. The more interesting flow pattern for accident deliberations is dispersed
flow, where small liquid droplets are transported in a gas core.
Experiments and theories in the literature show a simple relationship between the Nusselt- or Sherwood-number and the flow conditions, as demonstrated in Figure 23.19. The heat and mass transfer
is quite moderate especially for small droplets and even if under
transient conditions the Nusselt- or Sherwood-numbers would be
greater by one order of magnitude compared to the values in Figure
23.19, the time would be too short for a considerable heat transport or chemical reaction.

References

Behringer, P. (1932) "Steiggeschwindigkeit von dampfblasen in kesselrohren."
VDI-Forschungsheft 356.

Chan, C K, Lee, B and Simpson, M, (1977/8) "Steam chugging in pressure suppression containment." Annual Report prepared for the US Nuclear Regulatory Comm.

Fauske, H K, (1963) "Critical two phase steam water flows." Dissertation
University of Trondheim, Norway.

Gaertner, D. et al, (1978) "Einfluss thermodynamischer und fluiddynamischer vorgaenge auf die notentspannung chemischer reaktoren." *Chem.-Ing.-Tech.* **50**, Nr. 7, S. 503-510.

Henry, R E, (1979) "Two-phase compressible flow," *EPRI Workshop on Basic Two-Phase Flow Modelling in Reactor Safety and Performance,* Tampa, Flor. Published in *International Journal of Multiphase Flow.*

Henry, R E and Fauske, H K, (1970) "Two phase critical flow at low qualities", Part II, *Nucl. Sciences Engng.* **41**, S. 92-98.

Langner, H, (1978) "Untersuchungen des entrainment-verhaltens in stationaeren und transienten zweiphasigen ringstroemungen." Dissertation Universitaet Hannover.

Margulova, T H, (1953) "An experimental investigation of the relative velocity of vapour in bubbling through a layer of water at high pressure." *Trans. of the Power Inst.,* M V Molotow, **11**, Moscow.

Mayinger, F, (1978) "Heat and mass transfer at liquid gas interface". Lecture at International Centre for Heat and Mass Transfer International Seminar, Dubrovnik, Yugoslavia.

Moody, F J, (1965) "Maximum two phase vessel blowdown from pipes." APED - 4827 General Electric Comp., San Jose, Cal.

Olson, D H, (1974) "Experimental data report for single and two-phase steady state tests of the 1 1/2 loop MOD-1 semiscale system pump." ANCR-1150.

Sozzi, G L and Sutherland, W A, (1975) "Critical flow measurements of saturated and subcooled water at high pressure." NEDO-13418.

Viecenz, H-J, (1980) "Blasenaufstieg und phasenseparation in behaeltern bei dampfeinleitung und druckentlastung." Dissertation Universitaet Hannover.

Wilson, J P et al., (1961) "Steam volume fraction in a bubbling two phase mixture." *Trans. Am. Nucl. Soc.* **4**, No. 2.

Chapter 24

Two-Phase Flow and Heat Transfer, 1756–1980

A. E. BERGLES

24.1 Introduction

Since past is prologue, it is appropriate to review historical developments in two-phase flow and heat transfer which have led to the present state of understanding described in the preceding chapters.

In this concluding chapter, the early beginnings, particularly Leidenfrost's classic publication, will be discussed. The five decades of modern progress, 1930 to the present, are then described. The material is based on the keynote lecture for the Two-Phase Flow and Heat Transfer Symposium-Workshop which was held in Fort Lauderdale, Florida, October 1976.

24.2 Early History

The application of two-phase processes to machines can be traced back to the steam cannon of Archimedes (born 287 B.C.). It was reported that iron balls were propelled through a copper tube by expanding steam (Matschoss (1908)). The aeolipile of that general period was a bronze boiler vessel with a relatively small opening through which steam would pass at high velocity. The whirling aeolipile shown in Figure 24.1 was described by Heron of Alexandria in the first century A.D. (De Camp, 1963). This machine was actually quite sophisticated, considering that the scientific bases needed to explain its operation (nucleate boiling, two-phase flow, and angular momentum) were completely lacking. I do mean sophisticated, since when one of my students constructed a working model for his history of technology class, he was unable to get the sphere to rotate without using Teflon bearings!

More than 1500 years passed before heat engines were developed for large-scale power production. The aeolipile principle was embodied in the first reaction turbines conceived in the early 1600's. The Savery pulsometer pump of 1698 was the first steam-operated pump. It was superceded by the safer Newcomen engine of 1712, which was the forerunner of all piston-type heat engines.

It was not until later in the 18th century that some of the necessary understanding of two-phase phenomena began to appear. As pointed out by Prof. K J Bell (1966), the first literature on boiling as a serious topic of study is by Johann Gottlob Leidenfrost, a German medical doctor, who in 1756 published "A Tract About Some

Figure 24.1: Heron's whirling aeolipile (De Camp, 1963)

Qualities of Common Water." In one section of this treatise, the boiling of small liquid masses on a hot surface is discussed. The particular case of film boiling described below has been termed the Leidenfrost Phenomenon.

> J G Leidenfrost. De Aquaea Communis Nonnulis Qualitatibus Tractatus (Leidenfrost, 1756).

> On the Fixation of Water in Diverse Fire
> XV 1. An iron spoon...is heated over glowing coals...To this glowing spoon...send...one drop of very pure distilled water...This water globule...delights in a very swift motion of turning...Moreover, this drop only evaporates very slowly ...When at last exceedingly diminished so that it can hardly any more be seen, with an audible crack, which with the ears one easily hears, it finishes its existence...
> XVI ...these observations suggest that water is changed into earth by a large fire, because always after the complete evaporation of the drop some terrestrial matter remains in the heated vessel...

It is noted from the latter excerpt that one of Leidenfrost's observations pertains to crud deposition. As shown in the next excerpt, the emphasis was on practical applications.

> XIX I show a new method by which the most perfect goodness of alcoholic wine can be determined...The spirit of wine is evaluated by measuring the quantity of superfluous water, burning it of course...It is necessary to arrange the thing so that at the same time that the spirit flies off the water is fixed and impeded in its evaporation...if in the spirit of wine a little water is mixed, this after the conflagration of the first is completed...can be seen under the form of little clear globes in the glowing spoon...the pure spirit will be

totally consumed so that no liquid remains. From this you
are certain that the spirit of wine is pure alcohol.

If someone was trying to pass off low proof wine or brandy, the hot
spoon method would ferret the scoundrel out. (The alcohol would
rapidly burn off and the telltale Leidenfrost droplets of water
would be left.)

Leidenfrost had a problem in getting heat transfer rates,
since the latent heat of evaporation was not shown to exist until
a few years later. We must credit him, however, with the first
publication in boiling.

The reciprocating steam engine occupied center stage throughout the Industrial Revolution, 1750-1850. In 1774 Smeaton improved
the efficiency of Newcomen's engine from 0.5 to over 1%. James Watt
invented the separate condenser in 1769, and subsequently made many
improvements; however, not much understanding of either boiling or
condensing was involved. His single-acting steam engine is shown
in Figure 24.2. One of the first steam engines used for locomotion
is shown in Figure 24.3.

Technically successful steamboats were built prior to 1800
in both England and the United States. In the early 1980's many
steamboats were put into service on the Mississippi river watershed.
The noncondensing engines frequently operated at 150 psi, and some
boats were powered by as much as 1200 hp. Great risks to life and
property were taken to achieve such concentrations of power. Many
of the fire-tube boilers exploded as a result of overheating at
the vessel bottom where the mud settled, or as a result of deactivated safety valves (Ferguson, 1967).

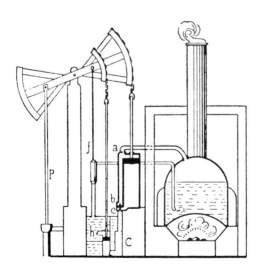

Figure 24.2: Watt's single-acting steam engine (Burstall, 1965)

TREVITHICK8,
PORTABLE STEAM ENGINE.

Catch me who can .

Mechanical Power Subduing
Animal Speed .

Figure 24.3: Locomotive developed by Trevithick (Burstall, 1965)

The Great Eastern of 1859 was one of the most ambitious engineering feats of the century: 680 ft long, 23,000 tons, and 2600 hp (paddle and screw engines). On her maiden voyage, as depicted in Figure 24.4, there was a problem resulting from over-pressure of the boiler feedwater heaters. I K Brunel, the chief engineer and promoter, died from the shock of the disaster; however, the ship was repaired and went into sporadic service for another 30 years.

Railroads had their problems too. The New York, New Haven, and Hartford engine met its demise due to a boiler explosion on December 19, 1890, at Wallingford, Connecticut. Figure 20.5 represents a picture from the report by Scientific American; however, it is fair to say that not much science was yet involved in boiling heat transfer.

Regarding the scientific aspects, it was first necessary to clear up the controversy about the "energy doctrines," the laws of thermodynamics. Carnot, Joule, Clausius, and Rankine had to provide today's standard textbook fare. We do find a few heat transfer papers of interest during the latter part of the 19th century. In 1861, Joule wrote "On the Surface Condensation of Steam" describing experiments on vertical in-tube condensation. He also proposed some augmentation devices for the water flowing in the concentric cooling jacket. We then find in the 1890's a series of papers on heat transfer and circulation rates in water tube boilers for marine engines, eg, see J T Thorneycroft (1895). There were even some complaints of flow oscillations. Critical flow of steam-water mixtures was noted by Sauvage (1892) and Adam (1906).

Not withstanding Joule's earlier contribution, it was Nusselt who began the modern understanding of the film condensation process when he published his classic paper in 1916. In 1917, Lord Rayleigh published his very important work on vapor bubble collapse.

Let us now move on to the past five decades, the modern era

Figure 24.4: The Great Eastern's ill-fated maiden voyage (Pudney, 1974)

Figure 24.5: Reconstruction of a locomotive explosion in 1890 (Reed, 1968)

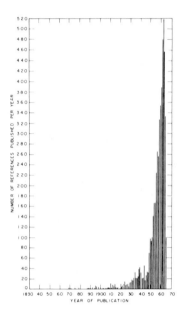

Figure 24.6: References on two-phase gas-liquid flow (Gouse, 1966)

of two-phase flow and heat transfer. Looking at Figure 24.6,
which summarizes the literature citations in two-phase gas-liquid
flow collected by Gouse (1966), 1930 suggests the point of "significant literature generation" - analogous to the point of net vapor
generation associated with the void profile in subcooled forced-
convection boiling. I shall attempt to characterize each of the
decades by noting some of the major investigators and their major
areas of activity, as well as the primary applications of the new
knowledge. Consider first the period 1930-1940.

24.3 The Period 1930-1940

Table 24.1

Two-Phase Flow and Heat Transfer in the 1930's

Investigators	Area of Activity
Jakob & Fritz (1931, 1936)	Pool boiling
Nukiyama (1934)	Pool boiling
Fritz (1935)	Pool boiling
Ledinegg (1938)	Static instability

In reminiscing about the 1930's, one of the significant
events which is still cited heavily is the clarification of the

boiling curve for nucleate pool boiling. Jakob and Fritz (1931, 1936) demonstrated at a very early stage that nucleate boiling is beset by uncertainties regarding the nature of the surface as well as the history of operation. The effects of "nuisance variables," such as those depicted in Figure 24.7, still cannot be quantitatively predicted. Nukiyama (1934) established the complete characteristics of the boiling curve using an electrically heated wire. As shown in Figure 24.8, he dramatized the first-order transitions from nucleate to film boiling, and from film boiling back to nucleate boiling. Detailed studies of bubble dynamics were carried out by Jakob and also by Fritz (1935).

Following the lead of a number of earlier investigators, eg Schnackenberg, 1937, Ledinegg (1938) interpreted the static instability that contributed to failure of a number of steam generators. The first-order instability was governed by unfavorable intersection of the boiler tube and pump head-flow characteristics. It is interesting to note that Ledinegg published an article recently (37 years after his original article) clarifying further aspects of this supposedly simple phenomenon (Ledinegg and Wiesenberger, 1975).

24.4 **The Period 1940-1950**

The 1940's were characterized by further fundamental studies of nucleate boiling experiments on subcooled flow boiling, and the first of many models concerned with two-phase pressure drop. Much of the work was spurred on by the necessity to have better design information for the chemical and process industries.

Bonilla and co-workers (1941) and Bromley (1950) were associated with important nucleate pool boiling studies. McAdams and

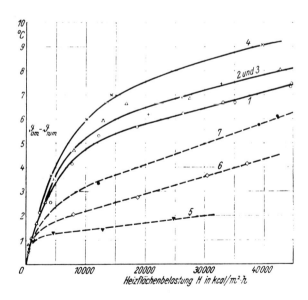

Figure 24.7: Influence of surface roughness and history on nucleate boiling of water (Jakob and Fritz, 1931)

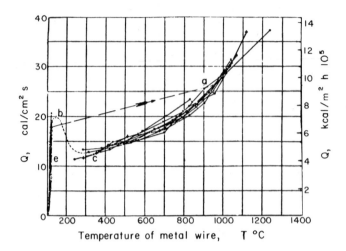

Figure 24.8: Nukiyama's boiling curves for a platinum wire (Nukiyama, 1934)

his students (McAdams et al, 1949) clarified the boiling curve for
fully developed subcooled nucleate boiling, demonstrating that the
effects of subcooling and velocity did not affect the boiling curve
to a significant extent (Figure 24.9). Martinelli and co-workers
(Lockhart and Martinelli, 1948, Martinelli and Nelson, 1949)

TABLE 24.2

Two-Phase Flow and Heat Transfer in the 1940's

Investigators	Area of Activity
Bonilla (1941)	Pool nucleate boiling
Bromley (1950)	Film boiling
McAdams (1949)	Subcooled flow boiling
Martinelli (1948, 1949)	Two-phase pressure drop

proposed the classic model for pressure drop in separated two-
phase flow. The concept of multipliers given as a function of the
Martinelli parameter shown, for example, in Figure 24.10, is still
with us today, and the proposed correlating curves are recommended
in most texts and references. The discussion of Lockhart and
Martinelli (1948) pointed out that the recommended curves ignored
a conspicuous velocity effect shown by some of the data. It seems,
however, that subsequent attempts to modify the multipliers to
account for such secondary effects demonstrate that accuracy is
achieved only with a substantial increase in complexity (Baroczy,
1965 and Chisholm, 1968).

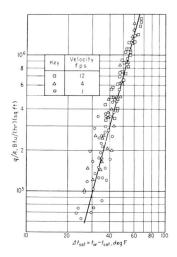

Figure 24.9: Approximate correlation of subcooled boiling of water (McAdams et al, 1949)

24.5 The Period 1950-1960

The 1950's brought about a near explosion of activity in the two-phase flow and heat transfer area. The stimulus was provided to a large extent by requirements in the aerospace industry and by the development of commercial nuclear power.

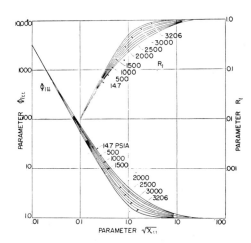

Figure 24.10: Correlation of frictional pressure drop and holdup for steam-water mixtures (Martinelli and Nelson, 1948)

TABLE 24.3

Two-Phase Flow and Heat Transfer in the 1950's

Investigators	Area of Activity
Rohsenow (1952)	Pool nucleate boiling
Nishikawa (1956)	Pool nucleate boiling
Kutateladze (1951)	Pool boiling burnout
Borishanskii (1956)	Pool nucleate boiling
Zuber (1958)	Pool boiling burnout
Doroschchuk (1959)	Flow boiling burnout
Zenkevich (1959)	Flow boiling burnout
Baker (1954)	Flow patterns
Westwater (1954-66)	Photographic studies
Katz et al (1955)	Industrial heat exchangers
Gambill et al (1958, 1961)	High Heat fluxes
Isbin (1957)	Critical flows
Gregorig (1954)	Condensation augmentation

Rohsenow (1952) modelled fully developed nucleate pool boiling in order to clarify the heat transfer mechanism and to correlate the pressure effect. Using the analogy of a forced flow of liquid brought about by bubble pumping, the Nusselt number and Reynolds number were defined using bubble departure diameter as a characteristic dimension. The Prandtl number was that of the liquid being "pumped." As shown in Figure 24.11 only one experimentally determined constant for a particular surface-fluid combination was required to correlate data for various pressure (assuming that the exponents are "universal"). Fundamental contributions to the understanding of nucleate boiling were also made by Nishikawa (1956).

The hydrodynamic theory of nucleate boiling burnout was developed during this period. Starting with closely related work, by Kutateladze (1951) and Borishanskii (1956), Zuber (1958) suggested that the instability generated by the interaction between the upward moving vapor and downward moving liquid was responsible for the burnout or peak nucleate heat flux condition. With one constant, which was related to the distribution of vapor jets, the correlation was in reasonable agreement with data for a variety of fluids.

During this period there was intense activity in many laboratories in the Soviet Union (Kutateladze, 1951, Borishanskii, 1956,

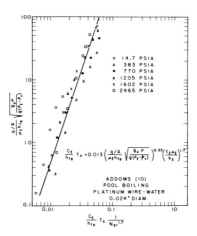

Figure 24.11: Rohsenow's correlation of the pressure effect for fully developed saturated nucleate pool boiling of water (Rohsenow, 1952)

Doroshchuk and Frid, 1959 and Zenkevich, 1959). Those of us who initiated our own studies in the 50's were introduced to the mysteries of Russian journals and those intriguing plots of burnout data, as in Figure 24.12 for subcooled forced-convection burnout. Even after going through all the trouble to learn Russian as one of our Ph.D. languages, so that we could at least read the captions, we were perplexed because the "cosmic" correlations were sometimes presented, as shown in Figure 24.12, without having the scales identified! Now, of course, most of the important foreign journals in the field are translated.

During this period there was much work on two-phase flow patterns. In the mind's eye, at least, a number of characteristic flow patterns can be identified and mapped in terms of independent variables. The flow regime map developed by Baker (1954) for horizontal, adiabatic, two-phase flow (Figure 24.13) has been clarified and modified; however, it still forms a basis for many two-phase studies today. This period was also characterized by attempts to photographically record boiling and condensing characteristics. The high speed motion pictures taken by Westwater and his colleagues are particularly noteworthy (1954-1966).

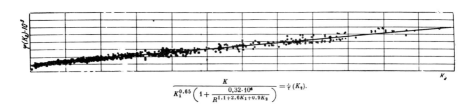

$$\frac{K}{K_1^{0.65}\left(1 + \dfrac{0.32 \cdot 10^8}{R^{1.1 + 2.6 K_1 + 0.9 K_2}}\right)} = \psi\,(K_2).$$

Figure 24.12: Correlation of forced convection subcooled boiling burnout (Zenkevich, 1959)

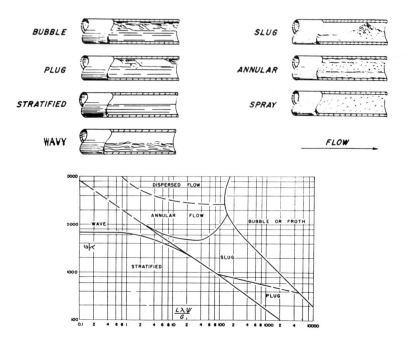

Figure 24.13: Definition and delineation of flow regimes in horizontal, adiabatic, two-phase flow according to Baker (1954)

Other important work during this decade involved study of boiling under conditions more nearly representing industrial heat exchangers (Katz et al, 1955), determination of maximum heat fluxes which can be accommodated by nucleate boiling (Gambill et al (1958, 1961), and modelling of two-phase critical flows (Isbin et al (1957). Gregorig (1954) developed the technique for obtaining thin-film condensation on the crests of ridged or rippled condensing surfaces. The type of surface shown in Figure 24.14 has found rather wide application in practical systems.

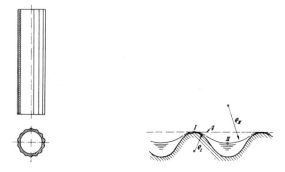

Figure 24.14: Vertical condenser tube with surface geometry developed by Gregorig (1954)

24.6 The Period 1960-1970

As we move into the 60's, it is more difficult to single out individual contributions. Work proceeded in response to needs in all of the areas previously mentioned, and there were new fundamental and applied studies associated with cooling of electronic equipment.

TABLE 24.4

Two-Phase Flow and Heat Transfer in the 1960's

Activity	Area of Activity
Gouse (1966)	Literature reviews
Han and Griffith (1965)	Nucleate boiling analysis
Merte and Clarke (1964)	Acceleration effects on pool boiling
Lienhard (1970)	Pool boiling burnout
Van Stralen (1966)	Pool boiling of mixtures
Grigull (1967-68)	Boiling near the critical point
Ornatskiy (1969)	Correlations of flow boiling burnout
Macbeth (1963, 1964)	Correlations of flow boiling burnout
Stevens and Kirby (1964)	Fluid-to-fluid modelling
Stenning and Veziroglu (1969)	Dynamic instabilities
Rose (1967)	Drop condensation
Dukler (1970)	Two-phase film behavior
Hewitt (1963)	Two-phase film behavior
Levy (1966)	Two-phase flow models
Bergles (1969)	Augmentation
Tong (1965)	Textbook
Wallis (1969)	Textbook

A model for predicting the nucleate boiling curve, which incorporated analytical representations of the basic phenomena - nucleation, growth, and departure - was developed by Han and Griffith (1965). A comparison of the prediction with experimental data is shown in Figure 24.15. The two "adjustable" constants, active nucleation site density and dynamic contact angle, represent probably the absolute minimum amount of information which

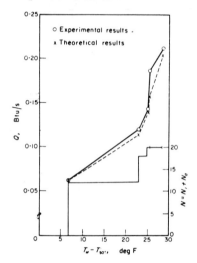

Figure 24.15: Comparison of theoretical predictions and experimental results
for nucleate boiling of water (Han and Griffith, 1965)

characterizes the surface and fluid of interest. In general, a
majority of the pool boiling studies emphasized field effects,
nonstandard geometries, and variations from pure coolants. A number
of studies (eg, Merte and Clark, 1964) produced data clarifying the
effects of acceleration, including fractional gravity, on pool
boiling burnout. Van Stralen (1966) continued his extensive
experimental and analytical studies of pool boiling of binary
mixtures. These studies held out the tantalizing prospect that a
small amount of an addition agent could produce dramatic improvements in the pool boiling critical heat flux, at least for small-
diameter test sections. Lienhard and his students (1970)
developed comprehensive extensions of burnout, according to hydrodynamic stability theory, to small test-section sizes. Pool boiling
experiments were also extended to conditions near the critical
point (Grigull and Abadzic, 1967-68).

The 60's were definitely the era of large scale experiments
in boiling heat transfer. Loops in the megawatt class sprang up
throughout the world. It was largely a matter of brute force,
as the desired geometry could be tested full size and at design
pressure. The laboratory report literature describing this work
became so voluminous that special glossaries were required to
identify the originating institutions: AE, AECL, AEEW, AERE, APED,
ANL, etc. Numerous correlations were developed from the extensive
data accumulated in these facilities. By the end of the decade,
for example, more than 20,000 burnout points were estimated to be
on record.

While some burnout researchers continued to hope for more

general correlations through use of dimensionless groups, the best accuracy was achieved by using graphical presentations (Ornatskiy, 1969) or dimensional correlations (Macbeth, 1963, Thompson and Macbeth, 1964) for a specific fluid. For example, for a forced convection boiling of high pressure water, Macbeth (1963) achieved good accuracy by using six or twelve (Thompson and Macbeth, 1964) adjustable constants derived from "world" data. There were, of course those who felt the need to develop fluid-to-fluid modelling parameters, particularly when not even the large powers available could handle the massive rod bundles which were to be tested. Scaling parameters were developed to relate refrigerant data to water data, so that the less expensive refrigerant facilities could be utilized (Stevens and Kirby, 1964). The good accuracy achieved by use of appropriate scaling parameters is shown in Figure 24.16.

Dynamic instabilities were the subject of much attention. After the work of Stenning and Veziroglu (1967), among others, "density wave" and "pressure drop" oscillations joined the vocabulary. Typical results are shown in Figure 24.17.

Condensation was still of interest, particularly the promotion of dropwise condensation. It is evident from Figure 24.18 that the type of promoter and probably a few other ill-defined variables have an effect on the heat transfer coefficient which is ultimately obtained. Detailed experiments and analyses were carried out to explain the high coefficients obtainable with drop condensation, eg, Rose (1967).

Experiments in annular two-phase flow were established to clarify the detailed nature of the annular liquid film and vapor core. The wavy interface studies of Dukler and students (eg Telles and Dukler, 1970) and the film flow rate experiments of Hewitt and his associates (1963) are typical of this activity. Very careful

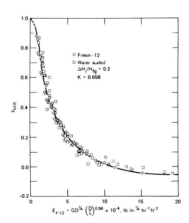

Figure 24.16: Fluid-to-fluid modelling of burnout in forced convection boiling (Stevens and Kirby, 1964).

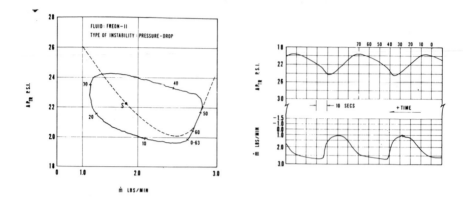

Figure 24.17: Characteristics of pressure-drop oscillations (Stenning et al,
          (1969)

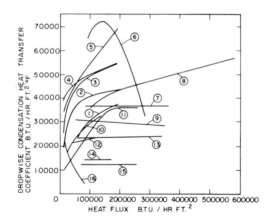

Figure 24.18: Dropwise condensation from a variety of sources (Griffith, 1967)

experiments provided the framework for models of specific flow
regimes, such as that of Levy (1966) for annular flow.

     Heat transfer augmentation emerged as a major speciality area
during the 1960's. "Enhanced" heat exchanger surfaces were studied
for desalination applications and for a variety of other two-phase
heat exchangers. Rough surfaces, twisted-tape inserts, additives,
surface vibration, fluid vibration, and other techniques were
studied (Bergles, 1969).

     As might be expected this very intense activity, the genera-
tion of literature accelerated. Gouse's analysis of the data on
citations (Figure 24.6) testifies to the growth of publications.

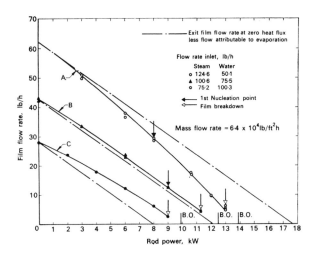

Figure 24.19: Demonstration that film flow rate in annular flow is essentially
zero at burnout point (Hewitt et al, 1963)

Efforts to disseminate and consolidate this information were also
evident. Reference compilations such as those of Gouse (5,253
citations) (1966) and Kepple and Tung (2,843 citations (1963)) are
valuable resources for beginning investigators. There were many
short courses and international symposia. University graduate
courses were initiated, and the first textbooks dealing exclusively
with two-phase, gas-liquid flow and heat transfer (Tong, 1965,
Wallis, 1969) appeared.

24.7 The Period 1970-1980

Historians would be quick to point out that it is too early to
assess the 70's. I will, however, note some of the trends which
have been evident in the past decade.

There is first of all considerable activity in developing the
equations of two-phase flow, eg (Delhaye, 1973, Ishii, 1975, Bouré
and Latrobe, 1976, and Gidaspow, 1976). These equations are
complex, but the activity has started in earnest as indicated by
periodic confrontations among proponents of various formulations.

Vapor explosions are the subject of considerable contemporary
interest (Henry and Fauske, 1975), and transient two-phase flows
(eg, transient pressure loss in piping components) are being studied
in detail (Weisman, 1977).

Perhaps the dominant activity has been associated with the
thermal-hydraulic aspects of nuclear reactor safety. This activity
includes large scale experiments, but to a large extent the effort
is concentrated on paper studies. The hypothetical double-ended
break is postulated to occur, and enormously complicated computer
codes are utilized to assess the thermal and hydraulic behavior
during blowdown and reflood. We are in an age of jargon: to

calculate the consequences of a large PWR break, the procedure
outlined in Figure 24.20 employs SATAN-V, CHIK-KIN, LOCTA-R2,
THINC, and FLECHT (Ybarrondo et al, 1972). The arsenal of programs
also includes BLODWN, BLOW, BRUCH, DYNAM, FABLE, FLASH, FLOOD,
HYDNA, MOXY, NEMI, RAMONA, RELAP, SABRE, SLUG, STABLE, and STUBE.

TABLE 24.5

Two-Phase Flow and Heat Transfer in the 1970's

Investigators	Area of Activity
Delhaye (1973)	Equations of two-phase flow
Ishii (1975)	Equations of two-phase flow
Bouré (1976)	Equations of two-phase flow
Gidaspow (1978)	Equations of two-phase flow
Henry and Fauske (1975)	Vapor explosions
Weisman (1977)	Transient two-phase flow
Prosching et al (1969)	Nuclear safety analysis code, FLASH-4
Moore and Rettig (1973)	Nuclear safety analysis code, RELAP 4

In the non-nuclear area, much effort is being directed toward
developing scale-up information or testing full-size components.
Noteworthy is the study conducted by Heat Transfer Research, Inc.
of boiling on tube bundles of kettle reboilers (Palen and Taborek,
1972). The bundle-average performance is not the same as that
which would be inferred from single-tube tests. Two discussions
of two-phase pressure drop across tube banks were recently
presented (Grant and Chisholm, 1977 and Ishihara et al, 1977).
Much more experimental information of this type will be required
for design of the enormous boilers and condensers planned for
ocean thermal energy conversion (OTEC) systems (Dugger, 1975).
With the increasing emphasis on more efficient energy utilization
and alternate systems of energy conversion, it is expected that
new two-phase flow and heat transfer problems will receive
attention in the 1980's.

New books published include those by Hewitt and Hall-Taylor
(1970), Collier (1972), Govier and Aziz (1972), Hsu and Graham
(1976), Hahne and Grigull (1977), Lahey and Moody (1977),
Butterworth and Hewitt (1977), Jones and Bankoff (1977), and
Ginoux (1978). In addition, a number of comprehensive surveys on
various two-phase phenomena have appeared recently (Bouré et al,
1973, Bouré and Hewitt, 1974, Bergles, 1975, Hewitt and Lovegrove,
1976, Tong and Bennett, 1977, Bergles, 1977, Elias and Yadigaroglu,
1977, Bergles, 1978, and Hsu, 1978. Since 1974 the
International Journal of Multiphase Flow has provided a major forum

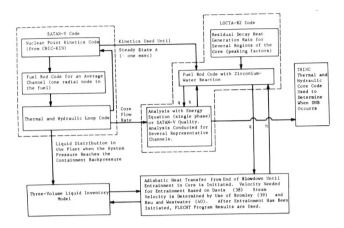

Figure 24.20: Analysis procedure for loss-of-coolant accient (Ybarrondo et al, 1972)

for work in the area. The proceedings of two recent international conferences, the NATO Advanced Study Institute on Two-Phase Flow and Heat Transfer, held in Istanbul, Turkey, August 1976, and the Two-Phase Flow and Heat Transfer Symposium-Workshop, held in Fort Lauderdale, Florida, October 1976, and again in Miami Beach, Florida, April 1979, also contain valuable survey papers as well as contributed papers. In the early 1970's it appeared that there was a slowdown in publications devoted to two-phase flow and heat transfer; this trend has clearly been reversed.

24.8   Concluding Thoughts

The 70's have demonstrated that there is still no lack of problems or lack of interest in two-phase flow and heat transfer. I am sure that this book will contribute further to the flow of information. Communication and consolidation of information is extremely important, and we find that we can now devote a good-sized book shelf to major publications and books in this area.

These contributions are a reminder that the progress and understanding of two-phase flow and heat transfer has been made by individuals - people with distinct temperaments and idiosyncrasies who have occasionally had flashes of brilliant insight. Looking into the future, there will be much more emphasis on computer solutions of two-phase phenomena. We must remember, however, that the computer codes are only as good as the basic thermal-hydraulic input. We must remember that our understanding is imperfect in many areas of two-phase flow and heat transfer, and that we will always be looking for the individual contributions to advance the state of the art or science of two-phase flow, as was done by Leidenfrost in 1756 when he told us about "the fixation of water in diverse fire".

Acknowledgement

During the past two decades I have had the pleasure of becoming acquainted with most contemporary investigators referred to in this review. I have also been privileged to associate closely with a number of them. I regret that space does not permit more than passing mention of their contributions to two-phase flow and heat transfer. I also regret that it is not possible to mention more of the growing number of worldwide contributors to this fascinating field.

References

Adam, J, (1906) "Der ausfluss von heissem wasser," *Zeitschrift des VDI*, 50, 1143-1150, 1269-1273.

Baker, O, (1954) "Simultaneous flow of oil and gas". *Oil and Gas Journal*, **53** (7), 185-195.

Baroczy, C J, (1965) "A systematic correlation for two-phase pressure drop," *Chemical Engineering Progress Symposium Series*, 62, No. 64, 232-249.

Bell, K J, (1966) "Introduction to work of J G Leidenfrost". *International Journal of Heat and Mass Transfer*, **9**, 1153-1154.

Bergles, A E, (1969) "Survey and evaluation of techniques to augment convective heat transfer". *Progress in Heat and Mass Transfer*, *Pergamon Press, Oxford*, **1**, 331-424.

Bergles, A E, (1975) "Burnout in boiling heat transfer. Part I: Pool-boiling systems". *Nuclear Safety*, **16**, 29-42.

Bergles, A E, (1977) "Burnout in boiling heat transfer. Part II: Subcooled and low-quality forced-convection systems". *Nuclear Safety*, 18, 154-167.

Bergles, A E, (1978) "Survey of augmentation of two-phase heat transfer". *Two-Phase Transport and Reactor Safety*, *II, Hemisphere, Washington, D.C.*, 457-477.

Bonilla, C F and Perry, C H, (1941) "Heat Transmission to boiling binary liquid mixtures". *Transactions AIChE*, **37**, 685-705.

Borishanskii, V M, (1956) "An equation generalizing experimental data on the cessation of bubble boiling in a large volume of liquid", *Zhurnal Tekhnicheskoi Fiziki*, 26, 452-456.

Bouré, J A, Bergles, A E and Tong, L S, (1973) "Review of two-phase flow instability". *Nuclear Engineering and Design*, 25, 165-192.

Bouré, J A and Hewitt, G F, (1974) "Some recent results and developments in gas-liquid flow: A review". *International Journal of Multiphase Flow*, 1, 139-172.

Bouré, J A and Latrobe, A, (1976) "On well-posedness of two-phase flow problems". *Sixteenth National Heat Transfer Conference, St. Louis, Missouri*, Paper No. 76-CSME/CSChE-9.

Bromley, L A, (1950) "Heat transfer in stable film boiling". *Chemical Engineering Progress*, **46** (5), 221-227.

Burstall, A F, (1965) "A history of mechanical engineering". *The MIT Press, Cambridge, Massachusetts*.

Butterworth, D and Hewitt, G F, editors (1977) "Two-phase flow and heat transfer", *Oxford University Press, Oxford*.

Chisholm, D, (1968) "The influence of mass velocity on friction pressure gradients during steam-water flow." *Proceedings Institution of Mechanical Engineers*, London, 182, Pt. 3H, 336-341.

Collier, J G, (1972) "Convective boiling and condensation". *McGraw-Hill Book Company (UK), Maidenhead*.

De Camp, L S, (1963) "The ancient engineers". *Doubleday & Company, Garden City, New York*, 242.

Delhaye, J M, (1973) "Jump conditions and entropy sources in two-phase systems: Local instant formulation". *International Journal of Multiphase Flow*, 1, 395-409.

Doroshchuk, V Y and Frid, F P, (1959) "About the effect of throttling the flow and the heated section of a pipe on critical thermal loadings". *Teploenergetika*, **6** (9), 74-79.

Dugger, G L, Editor, (1975) "Proceedings, Third Workshop on Ocean Thermal Energy Conversion (OTEC)". *APL/JHU SR 75-2*.

Elias, E and Yadigaroglu, G, (1977) "Rewetting and liquid entrainment during reflooding - State of the art". *Electric Power Research Institute Report*, EPRI NP-435.

Ferguson, E S, (1967) "Steam transportation". in *Technology in Western Civilization*, 1, M Kranzberg and C W Pursell, Jr, editors, *Oxford University Press, New York*, 284-302.

Fritz, W, (1935) "Berechnung des maximal volumens von dampfblasen". *Physikalisches Zeitschrift*, **36**, 379-384.

Gambill, W R and Greene, N E, (1958) "Boiling burnout with water in vortex flow". *Chemical Engineering Progress*, **54** (10), 68-76.

Gambill, W R, Bundy, R D and Wansbrough, R W, (1961) "Heat transfer, burnout and pressure drop for water in swirl flow through tubes with internal twisted tapes". *Chemical Engineering Progress Symposium Series*, **57** (32), 127-137.

Gidaspow, D, (1978) "Hyperbolic compressible two-phase flow equations based on stationary principles and the Fick's Law". *Two-Phase Transport and Reactor Safety*, Vol. I, *Hemisphere, Washington, D.C.*, 283-298.

Ginoux, J J, editor (1978) "Two-phase flows and heat transfer with application to nuclear reactor design problems", *Hemisphere, Washington, DC*.

Gouse, S W, Jr, (1966) "An index to the two-phase gas-liquid flow literature", *The MIT Press, Cambridge, Massachusetts.*

Govier, G W and Aziz, K, (1972) "The flow of complex mixtures in pipes". *Van Nostrand Reinhold, New York.*

Grant, I D R and Chisholm, D, (1977) "Two-phase flow on the shell-side of a segmentally baffled shell-and-tube heat exchanger". ASME *Paper No* 77-WA/HT-22.

Gregorig, R, (1954) "Hautkondensation an feingewellten oberflachen bei berück-sichtigung der oberflächenspannungen". *Zeitschrift für Angewandete Mathematik und Physik,* **5**, 36-49.

Griffith, P, (1973) "Dropwise condensation". *Notes for Special Summer Program Two-Phase Gas-Liquid Flow and Heat Transfer, MIT, July 10-21 1967.* Also in *Handbook of Heat Transfer, McGraw-Hill, New York,* 12-35.

Grigull, U and Abadzic, E, (1967-68) "Heat transfer from a wire in the critical region". *Proceedings, Institution of Mechanical Engineers, London,* **182**, Pt. 3I, 52-57.

Hahne, E and Grigull, U, Editors, (1977) "Heat transfer in boiling". *Hemisphere Washington, DC.*

Han, C-Y and Griffith, P, (1965) "The mechanism of heat transfer in nucleate pool boiling - Part II". *International Journal of Heat and Mass Transfer,* **8**, 905-914.

Henry, R E and Fauske, H K, (1975) "Energetics of vapor explosions", ASME *Paper No.* 75-HT-66.

Hewitt, G F and Hall-Taylor, N S, (1970) "Annular two-phase flow". *Pergamon Press, Oxford.*

Hewitt, G F, Kearsey, H A, Lacey, P M C and Pulling, D J, (1963) "Burnout and nucleation in climbing film flow", AERE-R4374.

Hewitt, G F and Lovegrove, P C, (1976) "Experimental methods in two-phase flow studies". *Electric Power Research Institute Report* EPRI NP-118.

Hsu, Y Y, Editor, (1978) "Two-phase flow instrumentation review group meeting". NUREG-0375.

Hsu, Y Y and Graham R W, (1976) "Transport processes in boiling and two-phase systems including near critical fluids". *Hemisphere, Washington, DC.*

Isbin, H S, Moy, J E and DaCruz, A J R, (1957) "Two-phase, steam-water critical flow". AIChE *Journal,* **3**, 361-365.

Ishihara, K, Palen, J W and Taborek, J, (1977) "Critical review of correlations for predicting two-phase pressure drop across tube banks". ASME *Paper No.* 77-WA/HT-23.

Ishii, M, (1975) "Thermo-fluid dynamic theory of two-phase flow". Monograph No. 22, *Collection de la Direction des Etudes et Recherches D'Électricité de France, Eyrolles, Paris.*

Jakob, M, (1936) "Heat transfer in evaporators and condensation - I," *Mechanical Engineering*, 643-660.

Jakob, M and Fritz, W, (1931) "Versuche über den verdampfungsvorgang". *Forschung auf dem Gebiete des Ingenieurwesens*, **2**, 435-447.

Jones, O C, Jr and Bankoff, S G, Editors, (1977) "Thermal and hydraulic aspects of nuclear reactory safety". 1, Light Water Reactors; 2, Liquid metal fast breeder reactors. *ASME, New York*.

Joule, J P, (1861) "On the surface-condensation of steam". *Philosophical Transactions of the Royal Society of London*, **151**, 133-160.

Katz, D L, Myers, J E, Young, E H and Bakkjian, G, (1955) "Boiling outside finned tubes". *Petroleum Refiner*, **34** (2), 113-116.

Keppel, R R and Tung, T V, (1963) "Two-phase (gas-liquid) system: Heat transfer and hydraulics, an annotated bibliography". ANL-6734.

Kutateladze, S S, (1951) "A hydrodynamic theory of change in the boiling process under free convection conditions", *Izvestiya Akademii Nauk SSSR, Otdelenie Tekhnicheskikh Nauk*, (4), 529-536.

Lahey, R T, Jr and Moody, F J, (1977) "The thermal-hydraulics of a boiling water nuclear reactor". *The American Nuclear Society, Hinsdale, Illinois*.

Ledinegg, M, (1938) "Unstabilität der Strömung bei natürlichem und Zwangumlauf". *Die Wärme*, **61**, 891-898, AEC-tr-1861.

Ledinegg, M and Wiesenberger, J, (1975) "Strömung in beheizten rohrregistern". *VGB Kraftwerks-Technik*, **55**, 581-589.

Leidenfrost, J G, (1756) "On the fixation of water in diverse fire". Translation in *International Journal of Heat and Mass Transfer*, **9**, 1154-1166.

Levy, S, (1966) "Prediction of two-phase annular flow with liquid entrainment". *International Journal of Heat and Mass Transfer*, **9**, 171-188.

Lockhart, R W and Martinelli, R C, (1949) "Proposed correlation of data for isothermal two-phase two-component flow in pipes". *Chemical Engineering Progress*, **45** (1), 39-48.

Macbeth, R V, (1963) "Burn-out analysis. Part 4. Application of a local condition hypothesis to world data for uniformly heated round tubes and rectangular channels", AEEW-R267.

Martinelli, R C and Nelson, D B, (1948) "Predictions of pressure drop during forced-circulation boiling of water". *Transactions of the ASME*, **70**, 695-702.

Merte, H, Jr and Clark, J A, (1964) "Boiling heat transfer with cryogenic fluids at standard, frictional, and near-zero gravity". *Journal of Heat Transfer*, **86**, 351-359.

McAdams, W H, Kennel, W E, Mindon, C S, Carl, R, Picornel, P M and Dew, J E, (1949) "Heat transfer to water with surface boiling", *Industrial and Engineering Chemistry*, **41**, 1945-1953.

Matschoss, C, (1908) "Die entwicklung der Dampfmaschine." Springer, Berlin.

Moore, K V and Rettig, W H, (1973) "RELAP4: A computer program for transient thermal-hydraulic analysis", USAEC Report ANCR-1127, Aerojet Nuclear Company.

Nishikawa, K, (1956) "Heat transfer in nucleate boiling". Memoirs of the Faculty of Engineering, Kyushu University, **16** (1), 1-28.

Nukiyama, S, (1934) "The maximum and minimum values of the heat Q transmitted from metal to boiling water under atmospheric pressure". Journal Japan Society of Mechanical Engineers, **37**, 367-374. Translation in International Journal of Heat and Mass Transfer, 9, 1419-1433.

Nusselt, W, (1916) "Die oberflächen kondensation des wasserdampfes". Zeitschrift des Vereines Deutscher Ingenieurs, **60**, 541-546, 569.

Ornatskiy, A, (1969) "The effect of basic regime parameters and channel geometry on critical heat fluxes in forced convection of subcooled water". Heat Transfer - Soviet Research, **1** (3), 17-22.

Palen, J W and Taborek, J, (1972) "Characteristics of boiling outside large scale multitube bundles". AIChE Symposium Series, **68** (118), 50-61.

Prosching, T A, et al, (1969) "FLASH-4: A fully implicit FORTRAN-IV program for the digital simulation of transients in a reactor plant". USAEC Report WAPD-TM-840, Bettis Atomic Power Laboratory.

Pudney, J, (1974) "Brunel and his world", Thames and Hudson, London, 107-108.

Lord Rayleigh, (1917) "On the pressure developed in a cavity during the collapse of a spherical cavity". Philosophical Magazine, **34**, 94-99.

Reed, R C, (1968) "Train wrecks", Bonanza Books, New York, 179.

Rohsenow, W M (1952) "A method of correlating heat transfer data for surface boiling of liquids". Transactions of the ASME, **74**, 969-976.

Rose, J W, (1967) "On the mechanisms of dropwise condensation". International Journal of Heat and Mass Transfer, **10**, 755-762.

Schnackenberg, H, (1937) "Die wasserverteilung in zwanglaufheizflächen, insbesondere in speisewasservorwärmern," Die Wärme, 60, 481-484.

Sauvage, E, (1892) "Ecoulement de l'eau des chaudieres," Annales des Mines, 11, 192-202.

Stenning, A H, Veziroglu, T N and Callahan, G M, (1969) "Pressure-drop oscillations in forced convection flow with boiling". Proceedings of the Symposium on Two-Phase Flow dynamics, **1**, The Commission of the European Communities, Brussels, 406-427.

Stevens, G F and Kirby, G J, (1964) "A quantitative comparison between burnout data for water at 1000 psia and Freon-12 at 155 psia, uniformly heated round tubes, vertical flow". AEEW-R327.

Sun, K H and Lienhard, J H, (1970) "The peak pool boiling heat flux on horizontal cylinders", International Journal of Heat and Mass Transfer, **13**, 1425-1439.

Telles, A S and Dukler, A E, (1970) "Statistical characteristics of thin, vertical, wavy, liquid films", *Industrial and Engineering Chemistry Fundamentals*, **9**, 412-421.

Thompson, B and Macbeth, R V, (1964) "Boiling water heat transfer - burnout in uniformly heated round tubes: A compilation of world data with accurate correlations", *AEEW-R356*.

Thorneycroft, J T, (1895) "The influence of circulation on evaporative efficiency of water-tube boilers". *Transactions of the Institution of Naval Architects*, **36**, 40.

Tong, L S, (1965) "Boiling heat transfer and two-phase flow. *John Wiley & Sons, New York*.

Tong, L S, (1972) "Boiling crisis and critical heat flux". *USAEC Report TID-25887, Westinghouse Electric Corp*.

Tong, L S and Bennett, G L, (1977) "NRC Water-Reactor Safety-Research Programs". *Nuclear Safety*, **18**, 1-40.

van Stralen, S J D, (1966) "The mechanism of nucleate boiling in pure liquids and in binary mixtures - Part I, Part II". *International Journal of Heat and Mass Transfer*, **9**, 995-1046.

Wallis, G B, (1969) "One-dimensional two-phase flow". *McGraw-Hill, New York*.

Westwater, J W, et al, (1954-1966) Films for Engineering Societies Heat Transfer Film Library. Nos. W-4 - W-19.

Weisman, J, (1977) "Evaluation of pressure drop across area changes during blowdown". *NUREG-0047-6, NRC-2*.

Ybarrondo, L J, Solbrig, C W and Isbin, H S, (1972) "The calculated loss-of coolant accident: A review". *AIChE Monograph Series No 7, American Institute of Chemical Engineers*.

Zenkevich, B A, (1959) "The generalization of experimental data on critical heat fluxes in forced convection of subcooled water". *Atomnaya Energiya*, **6**, 169-173.

Zuber, N, (1958) "On the stability of boiling heat transfer". *Transactions of the ASME*, **80**, 711-720.

Index